Uwe Kiencke

Lars Nielsen

**Automotive Control Systems**

For Engine, Driveline, and Vehicle

Uwe Kiencke

Lars Nielsen

# Automotive Control Systems

## For Engine, Driveline, and Vehicle

Second edition

With 345 figures and 13 tables

 Springer

**Prof. Dr.-Ing. Uwe Kiencke**
Universität Karlsruhe (TH)
Department of Electrical Engineering
76187 Karlsruhe
Germany
*kiencke@iiit.uni-karlsruhe.de*

**Prof. Dr. Lars Nielsen**
Division of Vehicular Systems
Department of Electrical Engineering
Linköping University
581 83 Linköping
Sweden
*lars@isy.liu.se*

ISBN 978-3-642-06211-7          e-ISBN 978-3-540-26484-2

**Springer is a part of Springer Science+Business Media**
springeronline.com
© Springer-Verlag Berlin Heidelberg 2010
Printed in Germany

Final processing by PTP-Berlin Protago-T$_E$X-Production GmbH, Germany
Cover-Design: Medionet AG, Berlin
Printed on acid-free paper          62/3141/Yu – 5 4 3 2 1 0

*To Margarete and Ingrid*

# Preface to the second edition

Since the first edition of this book was published, already five years have passed, and during this period research on automotive control has flourished, and at the same time, the amount of industrial applications and products have also prospered. This means that there was a rich flora of possibilities regarding the selection of material when planning the second edition. In this process new topics have been added, important fields have been deepened, and in order to keep the number of pages down, sections of limited interest have been eliminated.

In the selection of the new material we have concentrated on subjects that are both of current interest and importance, but at the same time are subjects that also contribute to a better understanding of basic processes and theories. The new material includes two completely new chapters: Diesel Engine Modeling and Diagnosis.

In driveline control a new section on Anti-Jerk Control has been added. Large parts of Vehicle Dynamics and Control have been rewritten, which significantly improves the presentation of that material. Further, in this second edition, we have hopefully corrected most of the errors in the first edition, reviewed the nomenclature, and in order to facilitate to work with this book, added an index.

The level of presentation has been thought trough to be suited for students at late undergraduate level or at early graduate level. The so called Bologna process is influencing universities in Europe at this moment, and in that perspective this book should be well suited for a course at the two year Masters level.

We like to thank Dr. Dara Torkzadeh and Thomas Rambow for their contributions in Diesel Engine Modeling, Dr. Mattias Nyberg for his work in Diagnosis, Julian Baumann for his participation in Anti-Jerking Control and Dr. Marcus Hiemer as well as Jörg Barrho for their work in Vehicle Dynamics and reviewing this book.

November 2004

Uwe Kiencke            Lars Nielsen

# Preface to the second edition

# Preface to the first edition

Automotive control has become a driving factor in automotive innovation over the last twenty five years. In order to meet the enhanced requirements for lower fuel consumption, lower exhaust emissions, improved safety as well as comfort and convenience functions, automotive control had to be applied.

In any area of technology, control design is an interplay between reality, physics, modeling, and design methods. This is also true in automotive control, and there has been extensive work done in research and development leading to a number of descriptions, models, and design methodologies suited for control.

## Goal of the book

Our purpose of writing a book on Automotive Control is to present this interplay between thermodynamics, basics of engine operation, vehicle mechanics as well as parameter estimation and automotive control approaches.

There are several good books available on the separate disciplines (some of the major references are in German). However, up until now there has not been a text available that explores more deeply the connections between reality, measurements, models and control design.

It has been natural for us to treat all the major aspects of automotive control in the same book. This means that we cover engine, driveline, and complete vehicle. One reason is that there are similarities in methodology when analyzing and designing automotive control systems. This includes the point of view of finding models of suitable complexity and expressiveness. Another, perhaps more important, reason is that there is a strong trend that engine control, driveline control, and vehicle control rather than being separate will be more and more integrated, so that overall vehicle optimization is possible.

It has also been important to us to show real measurements. This gives a reader the possibility to see how models are approximations of reality, and to judge the modeling assumptions. A consequence of this approach is that we have selected to treat systems that are close to some of those utilized in actual vehicles, rather than discussing speculative systems or presenting purely theoretical results.

## Intended readers

This book should enable control engineers to understand engine and vehicle models necessary for controller design and should introduce mechanical engineers into vehicle-specific signal processing and automatic control.

In fact, our inspiration to write the book came from this. We are both members of the IFAC technical committee on Automotive Control (with the first author being the chairman). We met there and also at SAE meetings, and we saw the potential value of bridging a gap that was obvious to us. However, even more important to us is to share some of the fun and excitement that goes into the area of Automotive Control Systems and thus give it the attention it deserves.

## Organization of the book

The outline of the book starts with engines, continues with drivelines, and finally deals with the vehicle.

Chapters 2 to 4 treat engines with regard to basics, thermodynamics, models, control, and advanced concepts. All the major control systems and their design are treated. The thermodynamic models in Chapter 2 deal with parameters that vary under one cycle and the resolution of interest is typically one crank angle degree, whereas the time scales of mean value models are in the order of 1 to several engine cycles, and the variation in variables that are considered are also averaged over one or several cycles. These models form the basis for understanding the complex phenomena that influence the engine operation, efficiency and emissions. They also serve the purpose of describing the properties influencing control design and performance in Chapters 3 and 5.

The driveline (engine, clutch, transmission, shafts, and wheels) which is a fundamental part of a vehicle is the topic in Chapter 7. Since the parts are elastic, mechanical resonances may occur. The handling of such resonances is basic for functionality and driveability, but is also important for reducing mechanical stress and noise. Two important modes of driveline control that are treated are driveline speed control and driveline torque control, having their applications in cruise control and automatic gear shifting control.

Vehicle dynamics control systems help the driver to perform the task of keeping the vehicle on the road in a safe manner. These systems are thus often safety-oriented, which means that they only interact in situations where they can reduce the possibility of an accident, but then they affect the immediate behavior of the vehicle within fractions of a second. Some systems are also used for improving the comfort of the driver. The performance of a vehicle, regarding the motions coming from accelerating, braking, cornering, or ride, is mainly a response to the forces imposed on the vehicle from the tire-road contact. Much of study of vehicle dynamics is a study on why and how these forces are produced and how they can be effectively understood and treated in simplified models.

The basics of these models and some associated control systems are presented in Chapters 8 to 10.

Chapter 11 is the exception from that all the systems and principles in this book is close to some of those utilized in actual vehicles. The reason is that road and driver modeling is part of simulation design rather than part of a vehicle. Nevertheless, it is important to realize that road and driver models are important parts in the design cycle of automotive systems design due to the importance of advanced simulation.

## Background and use of the book

The material in this book has been used in courses at the universities of Karlsruhe, Germany and Linköping, Sweden. It is well suited for the later stages (third or fourth year) of the engineering programs at our technical institutes ("Diploma-engineer", "Master of Science").

The book, to a large extent, covers the basic material needed, but of course it is advantageous to have a background from basic undergraduate courses in automatic control, signals and systems, mechanics, and physics.

The course lay-out includes problem-solving sessions and laboratory experiments. The laboratory assignments typically include measurements, building models of the type treated in the book, and finally designing controllers and simulating them. Here students with more background, for example in modern control, can do more elaborate designs. This is also the case when the book is used in an introductory graduate course.

## The authors

Dr. Kiencke's experience in this field started in the early nineteen seventies when developing adaptive lambda control and knock control at Robert Bosch Corporation. In the following years more complex approaches for engine modelling [2], [22] and controller design [63] were published. At that time he headed a team that developed the vehicle communication network "Controller Area Network (CAN)" [67]. Networking allowed to combine formerly stand-alone control schemes into an integrated vehicle control system. In the early nineteen nineties he joined the University of Karlsruhe in Germany where he could intensify engine and vehicle dynmaic control research.

Dr. Nielsen has more than fifteen years background in academic mechatronics research (obtaining a good start at the Department of Automatic Control in Lund, Sweden). He has during that time continuously collaborated with industry, and has lead joint research projects with Scania AB, Mecel AB, Saab Automobile AB, Volvo AB, and DaimlerChrysler. He is since 1992 holder of the chair Sten Gustafsson professor of vehicular systems at Linköping University in Sweden.

## Acknowledgments

The control systems presented were mostly developed within a team. Therefore the first author would like to thank especially the following cooperation partners: Dr. Martin Zechall in lambda (air-fuel ratio) control, Dr. Böning in knock control and engine map optimization, Alfred Schutz in engine idle speed control, Heinz Leiber in ABS braking control, Dr. Michael Henn in misfire detection, Dr. Achim Daiss in vehicle modelling and identification and Dr. Rajjid Majjad in road and driver modelling. It was a great pleasure to cooperate with these people and it created many friendships. The second author is especially indebted to Magnus Petterssson for joint work in driveline control, and to Lars Eriksson for joint work in engine modelling and control. Also Lars-Gunnar Hedström, Jan Nytomt, and Jan Dellrud deserves special mentioning as research dedicated industrial partners.

Furthermore we both thank Christopher Riegel, Jochen Schöntaler, Dara Torkzadeh, and Dr. Tracy Dalton for their tremendous effort to translate and revise parts of the book, as well as Dr. Dietrich Merkle as a publisher.

Last but not least we to thank our families and especially our wives Margarete and Ingrid for tolerating that so much weekend and vacation time was dedicated to this book.

Being in November 1999 looking forward to the next millennium, we hope that readers will share some of the excitement that comes along with Automotive Control Systems.

Uwe Kiencke                    Lars Nielsen

# Contents

# 1 Introduction

Vehicles are now computerized machines. This fact has had an enormous effect on the possibilities for functionality of vehicles, which together with needs and requirements from customers and from society have created vigorous activities in development.

## 1.1 Overall demands

The overall demands on a vehicle are that it should provide safe and comfortable transportation together with good environmental protection and good fuel economy. This means that there are three main objectives for automotive control systems:

- Efficiency, which leads to lower fuel consumption.

- Emissions should be low to protect the environment.

- Safety is of course a key issue.

There are a number of additional objectives like comfort, driveability, low wear, availability, and long term functionality.

## 1.2 Historic remark

Many of the technologies that today are considered advanced, sometimes even new, have been around for a long time. It is therefore interesting to ask ourselves why these technologies are surfacing now as commercial products. Direct injection of gasoline engines is one example. These concepts are not new, even if sometimes presented so, but the novelty is instead that they now with proper **control** can achieve competitive functionality and performance.

It is thus the breakthrough of computer control that is a driving factor. A good example is ABS (Anti-lock Braking Systems) which is an old idea, but it was not functional enough using mechanical solutions or analog electronics. Now these systems are readily available and widely spread.

## 1.3   Perspectives

Looking at the future, the three overall objectives above will be in focus. The demands of reduced emissions and advanced diagnosis functionality are steadily increased by legislators and customers. The key areas that help meeting the increased demands are the development of control and diagnosis functions in the control units. Further, this functionality will have to be obtained not only when the car is new, but over a sustained period of time. Other examples are improved stability due to handling control, and improved driveability due to driveline torque control, which also can be used to e.g. reduce clutch wear.

Regarding methodology development, mathematical models will play an important role. They will be used for model based control and diagnosis. They will also be the basis for e.g. sensor fusion, adaptive control, and supervision.

### Co-design

Automotive control will not only improve existing vehicle designs. It will also to a large extent change the view of vehicular systems design leading to:

- New mechanical designs. These new designs are made possible by, and rely on, the existence of a control system.

Design of vehicles is thus evolving into co-design of mechanics and control. The goals for this development can be set high, and the perspectives on automotive control systems are therefore concluded with an inspiring mind teaser:

### A mind teaser

It can not be ruled out that a car can function as an air cleaner for usual town air. In a typical town in the industrial world, the air is typically somewhat polluted from many sources, including cars, but also due to house heating and industries. Existing and upcoming technology lowering exhaust emissions are such that the concentrations in the exhaust after the catalyst can be lower than in town air. This means that the originally available pollutants have been combusted or have been collected in the catalyst.

Since such a perfect combustion produces only water and carbon dioxide, the problem of carbon dioxide is then a matter of less fuel consumption.

If at all possible, such a development will rely heavily on automotive control systems since the car has to function as an air cleaner under all possible circumstances, load variations, and driving styles. This is a mind teaser, but understanding the material in this book is a first step.

# 2 Thermodynamic Engine Cycles

In this chapter, the thermodynamic characteristics of basic engine cycles are explained. For each concept, the thermal efficiency is derived from thermodynamic equations.

## 2.1 Introduction to Thermodynamics

A short introduction to thermodynamics is given in this section. It is essential to understand the thermodynamic processes for the different types of engine models explained in this book.

### 2.1.1 First Thermodynamic Law

Ideal gases are always in a gaseous state and they behave according to the *ideal gas equation*:

$$pV = mR\vartheta \qquad (2.1)$$

where:

- $p$    is the pressure in $N/m^2$
- $V$    is the volume in $m^3$
- $m$    is the mass in $kg$
- $R$    is the gas constant $R = 287.4\,m^2/(s^2\,K)$
- $\vartheta$    is the absolute temperature in $K$

The expression $pV$ has the unit *Joule* and it depends on the mass and temperature of the gas.

The thermal energy is defined as:

$$q = cm\vartheta \qquad (2.2)$$

where:

$q$    is the thermal energy in $J$
$c$    is the specific heat constant in $J/(kg\ K)$

This equation is valid for all kind of materials in solid, liquid and gaseous state. The specific heat constant is material depending. In gases, different state changes have to be distinguished which will be discussed in the next section.

### First Thermodynamic Law

$$dq = du + dw \qquad (2.3)$$

$dq$    is the differential change of thermal energy. Energy which is brought into the gas is **positive**.

$du$    is the differential change of internal energy. Internal energy which is brought into the gas **is positive**.

$dw$    is the differential change of mechanical work. Mechanical work which is brought into the gas is **negative**.
Please note, that this may be defined differently in other books.

The thermal energy $dq$ brought into a gas of constant mass results in a rise in it's internal energy $du$ and/or is transformed into mechanical work outside the gas. Thus, energy is neither created nor destroyed. This equation is solved by integration to compute the energies in state changes.

### Volume Change

There are two basic principles how mechanical work can be delivered to or from the system: By a change of volume or a change of pressure.

In engines, a limited amount of compressed gas expands in a cylinder. The differential mechanical work $dw_v$ depends on the force $F$ upon the piston and the differential piston stroke $ds$:

$$dw_v = F\ ds = p\,A\ ds$$

where $A$ is the cross-section of the cylinder and $p$ the pressure. More general, this yields to:

$$dw_v = p\ dV \qquad (2.4)$$

The work $dw_v$ is equivalent to the kinetic energy brought into the mechanical system due to the expansion of the gas in the cylinder. Mechanical energy brought into the gas is negative: $dV < 0$ (compression) and therefore $dw_v < 0$. Output of mechanical work from the gas is positive, $dV > 0$ (expansion), $dw_v > 0$. This can also be seen in Figure 2.1.

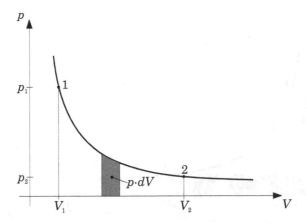

**Figure 2.1** Mechanical work due to volume change.

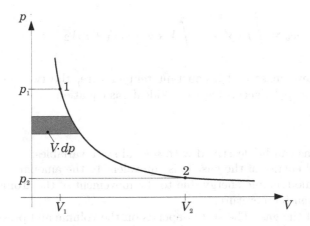

**Figure 2.2** Mechanical work due to pressure change

### Pressure Change

In turbines, kinetic energy is transmitted by a continuous gas flow between two locations of different pressure:

$$dw_p = -V \, dp \tag{2.5}$$

The work $dw_p$ is equivalent to the loss of potential energy of the gas. Figure 2.2 graphically explains the state change in the pV-diagram. Output of mechanical work from the gas is positive because the pressure drop is negative: $dp < 0$ (expansion), $dw_p > 0$.

The two different types of work are graphically explained in Figure 2.3 and they are linked by the following equation:

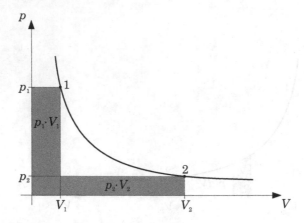

**Figure 2.3**  Relationship between the two mechanical works.

$$w_v = \int_1^2 p \, dV = - \int_1^2 V \, dp - p_1 V_1 + p_2 V_2 \qquad (2.6)$$

Assuming a constant mass and a constant temperature, the two energies are equivalent as $p_1 V_1 = p_2 V_2$ according to the ideal gas equation.

## Enthalpy

The state of the gas can be described with several state variables:

$u$     Internal energy of the gas. It is equivalent to the amount of thermodynamic energy due to the movement of the atoms at a given temperature.

$pV$    State of the gas. The state depends on the volume and pressure of the gas. It is related to the internal energy of the gas.

A state change can be caused by a change of pressure, a change of volume or both:

$$d(pV) = p \, dV + V \, dp \qquad (2.7)$$

$p \, dV$ Kinetic energy caused by a change of volume.

$V \, dp$ Potential energy caused by a change of pressure.

The enthalpy is another state variable of the gas:

$$h = u + pV \qquad (2.8)$$

The differential is:

$$dh = du + p \, dV + V \, dp \qquad (2.9)$$

Compared to the first thermodynamic law in Equation 2.3

$$dq = du + dw$$

the thermal energy can be written as:

$$dq = dh - V\,dp = du + p\,dV \tag{2.10}$$

For isobaric expansion, we have $dw_v = p\,dV$ and $V\,dp = 0$ and therefore $dq = dh$. This yields to:

$$dq = du + p\,dV \tag{2.11}$$

For isochoric pressure drop, we have $dw_p = -V\,dp$ and $p\,dV = 0$ and therefore $dq = du$. This results in:

$$dq = dh - V\,dp \tag{2.12}$$

The enthalpy $h$ can be very useful as a state variable when dealing with gas turbine engines. The internal energy $u$ is useful when dealing with piston engines.

## 2.1.2 Specific Heat Constant

### Input of Thermal Energy at Constant Volume

A temperature rise can be observed when increasing the thermal energy $q$ of a constant mass of gas at a **constant volume**. The specific heat constant $c_v$ is defined as:

$$c_v = \frac{1}{m}\left(\frac{dq}{d\vartheta}\right)_V \tag{2.13}$$

The differential of the internal energy is

$$du = \left(\frac{\partial u}{\partial V}\right)_\vartheta dV + \left(\frac{\partial u}{\partial \vartheta}\right)_V d\vartheta \quad,$$

Inserting this into the first thermodynamic law 2.11yields:

$$\begin{aligned}
dq &= \left(\frac{\partial u}{\partial V}\right)_\vartheta dV + \left(\frac{\partial u}{\partial \vartheta}\right)_V d\vartheta + p\,dV \tag{2.14}\\
&= \left(\frac{\partial u}{\partial \vartheta}\right)_V d\vartheta + \left(\left(\frac{\partial u}{\partial V}\right)_\vartheta + p\right) dV \tag{2.15}\\
&= \left(\frac{\partial u}{\partial \vartheta}\right)_V d\vartheta \tag{2.16}
\end{aligned}$$

At a constant volume we have $dV = 0$. After division by $d\vartheta$:

$$\left(\frac{dq}{d\vartheta}\right)_V = \left(\frac{\partial u}{\partial \vartheta}\right)_V$$

Equation 2.13 can be written as:

$$c_v = \frac{1}{m}\left(\frac{\partial u}{\partial \vartheta}\right)_V \tag{2.17}$$

We have shown that a change in the thermal energy $dq$ at constant volume is equivalent to a change in the internal energy $du$ of the gas.

## Input of Thermal Energy at Constant Pressure

A temperature rise can also be observed when increasing the thermal energy of a constant mass of gas at a **constant pressure**. Please note the difference to the previous section, as the specific heat constant $c_p$ is now defined as:

$$c_p = \frac{1}{m}\left(\frac{dq}{d\vartheta}\right)_p \tag{2.18}$$

The differential of the enthalpy is:

$$dh = \left(\frac{\partial h}{\partial \vartheta}\right)_p d\vartheta + \left(\frac{\partial h}{\partial p}\right)_\vartheta dp$$

This relationship can be inserted into 2.12 and then be simplified because $dp = 0$:

$$dq = \left(\frac{\partial h}{\partial \vartheta}\right)_p d\vartheta + \left(\frac{\partial h}{\partial p}\right)_\vartheta dp - V\,dp \tag{2.19}$$

$$= \left(\frac{\partial h}{\partial \vartheta}\right)_p d\vartheta + \left(\left(\frac{\partial h}{\partial p}\right)_\vartheta - V\right) dp \tag{2.20}$$

$$= \left(\frac{\partial h}{\partial \vartheta}\right)_p d\vartheta \tag{2.21}$$

The last equation is divided by $d\vartheta$:

$$\left(\frac{dq}{d\vartheta}\right)_p = \left(\frac{\partial h}{\partial \vartheta}\right)_p$$

Similar to the above section, Equation 2.18 can also be written as:

$$c_p = \frac{1}{m}\left(\frac{\partial h}{\partial \vartheta}\right)_p \tag{2.22}$$

The enthalpy of the system is changed. The input of thermal energy $dq$ at constant pressure not only increases the internal energy $du$, but also produces an output of work $p\,dV$.

## Gas Constant and Adiabatic Coefficient

The relationship between $c_p$ and $c_v$ can be seen by using the enthalpy (Equation 2.8) and the ideal gas equation (Equation 2.1):

$$h = u + pV \tag{2.23}$$
$$dh = du + d(pV) \tag{2.24}$$
$$m\,c_p\,d\vartheta = m\,c_v\,d\vartheta + m\,R\,d\vartheta \tag{2.25}$$
$$c_p = c_v + R \tag{2.26}$$

The ideal gas constant $R$ is equal to the difference between $c_p$ and $c_v$:

$$R = c_p - c_v \tag{2.27}$$

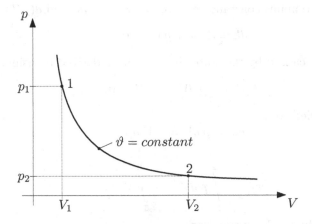

**Figure 2.4** Isothermic state change

The ratio of $c_p$ to $c_v$ is called the **adiabatic exponent**:

$$\kappa = \frac{c_p}{c_v} \qquad (2.28)$$

Under normal conditions its value for air is $\kappa_{air} = \frac{3.5}{2.5} = 1.4$.

## 2.1.3 State Changes of Ideal Gases

**Isothermal Change:** $\vartheta = const$

Assuming a constant temperature, the ideal gas equation $pV = mR\vartheta$ yields:

$$p = \frac{mR\vartheta}{V} = \frac{const}{V}$$

where $pV = const$. The state change is equivalent to a hyperbolic curve in the $pV$-diagram which can be seen in Figure 2.4. The isothermal state change is characterized by using the ideal gas equation:

$$p_1 V_1 = p_2 V_2 \qquad (2.29)$$

The slope of the curve in the $pV$-diagram is:

$$\frac{dp}{dV} = -\frac{p}{V} \qquad (2.30)$$

The internal energy of the gas remains constant:

$$du = m\, c_v \, d\vartheta = 0$$

Hence, the first thermodynamic law can be written as:

$$dq = du + dw = dw$$

The enthalpy also remains constant because of $pV = const$ and $d(pV) = 0$:

$$dh = du + d(pV) = 0$$

The kinetic energy caused by the state change can be derived by using:

$$d(pV) = p\,dV + V\,dp = 0$$

Hence, it is simplified to:

$$dw = p\,dV = -V\,dp$$

and integrated:

$$w_{1,2} = \int_1^2 p\,dV = -\int_1^2 V\,dp$$

Finally, by using the ideal gas equation:

$$w_{1,2} = q_{1,2} = m\,R\,\vartheta \int_1^2 \frac{1}{V}\,dV = m\,R\,\vartheta \ln \frac{V_2}{V_1}$$

Work $w_{1,2}$ is generated from the system by expanding the gas. The same amount of thermal energy $q_{1,2}$ must be brought into the system in order to compensate for the work and to maintain the constant temperature.

**Isobaric Change:** $p = const$

A state change at constant pressure is described by the ideal gas equation:

$$p = \frac{m\,R\,\vartheta}{V} = const$$

and therefore:

$$\frac{\vartheta_1}{\vartheta_2} = \frac{V_1}{V_2} \tag{2.31}$$

The state change is equivalent to a horizontal line in the $pV$-diagram which can be seen in Figure 2.5. The isobaric state change $1 \rightarrow 2$ results in an output of mechanical work $w_{1,2}$ due to the expansion of the gas, and at the same time it results in an increased internal energy $u$ due to the temperature rise. The input of thermal energy $q$ is equivalent to the increase of the enthalpy $h$.

In this case, the first thermodynamic law can be written as:

$$dq = dh - V\,dp = dh$$

since pressure change vanishes: $dp = 0$. The differential change of thermal energy is given by:

$$dq = m\,c_p\,d\vartheta$$

Integration leads to:

$$q_{1,2} = m\,c_p \int_1^2 d\vartheta = m\,c_p(\vartheta_2 - \vartheta_1) \tag{2.32}$$

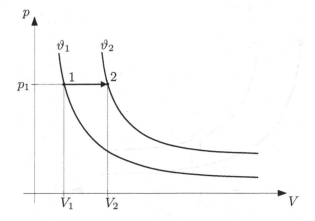

**Figure 2.5** Isobaric state change

The mechanical work can be derived by using the fact that $dq = dh$ and $dh = du + p\,dV$:

$$dq = du + p\,dV$$

Because of $dw_v = p\,dV$, the change of kinetic energy is:

$$
\begin{aligned}
dw_v &= p\,dV = dh - du & (2.33)\\
&= m\,c_p\,d\vartheta - m\,c_v\,d\vartheta & (2.34)\\
&= m\,R\,d\vartheta & (2.35)
\end{aligned}
$$

Integration leads to:

$$w_{1,2} = m\,R\,\vartheta \int_{1}^{2} d\vartheta = m\,R(\vartheta_2 - \vartheta_1)$$

Please note that only a small portion of thermal energy $q$ is converted into kinetic energy $w$. The ratio of the input of thermal energy to the output of mechanical work can be expressed:

$$\frac{dw}{dq} = \frac{m(c_p - c_v)\,d\vartheta}{m\,c_p\,d\vartheta} = \frac{c_p - c_v}{c_p} = \frac{\kappa - 1}{\kappa}$$

which is 0.29 for air.

**Isochoric Change:** $V = const$

In this case, the ideal gas equation is:

$$V = \frac{m\,R\,\vartheta}{p} = const$$

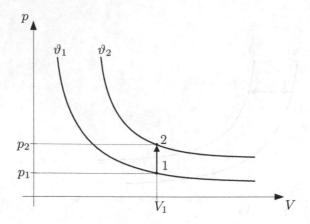

**Figure 2.6** Isochoric state change

and therefore:

$$\frac{\vartheta_1}{\vartheta_2} = \frac{p_1}{p_2} \tag{2.36}$$

The state change is equivalent to a vertical line in the pV-diagram illustrated in Figure 2.6. The mechanical work vanishes for an isochoric state change $1 \to 2$. The input of thermal energy $dq$ is equivalent to the increased internal energy $du$.

As $dq = du + p\,dV = du$, the thermal energy is given by:

$$dq \;=\; m\,c_v\,d\vartheta \tag{2.37}$$

$$q_{1,2} \;=\; m\,c_v \int_1^2 d\vartheta = m\,c_v(\vartheta_2 - \vartheta_1) \tag{2.38}$$

The enthalpy is increased more than the internal energy:

$$dh = du + p\,dV + V\,dp$$

where $p\,dV = 0$ and $V\,dp = m\,R\,d\vartheta$. The change of enthalpy is therefore:

$$dh = m\,c_v\,d\vartheta + m\,R\,d\vartheta = m\,c_p\,d\vartheta$$

The internal energy is $du = dh - V\,dp$. It has the same value as the thermal energy:

$$du = m(c_p - R)\,d\vartheta = m\,c_v\,d\vartheta = dq$$

**Isentropic or Adiabatic Change:** $q = const$

Assuming an insulated gas volume, the thermal energy of the gas remains constant. In an isentropic state change $1 \to 2$, the gas expands and mechanical work is delivered from the gas while its internal energy is reduced. The gradient of the

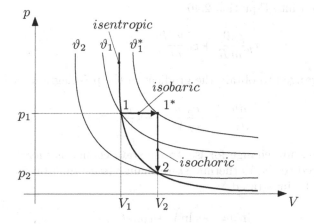

**Figure 2.7** Isentropic state change

adiabatic change in the $pV$-diagram is steeper than the hyperbolic curve of the isothermic change. This is due to the fact that no thermal energy is brought into the gas.

$$dq = du + p\,dV = 0$$

Hence, the first thermodynamic law can be written as:

$$-du = p\,dV$$

This means that the mechanical work is equivalent to the loss of internal energy. The change of enthalpy is

$$dh = du + p\,dV + V\,dp = V\,dp$$

because of $dq = du + p\,dV = 0$. The isentropic change can be considered as the sequence of two state changes:

- An isobaric change where $dq_p = m\,c_p\,d\vartheta_p = dh$.

- An isochoric change where $dq_v = m\,c_v\,d\vartheta_v = du$.

The summation of exchanged thermal energies is supposed to be zero:

$$
\begin{aligned}
dq = dq_p + dq_v &= 0 & \text{(2.39)}\\
c_p\,d\vartheta_p + c_v\,d\vartheta_v &= 0 & \text{(2.40)}
\end{aligned}
$$

From the isobaric and the isochoric state changes we get:

$$
\begin{aligned}
isobaric : d\vartheta_p &= \frac{p\,dV}{mR} & \text{(2.41)}\\
isochoric : d\vartheta_v &= \frac{V\,dp}{mR} & \text{(2.42)}
\end{aligned}
$$

This can be inserted into Equation 2.40

$$c_p \frac{p \, dV}{m \, R} + c_v \frac{V \, dp}{m \, R} = 0 \quad ,$$

which can be rearranged to obtain the gradient in the pV-diagram:

$$\frac{dp}{dV} = -\frac{c_p}{c_v} \frac{p}{V} = -\kappa \frac{p}{V} \qquad (2.43)$$

The differential pressure change for isentropic state change is increased by the factor of $\kappa$ compared to the isothermic state change (see Equation 2.30). Equation 2.43 can be integrated:

$$\ln p = -\kappa \ln V + const$$

This yields:

$$p V^\kappa = const \qquad (2.44)$$

An isentropic change $1 \rightarrow 2$ is characterized by:

$$p_1 V_1^\kappa = p_2 V_2^\kappa$$

The kinetic energy $w_{1,2}$ is obtained by integration:

$$w_{1,2} = \int_1^2 p \, dV$$

Exchanging $p$ with $p = const \, V^{-\kappa}$ leads to:

$$w_{1,2} = const \int_1^2 V^{-\kappa} \, dV = \frac{const}{1 - \kappa} (V_2^{1-\kappa} - V_1^{1-\kappa})$$

Replacing $const = p_1 V_1^\kappa$:

$$w_{1,2} = p_1 V_1^\kappa \frac{1}{1 - \kappa} (V_2^{1-\kappa} - V_1^{1-\kappa}) = p_1 V_1 \frac{1}{1 - \kappa} \left( \left( \frac{V_1}{V_2} \right)^{\kappa-1} - 1 \right)$$

Further simplifications:

$$\left( \frac{V_1}{V_2} \right)^{\kappa-1} = \frac{p_1 V_1^\kappa}{p_2 V_2^\kappa} \frac{p_2 V_2}{p_1 V_1} = \frac{p_2 V_2}{p_1 V_1} = \frac{\vartheta_2}{\vartheta_1}$$

The work is given by:

$$w_{1,2} = p_1 V_1 \frac{1}{1 - \kappa} \left( \frac{\vartheta_2}{\vartheta_1} - 1 \right)$$

By inserting the ideal gas equation $p_1 V_1 = m R \vartheta_1 = m(c_p - c_v)\vartheta_1$ and the fact that $\kappa = c_p/c_v$, which yields:

$$w_{1,2} = m(c_p - c_v)\vartheta_1 \frac{1}{\kappa - 1}\left(1 - \frac{\vartheta_2}{\vartheta_1}\right) \tag{2.45}$$

$$= m(c_p - c_v)\frac{1}{\frac{c_p}{c_v} - 1}(\vartheta_1 - \vartheta_2) \tag{2.46}$$

and finally, the mechanical work for an isentropic state change is given by:

$$w_{1,2} = m\, c_v (\vartheta_1 - \vartheta_2) \tag{2.47}$$

**Polytropic Change**

Polytropic state changes are necessary to model a realistic isentropic process with insufficient insulation. Therefore, the polytropic exponent $n$ has values smaller than the isentropic exponent $\kappa$:

$$p V^n = const$$

with $n < \kappa$.

## 2.1.4   Thermodynamic Cycles

**Entropy**

Two different types of state changes have to be distinguished: reversible and irreversible state changes. They can be described by the following characteristics.
Reversible state changes:

- enable the system to return to the initial state

- do not need energy from outside the system when restored to their initial state.

- do not leave a permanent state change in a closed system after restoration.

Irreversible state changes:

- go into **one** direction. For example, expansion of molecules when a larger volume is available to them

- need energy from outside which is transformed into thermal energy. For example, a falling stone.

- do not return to the initial state.

The entropy is defined as:

$$S = \frac{q}{\vartheta} \tag{2.48}$$

An example is given to explain the entropy:

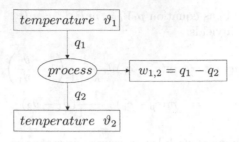

**Figure 2.8** Ideal cycle process

There are two bodies at two different temperatures $\vartheta$. Thermal energy $q$ is exchanged by conduction from the hot body at temperature $\vartheta_1$ to the cold body at temperature $\vartheta_2$. The change of entropy within the system after the transfer of thermal energy can be described by:

$$\Delta S = \frac{q_2}{\vartheta_2} - \frac{q_1}{\vartheta_1} \tag{2.49}$$

Even when $q_1=q_2$ there is a positive change of entropy $\Delta S$ due to the irreversible process of energy flow from the hot to the cold body. Thermal energy can only be transferred from hot to cold media and not in the opposite direction. The entropy of the cold body 2 is larger than the entropy of the hot body 1 as the entropy is inversely proportional to the temperature.

A reversible state change is characterized by:

$$\Delta S = 0 \tag{2.50}$$

and an irreversible state change is characterized by:

$$\Delta S > 0 \tag{2.51}$$

which is also called the second thermodynamic law. It means that the entropy within the system always increases for real state changes.

## Ideal Cycle Process

Assuming reversible state changes, the following ideal cycle process can be described: The gas absorbs thermal energy $q_1$ in an isothermal and reversible state change at high temperature $\vartheta_1$. The gas then loses the thermal energy $q_2$ in an isothermal and reversible state change at low temperature $\vartheta_2$. Kinetic energy $w_{1,2}$ is delivered by such an ideal engine which is equivalent to the difference of the thermal energies: $w_{1,2} = q_1 - q_2$. The entropy change shall be zero:

$$\Delta S = \frac{q_2}{\vartheta_2} - \frac{q_1}{\vartheta_1} = 0$$

and therefore:

$$\frac{q_1}{\vartheta_1} = \frac{q_2}{\vartheta_2} \tag{2.52}$$

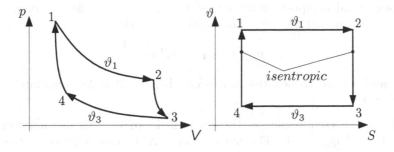

**Figure 2.9** $pV$-diagram and $\vartheta S$-diagram of Carnot Cycle.

The thermal efficiency of this ideal cycle process is equivalent to the ratio of mechanical work $w_{1,2}$ to the absorbed thermal energy $q_1$:

$$\eta = \frac{w_{1,2}}{q_1} = \frac{q_1 - q_2}{q_1} = \frac{\vartheta_1 - \vartheta_2}{\vartheta_1} = 1 - \frac{\vartheta_2}{\vartheta_1} \qquad (2.53)$$

It can be seen that the thermal efficiency $\eta$ only depends on the ratio of the absolute temperatures $\vartheta_1$ to $\vartheta_2$. $\eta$ is always smaller than 1. Please note that in non-ideal cycle processes the entropy change is always non-zero, i. e. $\Delta S > 0$.

**Carnot Cycle**

The Carnot cycle is characterized by four state changes. At the end of the process, the initial state is reached again which is the case for all periodic processes in engines. Two different cycles have to be distinguished:

**Reaction within a closed combustion chamber:** Engines with periodic combustion cycles and emission of mechanical work.

**Gas flow through an open combustion chamber:** Turbines with continuous combustion and emission of mechanical work.

The Carnot cycle is used as reference model for thermodynamic processes. It is illustrated in the $pV$-diagram of Figure 2.9:

$1 \rightarrow 2$ Isothermal expansion of the gas from $V_1$ to $V_2$ while the temperature remains constant $\vartheta_1 = \vartheta_2$.

$$w_{1,2} = q_{1,2} = m\,R\,\vartheta_1 \ln \frac{V_2}{V_1}$$

The mechanical work $w_{1,2}$ is delivered since the gas expands. Thermal energy $q_{1,2}$ of the same amount has to be brought into the system in order to keep the temperature constant.

$2 \rightarrow 3$ Isentropic expansion of the gas from $V_2$ to $V_3$. No thermal energy is exchanged $q_{2,3} = 0$. Output of kinetic energy:

$$w_{2,3} = m\,c_v(\vartheta_2 - \vartheta_3)$$

$3 \to 4$ Isothermal compression of the gas from $V_3$ to $V_4$ while the temperature remains constant $\vartheta_3 = \vartheta_4$. The mechanical work

$$w_{3,4} = q_{3,4} = m\,R\,\vartheta_3 \ln \frac{V_4}{V_3}$$

is needed which is negative in this case because it is delivered into the gas $(V_4 < V_3)$.

$4 \to 1$ Isentropic compression of the gas from $V_4$ to $V_1$. No thermal energy is exchanged $q_{2,3} = 0$. The mechanical work is also negative in this case $(\vartheta_1 > \vartheta_4)$:

$$w_{4,1} = m\,c_v(\vartheta_4 - \vartheta_1)$$

Now, all kinetic energies of the Carnot cycle have to be added to derive the thermodynamic efficiency:

$$
\begin{aligned}
w &= w_{1,2} + w_{2,3} + w_{3,4} + w_{4,1} & (2.54) \\
&= m\,R\,\vartheta_1 \ln \frac{V_2}{V_1} + m\,c_v(\vartheta_2 - \vartheta_3) + m\,R\,\vartheta_3 \ln \frac{V_4}{V_3} + m\,c_v(\vartheta_4 - \vartheta_1) & (2.55)
\end{aligned}
$$

The work $w_{2,3}$ and $w_{4,1}$ of the isentropic state changes compensate each other because of $\vartheta_1 = \vartheta_2$ and $\vartheta_3 = \vartheta_4$:

$$w = m\,R(\vartheta_1 \ln \frac{V_2}{V_1} + \vartheta_3 \ln \frac{V_4}{V_3})$$

To simplify this equation, the characteristic equations of the isentropic process $1 \to 2$ and $3 \to 4$ are used:

$$pV^\kappa = pV \cdot V^{\kappa-1} = m\,R\,\vartheta \cdot V^{\kappa-1} = const. \qquad (2.56)$$

$$\vartheta_2 V_2^{\kappa-1} = \vartheta_3 V_3^{\kappa-1} \quad , \quad \vartheta_1 V_1^{\kappa-1} = \vartheta_4 V_4^{\kappa-1} \qquad (2.57)$$

$$\left(\frac{V_4}{V_1}\right)^{\kappa-1} = \frac{\vartheta_1}{\vartheta_4} \qquad (2.58)$$

Because of $\vartheta_1 = \vartheta_2$ and $\vartheta_3 = \vartheta_4$ we get

$$\left(\frac{V_2}{V_1}\right)^{\kappa-1} = \left(\frac{V_3}{V_4}\right)^{\kappa-1} \qquad (2.59)$$

$$\frac{V_2}{V_1} = \frac{V_3}{V_4} \qquad (2.60)$$

$$\ln \frac{V_4}{V_3} = -\ln \frac{V_2}{V_1} \qquad (2.61)$$

and therefore:

$$w = m\,R(\vartheta_1 - \vartheta_3) \ln \frac{V_2}{V_1}$$

This result is now used for the thermodynamic efficiency:

$$\eta = \frac{w}{q_{1,2}} = \frac{m\,R(\vartheta_1 - \vartheta_3) \ln \frac{V_2}{V_1}}{m\,R\,\vartheta_1 \ln \frac{V_2}{V_1}} \qquad (2.62)$$

$$= \frac{\vartheta_1 - \vartheta_3}{\vartheta_1} = 1 - \frac{\vartheta_3}{\vartheta_1} \qquad (2.63)$$

The efficiency depends only on the temperature ratio of $\vartheta_3$ to $\vartheta_1$. Using the compression ratio $\varepsilon$ :

$$\varepsilon = \frac{V_4}{V_1} \tag{2.64}$$

the efficiency is ($\vartheta_3 = \vartheta_4$):

$$\eta = 1 - \frac{\vartheta_4}{\vartheta_1} = 1 - \left(\frac{V_1}{V_4}\right)^{\kappa - 1} \tag{2.65}$$

$$= 1 - \frac{1}{\varepsilon^{\kappa - 1}} \tag{2.66}$$

The thermal efficiency can be explained graphically in the $pV$-diagram in Figure 2.9: It is equivalent to the ratio of the integral within the cycle to the area which is produced by integrating $q_{1,2}$. The efficiency is improved by increasing the area within the cycle. This can only be achieved by a rise of the compression ratio $\varepsilon$ or the temperature ratio.

To give an example: Assuming absolute temperatures:

$$\vartheta_1 = 2800\,°K \tag{2.67}$$
$$\vartheta_3 = 300\,°K \tag{2.68}$$

the thermal efficiency would be:

$$\eta = 0.89$$

and the compression ratio:

$$\varepsilon = 266$$

supposing a $\kappa = 1.4$. Such a compression ratio can hardly be produced by real engines. Typical compression ratios are at least ten times smaller.

## 2.2 Ideal Combustion Engines

Commonly used combustion engines in cars are four-stroke engines. They have two intermittent cycles: the gas is compressed, combusted and expanded in the first cycle, and the gas is exchanged in the second cycle. In this section the second (or passive) cycle will not be considered to simplify the mathematical derivations. The processes related to the second cycle will be discussed in Chapter 3.

Two different types of combustion engines have to be distinguished:

1. Spark-ignited (SI) Engine: Combustion caused by an electric spark-ignition.

2. Diesel Engine: Combustion caused by self inflammation due to compressional heat.

In most sections, $p$ represents the in-cylinder pressure, $V$ the cylinder volume, $\vartheta$ the in-cylinder temperature, $S$ the entropy, $q$ the thermal energy of the gas, $u$ its internal energy and $h$ its enthalpy.

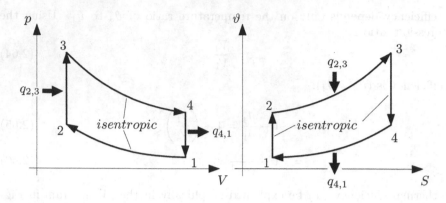

**Figure 2.10** $pV$-diagram (left) and $\vartheta S$-diagram (right) of the SI engine process

## 2.2.1   Spark-ignited (SI) Engine

The first SI engine was presented by Nikolaus Otto in 1862. The combustion process can be modeled as an **isochoric process** where the gas volume is considered to be constant. The $pV$-diagram in Figure 2.10 illustrates that the gas volume does not change between step 2 and step 3. The ratio of maximum to minimum volume is given by:

$$\varepsilon = \frac{V_1}{V_2} \tag{2.69}$$

This ratio $\varepsilon$ is called the **compression ratio** of the engine. The different steps for a complete cycle in the $pV$-diagram and in the $\vartheta S$-diagram can be seen in Figure 2.10. Mathematically they can be described as followed:

$1 \rightarrow 2$ : Isentropic compression, $dq = 0$:

$$
\begin{aligned}
dq &= du + dw = 0 \\
q_{1,2} &= 0 \\
dw &= -du = -m\,c_v\,d\vartheta \\
w_{1,2} &= -\int_1^2 m\,c_v\,d\vartheta = -m\,c_v(\vartheta_2 - \vartheta_1)
\end{aligned}
$$

The work $w_{1,2}$ is used to compress the gas and therefore, it is negative.

$2 \rightarrow 3$ : Isochoric input of thermal energy, $dV = 0$:

$$dw = p\,dV = 0$$

$$w_{2,3} = \int_2^3 p\,dV = 0$$

$$dq = du = m\,c_v\,d\vartheta$$

$$q_{2,3} = m\,c_v \int_2^3 d\vartheta = m\,c_v(\vartheta_3 - \vartheta_2)$$

The increased thermal energy $q_{2,3}$ is caused by combustion of the gas.

$3 \rightarrow 4$ : Isentropic expansion, $dq = 0$:

$$q_{3,4} = 0$$

$$dw = -du = -m\,c_v\,d\vartheta$$

$$w_{3,4} = -\int_3^4 m\,c_v\,d\vartheta = -m\,c_v(\vartheta_4 - \vartheta_3)$$

This state change describes the power stroke of the engine where $w_{3,4}$ is the output of kinetic energy from the gas, which is positive ($\vartheta_4 < \vartheta_3$).

$4 \rightarrow 1$ : Isochoric heat loss, $dV = 0$:

$$dw = p\,dV = 0$$

$$w_{4,1} = \int_4^1 p\,dV = 0$$

$$dq = du + dw = m\,c_v\,d\vartheta$$

$$q_{4,1} = m\,c_v \int_4^1 d\vartheta = m\,c_v(\vartheta_1 - \vartheta_4)$$

The loss of thermal energy $q_{4,1}$ is due to the gas exchange: The burnt hot gas is pumped into the exhaust and the combustion chamber is filled with a cold mixture of unburnt fuel vapor and air ($q_{4,1}$ is negative because of $\vartheta_1 < \vartheta_4$).

The thermal efficiency of the engine is equivalent to the ratio of all the kinetic energies to the input of thermal energy $q_{2,3}$ at the combustion of a complete cycle:

$$
\begin{aligned}
\eta_{th} &= \frac{w_{1,2} + w_{2,3} + w_{3,4} + w_{4,1}}{q_{2,3}} \\
&= \frac{m\,c_v(-\vartheta_2 + \vartheta_1 - \vartheta_4 + \vartheta_3)}{m\,c_v(\vartheta_3 - \vartheta_2)} \\
&= 1 - \frac{\vartheta_4 - \vartheta_1}{\vartheta_3 - \vartheta_2} \\
&= 1 - \frac{\vartheta_1}{\vartheta_2}\frac{\vartheta_4/\vartheta_1 - 1}{\vartheta_3/\vartheta_2 - 1}
\end{aligned}
$$

The relationship for isentropic changes $1 \to 2$ and $3 \to 4$ can be used to simplify the equation:

$$
\frac{\vartheta_4}{\vartheta_3} = \left(\frac{V_3}{V_4}\right)^{\kappa-1} = \frac{1}{\varepsilon^{\kappa-1}} = \frac{\vartheta_1}{\vartheta_2} \tag{2.70}
$$

This yields:

$$
\eta_{th} = 1 - \frac{1}{\varepsilon^{\kappa-1}} \tag{2.71}
$$

Please note that the thermal efficiency $\eta_{th}$ does not depend on the absolute temperature values. It mainly depends on the compression ratio $\varepsilon$. Example: For a compression ratio of $\varepsilon = 11$ and an adiabatic coefficient of $\kappa = 1.4$ the theoretical thermal efficiency $\eta_{th}$ is:

$$
\eta_{th} = 0.617
$$

## 2.2.2   Diesel Engine

Rudolf Diesel developed this engine from 1893 to 1897. In a diesel engine, the combustion takes place in an **isobaric state change** during the downward movement of the piston. At the beginning of this process the combustion is controlled by the injection of fuel to maintain a constant pressure at the expansion from 2 to 3. The isobaric state change is indicated between steps 2 and 3 in the $pV$-diagram in Figure 2.11. The more fuel is injected, the longer the distance between steps 2 and 3 and the larger the volume ratio:

$$
\rho = \frac{V_3}{V_2} = \frac{\vartheta_3}{\vartheta_2} \quad . \tag{2.72}
$$

This ratio is called **injection ratio** or **load**. The injection ratio $\rho$ has an impact on the thermodynamic efficiency which is derived after explaining the different parts of the cycle:

$1 \to 2$ : Isentropic compression, $dq = 0$:

$$
\begin{aligned}
dq &= du + dw = 0 \\
q_{1,2} &= 0 \\
dw &= -du = -m\,c_v\,d\vartheta \\
w_{1,2} &= -m\,c_v(\vartheta_2 - \vartheta_1)
\end{aligned}
$$

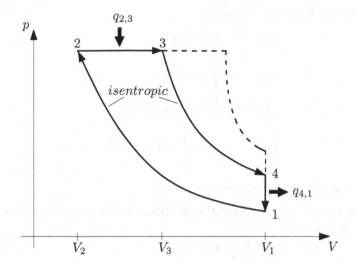

**Figure 2.11** pV-diagram for Diesel Engine

The mechanical work $w_{1,2}$ is used to compress the gas (equivalent to the SI engine). It is negative.

$2 \rightarrow 3$ : Isobaric gain of thermal energy, $dp = 0$:

$$
\begin{aligned}
dq &= dh - V\, dp = m\, c_p\, d\vartheta \\
q_{2,3} &= m\, c_p(\vartheta_3 - \vartheta_2) \\
dw &= p\, dV = m\, R\, d\vartheta \\
w_{2,3} &= m\, R(\vartheta_3 - \vartheta_2)
\end{aligned}
$$

In this process, the combustion generates the thermal energy $q_{2,3}$ and produces the kinetic energy $w_{2,3}$.

$3 \rightarrow 4$ : Isentropic expansion, $dq = 0$:

$$
\begin{aligned}
q_{3,4} &= 0 \\
dw &= -du = -m\, c_v\, d\vartheta \\
w_{3,4} &= -m\, c_v(\vartheta_4 - \vartheta_3)
\end{aligned}
$$

Note that $w_{3,4}$ is positive since $\vartheta_4 < \vartheta_3$.

$4 \rightarrow 1$ : Isochoric heat loss, $dV = 0$:

$$dw = p\,dV = 0$$

$$w_{4,1} = \int_4^1 p\,dV = 0$$

$$dq = du + dw = m\,c_v\,d\vartheta$$

$$q_{4,1} = m\,c_v \int_4^1 d\vartheta = m\,c_v(\vartheta_1 - \vartheta_4)$$

Note that $q_{4,1}$ is negative since $\vartheta_1 < \vartheta_4$.

With $\kappa = \frac{c_p}{c_v}$ and $R = (c_p - c_v)$ the thermodynamic efficiency of the diesel engine can now be calculated:

$$
\begin{aligned}
\eta_{th} &= \frac{w_{1,2} + w_{2,3} + w_{3,4} + w_{4,1}}{q_{2,3}} \\
&= \frac{-m\,c_v(\vartheta_2 - \vartheta_1) + m(c_p - c_v)(\vartheta_3 - \vartheta_2) - m\,c_v(\vartheta_4 - \vartheta_3)}{m\,c_p(\vartheta_3 - \vartheta_2)} \\
&= 1 - \frac{1}{\kappa}\frac{\vartheta_1}{\vartheta_2}\frac{\vartheta_4/\vartheta_1 - 1}{\vartheta_3/\vartheta_2 - 1}
\end{aligned}
$$

This equation can be simplified by using the relationship for the isentropic process (Equation 2.70) and the relationship for the isobaric process (Equation 2.72). Additionally, the following relationship is used for the isochoric heat loss:

$$\frac{\vartheta_4}{\vartheta_1} = \frac{p_4}{p_1} = \frac{p_4\,p_2}{p_3\,p_1} = \frac{V_3^\kappa}{V_4^\kappa}\frac{V_1^\kappa}{V_2^\kappa}$$

In that, we have $p_2 = p_3$.

$$\frac{\vartheta_4}{\vartheta_1} = \left(\frac{\vartheta_4}{\vartheta_3}\right)^{\kappa/\kappa-1}\left(\frac{\vartheta_2}{\vartheta_1}\right)^{\kappa/\kappa-1} = \left(\frac{\vartheta_4}{\vartheta_1}\right)^{\kappa/\kappa-1}\left(\frac{\vartheta_2}{\vartheta_3}\right)^{\kappa/\kappa-1}$$

This results in: $\vartheta_4/\vartheta_1 = (\vartheta_3/\vartheta_2)^\kappa = \rho^\kappa$, which yields the thermodynamic efficiency of the Diesel engine:

$$\eta_{th} = 1 - \frac{1}{\varepsilon^{\kappa-1}}\frac{1}{\kappa}\frac{\rho^\kappa - 1}{\rho - 1} \tag{2.73}$$

It can be seen that the efficiency $\eta_{th}$ decreases as the load $\rho$ is increased. At high loads, the diesel engine has a lower efficiency compared to the SI engine, supposing the same compression ratio $\varepsilon$ for both (see Figure 2.13). The compression ratio for Diesel engines is however much higher than for SI engines to improve the thermodynamic efficiency.

## 2.2.3   Seiliger Process

The Seiliger process models the thermodynamic process in automotive engines much better than the previously described models of SI and Diesel engines. Figure 2.12 shows that the combustion is now divided into two parts: In the first

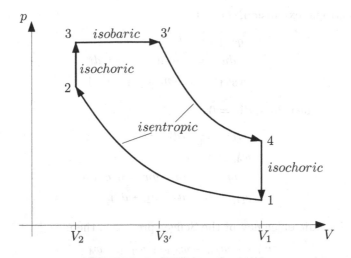

**Figure 2.12** pV-diagram of Seiliger process

part, the gas is heated in an **isochoric process** between step 2 and step 3. In the second part the gas is expanded in an **isobaric state change** between step 3 and step 3'. The cycle is characterized by the **compression ratio** $\varepsilon = V_1/V_2$ , the **injection ratio** $\rho = V_{3'}/V_3$, and the **pressure ratio**:

$$\chi = \frac{p_3}{p_2} \tag{2.74}$$

The different steps of the cycle are the following:

$1 \to 2$ : Isentropic compression, $dq = 0$:

$$
\begin{aligned}
dq &= du + dw = 0 \\
q_{1,2} &= 0 \\
dw &= -du = -m\, c_v\, d\vartheta \\
w_{1,2} &= -m\, c_v(\vartheta_2 - \vartheta_1)
\end{aligned}
$$

$2 \to 3$ : Isochoric input of thermal energy, $dV = 0$:

$$
\begin{aligned}
dw &= p\, dV = 0 \\
w_{2,3} &= 0 \\
dq &= du = m\, c_v\, d\vartheta \\
q_{2,3} &= m\, c_v(\vartheta_3 - \vartheta_2)
\end{aligned}
$$

$3 \to 3'$ : Isobaric input of thermal energy, $dp = 0$:

$$
\begin{aligned}
dq &= dh - V\, dp = m\, c_p\, d\vartheta \\
q_{3,3'} &= m\, c_p(\vartheta_{3'} - \vartheta_3) \\
dw &= p\, dV = m\, R\, d\vartheta \\
w_{3,3'} &= m\, R(\vartheta_{3'} - \vartheta_3)
\end{aligned}
$$

$3' \rightarrow 4$ : Isentropic expansion, $dq = 0$:

$$
\begin{aligned}
q_{3',4} &= 0 \\
dw &= -du = -m\,c_v\,d\vartheta \\
w_{3',4} &= -m\,c_v(\vartheta_4 - \vartheta_{3'})
\end{aligned}
$$

$4 \rightarrow 1$ : Isochoric heat loss, $dV = 0$:

$$
\begin{aligned}
dw &= p\,dV = 0 \\
w_{4,1} &= 0 \\
dq &= du + dw = m\,c_v\,d\vartheta \\
q_{4,1} &= m\,c_v(\vartheta_1 - \vartheta_4)
\end{aligned}
$$

The thermodynamic efficiency of the Seliger process is then:

$$
\begin{aligned}
\eta_{th} &= \frac{w_{1,2} + w_{2,3} + w_{3,3'} + w_{3',4} + w_{4,1}}{q_{2,3}} \\
&= 1 - \frac{\vartheta_4/\vartheta_1 - 1}{\vartheta_3/\vartheta_1 - \vartheta_2/\vartheta_1 + \kappa(\vartheta_{3'}/\vartheta_1 - \vartheta_3/\vartheta_1)}
\end{aligned}
$$

The isentropic process is characterized by:

$$
\frac{\vartheta_2}{\vartheta_1} = \varepsilon^{\kappa-1}
$$

which yields:

$$
\begin{aligned}
\frac{\vartheta_4}{\vartheta_{3'}} &= \left(\frac{V_{3'}}{V_4}\right)^{\kappa-1} \\
&= \left(\frac{V_{3'}}{V_3}\frac{V_3}{V_4}\right)^{\kappa-1} = \left(\frac{V_{3'}}{V_3}\frac{V_2}{V_1}\right)^{\kappa-1} \\
&= \left(\frac{\rho}{\varepsilon}\right)^{\kappa-1}
\end{aligned}
$$

In the isochoric process, the relationship

$$
\frac{\vartheta_3}{\vartheta_2} = \frac{p_3}{p_2} = \chi
$$

can be used, and the temperature ratio in the isobaric process is given by:

$$
\frac{\vartheta_{3'}}{\vartheta_3} = \frac{V_{3'}}{V_3} = \rho \quad .
$$

Therefore, the following temperature ratios may be expressed as:

$$
\begin{aligned}
\frac{\vartheta_3}{\vartheta_1} &= \frac{\vartheta_3}{\vartheta_2}\frac{\vartheta_2}{\vartheta_1} = \chi\,\varepsilon^{\kappa-1} \\
\frac{\vartheta_{3'}}{\vartheta_1} &= \frac{\vartheta_{3'}}{\vartheta_3}\frac{\vartheta_3}{\vartheta_1} = \rho\,\chi\,\varepsilon^{\kappa-1} \\
\frac{\vartheta_4}{\vartheta_1} &= \frac{\vartheta_4}{\vartheta_{3'}}\frac{\vartheta_{3'}}{\vartheta_1} = \left(\frac{\rho}{\varepsilon}\right)^{\kappa-1}\rho\,\chi\,\varepsilon^{\kappa-1} = \chi\,\rho^{\kappa}
\end{aligned}
$$

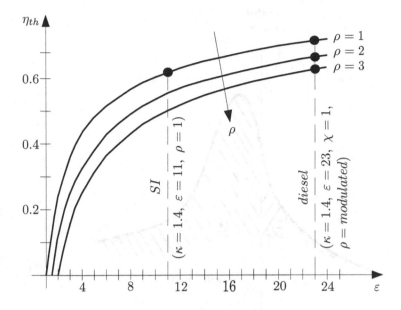

**Figure 2.13** Thermodynamic efficiency $\eta_{th}$ depending on compression ratio $\varepsilon$ and load $\rho$ (Equation 2.75)

The thermodynamic efficiency of the Seliger process can be simplified to:

$$\eta_{th} = 1 - \frac{1}{\varepsilon^{\kappa-1}} \frac{\chi \, \rho^{\kappa} - 1}{\chi - 1 + \kappa \, \chi \, (\rho - 1)} \qquad (2.75)$$

It can be seen that the thermodynamic efficiency of the SI engine is obtained when $\rho = 1$ and that of the Diesel engine when $\chi = 1$ (Figure 2.13).

## 2.2.4 Comparison of Different Engine Concepts

The in-cylinder pressure during combustion is plotted over the crankshaft angle in Figure 2.14. The compression ratio $\varepsilon$ for the SI engine is limited by the maximum allowable pressure $p_3$ during the combustion process. Under part load conditions, the maximum pressure of a cycle is far below this limit, since the SI engine power output is modulated by throttling the air intake thus modulating $p_1$ (Figure 2.15). A low compression ratio $\varepsilon$ is also helpful to reduce knocking of the engine and to meet material demands. In contrast to SI engines, the maximum pressure for Diesel engines is closer set to the maximum allowable pressure $p_3$. As the Diesel engine is unthrottled (modulation of $\rho$), it can afford higher compression ratios $\varepsilon = V_1/V_2$ than the SI engine (Figure 2.16).

The four-stroke engine works intermittently: A hot combustion cycle is always followed by a cool gas exchange cycle. In the first cycle, peak temperatures of $2500 - 2800\,°C$ occur. Compared to that, the maximum temperature of gas turbines must be kept much lower ($1300\,°C$), since the combustion process is continuous. Diesel engines have a higher thermodynamic efficiency than SI engines

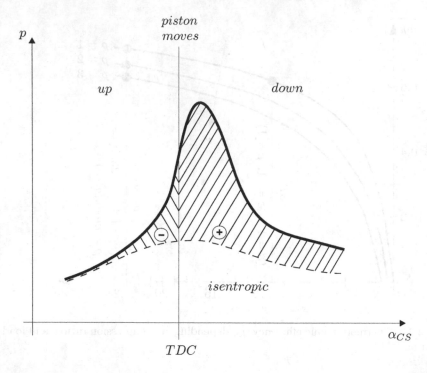

**Figure 2.14** In-cylinder pressure over crankshaft angle during combustion process

at low and medium power (Figure 2.13). At turbocharged Diesel engines also $p_1$ is modulated. In order to stay below maximum temperatures, the compression ratio $\varepsilon$ is reduced at turbocharged Diesel engines. The absolute effective power output of combustion engines depends on the displacement volume $V_d$, the specific work output per power-stroke $w_e$ and the number of crankshaft revolutions $n$:

$$P_e = w_e \, V_d \, \frac{n}{2} \tag{2.76}$$

where:

$P_e$ — is the absolute effective power output
$w_e$ — is the specific work referred to the displacement volume $V_d$
$V_d$ — is the displacement volume $V_1 - V_2$
   (here, only one cylinder is regarded)
$n$ — is the number of crankshaft revolutions per minute

The factor $1/2$ is necessary in determining the power output of four-stroke engines since only every second cycle contributes power. The mean piston velocity $\bar{s}$ depends on the number of crankshaft revolutions per minute $n$ and the maximum piston stroke $2\,r$ from top dead to bottom dead center. The mean piston velocity is:

$$\bar{s} = 4\,n\,r \tag{2.77}$$

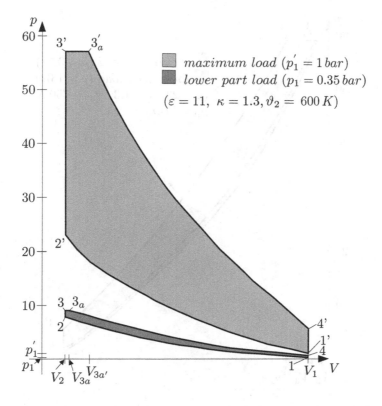

**Figure 2.15** Load behavior of SI engine (Modulation of $p_1$), $1\,bar = 10^5\,Pa$

A typical value in car engines is $\bar{s} = 15\,m/s$ at maximum power output. Equation 2.76 can now be written as:

$$\frac{P_e}{V_d} = w_e \frac{\bar{s}}{8\,r} \qquad (2.78)$$

Please note that the specific power $P_e/V_d$ is inversely proportional to the maximum piston stroke $2\,r$ ($r$ is the crankshaft radius). Table 2.1 gives some examples of engine types and their characteristic values. It can be seen that the specific power decreases with increasing dimensions $r$ at large engines. They are mainly used in ships because of their thermodynamic efficiency. When size and weight are major concerns at high power levels, gas turbines are preferred e.g. in aircraft.

## 2.3 Alternative Combustion Engines

### 2.3.1 Gas Turbine

In the gas turbine engine fuel is combusted in a continuous process. The air has to be compressed before it flows into the combustion chamber. There the fuel

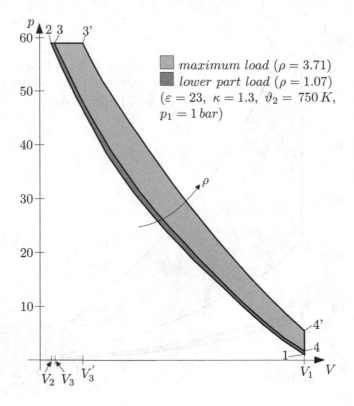

**Figure 2.16** Load behavior of Diesel engine (Modulation of $\rho$), $1\,bar = 10^5\,Pa$

is injected and burned. The resulting expansion of the gas is used to turn the turbine shaft. Therefore, the effective power is the power of the turbine minus the power used for the compressor. The process can be modeled as a **Brayton cycle** (or Joule process) which is shown in Figure 2.17. The different steps of the Brayton cycle:

$1 \rightarrow 2$ : Isentropic compression, $dq = 0$:

$$
\begin{aligned}
dq &= du + dw = 0 \\
q_{1,2} &= 0 \\
dw &= -dh = -m\,c_p\,d\vartheta \\
w_{1,2} &= -m\,c_p(\vartheta_2 - \vartheta_1)
\end{aligned}
$$

The kinetic energy $w_{1,2}$ is provided by the compressor.

$2 \rightarrow 3$ : Isobaric input of thermal energy, $dp = 0$:

$$
\begin{aligned}
dq &= dh - V\,dp = m\,c_p\,d\vartheta \\
q_{2,3} &= m\,c_p(\vartheta_3 - \vartheta_2) \\
dw &= p\,dV = m\,R\,d\vartheta \\
w_{2,3} &= m\,R(\vartheta_3 - \vartheta_2)
\end{aligned}
$$

**Table 2.1** Specific power output of various combustion engines

| Engine type | $\bar{\dot{s}}$ | $w_e$ | $2\,r$ | $P_e/V_d$ |
| --- | --- | --- | --- | --- |
| | $[m/s]$ | $[J/cm^3]$ | $[cm]$ | $[kW/dm^3]$ |
| Car engine | 15 | 1.6 | 8 | 75 |
| Truck engine | 11 | 1.9 | 14 | 37 |
| Big four-stroke diesel | 9 | 2.0 | 32 | 14 |
| Big two-stroke diesel | 7 | 1.8 | 60 | 5 |

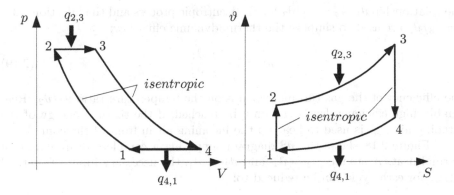

**Figure 2.17** pV-diagram (left) and $\vartheta$S-diagram (right) of the Brayton cycle

$3 \rightarrow 4$ : Isentropic expansion, $dq = 0$, generation of work:

$$
\begin{aligned}
q_{3,4} &= 0 \\
dw &= -dh = -m\,c_p\,d\vartheta \\
w_{3,4} &= -m\,c_p(\vartheta_4 - \vartheta_3)
\end{aligned}
$$

$4 \rightarrow 1$ : Isobaric heat loss, $dp = 0$:

$$
\begin{aligned}
dq &= dh - V\,dp = m\,c_p\,d\vartheta \\
q_{4,1} &= m\,c_p(\vartheta_1 - \vartheta_4) \\
dw &= m\,R\,d\vartheta = 0 \\
w_{4,1} &= m\,R(\vartheta_1 - \vartheta_4)
\end{aligned}
$$

The thermal energy $q_{4,1}$ and mechanical work $w_{4,1}$ are both negative because of $\vartheta_1 < \vartheta_4$.

The thermal efficiency of the gas turbine is:

$$
\begin{aligned}
\eta_{th} &= \frac{q_{2,3} + q_{4,1}}{q_{2,3}} \\
&= \frac{m\,c_p(\vartheta_3 - \vartheta_2) + m\,c_p(\vartheta_1 - \vartheta_4)}{m\,c_p(\vartheta_3 - \vartheta_2)} \\
&= 1 - \frac{\vartheta_4}{\vartheta_3}\frac{1 - \vartheta_1/\vartheta_4}{1 - \vartheta_2/\vartheta_3}
\end{aligned}
$$

As pointed out before, the isentropic process is characterized by:

$$
\frac{\vartheta_1}{\vartheta_2} = \left(\frac{V_2}{V_1}\right)^{\kappa-1} = \frac{1}{\varepsilon^{\kappa-1}} = \frac{\vartheta_4}{\vartheta_3}
$$

The relationship $\vartheta_1/\vartheta_4 = \vartheta_2/\vartheta_3$ for the isentropic process and the injection ratio $\rho = \vartheta_3/\vartheta_2$ are used to simplify the thermodynamic efficiency:

$$
\eta_{th} = 1 - \frac{1}{\rho}\frac{\vartheta_4}{\vartheta_2} \tag{2.79}
$$

The efficiency of the gas turbine depends on the temperature ratio $\vartheta_4/\vartheta_2$. Reasonably high efficiency levels can only be reached, if the thermal energy of the outgoing air $q_{4,4'}$ is used to heat up the incoming air in front of the compressor $q_{2,2'}$. Figure 2.18 shows the $\vartheta S$-diagram supposing a complete recirculation of thermal heat. Assuming $\vartheta_2 = \vartheta_{4'}$ and $\vartheta_{2'} = \vartheta_4$, the necessary input of thermal energy for each cycle can be reduced to:

$$
q_{2',3} = m\,c_p(\vartheta_3 - \vartheta_{2'}) = m\,c_p(\vartheta_3 - \vartheta_4) \tag{2.80}
$$

and the heat loss is also reduced to:

$$
q_{4',1} = m\,c_p(\vartheta_1 - \vartheta_{4'}) = m\,c_p(\vartheta_1 - \vartheta_2) \tag{2.81}
$$

Equation 2.79 can be modified:

$$
\begin{aligned}
\eta_{th} &= \frac{q_{2',3} + q_{4',1}}{q_{2',3}} \\
&= 1 - \frac{\vartheta_2}{\vartheta_3}\frac{1 - \vartheta_1/\vartheta_2}{1 - \vartheta_4/\vartheta_3}
\end{aligned}
$$

Finally, by using the same relationships as before, the thermodynamic efficiency of the gas turbine with complete heat recirculation is:

$$
\eta_{th} = 1 - \frac{1}{\rho} \tag{2.82}
$$

The thermal efficiency depends only on the load $\rho = \vartheta_3/\vartheta_2$. The higher the load the better the efficiency. The maximum temperature $\vartheta_3$ is mainly limited by the material from which the gas turbine is constructed. Equation 2.82 is not valid for extremely small loads as the heat recirculation does not work properly under these conditions. The main advantages of gas turbines are the following:

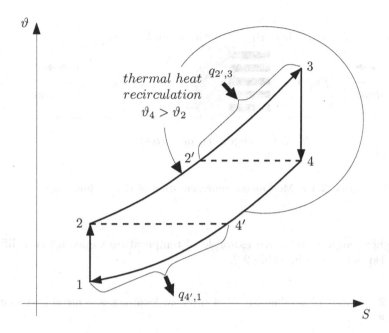

**Figure 2.18** $\vartheta$S-diagram of the Joule process with heat recirculation

- Gas turbines are much smaller and lighter compared to four-stroke piston engines: The weight per power output in gas turbines is $0.3 - 0.5\,kg/kW$ and for four-stroke piston engines it is $1.0 - 1.5\,kg/kW$.

- High torque can be generated even at very low revolutions.

- Different types of fuel can be used for combustion: multi-fuel capability.

- Gas turbines are easy to start even at low temperatures.

- Low vibrations because of a continuous combustion and the rotary motion.

- Long service intervals between required maintenance.

- Reduced emissions of noxious exhaust gases. ECE test results:
  $CO = 20\,g$, $HC = 0.8\,g$, $NO_x = 2 - 3\,g$ per test.

However, there are some disadvantages:

- Low efficiency for low loads.

- Poor dynamic behavior during transients.

Gas turbines are mainly used in air planes because of their low weight. Their efficiency can be increased by raising the maximum allowable temperature. Ceramics like $Al_2O_3$ or laminated silicon-carbon materials are used for the construction to

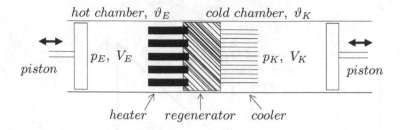

**Figure 2.19** Mechanical representation of the Stirling engine

allow higher temperatures. An example of temperature values for two different turbines types is given in Table 2.2.

**Table 2.2** Temperature values in $^\circ C$ at different locations of a metal and a ceramic gas turbine

| Location of measurement | Metal gas turbine | Ceramic gas turbine |
|:---:|:---:|:---:|
| Compressor inlet | 230 | 250 |
| Heat exchanger outlet (air) | 700 | 950 |
| Combustion chamber outlet | 1000-1100 | 1250-1350 |
| Heat exchanger inlet (gas) | 750 | 1000 |
| Heat exchanger outlet (gas) | 270 | 300 |

## 2.3.2   Stirling Engine

The Stirling engine is a piston engine which uses a continuous heat supply. It was invented by Robert Stirling in 1816. The cycle process has a high efficiency, comparable to that of the reference Carnot cycle. Even though the engine was redesigned by Philips Corporation (in 1938) and recently by other companies, some mechanical problems remain unsolved. Figure 2.19 shows the mechanical representation of the Stirling engine. It consists of three main parts:

H   heater        the gas is heated from outside
R   regenerator   the regenerator stores thermal energy
C   cooler        the gas is cooled from outside

The heater, regenerator and cooler are located in the middle. The pistons on the left and right side are linked mechanically. The cycle consists of the following four steps that can be seen in Figure 2.20:

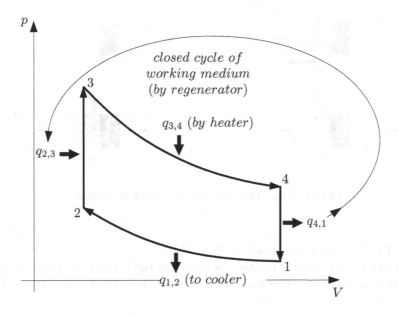

**Figure 2.20** pV-diagram of the Stirling cycle

$1 \rightarrow 2$ : Isothermal compression: $\vartheta_1 = \vartheta_2$
   The heat is absorbed by the cooler when the two pistons move to the left
   to compress the gas.

$$w_{1,2} = m R \vartheta_1 \ln \frac{V_2}{V_1}$$
$$q_{1,2} = w_{1,2}$$

The emission of heat $q_{1,2}$ is equivalent to the input of kinetic energy $w_{1,2}$.

$2 \rightarrow 3$ : Isochoric input of thermal energy: $V_2 = V_3$
   The two pistons move simultaneously to the left and the gas is heated by
   the regenerator. Both, pressure and temperature are increased.

$$q_{2,3} = m c_v (\vartheta_3 - \vartheta_2)$$

$3 \rightarrow 4$ : Isothermal expansion: $\vartheta_3 = \vartheta_4$
   The thermal energy is supplied by the heater, when the pistons move back
   to the right.

$$w_{3,4} = m R \vartheta_3 \ln \frac{V_4}{V_3}$$
$$q_{3,4} = w_{3,4}$$

The mechanical work $w_{3,4}$ is equivalent to the thermal energy $q_{3,4}$.

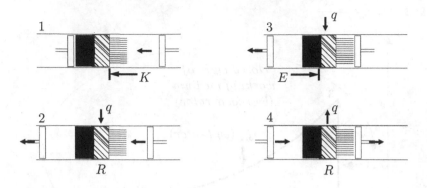

**Figure 2.21** The four steps of the Stirling cycle

$4 \rightarrow 1$ : Isochoric heat regeneration: $V_4 = V_1$

As the pistons move simultaneously to the right, thermal energy is stored in the regenerator. The pressure as well as the temperature drop to lower levels.

$$q_{4,1} = m \, c_v (\vartheta_1 - \vartheta_4)$$

By exploiting the fact that $\vartheta_1 = \vartheta_2$ and $\vartheta_3 = \vartheta_4$ for the isothermic processes, and $V_1 = V_4$ and $V_2 = V_3$ for the isochoric processes, the thermal efficiency of the Stirling engine is:

$$\eta_{th} = \frac{q_{1,2} + q_{3,4}}{q_{3,4}} = 1 - \frac{\vartheta_1}{\vartheta_3} \tag{2.83}$$

Hence, the efficiency depends only on the temperature ratio $\vartheta_1 / \vartheta_3$. For example, a temperature ratio of $80\,°C / 600\,°C$ results in a thermodynamic efficiency of:

$$\eta_{th} = 0.87 \tag{2.84}$$

which is very close to the efficiency of the Carnot cycle. The main advantages of the Stirling engine are:

- Engine is independent of the heat source. Instead of combusting fossil fuels, alternate heat sources such as solar heat could be employed.

- High (theoretical) efficiency.

- Very quiet.

- Reduced emission of noxious exhaust gases. ECE test results:
  $CO = 4 - 6\,g$, $HC = 0.5 - 2\,g$, $NO_x = 0.6 - 2.0\,g$ per test.

On the other hand, there are some disadvantages:

- Expensive to construct.

- Regenerator: conduction and storage of heat are difficult to combine.

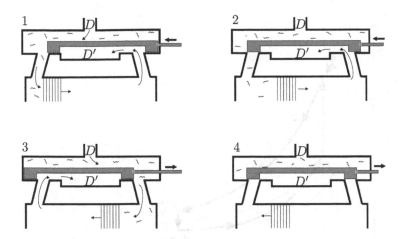

**Figure 2.22** Steam engine

- Heat resistant materials needed.

- A heat exchanger for the cooler is needed to increase the efficiency. It increases however volume and costs.

Experimental Stirling engines with temperatures of $40 - 80\,^{\circ}C/600 - 650\,^{\circ}C$ can reach an effective thermodynamic efficiency of

$$\eta_{eff} = 0.35 - 0.40 \tag{2.85}$$

which is much lower than the theoretical value (Equation 2.84).

### 2.3.3 Steam Engine

The steam engine introduced by James Watt is the oldest engine using continuous combustion. The steam is transferred from the boiler to the cylinder. The piston is moved by the expanding steam. The linear movement of the piston is translated into a rotation of the crankshaft by the connecting rod. The different steps are illustrated in Figure 2.22 and Figure 2.23.

The different steps of the steam engine cycle are:

$1 \rightarrow 2$ : Isochoric compression, followed by an isothermal expansion:

This process can be divided into two parts: First, hot steam is injected into the cylinder through the open valve at constant volume ($1 \rightarrow 1'$) where $V_{1'} = V_1$. Second, the gas expands at a constant temperature ($1' \rightarrow 2$) where $\vartheta_{1'} = \vartheta_2$:

$$w_{1,1'} = 0$$
$$q_{1,1'} = m\,c_v(\vartheta_{1'} - \vartheta_1) = m\,c_v(\vartheta_2 - \vartheta_1)$$
$$w_{1',2} = m\,R\,\vartheta_{1'} \ln\frac{V_2}{V_{1'}} = m\,R\,\vartheta_2 \ln\frac{V_2}{V_1}$$
$$q_{1',2} = w_{1',2}$$

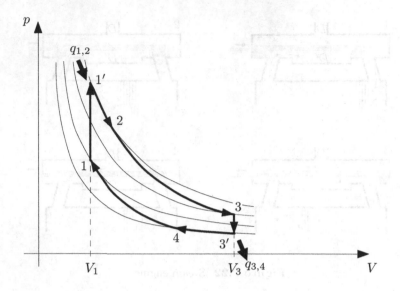

**Figure 2.23** pV-diagram of the steam engine

$2 \rightarrow 3$ : Isentropic expansion:

After the valve is closed, the expansion is continued until the maximum volume is reached.

$$q_{2,3} = 0$$
$$w_{2,3} = m\,c_v(\vartheta_2 - \vartheta_3)$$

$3 \rightarrow 4$ : Isochoric heat regeneration and isothermal compression:

This process can be divided into two steps: The pressure drops at constant volume after the valve is opened at $3 \rightarrow 3'$ where $V_3 = V_{3'}$. Second, the steam is compressed isothermally at $3' \rightarrow 4$ where $\vartheta_{3'} = \vartheta_4$.

$$w_{3,3'} = 0$$
$$q_{3,3'} = m\,c_v(\vartheta_{3'} - \vartheta_3) = m\,c_v(\vartheta_4 - \vartheta_3)$$
$$w_{3',4} = m\,R\,\vartheta_{3'} \ln\frac{V_4}{V_{3'}} = m\,R\,\vartheta_4 \ln\frac{V_4}{V_3}$$
$$q_{3',4} = w_{3',4}$$

$4 \rightarrow 1$ : Isentropic compression: After the valve is closed, the gas is compressed mechanically:

$$q_{4,1} = 0$$
$$w_{4,1} = m\,c_v(\vartheta_4 - \vartheta_1)$$

The mechanical work $w_{4,1}$ is negative.

The thermal efficiency of the steam engine is expressed by:

$$\eta_{th} = \frac{w_{1',2} + w_{2,3} + w_{3',4} + w_{4,1}}{q_{1,1'} + q_{1',2}}$$

$$= 1 - \frac{\vartheta_3 - \vartheta_4 + (\kappa - 1)\vartheta_4 \ln(V_3/V_4)}{\vartheta_2 - \vartheta_1 + (\kappa - 1)\vartheta_2 \ln(V_2/V_1)}$$

By inserting the compression ratio $\varepsilon = V_3/V_1$, the partial compression ratio $\rho = V_2/V_1 = V_3/V_4$ and the pressure ratio $\chi = p_{1'}/p_1$, this leads to:

$$\eta_{th} = 1 - \frac{1}{\varepsilon^{\kappa-1}} \frac{\rho^{\kappa-1}(\kappa - 1)(1 + \ln\rho)}{(\chi - 1) + (\kappa - 1)\chi \ln\rho} \qquad (2.86)$$

An example is given for $\rho = 2$ and $\chi = 10$:

$$\eta_{th} = 1 - \frac{1.065}{\varepsilon^{\kappa-1}} \qquad (2.87)$$

which is $\eta_{th} = 0.31$ for a $\kappa = 1.4$ and $\varepsilon = 3$.

Advantages of the steam engine:

- Engine is independent of the heat source: multi-fuel capability.

- Noxious exhaust emissions are low because of continuous combustion.

- High torque at low revolutions.

Disadvantages of the steam engine:

- Heavy weight.

- Poor thermodynamic efficiency.

- The water in the boiler needs to be heated before the engine can be started.

### 2.3.4 Potential of Different Fuels and Propulsion Systems

Table 2.3 illustrates that a constant amount of stored energy varies considerably in its volume and weight. Standard *lead batteries* are much too heavy. Other types of batteries are lighter, but they are still not comparable to the weight of ordinary fuel. Power is dissipated in the charging and discharging processes of the battery, reducing the overall efficiency. Eventually, battery driven vehicles with a reduced buffer size may be used in special applications at short distances. Another promising approach is that of hybrid vehicles, where an internal combustion engine is combined with an electrical motor. The electrical motor may be activated to smooth out transients of the combustion engine and the driveline, contributing to reduced noxious emissions. Under partial load conditions the combustion engine can also charge the battery, so that battery volume and weight are significantly reduced. *Hydrogen $H_2$ gas* is too voluminous to be used as adequate energy source. It can be stored either at an extremely cold temperature of $20\,K$ or at relatively high pressure at room temperature. Over long time

**Table 2.3** Typical storage volumes and weights of different energy sources with an energy of $1000\,kWh$.

| Source | Volume $V$ in $[l]$ | Mass $m_1$ in $[kg]$ | Tank $m_2$ in $[kg]$ | Mass+Tank $m_1 + m_2$ in $[kg]$ |
|---|---|---|---|---|
| Fuel | 117 | 83 | 21 | 104 |
| Diesel | 102 | 85 | 17 | 102 |
| Methanol | 224 | 180 | 41 | 221 |
| Liquid gas | 153 | 78 | 90 | 168 |
| Methane | 259 | 72 | 500 | 570 |
| $H_2$, liquid | 426 | 30 | 142 | 172 |
| $H_2$, hydride buffer | 200 | 30 | 970 | 1000 |
| Battery (lead) | 5000 | 0 | 10000 | 10000 |

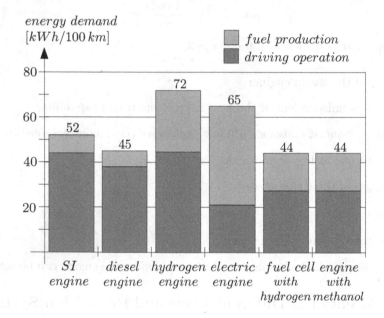

**Figure 2.24** Energy demand of different engine concepts [74]

periods, $H_2$ leaks through even thick walled steel tanks. In hydride buffers, $H_2$ is chemically bound. Since hydrogen burns at high combustion temperatures, emissions of nitrogen oxide $(NO_x)$ become a problem.

*Fuel cells* produce electrical energy directly at low temperatures. Thermal efficiencies of 70 % are reached for the synthesis of $H_2$ and $O_2$. The storage of hydrogen is again the problem. If $H_2$ must be however generated from natural gas or from methanol, efficiencies become much lower. The task is to generate the exact amount of hydrogen from e.g. methanol even under realtime transient engine conditions. For this the fuel conversion process can be modeled, and the actual masses reacting in the conversion process be estimated in realtime, as a basis for state space control. *Fuel cells* appear to be a promising alternative to combustion engines. In Figure 2.24, the relative energy requirements to move a vehicle by $100\,km$ are shown for different propulsion systems.

# 3 Engine Management Systems

## 3.1 Basic Engine Operation

### 3.1.1 Effective Work

Four-stroke engines are characterized by two alternate cycles: In the first cycle, equivalent to the first and second strokes, the gas is compressed, combusted and expanded. In the second cycle, equivalent to the third and fourth strokes, the gas is transferred to the exhaust pipe and the cylinder is filled with fresh air from the intake manifold. Figure 3.1 shows the two cycles. The crankshaft is turned 360° per cycle. SI and diesel engines are controlled differently: In diesel engines, fuel is directly injected into the combustion chamber. The amount of injected fuel per stroke is then proportional to engine torque. The amount of air is almost constant at a given speed. In SI engines, the amount of fuel as well as air is controlled. When the fuel is injected into the intake manifold, a homogeneous air-fuel mixture is sucked into the cylinders. The mechanical work generated in the combustion cycle can be obtained by integration in the $pV$-diagram. The mechanical work can be normalized when relating it to the displacement volume $V_d$:

$$w_i = \frac{1}{V_d} \sum_{j=1}^{CYL} \oint \Big( p_j(V_j) - p_0 \Big) dV_j \quad , \tag{3.1}$$

where:

| | |
|---|---|
| $V_d = CYL \cdot (V_1 - V_2)$ | is the displacement volume of all cylinders |
| $CYL$ | is the number of cylinders |
| $w_i$ | is the (normalized) **indicated specific work**. |

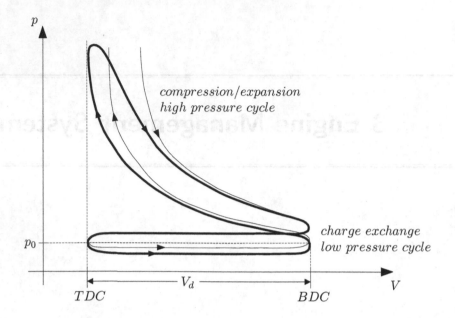

**Figure 3.1** pV-diagram of four-stroke combustion engine

The value of $w_i$ can be determined by measuring the in-cylinder pressure during a cycle. An indicated specific work of $1\,J/cm^3$ is equivalent to a mean pressure of $\bar{p} = 10\,bar\;(= 10^6\,Pa)$. The transfer of the combustion torque to the engine torque available at the crankshaft can be calculated from the following motion equations.

The piston stroke from Top Dead Center (TDC) is

$$s(\alpha_{CS}) = l(1 - \cos\beta) + r(1 - \cos\alpha_{CS}) \quad .$$

From Figure 3.2 we get

$$l\sin\beta \;=\; r\sin\alpha_{CS} \quad ,$$

$$\cos\beta \;=\; \sqrt{1 - \frac{r^2}{l^2}\sin^2\alpha_{CS}} \quad , \tag{3.2}$$

which yields the piston stroke as

$$s(\alpha_{CS}) = r\left(1 - \cos\alpha_{CS} + \frac{l}{r}\left(1 - \sqrt{1 - \frac{r^2}{l^2}\sin^2\alpha_{CS}}\right)\right) \quad . \tag{3.3}$$

At Top Dead Center, we have $\alpha_{CS} = 0$, $s(\alpha_{CS}) = 0$, and at Bottom Dead Center $\alpha_{CS} = \pi$, $s(\alpha_{CS}) = 2r$ respectively. The derivatives of the piston stroke are

$$\frac{ds}{d\alpha_{CS}} \;=\; r\left(\sin\alpha_{CS} + \frac{r}{l}\cdot\frac{\sin\alpha_{CS}\cos\alpha_{CS}}{\sqrt{1 - \frac{r^2}{l^2}\sin^2\alpha_{CS}}}\right) \tag{3.4}$$

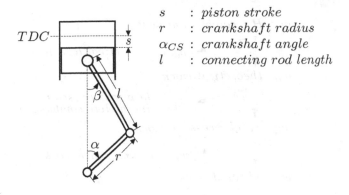

$s$ : *piston stroke*
$r$ : *crankshaft radius*
$\alpha_{CS}$ : *crankshaft angle*
$l$ : *connecting rod length*

**Figure 3.2** Piston and crankshaft motion

and

$$\frac{d^2 s}{d\alpha_{CS}^2} = r \left( \cos \alpha_{CS} + \frac{\frac{r}{l}(\cos^2 \alpha_{CS} - \sin^2 \alpha_{CS}) + \frac{r^2}{l^2}\sin^4 \alpha_{CS}}{\left(\sqrt{1 - \frac{r^2}{l^2}\sin^2 \alpha_{CS}}\right)^3} \right) \quad . \tag{3.5}$$

These derivatives with respect to crankshaft angle can be related to the derivatives with respect to time as follows:

$$\dot{s} = \frac{ds}{dt} = \frac{ds}{d\alpha_{CS}} \cdot \frac{d\alpha_{CS}}{dt} = \frac{ds}{d\alpha_{CS}} \cdot \dot{\alpha}_{CS}$$

$$\ddot{s} = \frac{d^2 s}{dt^2} = \frac{d}{dt}\left(\frac{ds}{d\alpha_{CS}} \cdot \frac{d\alpha_{CS}}{dt}\right) = \frac{d}{dt}\left(\frac{ds}{d\alpha_{CS}}\right) \cdot \frac{d\alpha_{CS}}{dt} + \frac{ds}{d\alpha_{CS}} \cdot \frac{d^2\alpha_{CS}}{dt^2}$$

$$= \frac{d^2 s}{d\alpha_{CS}^2} \cdot \dot{\alpha}_{CS}^2 + \frac{ds}{d\alpha_{CS}} \cdot \ddot{\alpha}_{CS} \tag{3.6}$$

The indicated specific work can be written as

$$w_i = \frac{1}{V_d} \oint \sum_{j=1}^{CYL} (p_j(\alpha_{CS}) - p_0)\, A_p\, \frac{ds_j(\alpha_{CS})}{d\alpha_{CS}} d\alpha_{CS}$$

$$= \frac{1}{V_d} \oint T_{comb}(\alpha_{CS}) d\alpha_{CS} \quad . \tag{3.7}$$

The combustion torque at the crankshaft is thus defined as

$$T_{comb}(\alpha_{CS}) = \sum_{j=1}^{CYL} (p_j(\alpha_{CS}) - p_0)\, A_p\, \frac{ds_j}{d\alpha_{CS}} \quad . \tag{3.8}$$

The piston strokes in different cylinders are shifted by phase.

$$s_j(\alpha_{CS}) = s\left(\alpha_{CS} - (j-1) \cdot \frac{4\pi}{CYL}\right), \quad j = 1,...,CYL \tag{3.9}$$

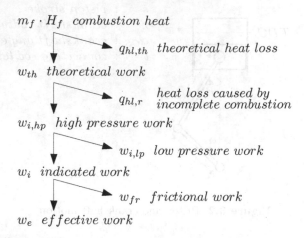

$m_f \cdot H_f$ *combustion heat*

$q_{hl,th}$ *theoretical heat loss*

$w_{th}$ *theoretical work*

$q_{hl,r}$ *heat loss caused by incomplete combustion*

$w_{i,hp}$ *high pressure work*

$w_{i,lp}$ *low pressure work*

$w_i$ *indicated work*

$w_{fr}$ *frictional work*

$w_e$ *effective work*

**Figure 3.3** The effective work delivered by the engine is much lower than the thermal energy caused by combustion.

The average combustion torque is

$$
\begin{aligned}
\bar{T}_{comb} &= \frac{1}{4\pi} \oint T_{comb}(\alpha_{CS}) d\alpha_{CS} \\
&= \frac{P_i}{\dot{\alpha}_{CS}} \quad,
\end{aligned}
\tag{3.10}
$$

where $P_i$ is the mean indicated power. The total indicated work $w_i V_d$ can now be written at stationary engine operation as

$$
w_i V_d = 4\pi \bar{T}_{comb} = 4\pi \frac{P_i}{\dot{\alpha}_{CS}} = \frac{4\pi P_i}{2\pi n} = \frac{2P_i}{n} \quad,
$$

and the normalized work

$$
w_i = \frac{2P_i}{V_d n} \quad,
\tag{3.11}
$$

where $n = \dot{\alpha}_{CS}/(2\pi)$ is the engine speed. In reality, the effective work $w_e$ per volume is much lower than the indicated work $w_i$ (see Figure 3.3). The effective thermodynamic efficiency $\eta_e$ is at constant fuel flow

$$
\eta_e = \frac{P_e}{\dot{m}_f H_f} = \frac{w_e V_d n}{2m_f n H_f} \cdot \frac{2}{CYL} = \frac{w_e}{m_f H_f} \cdot \frac{V_d}{CYL} \quad .
\tag{3.12}
$$

where:

$P_e$    is the effective power in $W$
$w_e$    is the effective specific work per cycle in $J/cm^3$
$m_f$    is the mass of fuel measured per cylinder in $kg$
$\dot{m}_f$    is the fuel flow in $kg/s$
$H_f$    is the specific energy of the fuel released in the combustion $J/kg$
$V_d$    is the total displacement volume in $dm^3$
     ($V_d/CYL$ displacement volume per cylinder)

The indicated thermodynamic efficiency (friction not considered) is:

$$\eta_i = \frac{w_i}{2m_f H_f} \cdot \frac{V_d}{CYL} \tag{3.13}$$

Some examples of typical values for the indicated efficiency are given in Table 3.1.

**Table 3.1** Indicated specific work $w_i$, theoretical heat loss $q_{hl,th}$, and realistic heat loss $q_{hl,r}$ for different engine types, related to fuel combustion heat.

| Engine Type | SI | Diesel | Big Diesel |
|:---:|:---:|:---:|:---:|
| $w_i$ | 33-35% | 40-43% | 45-48% |
| $q_{hl,th}$ | 23-28% | 22-25% | 12-14% |
| $q_{hl,r}$ | 37-44% | 35-40% | 26-33% |

## 3.1.2 Air-Fuel Ratio

The ratio of air to fuel is very important for the combustion process of internal combustion engines. There are several effects that have an impact on the amount of air $m_a$ transferred to the cylinder: Throttling of the air flow by the throttle butterfly, aerodynamic resistance and resonances in the intake manifold, rebounding of already burned gases from the cylinder into the inlet pipes and other effects. The amount of air which would theoretically fit into a displacement volume $V_d$ under the normalized pressure $p_0 = 1.013\,bar$ and the normalized air density $\rho_0 = 1.29\,kg/m^3$ is expressed by $m_{a,th} = \rho_0 V_d$. The ratio of real to theoretical value is equivalent to the relative air supply:

$$\lambda_a = \frac{m_a}{m_{a,th}} \tag{3.14}$$

Similarly, the ratio of measured fuel mass $m_f$ to theoretical fuel mass $m_{f,th}$ is equivalent to the relative fuel supply:

$$\lambda_f = \frac{m_f}{m_{f,th}} \tag{3.15}$$

The theoretical fuel mass $m_{f,th}$ is equivalent to the mass of fuel needed for an ideal stoichiometric combustion with the oxygen. Under normal conditions the stoichiometric ratio for gasoline is:

$$L_{st} = \frac{m_{a,th}}{m_{f,th}} = 14.66 \tag{3.16}$$

The air-fuel ratio lambda is defined as:

$$\lambda = \frac{\lambda_a}{\lambda_f} \tag{3.17}$$

It can be extended:

$$\lambda = \frac{m_a}{m_f} \frac{m_{f,th}}{m_{a,th}} = \frac{1}{L_{st}} \cdot \frac{m_a}{m_f} \tag{3.18}$$

For an ideal stoichiometric combustion, this ratio is equivalent to one: $\lambda = 1$. The air-fuel ratio has an impact on the effective work $w_e$ and the effective thermodynamic efficiency $\eta_e$. The air-fuel ratio can be influenced in two different ways, by variation of $\lambda_a$ or of $\lambda_f$:

1. **Variation of $\lambda_f$ at a given $\lambda_a$:**
   Typical applications are SI engines operating around a stoichiometric air-fuel ratio. The relative air supply $\lambda_a$ is determined by the driver.

   **Lean operation** ($\lambda > 1$): Less fuel is injected than needed for stoichiometric combustion (reduced $\lambda_f$). Due to a reduced high pressure work $w_{i,hp}$ the effective work $w_{i,hp}$ decreases. In the range of $1 < \lambda < 1.1$, the thermodynamic efficiency $\eta_e$ increases however, caused by higher combustion peak temperatures. This results in high emissions of nitrogen oxides $NO_x$. If $\lambda$ is further increased, $\eta_e$ will decrease because of an even lower high pressure work $w_{i,hp}$ at a given low pressure work $w_{i,lp}$.

   **Rich operation** ($\lambda < 1$): More fuel is injected than needed for stoichiometric combustion (higher $\lambda_f$). The fuel surplus increases both high pressure work $w_{i,hp}$ and effective work $w_e$. Below $\lambda < 0.9$ incomplete combustion results in high emissions of hydrocarbon $HC$ in the exhaust gases and in a decreasing effective work $w_e$. At $\lambda < 1$ the thermodynamic efficiency $\eta_e$ is always decreased.

2. **Variation of $\lambda_a$ at a given $\lambda_f$:**
   Typical applications are lean-burn SI-engines at part load and Diesel engines. The relative fuel supply $\lambda_f$ is determined by the driver.

   **Lean operation** ($\lambda > 1$): More air is admitted than needed for stoichiometric combustion (increased $\lambda_a$). Therefore, high pressure work $w_{i,hp}$ is increased while low pressure work $w_{i,lp}$ remains constant. Both effective work $w_e$ and thermodynamic efficiency $\eta_e$ are increased. It should be mentioned that lean gas mixtures are less flammable. In SI-engines delays between spark ignition and complete combustion increase.

   Misfiring at SI engines must be avoided by e.g. direct injection of a fuel stratified charge into the cylinder, which forms an enriched mixture around the spark plug. This operation is quite similar to that of Diesel engines. Combustion is either triggered by a spark or by self-inflammation due to high compression ratios. The engine can only operate up to its maximum gas load (maximum $\lambda_a$). Due to lean operation its maximum power according to the displacement volume is, however, not reached.

   **Rich operation** ($\lambda < 1$): Less air is admitted than needed for stoichiometric combustion (decreased $\lambda_a$). This leads to a decrease in both

(1)  *SI engines at maximum power output*
(2)  *Stochiometric SI engines*
(3)  *Diesel and lean − burn engines*
(4)  *Moderately lean SI engines*

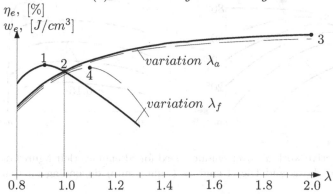

**Figure 3.4** Effective work $w_e$ and thermodynamic efficiency $\eta_e$ of combustion engines depending on variation of $\lambda_a$ or $\lambda_f$.

efficiency $\eta_e$ and effective work $w_e$. Incomplete combustion results in higher hydrocarbon $HC$ emissions and in a reduction in high pressure work $w_{i,hp}$.

Figure 3.4 shows the dependency of the effective work and effective thermodynamic efficiency over $\lambda$, assuming an optimal control of fuel injection and ignition timing.

Engines may use recirculated exhaust gas instead of fresh air to increase the relative air supply $\lambda_a$. As long as sufficient air is available for the combustion this is similar to a higher air-fuel ratio $\lambda$. Exhaust gas recirculation reduces the emission of $NO_x$ due to lower combustion peak temperatures.

### 3.1.3   Engine Concepts

The **SI engine** is controlled by the relative air supply $\lambda_a$. This is done by throttling the air flow into the engine. The relative fuel supply $\lambda_f$ is subsequently regulated by the engine management systems to maintain the desired air-fuel ratio $\lambda$. The range of $\lambda$ is limited by the ability to inflame air-fuel mixtures by spark ignition. Conventional SI engines operate on approximately homogeneous mixtures ($0.9 < \lambda < 1.3$). Lean-burn engines operate at very lean mixtures equivalent to Diesel engines. Combustion is ensured by directly injecting a stratified charge of rich air-fuel ratio around the spark plug.

The **Diesel engine** is controlled by the relative fuel supply $\lambda_f$. The intake manifold is not throttled. The relative air supply $\lambda_a$ is always at its maximum. Therefore, the air-fuel ratio $\lambda$ changes within a large range. The inflammation of extremely lean mixtures is still possible because of the non-homogeneous fuel distribution in the combustion chamber. Such inhomogeneous mixtures burn with

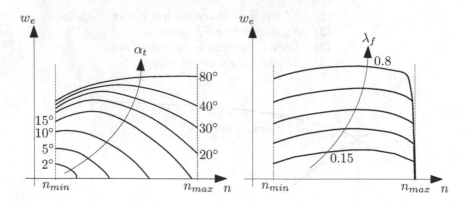

**Figure 3.5** Effective work $w_e$ over engine speed for SI engines (left figure) depending on the throttle angle $\alpha_t$ and Diesel engines (right figure) depending on relative fuel supply $\lambda_f$ .

a yellow flame. The average air-fuel ratio should not be below $\lambda = 1.3$ to avoid the generation of too much soot. Since the effective work $w_e$ is basically given by the amount of injected fuel, the fuel supply must be cut-off when reaching the maximum engine speed. Otherwise the engine power would continue to increase with speed resulting in a self-destruction of the engine. Fuel may be injected in several steps. A first small amount of fuel starts the combustion process more smoothly. The second main injection then results in lower peak pressures and temperatures, yielding lower $NO_x$ emissions and less combustion noise. An injection at the end of the combustion cycle heats up the exhaust pipe and exhaust gas treatment systems such as a soot filter. When Diesel engines are operated with very high exhaust gas recirculation rates, then they generate few noxious raw emissions [28].

**Lean-burn SI engines** are a compromise between diesel and stoichiometric SI engines. Driving at part load, the air-fuel ratio is very lean. By properly designing injection pressure, spray cone and air turbulence, an enriched stratified charge is assembled around the spark plug. The resulting combustion is equivalent to that of Diesel engines (inhomogeneous mixture, yellow flame). Lower noxious emissions can be achieved by separating into e.g. two injections. First, about 1/4 of all fuel is injected, which forms a lean homogeneous mixture in the combustion chamber. This lean mixture is also less sensitive to knocking. Second, 3/4 of fuel is injected as a stratified charge. After burning this rich mixture, the homogeneous lean mixture burns in a blue flame, resulting in a reduction of noxious emissions. At high loads, the operating conditions are shifted from very lean to stoichiometric mixture. At a given engine displacement volume, more fuel can be combusted from stoichiometric mixtures, increasing power output.

Figure 3.5 shows the dependency of the effective work $w_e$ over engine speed for SI and Diesel engines. The displacement volume of a naturally aspirated diesel or lean-burn SI engine must be 60 % higher than that of a stoichiometric SI engine to obtain the same maximum power output. Therefore diesel and lean-burn SI

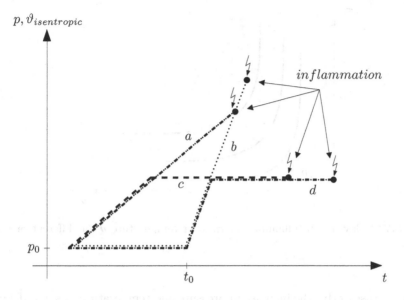

**Figure 3.6** Isentropic temperature $\vartheta_{isentropic}$ and pressure $p$ over time. The conditions for self inflammation (a, b, c ,d) are explained in the text below.

engines are often turbocharged which increases the relative air supply $\lambda_a$ at a given displacement volume $V_d$.

### 3.1.4 Inflammation of Air-Fuel Mixtures

The kinetic gas theory describes gases as a cloud of molecules with a given velocity distribution according to their temperature. The collision of different molecules will start a chemical chain reaction if their kinetic energy is over a certain activation energy $E$. The relative amount of effective collisions $A$ is expressed by the Arrhenius law

$$A = e^{-E/R\vartheta} \quad . \tag{3.19}$$

The activation energy $E$ is low for radicals (not saturated molecules). The probability for a collision is increased by the concentration of molecules and by the temperature. A chemical reaction must be started by a high temperature. During the chain reaction, more radicals are generated than destroyed. Under appropriate conditions, a spontaneous spark ignition is sufficient to start the combustion process at the location of the spark plug. The air-fuel mixture must be within a certain range ($0.9 < \lambda < 1.3$) and its pressure (or respective temperature) must be over a threshold for a certain period of time. The gas is compressed isentropically, neglecting heat conduction. Figure 3.6 shows how pressure and temperature courses over time influence self inflammation under different conditions.

1. In curve $a$, pressure and temperature rise immediately. In curve $b$, inflammation starts only when reaching a higher pressure level than in curve (a).

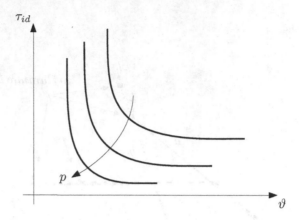

**Figure 3.7** Delay of self inflammation $\tau_{id}$ over temperature $\vartheta$ for different pressure values $p$.

In the case of (b), the increase in pressure and temperature was first delayed until $t_0$.

2. In curve $c$, temperature and pressure rise immediately, but level off at a lower level. Inflammation occurs after a longer time delay. In the case of (d) the rise is first delayed until time $t_0$. Self inflammation happens later than in (c).

It can be seen, that self inflammation depends on something like the integral of pressure or temperature over time. Many parameters have an impact on the time delay of self inflammation, like location of the spark plug within the combustion chamber, etc. Woschni [130] gives an empirical formula of the inflammation delay time, depending only on mean temperature $\vartheta$ and mean pressure $p$, without any integral portion.

$$\tau_{id} = 0.44\, e^{4650K/\vartheta} \left( \frac{p}{p_0} \right)^{-1.19} \tag{3.20}$$

The inflammation delay $\tau_{id}$ over temperature $\vartheta$ with the pressure $p$ as parameter can be seen in Figure 3.7. The time delay $\tau_{id}$ is reduced for high temperatures and high pressures. This is why e.g. turbo-charged Diesel engines operate with a start-of-injection angle which is approximately $10°$ (crankshaft angle) later than that of naturally aspirated Diesel engines.

### 3.1.5  Flame Propagation

The velocity $v_{fl}$ of flame propagation depends on two components:

1. **Combustion velocity:** The combustion propagates through the gas mixture, for example with a velocity of $1\,m/s$.

2. **Transport velocity:** The burning gas itself is swirled as the rising piston generates turbulence in the combustion chamber. The transport velocity

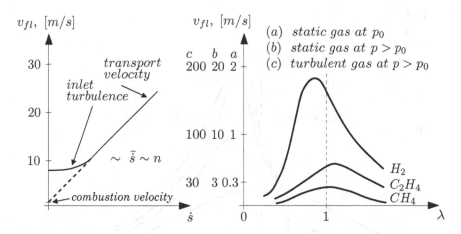

**Figure 3.8** Flame propagation velocity $v_{fl}$ over average piston velocity $\bar{\dot{s}}$ (left) and over air-fuel ratio $\lambda$ (right).

is approximately proportional to the piston velocity $\dot{s}$ which depends on the engine speed $n$. At low engine speeds, the transport velocity can be increased by a swirl inlet port generating a turbulent gas flow. The turbulences accelerate the combustion speed proportional to engine speed.

Figure 3.8 shows the flame propagation velocity $v_{fl}$ depending on piston velocity $\dot{s}$ (left) and air-fuel ratio $\lambda$ (right). The inflammation delay time $\tau_{id}$ must be considered in the engine control to position the combustion process right over the downward moving piston. The time delay $\tau_{id}$ must be convoluted to a crankshaft angle delay, increasing with the engine speed. The ignition angle $\alpha_i$ must therefore be advanced over engine speed.

Contrary to that, the position of the combustion process over the crankshaft angle is almost constant. This is due to the fact, that the flame propagation velocity at combustion is mostly determined by the transport velocity. Thus the angle position of the combustion process is independent of the engine speed. With higher engine speeds, flame transport velocity is increasing, speeding up combustion over time, leaving it however constant over engine speed.

## 3.1.6 Energy Conversion

The in-cylinder pressure can be plotted over time or over crankshaft angle $\alpha_{CS}$. The angle of 360° is equivalent to a complete high pressure cycle. Figure 3.9 shows the in-cylinder pressure over crankshaft angle.

The gas is compressed by the piston in an approximately isentropic process. With ignition at $\alpha_i$, the pressure rises only after time delay $\tau_{id}$. The maximum pressure varies from cycle to cycle. The inflammation delay $\tau_{id}$ depends on temperature, pressure, air-fuel ratio and self inflammation time as described in the previous section. It also depends on the type of fuel being used. Figure 3.10 shows some inflammation delays for different fuels over temperature. Oil compa-

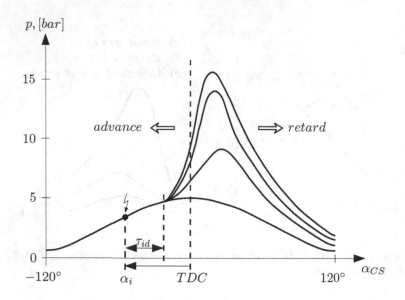

**Figure 3.9** In-cylinder pressure $p$ over crankshaft angle $\alpha_{CS}$.

nies adapt their fuel to weather conditions (summer, winter).

Turbulence caused by the upward moving piston has no impact on the time delay $\tau_{id}$. For a correct ignition angle, this delay must be considered. The time delay is convoluted to an angle delay, increasing proportional to engine speed. Contrary to that, the engine speed has almost no impact on the position of energy conversion as turbulences increase the transport velocity with higher engine speeds.

The energy conversion caused by combustion is shown in Figure 3.11 for different air-fuel ratios $\lambda$. In these curves, the isentropic pressure curves are suppressed. The differential output of thermal energy per angle $dQ/d\alpha_{CS}$ (its gradient) is normalized to the total combustion energy $Q_{comb}$. The shape of the relative energy conversion is therefore almost constant.

If the air-fuel ratio is increased e.g. to $\lambda = 1.2$ as shown in Figure 3.11, the ignition delay $\tau_{id}$ will rise. At a constant inflammation angle $\alpha_{i1}$ the energy conversion is then retarded. Therefore, the ignition angle must be advanced to $\alpha_{i2}$, to compensate for the increased delay. The energy conversion returns to its previous position. It should be mentioned that a high air-fuel ratio $\lambda$ increases also the variance of the time delay $\tau_{id}$.

The ignition angle $\alpha_i$ depends on $\lambda$ which can be seen in Figure 3.12. The angle is computed by averaging the energy conversion over 0.1 %, 1 %, 10 %, 50 %, 90 % points. The angles for $\alpha_{Q1\%}$ and higher are almost independent of the air-fuel ratio $\lambda$. In-cylinder pressure measurements can be used to control the ignition angle in a closed loop to maintain a constant position of energy conversion as shown in Figure 3.13. The angle of maximum pressure gradient $max(dp/d\alpha_{CS})$ may be used as a control variable. The controller time constant must be relatively large because of the high delay time variances between consecutive cycles. Thus

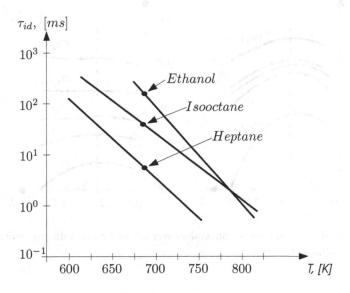

**Figure 3.10** Inflammation delay $\tau_{id}$ over temperature for different fuels.

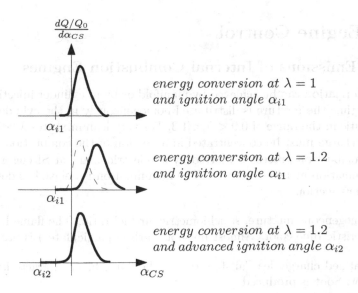

**Figure 3.11** Normalized energy conversion caused by combustion for different air-fuel ratios $\lambda$.

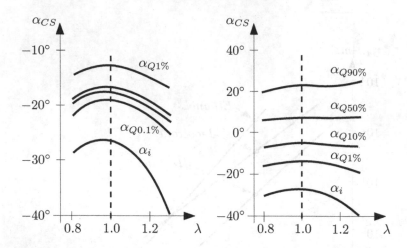

**Figure 3.12** Angle $\alpha_{CS}$ of energy conversion over air-fuel ratio $\lambda$ during ignition (left) and combustion (right) process.

closed loop ignition control may be too slow for the dynamic response of the engine.

The ignition angle is determined to find a compromise between fuel consumption, emissions or knocking (see Section 3.2.8). An equivalent procedure can be found for the fuel injection angle at Diesel engines.

## 3.2  Engine Control

### 3.2.1  Emissions of Internal Combustion Engines

Mixture formation can be achieved by manifold or by in-cylinder injection. With sufficient time the mixture is distributed homogeneously in the cylinder with an air-fuel ratio in the range of $0.9 < \lambda < 1.3$. For very lean mixtures $\lambda > 1.3$, a rich stratified charge must be concentrated in a portion of the combustion chamber.

The combustion process is started by an electric spark at SI engines and by self-inflammation at Diesel engines. The inflammation is delayed as described in the previous section.

- Homogeneous mixture, stoichiometric air-fuel ratio: The flame has a characteristic blue color. Almost no soot (carbon particulates) is produced.

- Stratified charge, lean air-fuel ratio: The flame has a characteristic yellow color. Soot is produced.

- Inflammation starts combustion from one location.

The inflammation process depends on pressure $p$, temperature $\vartheta$, air-fuel ratio $\lambda$ and activation energy $E$ of the fuel. For $\lambda < 1$ the exhaust gases are generated

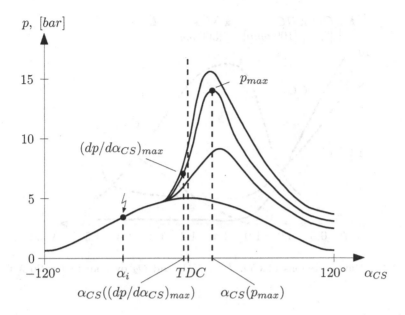

**Figure 3.13** Closed-loop control of ignition angle $\alpha_i$ to maintain a constant position of energy conversion.

according to the concentration ratio

$$k = \frac{n_{CO} \cdot n_{H_2O}}{n_{CO_2} \cdot n_{H_2}} . \tag{3.21}$$

This ratio is temperature dependant. A typical value for $\vartheta = 1850\,°K$ is $k = 3.6$.

The pollutant emissions like $CO$, $HC$, $NO_x$ depend strongly on the air-fuel ratio which is shown in Figure 3.14.

$\lambda < 1$: Increased emission of hydrocarbon $HC$ and carbon monoxide $CO$.

$\lambda = 1$: Stoichiometric combustion. Very low emissions after three way catalytic converter.

$\lambda \approx 1.1$: Highest nitrogen oxide $NO_x$ emissions due to highest combustion peak temperatures.

$\lambda > 1.1$: Decreasing nitrogen oxide $NO_x$ concentration and lower combustion temperatures. Increasing hydrocarbon $HC$ emissions at eventual misfires.

$\lambda > 1.5$: Lean operation. For very low emissions, a $NO_x$ reducing catalytic converter is required.

The concentration of oxygen $O_2$ in the exhaust gas can be used to determine the air-fuel ratio $\lambda$ for $\lambda \geq 1$ using a lambda-sensor.

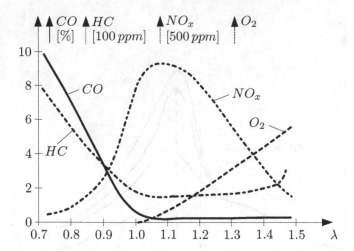

**Figure 3.14**  Raw emissions of $CO$, $HC$, $NO_x$ and $O_2$ over air-fuel ratio $\lambda$ for SI engines.

## 3.2.2   Fuel Measurement

The air-fuel ratio $\lambda$ is an important variable for fuel control which is based on different control concepts:

**rich mixture $\lambda < 1$:** Maximum power output per displacement volume because of increased relative fuel supply $\lambda_f$. It was used at high engine loads until 1970. Nowadays it is only used for cold engines during the warm-up phase. High noxious emission rates.

**stoichiometric mixture $\lambda = 1$:** Acceptable power output.  This ratio is required for proper operation of three-way catalytic converters. At high engine loads, a good compromise between power output and exhaust emissions is achieved.

**moderately lean mixture $1 < \lambda < 1.5$:** Good efficiency because of increased air supply $\lambda_a$, but high emissions of $NO_x$. This method was used at part loads until 1980.

**lean mixture $\lambda > 1.5$:** High efficiency because of high $\lambda_a$. $NO_x$ emissions are still high, so that catalytic converters for $NO_x$ reduction are required. This method is used in lean-burn engines at part loads and in Diesel engines. Maximum engine power output cannot be reached.

The reference torque desired by the driver controls either the relative air supply $\lambda_a$ via the throttle angle $\alpha_t$ at SI engines or the relative fuel supply $\lambda_f$ at Diesel engines. The amount of fuel being mixed with the air is regulated by the fuel control system to obtain a predefined air-fuel ratio $\lambda$. There are two different injection systems:

1. **Manifold injection:** The fuel is injected into the individual inlet pipes in front of the inlet ports. There is at least one inlet valve for each cylinder. Problems may occur at idling because of incomplete fuel evaporation due to a low air flow velocity into the cylinders. Additionally, the distribution of air flow into the different inlet pipes may vary. The amount of injected fuel is less accurate at idling because electromagnetic injection valves are time controlled: Errors due to different rise and fall times have a larger impact on the amount of fuel injected at small injection times. The advantage of manifold injection is the creation of a homogeneous fuel distribution in the cylinder at $\lambda = 1$. This burns with a blue flame. There are few restrictions for the design of the inlet pipes. The exchange of gas can be optimized without major effects on the injection system as it is located at the inlet valves. The inlet pipes are designed to produce acoustic resonances at low engine speed. This increases the relative air supply $\lambda_a$ and the effective work $w_e$ already without turbo charging. The inlet valves are cooled by the evaporating fuel. The reduced gas temperature lowers knocking and allows higher compression ratios to increase efficiency. The injection timing is phase-shifted for each cylinder. It is aimed to terminate the injection just before the inlet valve is opened to avoid the emission of soot. The injection can be controlled individually and fuel supply can be individually cut off for each cylinder: limitation of engine speed and vehicle speed, fuel cut-off at coasting or cylinder switch-off at multi-cylinder engines.

2. **In-cylinder injection:** The fuel is injected directly into the cylinders. At stoichiometric operation, a higher compression ration can be realized e.g. 12 . . . 13, due to the cooling effect of the evaporating fuel. At lean operation, the aim is to assemble a sufficiently rich mixture (e.g. stoichiometric) in a limited portion of the combustion chamber, e.g. at SI engines around the spark plug at the time of ignition. The amount of fuel, the injection pressure (thus fuel atomization), the injector spray angle (width and depth of injection) and the injection timing are adjusted in each engine operation point. The swirl is controlled by special geometry of the piston head.

   At part load, the aim is to burn very lean air-fuel mixtures. This can be done by multiple injection. By an early injection, a homogeneous lean mixture is developing throughout the combustion chamber until the time of ignition, due to the swirl. The follow-up injection creates the richer stratified charge, which is burning fast with reduced cycle-to-cycle variations. Due to the reduced amount of the fuel in the stratified charge, peak temperatures are also reduced. The early generated lean homogeneous mixture is burned afterwards. Since the mixture is very lean, it takes significant time, limiting maximum allowable engine speeds. In the combustion of the stratified charge, soot was generated. In the blue flame of the homogeneous mixture, this can be effectively reduced.

The total amount of injected fuel depends on the following parameters:

- aspirated air flow per time $\dot{m}_a$

- intake manifold pressure $p_m$ at SI engines

- throttle angle $\alpha_t$ and its derivative at SI engines

- engine speed $n$

- crankshaft angle $\alpha_{CS}$ and TDC signal of a reference cylinder

- engine temperature $\vartheta_e$

- ambient air temperature $\vartheta_a$

- battery voltage $U_b$ (indirectly)

Major functions of the fuel control:

- Control of injected fuel per time $\dot{m}_f$, following the aspirated air per time $\dot{m}_a$, depending on the desired air-fuel ratio $\lambda$.

- Enriched fuel injection in warm-up phase of the engine after cold start at SI engines.

- Increased relative air supply $\lambda_a$ or relative fuel supply $\lambda_f$ for the cold engine because of higher friction.

- Compensation of intake manifold dynamics at SI engines.

- Compensation of fuel film dynamics at manifold injection. This phenomenon is also temperature dependant.

- Fuel cutoff at coasting. This reduces overall fuel consumption by around 5 %.

- The measured air flow is eventually corrected for ambient air temperature $\vartheta_a$ and barometric pressure $p_0$ changes.

- Engine idle speed control.

- Maximum engine speed limitation by fuel cutoff.

- Lambda control of the air-fuel ratio.

- Exhaust gas recirculation control.

- Approximation of an engine torque signal, e.g. by means of a map for the thermodynamic efficiency $\eta_e$.

### 3.2.3  Intermittent Fuel Injection

Intermittent fuel injection has turned out to be more economical than continuous fuel injection, due to the different accuracies required in those systems. The power output of an engine varies within a wide range. Between idling power $P_{min}$ and maximum power $P_{max}$, the ratio is about $1 : 100$.

$$\frac{P_{max}}{P_{min}} = 100 \qquad (3.22)$$

The engine speed varies in the range of:

$$\frac{n_{max}}{n_{min}} = 10 \tag{3.23}$$

At stationary engine operation, the amount of injected fuel per time $\dot{m}_f$ is proportional to the effective power output $P_e$ of the engine which is expressed in Equation 3.12. In this consideration air pulsation in the intake pipes are neglected.

Supposing a **continuous fuel supply** $\dot{m}_f$, the relative error of the open loop fuel measurement at low loads should be

$$\frac{\Delta \dot{m}}{\dot{m}_{min}} < 3\% \quad . \tag{3.24}$$

This will cause an absolute error related to maximum fuel flow

$$\frac{\Delta \dot{m}}{\dot{m}_{max}} = \frac{\dot{m}_{min}}{\dot{m}_{max}} \frac{\Delta \dot{m}}{\dot{m}_{min}} = \frac{P_{min}}{P_{max}} \frac{\Delta \dot{m}}{\dot{m}_{min}} < 3 \cdot 10^{-4} \quad . \tag{3.25}$$

Hence, the absolute error at idling is 100 times smaller than the relative error. A continuous fuel injection system should be designed for the absolute error. It must be extremely accurate. High production costs are the consequence.

Fuel measurement can alternatively be achieved by **intermittent fuel injection**. For each combustion cycle, a certain amount of fuel $m_f$ is injected. The number of injections per second are proportional to the engine speed $n$. The amount of injected fuel per cylinder and combustion cycle is

$$m_f = \int_0^{\frac{2}{n \cdot CYL}} \dot{m}_f \, dt \quad , \tag{3.26}$$

where $CYL$ is the number of cylinders of the engine. The factor 2 is due to the fact that air is combusted only every second cycle in the four-stroke process. Supposing a constant fuel flow $\dot{m}_f$ at stationary operation, the integration leads to:

$$m_f = \frac{\dot{m}_f}{n} \cdot \frac{2}{CYL} \tag{3.27}$$

The ratio of maximum to minimum amount of injected fuel per cycle is given by Equation 3.12:

$$\frac{m_{max}}{m_{min}} = \frac{P_{max}}{P_{min}} \frac{n_{min}}{n_{max}} = 10 \tag{3.28}$$

Supposing again a relative error at minimum load of

$$\frac{\Delta m}{m_{min}} < 3\% \quad , \tag{3.29}$$

the absolute error related to maximum fuel per cycle is now

$$\frac{\Delta m}{m_{max}} < 3 \cdot 10^{-3} \quad , \tag{3.30}$$

i.e. 10 times larger. Compared to a continuous fuel supply, intermittent fuel injection systems can be implemented with significantly lower accuracy requirements and therefore can be produced at lower costs.

## 3.2.4   Injection Time Calculation

The fuel supply is controlled by the **injection time** $t_{inj}$ during which the injector valve is open. Therefore, the amount of fuel per injection into a cylinder can be calculated using the following relationship for constant air flow $\dot{m}_a = const.$

$$m_f = \frac{m_a}{L_{st}\,\lambda} = \frac{1}{L_{st}\,\lambda}\,\frac{\dot{m}_a}{n}\,\frac{2}{CYL} \quad , \tag{3.31}$$

where $L_{st} = 14.66$ (see Equation 3.16). The amount of injected fuel $m_f$ is proportional to the injection time $t_{inj}$ and the square root of the pressure difference $\Delta p$ between fuel rail and manifold at manifold injection or between fuel rail and combustion chamber at direct in-cylinder injection [69]. The fuel density $\varrho_f$ and the effective opening area of the valve $A_{eff}$ are assumed to be constant.

$$m_f \sim \varrho_f \cdot A_{eff} \cdot \sqrt{2\frac{\Delta p}{\varrho_f}} \cdot t_{inj} \tag{3.32}$$

At manifold injection, the pressure difference $\Delta p$ is around $5\,bar$. At direct in-cylinder injection, the pressure difference $\Delta p$ is up to $400\,bar$ for SI and up to $2000\,bar$ for Diesel engines.

The injection time at stationary engine operation is proportional to

$$t_{inj} \sim \frac{1}{\lambda}\,\frac{\dot{m}_a}{n}\,\frac{2}{CYL} \quad , \tag{3.33}$$

and for a reference air-fuel ratio $\lambda_0$ a reference injection time $t_0$ is proportional to

$$t_0 \sim \frac{1}{\lambda_0}\,\frac{\dot{m}_a}{n}\,\frac{2}{CYL} \quad . \tag{3.34}$$

For arbitrary air-fuel ratios $\lambda$ we get

$$t_{inj} \approx \frac{\lambda_0}{\lambda}\,t_0 \quad . \tag{3.35}$$

The injection time $t_{inj}$ per combustion cycle depends on the following values:

**Mass air flow** $\dot{m}_a$: Must be measured. Systematic measurement errors at some sensors can be reduced by taking air density and temperature into account.

**Air mass per stroke** $m_a$: is computed by implementation of Equation 3.39.

**Reference air-fuel ratio** $\lambda_0$: Must be determined, e.g. stoichiometric. A lookup table can be implemented to compensate for possible errors of sensors and actuators: $\lambda_0 = \lambda_0(\dot{m}_a, n)$.

**Actual air-fuel ratio** $\lambda$: Depends on several factors such as temperature depending enrichment during warm-up and correction for dynamic transients. At Diesel engines, lambda is always $\lambda > 1.3$.

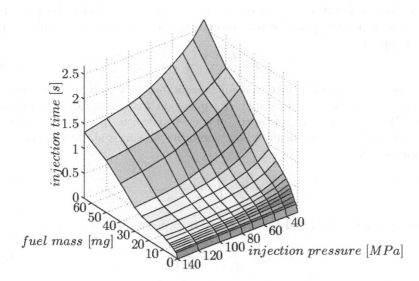

**Figure 3.15** Piezoelectric injector map ($1 MPa = 10 bar$)

**Battery voltage $U_b$:** It has an impact on the rise and fall times in electro-magnetically controlled injection nozzles. The effect can be compensated by adding a voltage dependant time correction $\Delta t(U_b)$. The compensated time is

$$t_{inj} + \Delta t(U_b) \quad . \tag{3.36}$$

Instead of direct measurement, the mass air flow $\dot{m}_a$ can be estimated from intake manifold pressure $p_m$ or the throttle angle $\alpha_t$ at throttled SI engines. The air flow into the cylinders also depends on the dynamic pressure changes in the intake manifold. It is a function of

$$\dot{m}_a = f_0(p_m, \dot{p}_m, n) \quad , \tag{3.37}$$

where $f_0$ must be measured for all possible $\dot{m}_a$ at stationary engine operation and corrected for dynamic pressure changes (Section 3.2.6).

Figure 3.15 shows a map for the required injection time $t_{inj}$ of a high-pressure piezoelectric injector over the desired injection fuel mass $m_f$ and the injection pressure difference $\Delta p$.

## 3.2.5 Air Mass per Combustion Cycle

The relative air supply $\lambda_a$ at low engine speeds can be increased by acoustic resonances in the inlet pipes to each cylinder. These resonances are exited by the periodic opening and closing of the inlet valves. The geometry of the inlet pipes is designed for resonances at lower engine speeds. It is intended that a pressure maximum from the resonance occurs at the inlet valve at the time when it is opened. Hence, more air flows into the combustion chamber and increases the relative air supply $\lambda_a$ and the effective work $w_e$. Typical resonant frequencies are between 2000 and 3000 $rpm$. For even lower resonant frequencies the geometric dimensions of the inlet pipes become too large. The frequency of air pulsation in the inlet pipe is

$$f_p = \frac{n \cdot CYL}{2} \quad . \tag{3.38}$$

The factor 2 is due to the fact that air is aspirated only every second cycle in a four stroke process. For example, the pulsation frequency for a six cylinder engine ($CYL = 6$) at an engine speed of $n = 6000\,rpm$ is $f_p = 300\,Hz$. The air mass per cylinder can be calculated by integrating the mass air flow $\dot{m}_a$ over one pulsation period.

$$m_a = \int_{t_a}^{t_b} \dot{m}_a \, dt \tag{3.39}$$

The limits (begin $t_a$ and end $t_b$) of integration are given by

$$t_b - t_a = \frac{1}{f_p} = \frac{2}{n \cdot CYL} \quad , \tag{3.40}$$

and therefore, the aspirated air for one cylinder per cycle is

$$m_a = \int_{0}^{\frac{1}{f_p}} \dot{m}_a \, dt \quad . \tag{3.41}$$

The air supply $m_a$ can be calculated by integration of the mass air flow signal. The sampling rate must be high enough to avoid aliasing, and therefore it is about $5 - 10$ times higher as the highest pulsation frequency. Eventual non-linear characteristics of the air flow meter must be compensated for before the integration. A linear characteristic can be obtained by e.g. multiplying the sensor characteristic with it's inverse. Thus an eventual bias introduced by the integration can be avoided [68].

The proper timing for integration ($t_a, t_b$) can be derived from the crankshaft angle $\alpha_{cs}$ signal. For example, if the crankshaft sensor has 60 teeth, the duration of $t_b - t_a = \frac{1}{f_p}$ is equivalent to $\Delta\alpha_{CS} = 120°$ in a six cylinder engine. This is given by the shift of 20 teeth of the crankshaft sensor. Unfortunately, mass air flow is synchronized to time and not to the crankshaft angle $\alpha_{CS}$. Since mass air flow $\dot{m}_a$ is not sampled at start and stop times $t_a$ and $t_b$, it must be interpolated:

$$\dot{m}_a(t_a) = \dot{m}_a(t_0)\frac{t_1 - t_a}{T_s} + \dot{m}_a(t_1)\left(1 - \frac{t_1 - t_a}{T_s}\right)$$

$$\dot{m}_a(t_b) = \dot{m}_a(t_n)\frac{t_{n+1} - t_b}{T_s} + \dot{m}_a(t_{n+1})\left(1 - \frac{t_{n+1} - t_b}{T_s}\right)$$

The integration is approximated e.g. by the trapezoidal rule:

$$m_a \approx \frac{T_s}{CYL}\left[(\dot{m}_a(t_a) + \dot{m}_a(t_1))\frac{t_1 - t_a}{T_s} + \dot{m}_a(t_1) + 2\dot{m}_a(t_2) + \cdots\right.$$

$$\left.\cdots + 2\dot{m}_a(t_{n-1}) + \dot{m}_a(t_n) + (\dot{m}_a(t_n) + \dot{m}_a(t_b))\left(1 - \frac{t_{n+1} - t_b}{T_s}\right)\right]$$

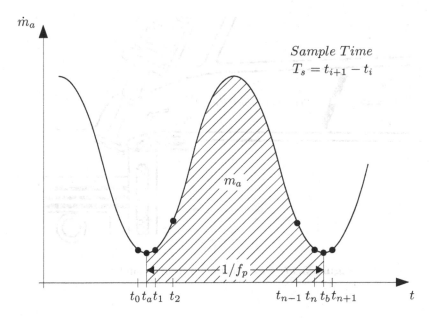

**Figure 3.16** Air mass $m_a$ obtained by integration of the air flow $\dot{m}_a$ over period $t_b - t_a$.

which is simplified to:

$$
m_a \approx \frac{T_s}{CYL}\left[\dot{m}_a(t_0)\frac{(t_1-t_a)^2}{T_s^2} - \dot{m}_a(t_1)\left(1 - \frac{(t_1-t_a)^2}{T_s^2}\right) + 2\sum_{i=1}^{n}\dot{m}_a(t_i) - \right.
$$
$$
\left. -\dot{m}_a(t_n)\frac{(t_{n+1}-t_b)^2}{T_s^2} + \dot{m}_a(t_{n+1})\left(1 - \frac{(t_{n+1}-t_b)^2}{T_s^2}\right)\right]
$$

### 3.2.6 Intake Manifold Dynamics

Figure 3.17 shows a cross-section of the intake manifold. The throttle angle controls the mass air flow $\dot{m}_{a,in}$ into the manifold. Diesel engines are either unthrottled or very moderately throttled in some operating points in order to ensure a sufficient exhaust gas recirculation. Exhaust gas recirculation is not considered in the following model.

The charge mass air flow out from the manifold into the cylinders $\dot{m}_{ac}$ depends on the pressure level in the intake manifold $p_m$ (and the pressure in the cylinder $p_c$). To control the air-fuel ratio $\lambda$ correctly also in transients, the injected amount of fuel must be adapted to the mass air flow into the cylinder $\dot{m}_{ac}$ rather than to the mass air flow $\dot{m}_{a,in}$ into the intake manifold.

The pressure oscillations shown in Figure 3.16 shall be neglected in the following deduction (averaged model). A change in mass air flow $\dot{m}_a$ results in a delayed change in manifold pressure $p_m$. The according differential equation is derived from an energy equilibrium: The change of the internal energy of the air mass in the intake manifold is equal to the sum of in- and outgoing energy flows

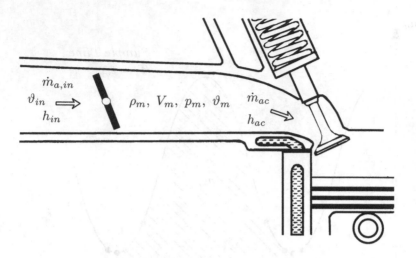

**Figure 3.17** Cross-section of intake manifold

plus the balance of energy changes of the gas due to the displacement work $pV$. By introducing the specific internal energy $u = U/m$ and the specific enthalpy $h = H/m$ the differential equation becomes:

$$\frac{d}{dt}(m_m u_m) = \dot{m}_{a,in} u_{in} - \dot{m}_{ac,air} u_{ac} + p_a \dot{V}_{in} - p_m \dot{V}_{ac} \qquad (3.42)$$

In this model, resonances within the manifold are neglected by assuming identical pressure $p_m$ at the throttle and the inlet pipes [1]. Additionally, the heat radiation from the engine is supposed to match the thermal heat required for the evaporation of the fuel. Therefore, no additional terms are added to Equation 3.42. The enthalpies at the inlet and outlet are equivalent to:

$$\dot{m}_{a,in} h_{in} = \dot{m}_{a,in} u_{in} + p_a \dot{V}_{in} \qquad (3.43)$$

$$\dot{m}_{ac,air} h_{ac} = \dot{m}_{ac,air} u_{ac} + p_m \dot{V}_{ac} \qquad (3.44)$$

which can be inserted in Equation 3.42:

$$m_m \dot{u}_m + \dot{m}_m u_m = \dot{m}_{a,in} h_{in} - \dot{m}_{ac,air} h_{ac} \qquad (3.45)$$

Using the specific heat coefficients $c_v = \partial u/\partial \vartheta$ and $c_p = \partial h/\partial \vartheta$ as well as the air density $\rho = m/V$ we get

$$\rho_m V_m c_v \dot{\vartheta}_m + c_v \vartheta_m \dot{\rho}_m V_m = \dot{m}_{a,in} c_p \vartheta_a - \dot{m}_{ac,air} c_p \vartheta_m \qquad (3.46)$$

and after division by $c_v V_m$

$$\rho_m \dot{\vartheta}_m + \dot{\rho}_m \vartheta_m = \frac{c_p}{c_v} \frac{\vartheta_a}{V_m} \left( \dot{m}_{a,in} - \frac{\vartheta_m}{\vartheta_a} \dot{m}_{ac,air} \right) \quad . \qquad (3.47)$$

---

[1]This relates to an averaged pressure model, which has proved to be sufficiently accurate for mass air flow transients.

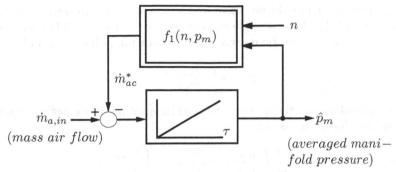

**Figure 3.18** Dynamic model of intake manifold

Inserting the adiabatic exponent $\kappa = c_p/c_v$ and the ideal gas equation $pV = mR\vartheta$, this yields the following equation for the pressure change:

$$\dot{p}_m = \frac{\kappa R \vartheta_a}{V_m} \left( \dot{m}_{a,in} - \frac{\vartheta_m}{\vartheta_a} \dot{m}_{ac,air} \right) \tag{3.48}$$

It is difficult to measure the charge mass air flow from the manifold into the cylinder $\dot{m}_{ac,air}$. Because the dynamic response of $\dot{m}_{ac,air}$ is much faster than that of the manifold pressure $p_m$, only the static behavior of $\dot{m}_{ac,air}$ shall be considered by a look-up table $f_1(n, p_m)$. The mass air flow $\dot{m}_{ac,air}$ depends on the engine speed $n$ and the manifold pressure $p_m$. At stationary engine operation, where the pressure derivative is $\dot{p}_m = 0$, it can be determined by $\dot{m}_{a,in}$.

$$\dot{m}_{ac,air}^* = \dot{m}_{ac,air} \frac{\vartheta_m}{\vartheta_a} = f_1(n, p_m) = \dot{m}_{a,in}, \quad \dot{p}_m = 0 \tag{3.49}$$

The pressure change in the intake manifold is given by:

$$\dot{p}_m = \frac{1}{\tau}(\dot{m}_{a,in} - f_1(n, p_m)) \tag{3.50}$$

with the integration constant $\tau$:

$$\tau = \frac{V_m}{\kappa R \vartheta_a} \tag{3.51}$$

The look-up table can be measured on an engine test bed at stationary operating points, where the derivative of the averaged manifold pressure $\dot{p}_m = 0$. Under these conditions the incoming mass air flow $\dot{m}_{a,in}$ is equal to the values in the look-up table for $\dot{m}_{ac,air}^*$. Figure 3.18 shows the block diagram of the pressure model of the intake manifold. By integration of Equation 3.50 the average manifold pressure can be estimated in realtime. This can be utilized at engine idle speed control (Section 5.2). Another application is to estimate the mass air flow $\dot{m}_{a,in}$ from the measured manifold pressure $p_m$ and engine speed $n$:

$$\dot{m}_{a,in} = \tau \cdot \dot{p}_m + f_1(n, p_m) . \tag{3.52}$$

The derivative $\dot{p}_m$ can only be calculated with some lag. In some applications, the derivative of the throttle angle $\dot{\alpha}_t$ is therefore used as an additional variable.

The time constant can be normalized at operating points $p_{m,0}$ and $\dot{m}_{a,0}$:

$$\tau_n = \frac{p_{m,0}}{\dot{m}_{a,0}} \tau \tag{3.53}$$

This normalized parameter $\tau_n$ represents the integration time constant in seconds. The pressure change $\dot{p}_m$ is then:

$$\tau_n \frac{d}{dt}\left(\frac{p_m}{p_{m,0}}\right) = \frac{\dot{m}_{a,in}}{\dot{m}_{a,0}} - \frac{f_1(n, p_m)}{\dot{m}_{a,0}} \tag{3.54}$$

The time constant $\tau_n$ depends mainly on the air flow $\dot{m}_{a,0}$ which can be seen in the following example: Assuming a manifold volume of $V_m = 4.25\,l$ at an ambient temperature of $\vartheta_a = 300\,^\circ K$ and an adiabatic coefficient $\kappa = 1.4$, the time constant $\tau_n$ at minimum and maximum power can be calculated:

- at maximum power, the manifold pressure is $p_{m,0} = 1\,bar$ and the mass air flow shall be $\dot{m}_{a,0} = 600\,kg/h$. This leads to a time constant of:

$$\tau_{n,1} = 21\,ms \tag{3.55}$$

- at minimum power (e.g. idling), the pressure is $p_{m,0} = 0.35\,bar$ and the mass air flow shall be $\dot{m}_{a,0} = 6\,kg/h$ which leads to:

$$\tau_{n,2} = 740\,ms \tag{3.56}$$

It can be seen that the dynamic behavior of the intake manifold has an impact on engine dynamics especially at low power such as idling.

## 3.2.7  Ignition Angle Control

Correct ignition timing over the entire engine operating range is very important as it has a major impact on fuel consumption as well as on emission rates. Combustion within the cylinder can be divided into two phases:

1. **Inflammation delay (Time Proportional)**
   In-cylinder pressure and temperature do not rise considerably within this time period. The inflammation delay time $\tau_{id}$ depends on the temperature, pressure and air-fuel ratio. The delay can be convoluted into an equivalent crankshaft angle which increases with engine speed.

2. **Combustion (Angle Proportional)**
   The equivalent crankshaft angle for the second phase is almost constant over the entire engine operating range. Induced by the piston movement, turbulences accelerate with engine speed, as well as the combustion process.

If combustion starts too late because of retarded ignition angles, the emission of hydrocarbons $HC$ will increase. High pressure amplitudes at advanced ignition angles increase the emission of $NO_x$. $NO_x$ can be reduced by delaying the ignition at the expense of a higher fuel consumption [2]. The following parameters are used to control the ignition angle:

- Intake manifold pressure $p_m$

- Mass air flow $\dot{m}_a$

- Engine speed $n$

- Throttle angle $\alpha_t$

- Air-fuel ratio $\lambda$

- Crankshaft angle $\alpha_{CS}$ and $TDC$ signal of a reference cylinder

- Ambient air temperature $\vartheta_a$

- Engine temperature $\vartheta_e$

- Battery voltage $U_b$

These values are the same as those needed for fuel control. The ignition angle $\alpha_i$ is dependent on many influences:

- The ignition angle $\alpha_i$ is a function of engine load approximated by injection time $t_{inj} \sim \dot{m}_a/(n \cdot \lambda)$ (see Equation 3.33), and of engine speed $n$. This can be described by a map $\alpha_i = f(t_i, n)$. The look-up table also covers the variation of inflammation depending on engine load and speed. Retarded ignition angles may be selected in order to compromise for reduced emissions and knocking. The ignition angle $\alpha_i$ is determined for each engine operating point by test bed experiments.

- Air-fuel ratio $\lambda$, which determines the inflammation delay $\tau_{id}$.

- Retarded ignition angle at high ambient air temperature $\vartheta_a$ to avoid knocking. A look-up table depending on $t_{inj}$ and $\vartheta_a$ may be used.

- Engine warm-up at low engine temperatures $\vartheta_e$. A retarded ignition angle $\alpha_i$ retards the energy conversion process to a phase where the exhaust valves are already opened. The exhaust pipe and the catalytic converter are then heated very fast.

- Engine speed stabilization at idling by advancing ignition angles at lower engine speeds, thus increasing torque.

- Engine speed limitation by retarding ignition angles in conjunction with fuel cut off.

- Retarded ignition angles during acceleration in order to avoid knocking.

- Closed-loop knock control

- The battery voltage $U_b$ has an impact on the electrical ignition energy.

Figure 3.19 shows an ignition angle map depending on engine speed and load. Angles $\alpha_i$ before Top Dead Center (TDC) are positive.

---

[2] Over all, the determination of the right ignition angle is a compromise between different objectives.

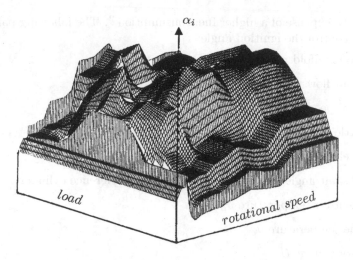

**Figure 3.19** Ignition angle map

## 3.2.8   Optimization of Engine Maps

The fuel amount and ignition angle are the two most important parameters that influence the fuel consumption as well as the emission of pollutants. This has already been shown in Sections 3.1.2 and 3.1.6. There is a conflict between minimizing either fuel consumption or emissions. This is shown in Figure 3.20.

If the ignition angle is chosen to minimize fuel consumption, the engine raw emission rates for $NO_x$ and $HC$ will be fairly high. On the other hand, if the ignition angle is selected to minimize emissions, the fuel consumption will be higher. A compromise must consider fuel consumption **and** emission levels at all engine operating points. Emission levels can be very high at some particular operating points. There, the optimization must focus on the emissions. Other operating points show acceptable emission rates. At these points the optimization must focus on fuel consumption.

Fuel consumption and emission levels are measured in special road driving cycles like the ECE-test or FTP-test. These tests specify the vehicle velocity over time. Translating vehicle to engine speeds, a test cycle is equivalent to a sequence of different engine operating points over time. Every operating point is defined by several control parameters including engine speed and load.

The fuel consumption can be described by the volume $\dot{V}$ of combusted fuel over time. The minimization criteria is the integral over the test cycle.

$$ V = \int\limits_0^T \dot{V}(t) \, dt \rightarrow min \qquad (3.57) $$

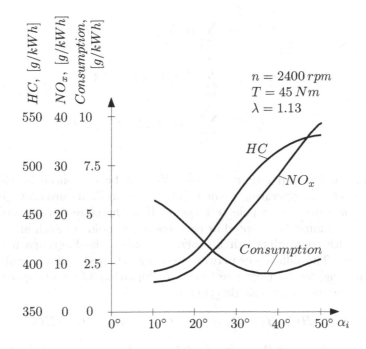

**Figure 3.20** Fuel consumption and emission levels over ignition angle $\alpha_i$.

The total fuel consumption $V$ for a test cycle time $T$ can also be obtained by a discrete summation over all visited engine operating points.

$$V = \sum_{i=1}^{N} \dot{V}_i(\alpha_i, \lambda_i) \, t_i \to min \qquad (3.58)$$

An analysis of the test cycle shows that most operating points are visited several times. The individual time periods where the engine stays in the same operating point $i$ can be summarized into a total time period $t_i$. The fuel consumption over time $\dot{V}_i$ can then be minimized independently for each operating point. The resulting values of $\alpha_i$ and $\lambda_i$ are stored into look-up tables $\alpha_i(t_{inj}, n)$ and $\lambda_i(t_{inj}, n)$ for every operating point.

When optimizing fuel consumption, the maximum allowable emission levels are treated as optimization constraints. The maximum emission rates are fixed by laws which specify the maximum integral masses of the different pollutants generated during a test cycle.

$$HC = \sum_{i=1}^{N} \dot{H}C(\alpha_i, \lambda_i)\, t_i \leq \hat{H}C \qquad (3.59)$$

$$CO = \sum_{i=1}^{N} \dot{C}O(\alpha_i, \lambda_i)\, t_i \leq \hat{C}O \qquad (3.60)$$

$$NO_x = \sum_{i=1}^{N} \dot{N}O_x(\alpha_i, \lambda_i)\, t_i \leq \hat{N}O_x \qquad (3.61)$$

The emission levels per time $\dot{H}C, \dot{C}O, \dot{N}O$ can be influenced by the values of $\alpha_i$ and $\lambda_i$ at each operating point $i$. The emission limits are only given for the integral mass over the whole test cycle. It is therefore not obvious which $\alpha_i$ and $\lambda_i$ values must be adopted at each operating point $i$. Such an optimization problem with constraints can be solved by using the Lagrange multiplication method [9]. The differences between actually achieved and acceptable emission levels are weighted by Lagrange factors $L$. Equation 3.58 and Equations 3.59 to 3.61 are combined into a single criteria.

$$W = V + L_{HC}(HC - \hat{H}C) + L_{CO}(CO - \hat{C}O) + L_{NO_x}(NO_x - \hat{N}O_x) \rightarrow \min \quad (3.62)$$

Now the cost function $W$ must be minimized. For example, if all emission rates were at the acceptable limits ($HC = \hat{H}C$, etc.), all terms except $V$ for fuel consumption would disappear. This would minimize $V$ as before. The value $W$ can be divided into two parts: a constant part $W_o$ which is independent of the operating points $i$ and a variable part influenced by $\alpha_i$ and $\lambda_i$.

$$
\begin{aligned}
W &= \sum_{i=1}^{N} \dot{V}_i(\alpha_i, \lambda_i) t_i + \\
&\quad \sum_{i=1}^{N} [L_{HC}\dot{H}C(\alpha_i, \lambda_i)t_i + L_{CO}\dot{C}O(\alpha_i, \lambda_i)t_i + L_{NO_x}\dot{N}O_x(\alpha_i, \lambda_i)t_i] \\
&\quad -L_{HC}\hat{H}C - L_{CO}\hat{C}O - L_{NO_x}\hat{N}O_x \\
W &= \sum_{i=1}^{N} Z(\alpha_i, \lambda_i)t_i - W_0
\end{aligned}
$$

where:
$$W_0 = L_{HC}\hat{H}C + L_{CO}\hat{C}O + L_{NO_x}\hat{N}O_x = const \qquad (3.63)$$

and:

$$
\begin{aligned}
Z(\alpha_i, \lambda_i) &= \dot{V}(\alpha_i, \lambda_i) \\
&\quad + L_{HC}\dot{H}C(\alpha_i, \lambda_i) \\
&\quad + L_{CO}\dot{C}O(\alpha_i, \lambda_i) \\
&\quad + L_{NO_x}\dot{N}O_x(\alpha_i, \lambda_i)
\end{aligned}
$$

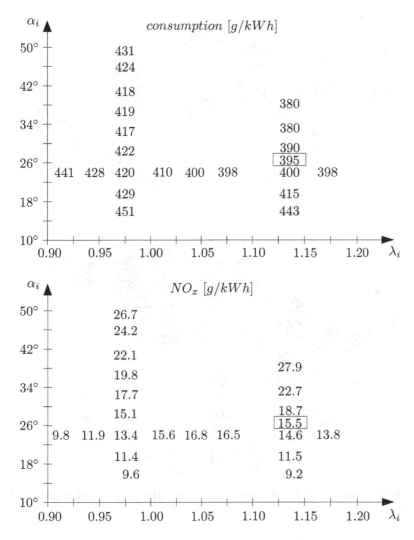

**Figure 3.21** Fuel consumption and $NO_x$ emissions over ignition angle $\alpha_i$ and air-fuel ratio $\lambda$ at an operating point $i$ with $L_{NO_x} = 3$.

The value of $W$ can be minimized by minimizing the function $Z(\alpha_i, \lambda_i)$ at each operating point:

$$Z_i(\alpha_i, \lambda_i) \rightarrow min \qquad (3.64)$$

Figure 3.21 shows $\dot{V}(\alpha_i, \lambda_i)$ and $\dot{NO}_x(\alpha_i, \lambda_i)$ for one operating point $i$ depending on ignition angle $\alpha_i$ and air-fuel ratio $\lambda_i$. The values were measured in an engine test bed run. The pair $\alpha_i, \lambda_i$ for minimum $Z(\alpha_i, \lambda_i)$ is marked by a rectangle.

It can be seen that fuel consumption $\dot{V}(\alpha_i, \lambda_i)$ at minimum $Z(\alpha_i, \lambda_i)$ is slightly higher than the absolute minimum in order to compromise with emissions $NO_x(\alpha_i, \lambda_i)$. The value $W$ is a minimum for the entire test cycle if correct values for the Lagrange factors $L_{HC}$, $L_{CO}$ and $L_{NO_x}$ were selected. In practice, several

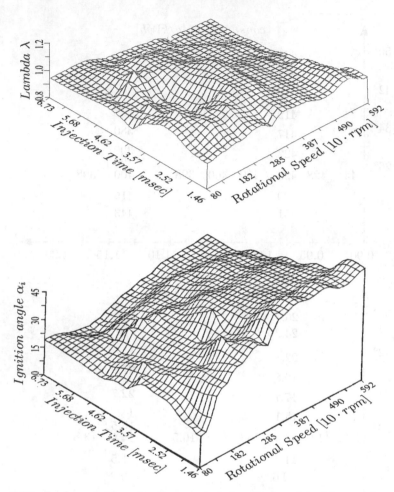

**Figure 3.22**  Look-up tables of optimum air-fuel ratio $\lambda$ and ignition angle $\alpha$ for an FTP test cycle.

iterations with modified Lagrange factors must be passed to obtain a minimum $W$ which also meets the legal emission constraints. For example, in a first iteration, low fuel consumption $V$ is attained. However, emissions $(HC, CO, NO_x)$ are still relatively close to or above the legal limits. The Lagrange weighting factors $L$ are then increased for the next iteration. Practical experience shows that only a few iterations are needed to minimize the function $Z$ at a small number of representative operating points $i$. Values for other operating points can be obtained by interpolation. Figure 3.22 shows resulting look-up tables for the ignition angle $\alpha$ and the air-fuel ratio $\lambda$. In engines at stoichiometric $\lambda = 1$ operation, only the ignition angle $\alpha_i$ is calculated. The optimization approach can also be extended to other variables such as exhaust gas recirculation rate.

# 4 Diesel Engine Modeling

In recent years engine modeling has become more and more important in the development process. In an early development stage new strategies can be examined without expensive test bench measurements. Further on, control strategies can be designed and optimized. Thereby new requirements for the models arise.

- Multiple direct fuel injection in diesel engines
  Direct fuel injection was a breakthrough for diesel engines. By means of multiple injections in common rail systems, the evaporation process and subsequently the combustion process can be directly influenced.

- Stratified charge in SI engines
  As explained in Section 5.1, SI engines work with a homogeneous air-fuel mixture at stoichiometric air-fuel ratios because the catalytic converter shows extremely good conversion rates under such conditions. At part load, directly injected SI engines work with lean air-fuel mixtures. In order to get a safe combustion a stratified charge is assembled around the spark plug. In such engines the charge exchange, turbulences inside the combustion chamber and the injection process must be modeled.

- Exhaust gas calculation
  One of the major problems is to model the exhaust gas generation fairly accurate. The maximum allowable emissions are determined by law. In order to meet such regulations two strategies are pursued in parallel. One is to minimize the raw emissions of noxious gases from the combustion process. The other is to clean the exhaust gases by means of catalytic after treatment systems. Both strategies need sophisticated control systems and dynamic models of the chemical processes involved.

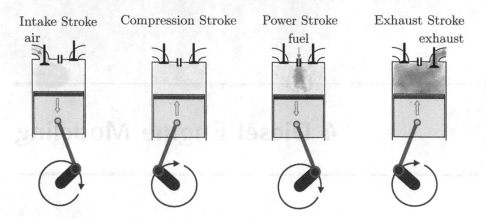

**Figure 4.1** Four stroke cycle diesel engine.

Because of the rising importance of diesel engines, models for flexible injection, energy conversion and evaporation processes will be presented in this chapter. In addition a model for soot accruement is introduced.

## 4.1   Four Stroke Cycle Diesel Engine

The movement of the piston up and down inside the cylinder is called its stroke. The crankshaft is connected to the piston by the connecting rod. By this means the linear motion of the piston is converted into rotational movement. The crank shaft connects all the pistons and the engine torque. Most diesel engines use a four stroke cycle because the piston has to travel up and down twice to complete a full power cycle turning the crank shaft through 720 degrees. As listed below the four strokes of the diesel cycle are illustrated in Figure 4.1.

- Intake stroke
  The intake stroke begins at top dead center. As the piston moves down, the intake valve opens. The downward movement of the piston draws fresh air through the intake valve into the combustion chamber.

- Compression stroke
  The compression stroke begins with the piston at bottom dead center. The piston is rising up to compress the in-cylinder charge. Since both the intake and exhaust valves are closed, the in-cylinder gas is compressed to a fraction of its original volume heating it up.

- Power stroke
  The next stroke is called power stroke. At the start of this second downward movement of the piston or at the end of the compression stroke fuel is sprayed into the cylinder. On contact with the hot in-cylinder charge the fuel inflames and the resulting combustion drives the piston down.

- Exhaust stroke

   The last stroke is called exhaust stroke. Here the exhaust gases are pushed out of the combustion chamber through the outlet valve.

In the following equations for the four stroke diesel engine model, the crankshaft angle $\alpha_{CS}$ is taken as an independent variable instead of time $t$. The relationship between these variables is

$$\alpha_{CS} = \omega \cdot t = 2\pi n \cdot t \ , \tag{4.1}$$

$$d\alpha_{CS} = \omega \cdot dt \ , \tag{4.2}$$

with $n$ as the engine speed.

## 4.2 Charge Exchange

During charge exchange the exhaust gases inside the combustion chamber are replaced with fresh air. The volumetric mass flow through a restrictor with the cross sectional area $A_{eff}$ into and out of the combustion chamber can be calculated according to [82] as

$$
\begin{aligned}
\frac{dm}{dt} &= \mu_{flow} \cdot A_{eff} \cdot \frac{p_0}{\sqrt{R \cdot \vartheta_0}} \cdot \sqrt{\frac{2 \cdot \kappa}{\kappa - 1} \left[ \left( \frac{p_1}{p_0} \right)^{\frac{2}{\kappa}} - \left( \frac{p_1}{p_0} \right)^{\frac{\kappa+1}{\kappa}} \right]} \\
&= \mu_{flow} \cdot A_{eff} \cdot \frac{p_0}{\sqrt{R \cdot \vartheta_0}} \cdot \Psi \ .
\end{aligned}
\tag{4.3}
$$

The outflow $\Psi$ depends on the pressure $p_0$ in front of and on the pressure $p_1$ behind the throttle as well as on the adiabatic exponent $\kappa$. In practice, the effective cross sectional area is smaller than the geometric cross sectional area. This fact is considered in the model with the flow ratio coefficient $\mu_{flow}$. The temperature before the throttle is $\vartheta_0$.

The geometric cross sectional area $A_{eff}$ depends on the valve movement and may be described as

$$A_{eff} = n_o \cdot \pi \cdot h_\nu \cdot \cos(\sigma) \cdot (d_\nu + h_\nu \cdot \sin(\sigma) \cdot \cos(\sigma)) \ , \tag{4.4}$$

depending on the number of input or output valves $n_o$, the valve stroke $h_\nu$, the valve seat angle $\sigma$ and the inner valve seat diameter $d_\nu$. Figure 4.2(a) shows the geometric parameters at the valve seat. The valve stroke $h_\nu$ during inlet and outlet can be approximated with sine curves, see Figure 4.2(b).

### 4.2.1 Flow into Exhaust Pipes

After the combustion, the exhaust gases are pumped into the exhaust pipes. At the beginning of the outlet phase the in-cylinder exhaust gas approximately consists of a homogeneous mixture of air and burnt fuel. The air mass $m_a$ and burnt fuel mass $m_{f,burnt}$ have to be adopted after completion of the combustion

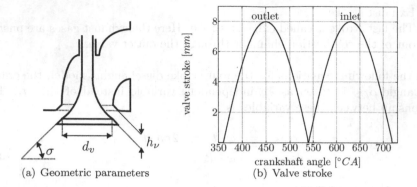

(a) Geometric parameters          (b) Valve stroke

**Figure 4.2** Cross sectional area at the valves.

process. The flow of exhaust gases $\frac{dm_{out}}{d\alpha_{CS}}$ through the outlet valves is depending
on crankshaft angle instead of time. It is calculated according to Equation 4.3.

$$\frac{dm_{out}}{d\alpha_{CS}} = \frac{10^5}{2\pi n} \cdot \mu_{flow,out} \cdot A_{eff\,out} \cdot \frac{p}{\sqrt{R \cdot \vartheta}} \cdot \sqrt{\frac{2 \cdot \kappa}{\kappa - 1} \left[ \left( \frac{p_{out}}{p} \right)^{\frac{2}{\kappa}} - \left( \frac{p_{out}}{p} \right)^{\frac{\kappa+1}{\kappa}} \right]} .$$
(4.5)

In this, $p_{out}$ is the pressure in the outlet pipe, $p$ the in-cylinder pressure and $\vartheta$
the in-cylinder temperature. Integration of Equation 4.5 over the outlet phase
yields $m_{out}$. Some remaining gases $m_{rem}$ will remain in the combustion chamber
after closing of the outlet valves.

$$m_{rem} = m_a + m_{f,burnt} - m_{out}$$
(4.6)

This remaining gas consists of air $m_{rem,air}$ (see Equation 4.14) and burnt fuel
$m_{rem,burnt}$ (see Equation 4.15). Since the exhaust gas is assumed to be a homogeneous mixture, the air-fuel ratio of the remaining gases is equal to that of the
gas mixture at the end of the previous combustion phase.

### 4.2.2   Flow into Combustion Chamber

The mass flow $\frac{dm_{ac}}{d\alpha_{CS}}$ into the combustion chamber from the intake manifold is
calculated with Equation 4.3 in dependency of the crankshaft angle $\alpha_{CS}$ and the
engine speed $n$ as

$$\frac{dm_{ac}}{d\alpha_{CS}} = \frac{10^5}{2\pi n} \cdot \mu_{flow,in} \cdot A_{eff\,in} \cdot \frac{p_m}{\sqrt{R \cdot \vartheta_m}} \cdot \sqrt{\frac{2 \cdot \kappa}{\kappa - 1} \left[ \left( \frac{p}{p_m} \right)^{\frac{2}{\kappa}} - \left( \frac{p}{p_m} \right)^{\frac{\kappa+1}{\kappa}} \right]} .$$
(4.7)

In this, $p_m$ is the intake manifold pressure and $p$ the in-cylinder pressure. Furthermore, the gas composition must be considered in the charge process, such as
in-cylinder remaining gases $m_{rem}$ and recirculated exhaust gases $m_{EGR}$ (exhaust
gas recirculation, EGR). By means of a given EGR-rate $X_{EGR}$, the charge air
mass from the manifold is

$$m_{ac,air} = m_{ac} \cdot (1 - X_{EGR})$$
(4.8)

and the mass of the recirculated gases is

$$m_{EGR} = m_{ac} \cdot X_{EGR} \ . \tag{4.9}$$

The charge mass of fresh air per stroke $m_{ac}$ is calculated by integrating Equation 4.7 over the inlet phase. Since the recirculated gas is composed of air and burnt fuel at the same proportion as the remaining gases at the end of the preceding outlet phase, the amount of recirculated air is

$$m_{EGR,air} = \frac{m_{rem,air}}{m_{rem}} \cdot X_{EGR} \cdot m_{ac} \ . \tag{4.10}$$

The resulting air mass $m_a$ inside the combustion chamber for the next combustion phase is the sum of the air mass from the manifold $m_{ac,air}$, the recirculated air mass $m_{EGR,air}$ and the air portion of the in-cylinder remaining gases $m_{rem,air}$ from the outlet phase

$$m_a = m_{ac} \cdot (1 - X_{EGR}) + m_{EGR,air} + m_{rem,air} \ . \tag{4.11}$$

The amount of recirculated burnt fuel $m_{EGR,burnt}$ is the complement to equation 4.10

$$m_{EGR,burnt} = X_{EGR} \cdot m_{ac} - m_{EGR,air} = X_{EGR} \cdot m_{ac} \cdot \frac{m_{rem,burnt}}{m_{rem}} \ . \tag{4.12}$$

The resulting charge mass of burnt fuel $m_{f,burnt}$ at the beginning of the combustion process is the sum of recirculated burnt fuel $m_{EGR,burnt}$ and the portion of the in-cylinder remaining gases $m_{rem,burnt}$ from the previous outlet phase.

$$m_{f,burnt} = m_{EGR,burnt} + m_{rem,burnt} \ . \tag{4.13}$$

The injected fuel is not yet burnt at the time of injection. Figure 4.3 shows all masses determined during charge exchange.

## 4.3   Air-fuel Ratio

The stoichiometry ratio, $L_{st}$ refers to an air-fuel ratio in which all combustible materials are used with no deficiencies or excesses. The air-fuel ratio $\lambda_c$ is the current ratio of air and burnt fuel masses divided by the stoichiometric constant. Local differences of the air-fuel ratio are neglected in this model. The gas is assumed to be homogeneous.

### 4.3.1   Exhaust Stroke

In this section of the diesel cycle, the air-fuel ratio $\lambda_c$ of the in-cylinder exhaust gases shall be identical to the air-fuel ratio at the end of the previous combustion process (section 4.3.3). Since we assume the exhaust gases to be homogeneous, $\lambda_c$ remains constant during the exhaust stroke. The resulting in-cylinder remaining air mass $m_{rem,air}$ and burnt fuel mass $m_{rem,burnt}$ are now derived as

$$m_{rem,burnt} = \frac{1}{\lambda_c \cdot L_{st} + 1} \cdot m_{rem} \tag{4.14}$$

$$m_{rem,air} = \frac{\lambda_c \cdot L_{st}}{\lambda_c \cdot L_{st} + 1} \cdot m_{rem} \ . \tag{4.15}$$

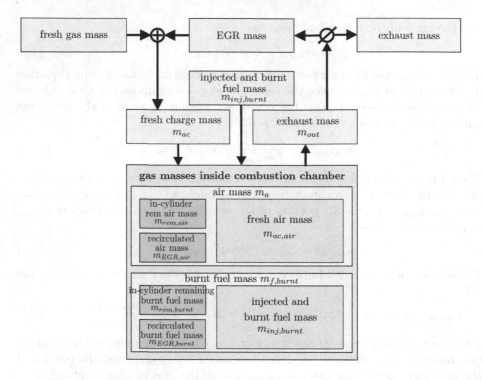

**Figure 4.3** Overview of all calculated masses during charge exchange.

## 4.3.2 Intake Stroke

Before fuel injection the combustion chamber is filled with fresh air and recirculated gases. Since the combustion chamber is charged with fresh air, the air-fuel ratio $\lambda_c$ is increasing. The resulting air-fuel ratio is

$$\lambda_c = \frac{m_a}{m_{f,burnt} \cdot L_{st}} = \frac{m_{ac} \cdot (1 - X_{EGR}) + m_{EGR,air} + m_{rem,air}}{(m_{EGR,burnt} + m_{rem,burnt}) \cdot L_{st}} \quad . \tag{4.16}$$

$m_{rem,air}$ and $m_{rem,burnt}$ are the remaining gas masses of the previous exhaust stroke (section 4.3.1). The change of the air-fuel ratio $\lambda_c$ during the intake stroke is

$$\frac{d\lambda_c}{d\alpha_{CS}} = \frac{1}{m_{f,burnt}^2 \cdot L_{st}} \cdot \left( \frac{dm_a}{d\alpha_{CS}} \cdot m_{f,burnt} - m_a \cdot \frac{dm_{f,burnt}}{d\alpha_{CS}} \right) \quad . \tag{4.17}$$

## 4.3.3 Compression and Combustion

The air-fuel ratio $\lambda_c$ is changing during combustion as well. Changes of the gas masses inside the combustion chamber result from the burning of injected fuel which causes the air-fuel ratio to decrease. The current ratio $\lambda_c$ between air and burnt fuel is calculated as

$$\lambda_c = \frac{m_a}{m_{f,burnt} \cdot L_{st}} = \frac{m_{ac} \cdot (1 - X_{EGR}) + m_{EGR,air} + m_{rem,air}}{(m_{EGR,burnt} + m_{rem,burnt} + m_{inj,burnt}) \cdot L_{st}} \quad . \tag{4.18}$$

$m_a$ represents the mass of fresh air and the mass of air contained in the combustion products inside the cylinder. After closing the inlet valves $m_a$ remains constant during the compression and the power stroke. The amount of burnt fuel $m_{f,burnt}$, consisting of the amount of burnt fuel in the remaining gases $m_{rem,burnt}$, the recirculated burnt fuel mass $m_{EGR,burnt}$ and the amount of injected and already burnt fuel $m_{inj,burnt}$ changes during the current combustion process. The change of the air-fuel ratio $\lambda_c$ in dependence of the crankshaft angle is

$$
\begin{aligned}
\frac{d\lambda_c}{d\alpha_{CS}} &= -\frac{\lambda_c}{m_{f,burnt}} \cdot \frac{dm_{inj,burnt}}{d\alpha_{CS}} , \\
&= -\frac{\lambda_c}{m_{EGR,burnt} + m_{rem,burnt} + m_{inj,burnt}} \cdot \frac{dm_{inj,burnt}}{d\alpha_{CS}} . \quad (4.19)
\end{aligned}
$$

The air-fuel ratio at the end of combustion is taken in Equation (4.14) and (4.15).

## 4.4 Mass Balance

A gas mass increment $dm$ inside the cylinder is the sum of the charge mass increment $dm_{ac}$ and the increment of the injected fuel that is already burnt $dm_{inj,burnt}$ minus the outlet mass increment $dm_{out}$. Blow-by effects at the valves and at the piston may be neglected for modern diesel engines.

$$
\frac{dm}{d\alpha_{CS}} = \frac{dm_{ac}}{d\alpha_{CS}} + \frac{dm_{inj,burnt}}{d\alpha_{CS}} - \frac{dm_{out}}{d\alpha_{CS}} . \quad (4.20)
$$

During the intake stroke, the gas mass derivate is

$$
\frac{dm}{d\alpha_{CS}} = \frac{dm_{ac}}{d\alpha_{CS}} . \quad (4.21)
$$

During the compression and power stroke, the valves are closed and the mass increment $dm$ is the combusted fuel mass increment $dm_{inj,burnt}$

$$
\frac{dm}{d\alpha_{CS}} = \frac{dm_{inj,burnt}}{d\alpha_{CS}} . \quad (4.22)
$$

During the exhaust stroke the mass derivative depends on the flow of exhaust gases out of the cylinder.

$$
\frac{dm}{d\alpha_{CS}} = -\frac{dm_{out}}{d\alpha_{CS}} . \quad (4.23)
$$

The liquid fuel inside the combustion chamber is ignored in this balance since only gases are relevant in the combustion process.

## 4.5 Fuel Injection

The energy conversion is determined by the injection process, i.e. how much fuel is injected over time. The course of injection is influenced by the injection pressure, the number of injections per combustion cycle, the duration of each injection, the geometry of the injector nozzle and the in-cylinder pressure.

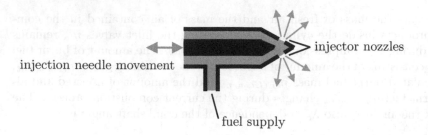

**Figure 4.4**  Schematic structure of an injector.

Figure 4.4 shows the schematic structure of an injector. The fuel inside the injector is under the same high pressure as inside the common rail. Because of the needle movement, the nozzles open and fuel flows into the cylinder according to the pressure gradient between common rail and combustion chamber. The flow of the injected fuel mass $\frac{dm_f}{d\alpha_{CS}}$ in dependency of the crankshaft angle $\alpha_{CS}$ can be calculated with the Bernoulli Equation for incompressible substances [98, 121] as

$$\frac{dm_f}{d\alpha_{CS}} = \frac{\mu_{inj}}{2 \cdot \pi \cdot n} \cdot A_n \cdot \sqrt{2 \cdot \rho_f \cdot (p_{rail} - p)} \; . \tag{4.24}$$

The injected fuel mass over rotation of the crankshaft depends on the effective cross sectional area of the nozzles $A_n$, the fuel density $\rho_f$ and the pressure difference between the injection pressure $p_{rail}$ and the in-cylinder pressure $p$. Furthermore, the effective cross sectional area $A_n$ of the nozzles depends on the needle movement. The factor $\mu_{inj}$ is required for fitting calculated fuel masses to actually measured fuel masses in each operating point. Because the needle movement is often not known, a model without needle movement is employed. According to [121] the fuel flow into the combustion chamber for direct injected diesel engines can be assumed as parabolic over rotation of the crankshaft.

The parabola

$$\frac{dm_f(\alpha_{CS})}{d\alpha_{CS}} = a \cdot \alpha_{CS}^2 + b \cdot \alpha_{CS} + c \tag{4.25}$$

has to be open downward, i.e. $a < 0$. The maximum point $(\alpha_{CS,max}, \frac{dm_{f,max}}{d\alpha_{CS}})$ is calculated as

$$\alpha_{CS,max} = -\frac{b}{2 \cdot a} \quad \text{and} \quad \frac{dm_{f,max}}{d\alpha_{CS}} = \frac{4ac - b^2}{4a}. \tag{4.26}$$

The injection flow can be written

$$\frac{dm_f(\alpha_{CS})}{d\alpha_{CS}} = a \cdot \alpha_{CS}^2 + b \cdot \alpha_{CS} \qquad 0 \le \alpha_{CS} \le \alpha_{CS,inj}, \tag{4.27}$$

if the injection starts at $\alpha_{CS} = 0$. The end of injection occurs at angle $\alpha_{CS,inj}$. The maximum fuel flow results from Equation 4.24 at maximum cross sectional area $A_{n,max}$

$$\frac{dm_{f,max}}{d\alpha_{CS}} = \frac{\mu_{inj}}{2 \cdot \pi \cdot n} \cdot A_{n,max} \cdot \sqrt{2 \cdot \rho_f \cdot (p_{rail} - p)}. \tag{4.28}$$

The crankshaft angle of maximum fuel flow is

$$\alpha_{CS,max} = \frac{\alpha_{CS,inj}}{2}. \tag{4.29}$$

The injection process is finished at $\alpha_{CS} = \alpha_{CS,inj}$. The parameter $a$ is calculated at the end of the injection, where $\frac{dm_f}{d\alpha_{CS}}$ is again zero in Equation 4.27

$$a = \frac{-b}{\alpha_{CS,inj}}. \tag{4.30}$$

Using the maximum flow $\frac{dm_{f,max}}{d\alpha_{CS}}$ in Equation 4.28, angle $\alpha_{CS,max}$ and Equations 4.27 and 4.30, the parameters of the parabolic equation are found as

$$a = -\frac{4}{\alpha_{CS,inj}^2} \cdot \frac{\mu_{inj}}{2 \cdot \pi \cdot n} \cdot A_{n,max} \cdot \sqrt{2 \cdot \rho_f \cdot (p_{rail} - p)} \tag{4.31}$$

$$b = \frac{4}{\alpha_{CS,inj}} \cdot \frac{\mu_{inj}}{2 \cdot \pi \cdot n} \cdot A_{n,max} \cdot \sqrt{2 \cdot \rho_f \cdot (p_{rail} - p)}. \tag{4.32}$$

Integrating Equation 4.24 over the injection angle segment $\alpha_{CS,inj}$ yields the total injected fuel mass $m_f$ for one combustion cycle

$$m_f = \frac{4}{\alpha_{CS,inj}} \cdot \frac{\mu_{inj}}{2\pi n} \cdot A_{n,max} \cdot \sqrt{2 \cdot \rho_f \cdot (p_{rail} - p)} \cdot \int\limits_{0}^{\alpha_{CS,inj}} \left( \alpha_{CS} - \frac{\alpha_{CS}^2}{\alpha_{CS,inj}} \right) d\alpha_{CS}. \tag{4.33}$$

The integration is done stepwise, generating a fuel package $m_{f,i}$ for each angle segment $\Delta\alpha_{CS}$ within limits $0 < \Delta\alpha_{CS} \cdot i \le \alpha_{CS,inj}$.

$$m_{f,i} = \frac{4 \cdot A_{n,max}}{\alpha_{CS,inj}} \cdot \frac{\mu_{inj}}{2\pi n} \cdot \sqrt{2 \cdot \rho_f \cdot (p_{rail} - p)} \cdot \int\limits_{\Delta\alpha_{CS} \cdot (i-1)}^{\Delta\alpha_{CS} \cdot i} \left( \alpha_{CS} - \frac{\alpha_{CS}^2}{\alpha_{CS,inj}} \right) d\alpha_{CS}. \tag{4.34}$$

If the entire injected fuel mass $m_f$ is given, the injection angle segment $\alpha_{CS,inj}$ is calculated as

$$\alpha_{CS,inj} = m_f \cdot \frac{3 \cdot \pi \cdot n}{\mu_{inj} \cdot A_{n,max} \cdot \sqrt{2 \cdot \rho_f \cdot (p_{rail} - p)}}. \tag{4.35}$$

To parameterize the fuel injection model, a map for $\mu_{inj}$ is determined for different operation points. Figure 4.5 shows a comparison of measured and modeled data of the injection time $t_{inj} = \frac{\alpha_{CS,inj}}{2 \cdot \pi \cdot n}$ for several pre- and main-injections.

## 4.6 Fuel Evaporation

The piezo electric injection technique allows to shape the course of injection and the progress of fuel evaporation. Many different injection curves are possible, see Figure 4.6. The pre-injection helps to reduce the typical hard diesel engine

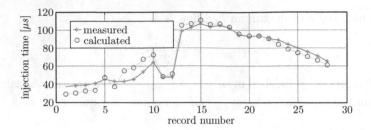

Figure 4.5  Comparison of measured and calculated injection durations for pre- and main-injection.

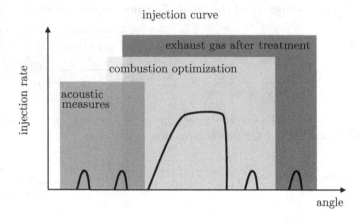

Figure 4.6  Possible injection curve of modern diesel injection systems.

combustion noise. The pressure gradient can be flattened with a small fuel injection at the beginning of the combustion cycle. The main injection and the shape of the main injection are used to optimize the power output of the engine. Finally, post-injections are useful for exhaust gas after treatment systems, e. g. to heat up the exhaust system and to burn soot in a catalytic converter inside the exhaust pipe.

For this reason the simulation model must be able to emulate this behaviour. Simple energy conversion models like the Vibe- or the Double-Vibe-functions [86, 98, 125] are not sufficient. Rather a phenomenological approach according to Constien [20] is employed in this book to model the energy conversion course.

During injection, the fuel sprays into the combustion chamber and evaporates depending on the injection pressure, the fuel density, injector nozzle geometry and the in-cylinder pressure. The air-fuel ratio is inhomogeneous inside the combustion chamber. Zones with lean and rich mixtures develop during injection.

The injected fuel evaporates into drops and droplets. Around the drops, a gaseous environment develops. The temperature is increasing from the center of each drop to its surface where it reaches the combustion chamber temperature. In addition to pressure, the self inflammation of evaporated fuel depends on the

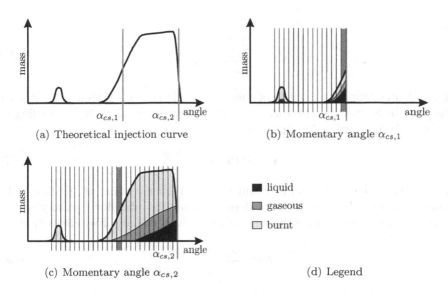

(a) Theoretical injection curve         (b) Momentary angle $\alpha_{cs,1}$

(c) Momentary angle $\alpha_{cs,2}$         (d) Legend

**Figure 4.7** Schematic sequence of the evaporation process.

temperature and the air-fuel ratio around the droplet. When the combustion starts, the droplets heat up and the oxygen concentration is decreasing. Around the droplets lambda values between zero and infinity occur.

According to [20, 122] the number of droplets, their surface and the amount of evaporated fuel inside the combustion chamber can be estimated. This information is needed to calculate the energy conversion process. Figure 4.7 shows how the evaporation process is modeled. An exemplary injection curve with pre- and main-injection is depicted in Figure 4.7(a). The fuel is assumed to be injected into the combustion chamber in discrete subsequent fuel mass packages $m_{f,i}$. For every fuel package the portion of liquid, gaseous and burnt fuel develops over time. With the crankshaft angle as the independent variable, the condition of each fuel mass package $m_{f,i}$ is therefore calculated over angle segments $\Delta\alpha_{CS} \cdot j$.

Figure 4.7(b) shows a snapshot of the fuel masses in the combustion chamber at the momentary crankshaft angle $\alpha_{CS,1}$. The perpendicular lines separate the different fuel mass packages $m_{f,i}$. Within each package, the three fuel portions are shown. Different gray tones mark the liquid, gaseous and burnt fractions of the injected fuel.

At the momentary crank angle $\alpha_{CS,2}$ almost all fuel has been injected, Figure 4.7(c). It can be seen that most of the fuel is already burnt, especially in the packages of the pre-injection. The fuel package singled out at $\alpha_{CS,1}$ is almost completely burnt at $\alpha_{CS,2}$. Only a little gaseous fuel remains.

The number of drops and the surface area of the drops shall now be calculated. Because it is impossible to analytically describe the multitude of all drop diameters, an average drop diameter $d_{32}$ in dependence of injection pressure and fuel package mass is adopted. According to the Sauter approach [20], it is calculated as the ratio of the entire drop volume divided by the entire drop surface area

$$d_{32} = \frac{\sum\limits_{r} d_{dr}^3 \cdot H_r}{\sum\limits_{r} d_{dr}^2 \cdot H_r} \quad . \tag{4.36}$$

The number of drops with the same diameter $d_{dr}$ is $H_r$. Since all $d_{dr}$ and $H_r$ are unknown, the Sauter diameter $d_{32}$ may alternatively be calculated according to the Varde/Popa/Varde [45] approach

$$d_{32} = 16.58 \left(Re \cdot We\right)^{-0.28} \quad , \tag{4.37}$$

where $Re$ is the Reynolds number and $We$ the Weber number of the flow. The Reynolds number $Re$ is calculated depending on the fuel flow velocity at the injector nozzle $v_n$, the nozzle diameter $d_n$ and the kinematic fuel viscosity $\vartheta_f$

$$Re = \frac{v_n \cdot d_n}{\vartheta_f} \quad . \tag{4.38}$$

The Weber number $We$ is calculated with the density of the combustion chamber charge $\rho_{charge}$ and the fuel surface tension $\sigma_f$ as

$$We = \frac{v_n^2 \cdot d_n \cdot \rho_{charge}}{\sigma_f} \quad . \tag{4.39}$$

To get the fuel flow velocity $v_n$ at the nozzle, the pressure difference $\Delta p$ between the injector and the combustion chamber as well as the fuel density $\rho_f$ are required

$$v_n = \sqrt{\frac{2 \cdot \Delta p}{\rho_f}} \quad . \tag{4.40}$$

Inserting Equations 4.38, 4.39 and 4.40 into Equation 4.37, the Sauter diameter $d_{32}$ can be written as

$$d_{32} = 12.392 \cdot \frac{d_n^{0.44} \cdot \rho_f^{0.42} \cdot (\sigma_f \cdot \vartheta_f)^{0.28}}{\Delta p^{0.42} \cdot \rho_{charge}^{0.28}} \quad . \tag{4.41}$$

The number of drops $N_{d,i}$ in a fuel package $i$ injected at the discrete crankshaft angle $\Delta\alpha_{CS} \cdot i$ is derived from the injected liquid fuel mass $m_{f,i}$

$$N_{d,i} = \frac{m_{f,i}}{\frac{\pi}{6} \cdot d_{32}^3 \cdot \rho_f} \quad . \tag{4.42}$$

$m_{f,i}$ is derived from the angle-discrete integration of Equation 4.34. The number of drops $N_{d,i}$ is calculated once for each fuel package. The portions of liquid, gaseous and burnt fuel in the package $i$ change over the discrete crankshaft angle progression $\Delta\alpha_{CS} \cdot j$. Also the drop diameter $d_{32,ij}$ changes over subsequent angles $\Delta\alpha_{CS} \cdot j$. The drop diameter in package $i$ is recalculated iteratively as

$$d_{32,ij} = \sqrt[3]{\frac{m_{f,ij}}{\frac{\pi}{6} \cdot N_{d,i} \cdot \rho_f}} \quad . \tag{4.43}$$

At the crank angle $\Delta\alpha_{CS} \cdot j = \Delta\alpha_{CS} \cdot i$, the liquid fuel mass has the initial value $m_{f,ij} = m_{f,i}$ (Equation 4.34). The surface area of all drops in package $i$ at the angle $\Delta\alpha_{CS} \cdot j$ is

$$A_{f,ij} = N_{d,i} \cdot \pi \cdot d_{32,ij}^2 \ . \tag{4.44}$$

The evaporated fuel mass of package $i$ at the angle $\Delta\alpha_{CS} \cdot j$ is

$$\Delta m_{f,ev,ij} = C_{diff} \cdot A_{f,ij} \cdot p^{m_p} \cdot \frac{\Delta\alpha_{CS}}{n \cdot d_{32,ij}} \ , \tag{4.45}$$

where $p$ is the combustion chamber pressure, $n$ the engine speed and $C_{diff}$ the diffusion constant which is determined in Section 4.8. By means of exponent $m_p$ the dependance of the evaporation process on the in-cylinder pressure can be fitted. Variable $\Delta\alpha_{CS}$ is the discrete crankshaft angle step size.

The liquid fuel portion of fuel package $i$ at the next step $j + 1$ is calculated by subtracting the evaporated fuel $\Delta m_{f,ev,ij}$ of Equation 4.45 from the liquid portion $m_{f,ij}$ of the last step $j$

$$m_{f,i\ j+1} = m_{f,ij} - \Delta m_{f,ev,ij} \ . \tag{4.46}$$

The drop diameter $d_{32,i\ j+1}$ and the surface area $A_{f,i\ j+1}$ are calculated from Equation 4.43 and 4.44 when increasing $j$ to $j + 1$.

In order to describe the transformation of evaporated into burnt fuel the inflammation delay time $\tau_{id,j}$ is determined at each step $j$. The inflammation delay time $\tau_{id,j}$ may be approximated by the empirical approach [20] depending on the average in-cylinder pressure $\bar{p}_j$ and the average temperature $\bar{\vartheta}_j$ between $\Delta\alpha_{CS} \cdot i$ and the momentary angle segment $\Delta\alpha_{CS} \cdot j$

$$\tau_{id,j} = 2,1 \cdot \bar{p}_j^{-1,02} \cdot e^{\frac{2100\ K}{\bar{\vartheta}_j}} \ . \tag{4.47}$$

After the inflammation delay $\tau_{id,j}$, the evaporated fuel mass $m_{f,ev,ij}$ of fuel package $i$ at the beginning of iteration step $j$ and the evaporated fuel mass $\Delta m_{f,ev,ij}$ during step $j$ are assumed to burn completely into

$$\Delta m_{f,burnt,ij} = m_{f,ev,ij} + \Delta m_{f,ev,ij} \ . \tag{4.48}$$

This occurs at the angle segment $\Delta\alpha_{CS} \cdot j \geq \Delta\alpha_{CS} \cdot i + 2\pi n \cdot \tau_{id,j}$. Before the next iteration step $j + 1$, $m_{f,ev,i\ j+1}$ is set to zero, because new fuel will evaporate at the step $j+1$. If the inflammation delay is not yet reached the overall evaporated fuel mass $m_{f,ev,i\ j+1}$ of package $i$ is updated at the beginning of the next iteration step $j + 1$

$$m_{f,ev,i\ j+1} = m_{f,ev,ij} + \Delta m_{f,ev,ij} \ . \tag{4.49}$$

The entire burnt fuel mass $m_{f,burnt,j}$ at iteration step $j$ is the sum of all burnt portions of the different injection packets $i$

$$m_{f,burnt,j} = \sum_i \Delta m_{f,burnt,ij} \tag{4.50}$$

and the mass of the injected and already burnt fuel $m_{inj,burnt}$ is the sum of the already burnt fuel masses

$$m_{inj,burnt} = \sum_j m_{f,burnt,j} \ . \tag{4.51}$$

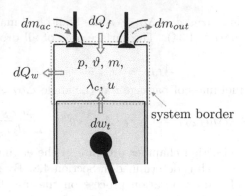

**Figure 4.8**  Charge and combustion model.

## 4.7   Cylinder Dynamics

The aim of modeling the combustion process is to get information about the in-cylinder conditions for different operating points of an engine. The development of engine control strategies and of the calibration procedure may be simplified with such models. Depending on the requirements, different models can be used. One application of the zero-dimensional model are for hardware-in-the-loop configurations. A more complex two-dimensional model can be found in [122].

Both models base on the ideal thermodynamic gas Equation 2.1. For the calculation of the combustion process itself, the thermodynamic balance

$$dw_t + dQ + \sum_r dm_r \cdot (h_r + e_r)$$

$$= dU + dE_{ext} \tag{4.52}$$

is regarded. $w_t$ is the technical work, $Q$ the external heat, $dm_r$ are different infinitesimal masses crossing the system border (combustion chamber) with the enthalpy $h_r$ and specific external energy $e_r$.

The right side of Equation 4.52 represents the energy stored inside the system, where $U$ is the internal energy and $E_{ext}$ the external energy e.g. kinetic or potential energy of the system.

Both approaches employ the phenomenological evaporation model in Section 4.6 and the fuel-injection model in Section 4.5 for flexible energy conversion rates.

### 4.7.1   Zero-Dimensional Modeling

In zero-dimensional modeling, the combustion chamber is regarded as a closed system with a single reaction zone. The gases inside are assumed to be ideally mixed. Local differences of the calculated variables, flows or eddies are neglected in the zero-dimensional model. Figure 4.8 gives an overview of the combustion chamber with some of the state variables.

In the following, the thermodynamic equations are formulated in differential notation. In order to calculate the energy balance in all four strokes of an engine cycle, the energy conversion during combustion including heat transfer, charge exchange and volume change is integrated.

## 4.7.2 Thermodynamic Equations

The ideal gas equation 2.1 is differentiated with respect to the crankshaft angle $\alpha_{CS}$

$$p \cdot \frac{dV}{d\alpha_{CS}} + V \cdot \frac{dp}{d\alpha_{CS}} = m \cdot R \cdot \frac{d\vartheta}{d\alpha_{CS}} + m \cdot \vartheta \cdot \frac{dR}{d\alpha_{CS}} + R \cdot \vartheta \cdot \frac{dm}{d\alpha_{CS}} \quad . \quad (4.53)$$

According to [98] and [111], a variation of the gas constant $R$ influences the thermodynamic state equations only for temperatures above $1800K$ or for air-fuel ratios below 1.2. In diesel engines the combustion temperatures are usually below $1800K$ and the air-fuel ratio is above 1.4. Hence, the gas constant $R$ is considered to be constant and $\frac{dR}{d\alpha_{CS}} = 0$.

## 4.7.3 Energy Balance

According to Equation 4.52 the internal energy $U$ inside the cylinder changes if there are changes of the external heat $dQ$, the technical work $dw_t$ or the enthalpy $dH$ crossing the system border. Because of its minimal impact, the influence of the external energy $dE_{ext}$ is neglected.

$$-\frac{dw_t}{d\alpha_{CS}} - \frac{dQ_w}{d\alpha_{CS}} - h_{out} \cdot \frac{dm_{out}}{d\alpha_{CS}} + h_{ac} \cdot \frac{dm_{ac}}{d\alpha_{CS}} + \frac{dQ_{comb}}{d\alpha_{CS}} = \frac{dU}{d\alpha_{CS}} \quad . \quad (4.54)$$

$\frac{dw_t}{d\alpha_{CS}}$ is the volumetric work of the piston. The heat loss through the wall is $\frac{dQ_w}{d\alpha_{CS}}$. $h_{out} \cdot \frac{dm_{out}}{d\alpha_{CS}}$ is the enthalpy change of the mass flow out of the combustion chamber. Equivalently the enthalpy change of the mass flow through the inlet channel into the cylinder is $h_{ac} \cdot \frac{dm_{ac}}{d\alpha_{CS}}$. $\frac{dQ_{comb}}{d\alpha_{CS}}$ is the released energy during combustion. All this sums up to the change of the internal energy of the gas $\frac{dU}{d\alpha_{CS}}$.

## 4.7.4 Volumetric Work

During the combustion, the combustion chamber is a closed system. The volumetric work drives the piston. Equation 2.4 relates volumetric changes to work. To calculate the volumetric work, the piston position is necessary. In Equation 3.4 the piston position $s$ in dependency of the crankshaft angle $\alpha_{CS}$ is given. At Top Dead Center (TDC), the volume is minimal, the so called rest volume $V_{TDC}$. The current volume is calculated from the piston surface $A_{piston}$, the total displacement volume $V_d$ and the piston position $s(\alpha_{CS})$

$$\begin{aligned} V(\alpha_{CS}) &= V_{TDC} + s(\alpha_{CS}) \cdot A_{piston} \\ &= \frac{V_d}{\varepsilon - 1} + s(\alpha_{CS}) \cdot \frac{V_d}{2 \cdot r} \quad . \end{aligned} \quad (4.55)$$

After differentiating Equation 4.55 with respect to the crankshaft angle $\alpha_{CS}$ the change of the combustion chamber volume is

$$\frac{dV}{d\alpha_{CS}} = V_d \cdot \left( \frac{\sin(\alpha_{CS})}{2} + \frac{r}{4 \cdot l} \cdot \frac{\sin(2 \cdot \alpha_{CS})}{\sqrt{1 - \frac{r^2}{l^2} \cdot \sin^2(\alpha_{CS})}} \right) \quad . \tag{4.56}$$

For a given combustion chamber pressure $p$ the volumetric work

$$\frac{dw_t}{d\alpha_{CS}} = p \cdot \frac{dV}{d\alpha_{CS}} \tag{4.57}$$

is calculated from Equation 2.4.

## 4.7.5   Heat Losses

The heat transfer between the in-cylinder gases and the combustion chamber wall depends on the gas charge and the gas movement within the combustion chamber. During combustion the heat transfer is at its highest level [116]. The heat transfer depends on the average combustion chamber temperature and pressure. It is caused by the following effects:

- Heat conduction
  Because of an inhomogeneous temperature distribution, thermal energy is transferred via molecular interaction.

- Convection
  Macroscopic particle movement from higher to lower temperatures.

- Heat radiation
  Electromagnetic waves are the carrier of the energy for heat radiation. Therefore, no physical matter is necessary and the heat radiation can also occur in a vacuum.

According to [98] the heat transfer between the combustion chamber and its wall may be described with the following equation

$$\frac{dQ_w}{d\alpha_{CS}} = \frac{dt}{d\alpha_{CS}} \cdot (\vartheta - \vartheta_w) \cdot \sum_{k=1}^{3} k_k \cdot A_k \quad . \tag{4.58}$$

The term $\vartheta - \vartheta_w$ is the temperature difference between the gas charge and the wall, $A_k$ the surface area of the part $k$ of the combustion chamber wall and $k_k$ the according heat release constant. The three surfaces are the piston head surface $A_{piston}$, the cylinder head surface $A_{CYL}$ and the side surface of the combustion chamber $A_{ch}$ depending on the piston position. The heat release constant $k_k$ is assumed to be the same for all surfaces. It depends on the heat transfer rate $\alpha_{ch}$ of the combustion chamber wall and the heat transfer rate $\alpha_{co}$ of the coolant

$$k_k = \frac{1}{\frac{1}{\alpha_{ch}} + \frac{1}{\alpha_{co}} + \frac{s_{ch}}{\lambda_{ch}}} \quad . \tag{4.59}$$

The thickness of the combustion chamber wall is $s_{ch}$ and the heat conductivity $\lambda_{ch}$.

$\alpha_{co}$ might be approximated by the value for water flowing around pipes [112]. Here, it is neglected. The heat transfer rate of the combustion chamber wall $\alpha_{ch}$ depends on convection and heat radiation

$$\alpha_{ch} = \alpha_{conv} + \alpha_{rad} \quad . \tag{4.60}$$

According to Hohenberg [98] $\alpha_{conv}$ may be calculated as

$$\alpha_{conv} = 130 \cdot V^{-0.06} \cdot p^{0.8} \cdot \vartheta^{-0.4} \cdot (v_p + 1,4)^{0.8} \quad . \tag{4.61}$$

Values for the different variables like volume $V$, temperature $\vartheta$, pressure $p$ and finally the piston velocity $v_p$ result from the process calculation.

$\alpha_{rad}$, $\alpha_{co}$ and $\frac{s_{ch}}{\lambda_{ch}}$ are neglected because of their low influence on the heat release constant $k_k$. Finally the heat release constant in Equation 4.59 simplifies to

$$k_k = \alpha_{conv} \quad . \tag{4.62}$$

## 4.7.6   Energy Conversion

The energy conversion during combustion $Q_{comb}$ is calculated from the burnt fuel mass $m_{inj,burnt}$. The energy content of fuels is based on the lower calorific value $H_{low}$. Taking the lower calorific value $H_{low}$ we get

$$Q_{comb} = H_{low} \cdot m_{inj,burnt} \quad , \tag{4.63}$$

and

$$\frac{dQ_{comb}}{d\alpha_{CS}} = H_{low} \cdot \frac{dm_{inj,burnt}}{d\alpha_{CS}} \quad . \tag{4.64}$$

As $m_{inj,burnt}$ is the sum of the angle-discrete burnt fuel masses $m_{f,burnt,j}$ (Equation 4.51), the energy conversion $\frac{dQ_{comb,j}}{d\alpha_{CS}}$ is calculated by means of the numerical derivative in each step $j$

$$\frac{dQ_{comb,j}}{d\alpha_{CS}} = H_{low} \cdot \frac{m_{f,burnt,j}}{\Delta\alpha_{CS}} \quad . \tag{4.65}$$

## 4.7.7   Enthalpy of Mass Flows

The change of enthalpy during charge exchange can be calculated with Equations 2.1 and 2.8. For the inlet and outlet follows

$$h_{ac} \cdot \frac{dm_{ac}}{d\alpha_{CS}} = (u_{ac} + R \cdot \vartheta_{ac}) \cdot \frac{dm_{ac}}{d\alpha_{CS}} \tag{4.66}$$

$$h_{out} \cdot \frac{dm_{out}}{d\alpha_{CS}} = (u + R \cdot \vartheta) \cdot \frac{dm_{out}}{d\alpha_{CS}} \quad . \tag{4.67}$$

The temperature $\vartheta$ and the internal energy $u$ of the in-cylinder gas may be used for calculation of the exhaust gas enthalpy $h_{out}$. This is necessary if e.g. a subsequent turbo charger shall be modeled.

## 4.7.8    Internal Energy of the Gas Charge

The internal energy of the gas charge depends on the pressure $p$, the temperature $\vartheta$ and the air-fuel ratio $\lambda_c$ inside the combustion chamber. It can be written in differential form as

$$
\begin{aligned}
\frac{dU}{d\alpha_{CS}} &= m \cdot \frac{du}{d\alpha_{CS}} + u \cdot \frac{dm}{d\alpha_{CS}} \\
&= m \cdot \frac{\partial u}{\partial \vartheta} \cdot \frac{d\vartheta}{d\alpha_{CS}} + m \cdot \frac{\partial u}{\partial \lambda_c} \cdot \frac{d\lambda_c}{d\alpha_{CS}} + m \cdot \frac{\partial u}{\partial p} \cdot \frac{dp}{d\alpha_{CS}} + u \cdot \frac{dm}{d\alpha_{CS}} .
\end{aligned}
$$
(4.68)

According to [58] the influence of in-cylinder pressure on the internal energy can be neglected and $\frac{\partial u}{\partial p}$ is set to zero. For numerical calculation Equation 4.68 is transformed into a Taylor-series around the reference temperature $\vartheta_0$ [58] ($\vartheta_0 = 273,15K$). The internal energy $u(\vartheta, \lambda_c)$ results as

$$
\begin{aligned}
u(\vartheta, \lambda_c) = 144,5 \cdot \Bigg[ \quad & - \left( 0,0975 + \frac{0,0485}{\lambda_c^{0,75}} \right) \cdot (\vartheta - \vartheta_0)^3 \cdot 10^{-6} \\
& + \left( 7,768 + \frac{3,36}{\lambda_c^{0,80}} \right) \cdot (\vartheta - \vartheta_0)^2 \cdot 10^{-4} \\
& + \left( 489,6 + \frac{46,4}{\lambda_c^{0,93}} \right) \cdot (\vartheta - \vartheta_0) \cdot 10^{-2} \\
& + \quad 1356,8 \Bigg] .
\end{aligned}
$$
(4.69)

After partial differentiation of Equation 4.69 the dependency of the internal energy $\frac{\partial u}{\partial \vartheta}$ on temperature and $\frac{\partial u}{\partial \lambda_c}$ on air-fuel ratio results.

## 4.7.9    Calculation of State Variables

Finally, the temperature $\vartheta$ inside the combustion chamber is calculated by means of the energy balance 4.54. Therefore we replace $\frac{dU}{d\alpha_{CS}}$ in Equation 4.68 taking into account the mass balance 4.20 and the assumption $\frac{\partial u}{\partial p} = 0$.

$$
\begin{aligned}
& -\frac{dw_t}{d\alpha_{CS}} - \frac{dQ_w}{d\alpha_{CS}} + \frac{dQ_{comb}}{d\alpha_{CS}} + h_{ac} \cdot \frac{dm_{ac}}{d\alpha_{CS}} - h_{out} \cdot \frac{dm_{out}}{d\alpha_{CS}} = \\
& = m \cdot \frac{\partial u}{\partial \vartheta} \cdot \frac{d\vartheta}{d\alpha_{CS}} + m \cdot \frac{\partial u}{\partial \lambda_c} \cdot \frac{d\lambda_c}{d\alpha_{CS}} + u \cdot \left( \frac{dm_{ac}}{d\alpha_{CS}} + \frac{dm_{inj,burnt}}{d\alpha_{CS}} - \frac{dm_{out}}{d\alpha_{CS}} \right) .
\end{aligned}
$$

With $\frac{dm_{inj,burnt}}{d\alpha_{CS}} = \frac{1}{H_{low}} \cdot \frac{dQ_{comb}}{d\alpha_{CS}}$ and the enthalpies $h_{ac} = u_{ac} + R \cdot \vartheta_{ac}$ and $h_{out} = u + R \cdot \vartheta$ of the mass flows in Equations 4.66 and 4.67 the temperature follows

$$\frac{d\vartheta}{d\alpha_{CS}} = \frac{1}{m \cdot \frac{\partial u}{\partial \vartheta}} \cdot \left[ \frac{dQ_{comb}}{d\alpha_{CS}} \cdot (1 - \frac{u}{H_{low}}) - \frac{dQ_w}{d\alpha_{CS}} - \frac{dw_t}{d\alpha_{CS}} \right.$$

$$- (R \cdot \vartheta) \cdot \frac{dm_{out}}{d\alpha_{CS}} + (u_{ac} - u + R \cdot \vartheta_{ac}) \cdot \frac{dm_{ac}}{d\alpha_{CS}}$$

$$\left. - m \cdot \frac{\partial u}{\partial \lambda_c} \cdot \frac{d\lambda_c}{d\alpha_{CS}} \right] . \qquad (4.70)$$

In order to set up an equation for the pressure progression $\frac{dp}{d\alpha_{CS}}$ from the differentiated ideal gas Equation 4.53

$$\frac{dp}{d\alpha_{CS}} = \frac{m \cdot R \cdot \frac{d\vartheta}{d\alpha_{CS}} + m \cdot \vartheta \cdot \frac{dR}{d\alpha_{CS}} + R \cdot \vartheta \cdot \frac{dm}{d\alpha_{CS}} - p \cdot \frac{dV}{d\alpha_{CS}}}{V} , \qquad (4.71)$$

the volume change (see also Equation 4.56) $\frac{dV}{d\alpha_{CS}}$, the temperature change $\frac{d\vartheta}{d\alpha_{CS}}$ (Equation 4.70) the mass change $\frac{dm}{d\alpha_{CS}}$ (Equation 4.20) and the assumption $\frac{dR}{d\alpha_{CS}} = 0$ must be inserted.

## 4.8  Fitting of Model Parameters

In this paragraph the fitting of important combustion model parameters is described. The model should be valid in different operating points. Geometric data of the modeled engine as well as measurement data from an engine test bench are necessary for different operating points .

The critical parameters of the energy conversion process are the inflammation delay time $\tau_{id}$ and the diffusion constant $C_{diff}$ [122]. An analytical approach shall be derived for these two parameters. The in-cylinder pressure and temperature are measured during combustion. Then we have $\frac{dm_{ac}}{d\alpha_{CS}} = 0$, $\frac{dm_{out}}{d\alpha_{CS}} = 0$ and $\frac{dm}{d\alpha_{CS}} = \frac{dm_{inj,burnt}}{d\alpha_{CS}}$. With the assumption $\frac{dR}{d\alpha_{CS}} = 0$ the differentiated ideal gas Equation 4.53 can be written as

$$p \cdot \frac{dV}{d\alpha_{CS}} + V \cdot \frac{dp}{d\alpha_{CS}} = m \cdot R \cdot \frac{d\vartheta}{d\alpha_{CS}} + R \cdot \vartheta \cdot \frac{dm_{inj,burnt}}{d\alpha_{CS}} . \qquad (4.72)$$

$\frac{d\vartheta}{d\alpha_{CS}}$ is now replaced by Equation 4.70.

$$p \cdot \frac{dV}{d\alpha_{CS}} + V \cdot \frac{dp}{d\alpha_{CS}} = \frac{R}{\frac{\partial u}{\partial \vartheta}} \cdot \left[ \frac{dQ_{comb}}{d\alpha_{CS}} \cdot (1 - \frac{u}{H_{low}}) - \frac{dQ_w}{d\alpha_{CS}} - \frac{dw_t}{d\alpha_{CS}} \right.$$

$$\left. - m \cdot \frac{\partial u}{\partial \lambda_c} \cdot \frac{d\lambda_c}{d\alpha_{CS}} \right] + R \cdot \vartheta \cdot \frac{dm_{inj,burnt}}{d\alpha_{CS}} . \quad (4.73)$$

Inserting $\frac{dw_t}{d\alpha_{CS}}$ from Equation 4.57, $\frac{d\lambda_c}{d\alpha_{CS}}$ from Equation 4.19 and $\frac{dm_{inj,burnt}}{d\alpha_{CS}}$ from

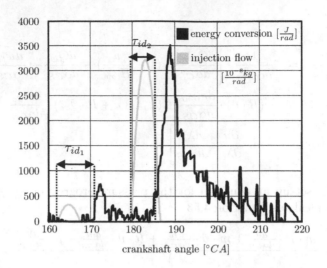

**Figure 4.9** Measured energy conversion and inflammation delay time

Equation 4.64 into Equation 4.73 we get

$$p \cdot \frac{dV}{d\alpha_{CS}} + V \cdot \frac{dp}{d\alpha_{CS}} m = \frac{R}{\frac{\partial u}{\partial \vartheta}} \cdot \left[ \frac{dQ_{comb}}{d\alpha_{CS}} \cdot (1 - \frac{u}{H_{low}}) - \frac{dQ_w}{d\alpha_{CS}} - p \cdot \frac{dV}{d\alpha_{CS}} \right.$$

$$\left. + m \cdot \frac{\partial u}{\partial \lambda_c} \cdot \frac{\lambda_c}{m_{f,burnt}} \cdot \frac{1}{H_{low}} \cdot \frac{dQ_{comb}}{d\alpha_{CS}} \right] + R \cdot \vartheta \cdot \frac{1}{H_{low}} \cdot \frac{dQ_{comb}}{d\alpha_{CS}} \ . \quad (4.74)$$

Therefore the energy conversion is

$$\frac{dQ_{comb}}{d\alpha_{CS}} = H_{low} \cdot \frac{\frac{dQ_w}{d\alpha_{CS}} + p \cdot \frac{dV}{d\alpha_{CS}} \left[ 1 + \frac{1}{R} \cdot \frac{\partial u}{\partial \vartheta} \right] + \frac{V}{R} \cdot \frac{\partial u}{\partial \vartheta} \cdot \frac{dp}{d\alpha_{CS}}}{H_{low} - u + \vartheta \cdot \frac{\partial u}{\partial \vartheta} + \frac{\lambda_c}{m_{f,burnt}} \cdot m \cdot \frac{\partial u}{\partial \lambda_c}} \ . \quad (4.75)$$

According to [98], the time delay between the start of injection and the start of energy conversion is the inflammation delay. In Figure 4.9 the calculated energy conversion $\frac{dQ_{comb}}{d\alpha_{CS}}$ as well as the related fuel injection flow are depicted. The approximation of the inflammation delay time $\tau_{id}$ in equation 4.47

$$\tau_{id} = a \cdot \overline{p}^b \cdot e^{\frac{c}{\vartheta}} \quad (4.76)$$

is extended by a term which also considers the injected fuel mass $m_f$

$$\tau_{id} = a \cdot \overline{p}^b \cdot e^{\frac{c}{\vartheta}} - d \cdot m_f^e \ . \quad (4.77)$$

The parameters $a, b, c, d$ and $e$ are fitted until Equation 4.77 is valid in all engine operation points. The Levenberg-Marquardt algorithm [36] and initial values found in [122] are used for the fitting procedure. Figure 4.10 shows the measured and calculated inflammation delay times for different engine operating points.

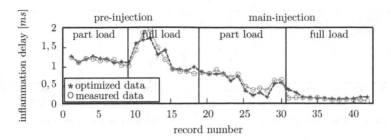

**Figure 4.10** Calculated and measured inflammation delay time for different operating points.

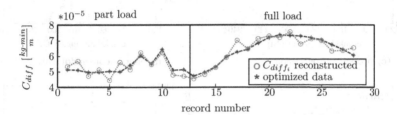

**Figure 4.11** Difference between reconstructed and analytical diffusion constant.

The second variable to be fitted is the diffusion constant $C_{diff}$ in Equation 4.45. An empirical formula depending on the average piston velocity $\bar{\bar{s}}$ and the air mass $m_a$ inside the combustion chamber is taken [82, 98].

$$C_{diff} = (a + b \cdot \bar{\bar{s}} \cdot m_a - c \cdot m_f) \cdot 10^{-5} \ , \qquad (4.78)$$

The diffusion constant cannot be directly measured. Therefore, two steps are necessary to fit the parameters $a, b, c$ in Equation 4.78 for the diffusion constant $C_{diff}$ to a real engine:

1. The diffusion constant $C_{diff}$ in Equation 4.45 is modified during the simulation until measured and calculated pressure data fit at different engine operation points. This yields a reconstructed diffusion constant at each engine operating point, which is adopted instead of the non-measurable one.

2. The parameters $a, b$ and $c$ in the analytical approach (Equation 4.78) are now fitted for all different operating points of the engine to the reconstructed diffusion constant of step 1. Initial values can be found in [20]. The Levenberg-Marquardt algorithm is also used for this procedure.

Figure 4.11 shows the results for the fitting of the diffusion constant.

## 4.8.1 Simulation results

In Figure 4.12 the calculated and measured in-cylinder pressure and the reconstructed energy conversion rates are compared for different engine operating

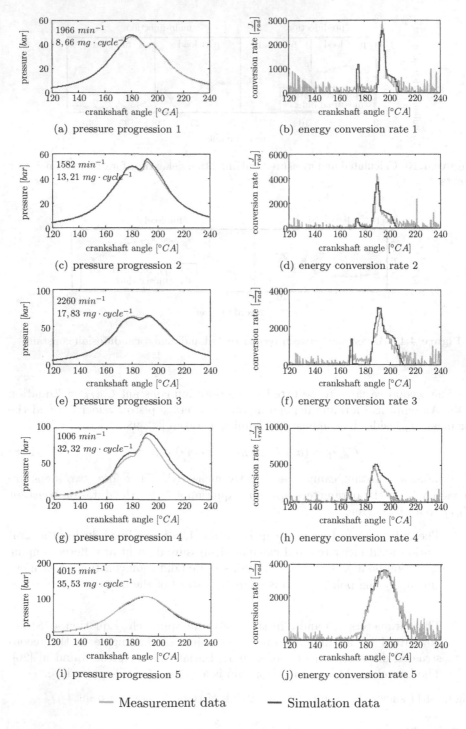

(a) pressure progression 1

(b) energy conversion rate 1

(c) pressure progression 2

(d) energy conversion rate 2

(e) pressure progression 3

(f) energy conversion rate 3

(g) pressure progression 4

(h) energy conversion rate 4

(i) pressure progression 5

(j) energy conversion rate 5

Measurement data          Simulation data

**Figure 4.12**  Comparison of measured pressure rates and simulated data.

points. In all operation points a pre- and a main-injection was used. The engine speed, engine load, injection pressure, cylinder charge and injected fuel mass are varied. The correlation between the measurement and analytical calculation data appears to be sufficiently accurate, for the pressure calculation as well as for the energy conversion. The deviation of the simulated from the real pressure is below 5% in almost all operating points. Only at low engine speed and large injected fuel masses, the deviation increases up to 8% (Figure 4.12(g)).

## 4.9 Soot Accruement

Soot is a major problem of diesel-engine combustion. 70 to 90% of particulate emissions of a diesel engine is soot. It is generated during incomplete combustion of fuel due to local oxygen deficiency in the inhomogeneous fuel/air-gas mixture. This condition mainly occurs when the fuel is injected directly into the flame [40]. Thereby, the molecules of the diesel fuel are disrupted because the combustion temperature increases and the concentration of oxygen decreases. Hydrogen is emitted and causes a polymerization. The combustion velocity of the polymerized molecules is low. Thus, they act as non-inflammable soot generators.

Soot accrues in rich mixtured zones within the combustion chamber, where very low $\lambda_c$-values and an inhomogeneous load are located. Therefore, soot accruement cannot be calculated in a simple way. We therefore turn to phenomenological approaches.

The phenomenological approach of Hiroyuki [50] and Boulouchos [113] combines the soot accruement and its oxidation. The balance

$$\frac{dm_{soot}}{d\alpha_{CS}} = \frac{dm_{soot,accrue}}{d\alpha_{CS}} - \frac{dm_{soot,oxidation}}{d\alpha_{CS}} \ . \tag{4.79}$$

yields the soot mass at the end of the combustion cycle (see Figures 4.13 and 4.14).

Soot accruement

$$\frac{dm_{soot,accrue}}{d\alpha_{CS}} = A_B \cdot \frac{dm_f}{d\alpha_{CS}} \left( \frac{p}{p_{ref}} \right)^{n_1} \cdot \exp \left( \frac{\vartheta_{soot,accrue}}{\vartheta} \right) \tag{4.80}$$

depends on the activation temperature $\vartheta_{soot,accrue}$ and the reference pressure $p_{ref}$ in the combustion chamber. $A_B$ is the constant of soot accruement.

The soot oxidation is

$$\frac{dm_{soot,oxidation}}{d\alpha_{CS}} = A_0 \cdot \frac{1}{\tau_{char}} \cdot m_{soot}^{n_2} \left( \frac{p_{O_2}}{p_{O_2,ref}} \right)^{n_3} \cdot \exp \left( \frac{\vartheta_{oxidation}}{\vartheta} \right) \tag{4.81}$$

where $A_0$ is the constant of soot oxidation and $\tau_{char}$ the characteristic mixture time. $p_{O_2}$ denotes the oxygen partial pressure and $p_{O_2,ref}$ the oxygen reference pressure. $\vartheta_{oxidation}$ is the activation temperature of soot oxidation. Low soot emissions can be achieved by proper control of the combustion process, where soot oxidation is enhanced during later part of the combustion.

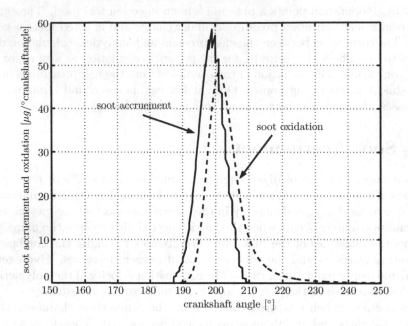

**Figure 4.13** Soot accruement ratio (solid line) and soot oxidation ratio (dashed line) over crankshaft angle

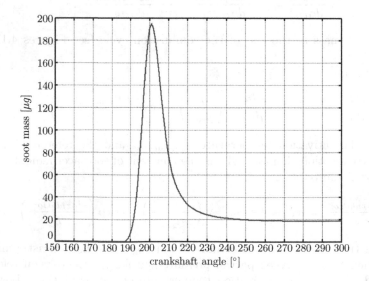

**Figure 4.14** Soot mass as difference between accruement and oxidation over crankshaft angle

# 5 Engine Control Systems

## 5.1 Lambda Control

In stoichiometric engine operation, emission levels heavily depend on how accurate the air-fuel ratio can be kept at $\lambda = 1$. Due to measurement and computational tolerances, sufficiently accurate stoichiometric operation requires a closed loop control.

### 5.1.1 Stoichiometric Operation of SI Engines

In SI engines, the air-fuel ratio $\lambda$ is either very lean at part load or stoichiometric at medium and high load. A stoichiometric ratio of $\lambda = 1$ should lead to an ideal combustion. Figure 5.1 shows the emissions at different air-fuel ratios. For $\lambda = 1$, the emissions of $HC$, $CO$ and $NO_x$ are relatively low. Due to turbulence and local inhomogeneity of the gas mixture, real combustion actually produces $HC$, $CO$ and $NO_x$ at the same time. By means of a catalytic converter, these raw emissions can be effectively reduced.

It can be seen in Figure 5.2 that the emission rates after the catalytic converter vary highly with the air-fuel ratio $\lambda$: A change of the average $\overline{\Delta\lambda} = 0.1\%$ would already double the emission rates. Therefore, it is important to have an accurate closed loop lambda control to guarantee an average air-fuel ratio within a window smaller than $0.1\%$ around $\lambda = 1$. When engine speed and torque change actual, lambda deviations of $2 - 3\%$ over a short period of time are allowed. If the average accuracy can be held, such deviations go into both directions. Within the volume of the catalytic converter excursions of the air-fuel ratio in one direction are compensated by those in the opposite direction. At the engine exhaust, short time lambda deviations of a few percent do not deteriorate the emissions after the catalytic converter.

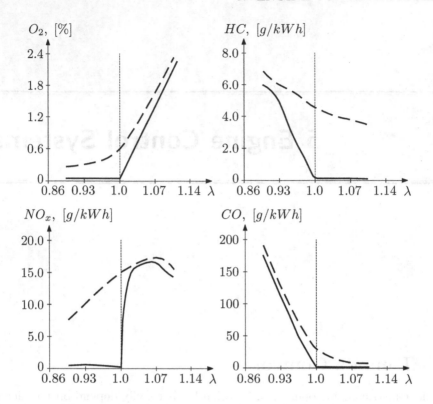

**Figure 5.1** Measurement of exhaust gases: oxygen $O_2$, hydrocarbon $HC$, nitrogen oxide $NO_x$ and carbon monoxide $CO$. The concentration before the catalytic converter are indicated by dotted and the concentrations after the catalytic converter by straight lines.

The block diagram of the lambda controlled SI engine is shown in Figure 5.3. The amount of injected fuel is controlled by the engine control unit which gets its feedback from the lambda sensor in the exhaust pipe as well as the mass air flow signal in the inlet pipe. Additional variables like engine speed and engine temperature are also used in the control scheme.

## Catalytic Converter

The catalytic aftertreatment reduces the emissions considerably (supposing a correct lambda control at $\lambda = 1$). Due to turbulences and flame propagation, the air-fuel mixture is still incompletely burned. Noxious gases like $HC$, $CO$ and $NO_x$ are converted to $CO_2$, $H_2O$ and $N_2$ by the catalytic converter. The converter is integrated into the exhaust pipe. It consists of a ceramic or metal carrier substrate covered by a wash coat with an extremely large surface which is again covered with a thin layer of platinum and rhodium as shown in Figure 5.4

The ratio of platinum to rhodium is approximately 2 to 1. Depending on the engine size about $1 - 3\,g$ of the precious metals are used. They both support the chemical reactions: Platin supports more the oxidation of $CO$ and $HC$ and

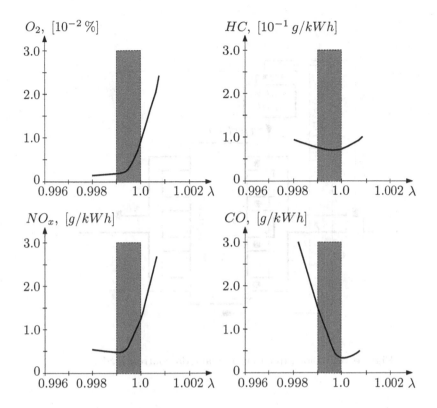

**Figure 5.2** Emission rates of an engine after the catalytic converter at a static operating point (engine speed $1800\,rpm$ and torque $T = 65\,Nm$). Average lambda (500 cycles) should be within the indicated window.

rhodium supports more the reduction of the nitrogen oxides $NO_x$.

Reduction and oxidation processes are simultaneously running in the catalytic converter. The conversion ratio is defined as the relative change of the gas concentration before and after the catalytic process.

$$c_r = \frac{c_{in} - c_{out}}{c_{in}} \tag{5.1}$$

The conversion ratio has typical values of $c_r > 90\,\%$. The most important chemical reactions are listed below:

**Oxidation of $HC$ and $CO$:**

$$H_nC_m + \left(m + \frac{n}{4}\right)O_2 \quad \rightarrow \quad m\,CO_2 + \frac{n}{2}H_2O \tag{5.2}$$

$$H_nC + 2H_2O \quad \rightarrow \quad CO_2 + \left(2 + \frac{n}{2}\right)H_2 \tag{5.3}$$

$$CO + \frac{1}{2}O_2 \quad \rightarrow \quad CO_2 \tag{5.4}$$

$$CO + H_2O \quad \rightarrow \quad CO_2 + H_2 \tag{5.5}$$

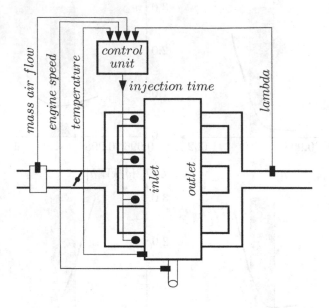

**Figure 5.3** Block diagram of a lambda controlled SI engine.

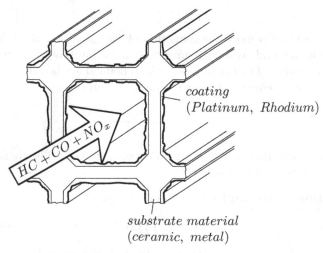

**Figure 5.4** Exploded view of a catalytic converter.

**Reduction of $NO_x$:**

$$CO + NO \rightarrow \frac{1}{2}N_2 + CO_2 \tag{5.6}$$

$$H_nC_m + 2\left(m + \frac{n}{4}\right)NO \rightarrow \left(m + \frac{n}{4}\right)N_2 + \frac{n}{2}H_2O + mCO_2 \tag{5.7}$$

$$H_2 + NO \rightarrow \frac{1}{2}N_2 + H_2O \tag{5.8}$$

**Other catalytic reactions:**

$$SO_2 + \frac{1}{2}O_2 \rightarrow SO_3 \tag{5.9}$$

$$SO_2 + 3H_2 \rightarrow H_2S + 2H_2O \tag{5.10}$$

$$\frac{5}{2}H_2 + NO \rightarrow NH_3 + H_2O \tag{5.11}$$

$$2NH_3 + \frac{5}{2}O_2 \rightarrow 2NO + 3H_2O \tag{5.12}$$

$$NH_3 + CH_4 \rightarrow HCN + 3H_2 \tag{5.13}$$

$$H_2 + \frac{1}{2}O_2 \rightarrow H_2O \tag{5.14}$$

The conversion ratio is influenced by the air-fuel ratio and the converter volume. Deviations of $\Delta\lambda < 3\%$ can be compensated for a short period of time. At stationary engine operation, the conversion ratio is high, even if the converter would be already partly damaged. During transients, excursions in the air-fuel ratio occur, leading to higher emissions. During the warm-up phase of the engine and the exhaust pipe, temperatures are too low for chemical reactions and the conversion ratio is poor. The catalytic converter has to reach temperatures beyond $300\,°C$ to be effective. There are several possibilities to accelerate engine warm-up.

- A fast heating of the exhaust pipe can be obtained by an ignition angle retard of e.g. $10° < \Delta\alpha_i < 20°$. The combustion is shifted to a phase of the thermodynamic cycle, where the exhaust valves are already opened.

- An additional start-up catalytic converter is mounted very close to the engine where the exhaust gases get hotter soon. After the warm-up period, this converter is bypassed.

- Fresh air is added to the exhaust gases by a secondary air pump. The engine runs with a rich mixture ($\lambda < 1$). The additional combustion process in the exhaust pipe heats up the catalytic converter.

- The catalytic converter is electrically heated. In order to reduce the required heating power, the heater is concentrated in the region of the converter where the exothermic reaction first starts.

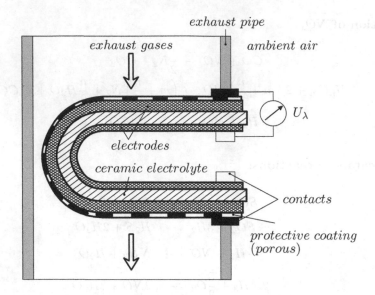

**Figure 5.5** Zirconium dioxide sensor

## 5.1.2   Oxygen Sensor

A lambda sensor is used to measure the concentration of oxygen $O_2$ in the exhaust pipe. The sensor is mounted in the collective exhaust pipe where the individual exhaust pipes from the cylinders end in. In engines with 6 or more cylinders two lambda sensors are used. In Figure 5.6 it can be seen, that the output voltage increases sharply at $\lambda = 1$. Thus the stoichiometric point can be determined.

### Zirconium Dioxide Sensor

The sensor consists of a solid ceramic electrolyte (zirconium dioxide), which conducts oxygen ions at temperatures above $250\,°C$. The outer electrode is covered with platinum. The oxygen partial pressure on the surface of the ceramic material is thus identical with the one inside the catalytic converter. The inner electrode has a direct contact with the ambient air. The exhaust gases flow around the outer electrodes. Figure 5.5 shows the construction of a Zirconium Dioxide Sensor.

Because of a difference in the partial oxygen pressure $p(O_2)$ inside and outside of the exhaust pipe, there is an electrolytic voltage between the electrodes:

$$U_\lambda = k\ \vartheta_{Sensor}\ ln\frac{p(O_2)_{ambient}}{p(O_2)_{exhaust}} \qquad (5.15)$$

The internal resistance ranges from $10^7\,\Omega$ at $200\,°C$ to $5\cdot10^3\,\Omega$ at $800\,°C$. Figure 5.6 shows a characteristic step in the sensor voltage curve close to $\lambda = 1$. This step is caused by the increase of the oxygen partial pressure over several orders of magnitude inside the exhaust pipe around $\lambda = 1$. Typical values for

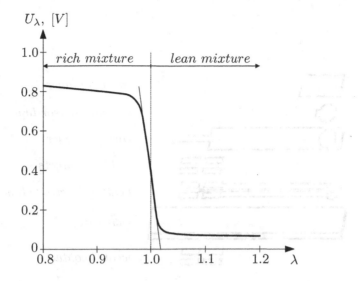

**Figure 5.6** Output voltage of zirconium dioxide sensor

the open circuit voltages are:

$$U_\lambda(rich) = 800 - 1000\,mV$$
$$U_\lambda(lean) = 50 - 200\,mV$$

The response time ranges from 15 to 30 $ms$.

**Strontium Titanate Sensor**

Strontium Titanate is a ceramic semiconductor material. Its conductivity depends on the material temperature and oxygen partial pressure. Conductivity in strontium titanate is less influenced by surface effects at high temperatures than in other materials. The dependance of the probe resistance from the temperature decreases at higher temperatures leaving the dependance on lambda only. As can be seen in Figure 5.7, the strontium titanate sensor has a planar structure.

The resistance characteristic of the sensor is shown in Figure 5.8

An advantage of the planar device is its short response time of a few milliseconds after lambda deviations. The protection pipe around the sensing device adds however further delays. Because of its operation at temperatures around 800 °$C$ it can be fitted closer to the engine. In the engine model (see Section 5.1.3) this leads to shorter time delays $T_{exh}$ between exhaust valve and lambda sensor.

## 5.1.3 Engine Model for Lambda Control

Figure 5.9 shows a suitable model of the engine for lambda control,

$CYL$ is the number of cylinders

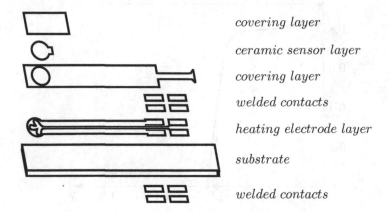

Figure 5.7  Planar structure of the strontium titanate sensor

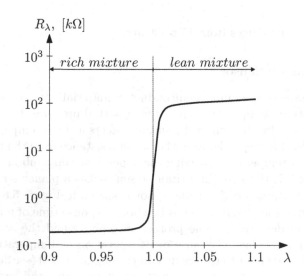

Figure 5.8  Resistance characteristic of the strontium titanate sensor

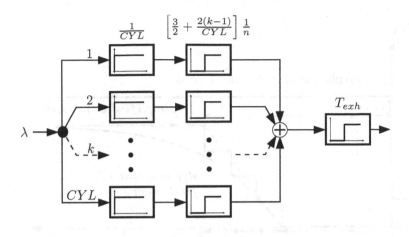

**Figure 5.9** Engine model for lambda control

$k$       is the respective cylinder $1, \ldots, CYL$

$\frac{1}{n}$      is the time needed for one crankshaft revolution

$T_{exh}$   is the time delay between exhaust valve and lambda sensor

Its simplified step response can be seen in Figure 5.10.

Fuel is injected into the intake manifold and sucked into the cylinders at phase-shifted time periods. This leads to the stair-step characteristic as a very simplified step response. For controller design the steps are approximated by a first-order lag element with the following structure (see also Figure 5.10):

$$\frac{K_{l,e}}{1 + T_{l,e}\,s} \tag{5.16}$$

The combustion can be modeled as a delay time $T_{burn}$ continuing until the opening of the exhaust valve. Another delay time $T_{exh}$ results from the time the exhaust gas needs to get to the lambda sensor.

$T_{exh}$   :   varies in dependence of the mass air flow
            between 20 and $500\,ms$

$T_{burn}$   :   time between opening of inlet and exhaust valves

$T_{l,e}$   :   the approximation delivers $\frac{2(CYL-1)}{n \cdot CYL}$

The delay times can be summed up to:

$$T_{d,e} = T_{exh} + T_{burn}$$

Figure 5.11 shows the simplified engine model containing only one lag time $T_{l,e}$ and only one delay time $T_{d,e}$.

Typical values of the parameters are:

$T_{d,e}$   :   $100\,ms \ldots 1.0\,s$

$T_{l,e}$   :   $50\,ms \ldots 0.5\,s$

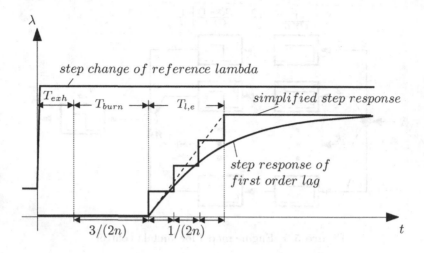

**Figure 5.10** Step response and its approximation by a first-order lag (CYL=4)

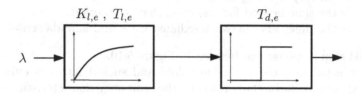

**Figure 5.11** Simplified dynamic portion of the engine model

Since the model parameters vary significantly with the operating conditions of the engine, the parameters of the lambda control are adapted in dependence of the engine operating point (feed forward adaptation). Each control parameter is stored in a map over the engine's operating points.

## 5.1.4 Lambda Control Circuit

The characteristic between the output voltage $U_\lambda$ and the air-fuel ratio $\lambda$ is non-linear. After several years of operation this characteristic slightly ages. Therefore the most stable measuring range of the characteristic is taken for control purposes. Figure 5.12 shows, that it is located in the steep linear range of the characteristic. The sensitivity factor in this range is $K_L$.

Outside the measurement range the characteristic is cut off. The center of the measurement range $\lambda_0$ is not at the desired reference value $\lambda_{ref}$ but is determined exclusively by the stability of the characteristic. The lambda reference value $\lambda_{ref}$ must however lie within the range $[\lambda_0 - \Delta\lambda_L, \lambda_0 + \Delta\lambda_L]$. The offset of $\lambda_0$ against the reference value $\lambda_{ref}$ can be compensated e.g. by a direction-dependant integral time constant of the PI controller.

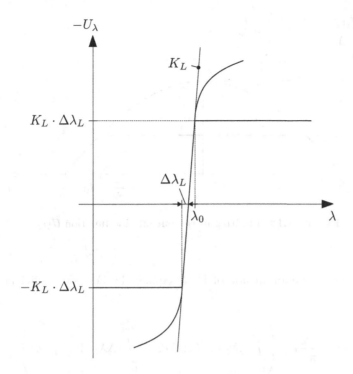

**Figure 5.12** Inverted $\lambda$-characteristic with limiting range $\pm\Delta\lambda_L$

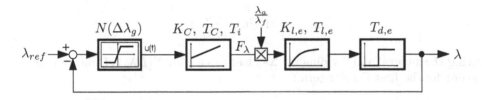

**Figure 5.13** Closed loop-control circuit of the lambda-control

To get the classical structure of a control loop, the sign of the characteristic voltage $U_\lambda(\lambda)$ is inverted. At the input of the controller a non-linear function is representing the range cut-off.

The closed loop-control circuit comprises a non-linear element and a delay time. Therefore it performs a limit cycle. For an analytic calculation the method of the harmonic balance [24] is used where the input of the non-linear element receives a sine function with the limit cycle amplitude $\Delta\lambda_g$:

$$\lambda(t) = \Delta\lambda_g \cdot \sin(\omega_g t) \tag{5.17}$$

From the output signal $U(t)$ only the first term $U_1(t)$ of a Fourier expansion is taken. This approach is justified by the fact that higher order oscillations in the Fourier expansions are much more damped in the control loop than the first

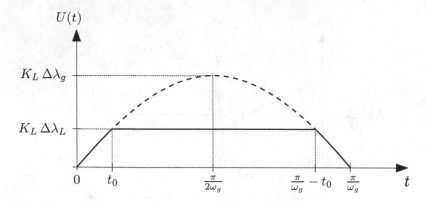

**Figure 5.14** Limiting of the output sine function $U(t)$

order oscillation. The amplitude of $U_1(t)$ equals the Fourier coefficient $U_1$ (see Figure 5.14).

$$U_1 = \frac{4\,\omega_g}{\pi} K_L \left( \int\limits_0^{t_0} \Delta\lambda_g \sin^2(\omega_g t)\, dt + \int\limits_{t_0}^{\frac{\pi}{2\omega_g}} \Delta\lambda_L \sin(\omega_g t)\, dt \right)$$

The time $t_0$ is given by the ratio of $\Delta\lambda_L$ to $\Delta\lambda_g$:

$$t_0 = \frac{1}{\omega_g} \arcsin\left( \frac{\Delta\lambda_L}{\Delta\lambda_g} \right)$$

Solving the integral and dividing by $\Delta\lambda_g$ leads to a gain $N(\Delta\lambda_g)$ of the nonlinear element for the first Fourier term:

$$N(\Delta\lambda_g) = \frac{U_1}{\Delta\lambda_g} = \frac{2}{\pi} K_L \left( \arcsin\left( \frac{\Delta\lambda_L}{\Delta\lambda_g} \right) + \frac{\Delta\lambda_L}{\Delta\lambda_g} \sqrt{1 - \left( \frac{\Delta\lambda_L}{\Delta\lambda_g} \right)^2} \right)$$

If the output range is strongly limited, meaning $\Delta\lambda_L \ll \Delta\lambda_g$, the gain can be approximated by:

$$N(\Delta\lambda_g) \approx \frac{4}{\pi} \left( \frac{\Delta\lambda_L}{\Delta\lambda_g} \right) K_L \qquad (5.18)$$

We assume the controller to be a PI-element with the following structure:

$$\frac{1 + T_C\, s}{T_i\, s} \cdot \frac{1}{K_C} \qquad (5.19)$$

The factor $\frac{1}{K_C}$ represents the steady-state relationship between controller variable and air-fuel ratio $\lambda$. First the time parameter $T_C$ is chosen to compensate the time lag $T_{l,e}$ of the engine:

$$T_C \approx T_{l,e} \qquad (5.20)$$

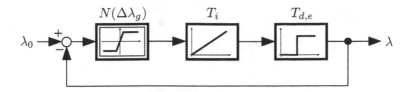

**Figure 5.15** Control loop after lag compensation

Because $T_{l,e}$ depends strongly on the operation point of the engine, $T_C$ needs to be adapted. Figure 5.15 shows the resulting structure of the control loop after compensation. The open-loop transfer function results to

$$G(s) = N(\Delta\lambda_g)\frac{1}{s\,T_i}\frac{1}{K_C} \cdot e^{-T_{d,e}s} \quad , \tag{5.21}$$

and the open-loop frequency response to

$$G(j\omega) = N(\Delta\lambda_g)\frac{1}{j\omega T_i}\frac{1}{K_C} \cdot [\cos\left(\omega T_{d,e}\right) - j\sin\left(\omega T_{d,e}\right)] \quad . \tag{5.22}$$

The stability limit of the closed-loop system is at

$$G(j\omega) = -1 \quad . \tag{5.23}$$

The frequency of the limit cycle $\omega_g$ is calculated from the imaginary part of $G(j\omega)$:

$$Im\{G(j\omega_g)\} = 0 \qquad \Rightarrow \qquad \cos(\omega_g T_{d,e}) = 0 \tag{5.24}$$

$$\Rightarrow \qquad \omega_g = \frac{\pi}{2 \cdot T_{d,e}} \tag{5.25}$$

With $\omega_g = \frac{\pi}{2 \cdot T_{d,e}}$ the real part yields:

$$Re\{G(j\omega_g)\} = -N(\Delta\lambda_g)\frac{2}{\pi}\frac{1}{K_C}\frac{T_{d,e}}{T_i} \tag{5.26}$$

The stability criterion of the control loop is given by (Figure 5.16):

$$|Re\{G(j\omega_g)\}| \le 1 \tag{5.27}$$

Inserting Equation 5.26 into Equation 5.27 leads to

$$T_i > N(\Delta\lambda_g)\frac{2}{\pi}\frac{1}{K_C}T_{d,e} \tag{5.28}$$

or with Equation 5.18

$$T_i > \frac{8}{\pi^2} \cdot \frac{\Delta\lambda_L}{\Delta\lambda_g} \cdot \frac{K_L}{K_C} \cdot T_{d,e} \quad . \tag{5.29}$$

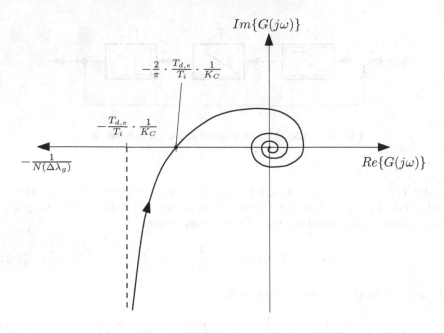

**Figure 5.16** Root locus diagram of lambda-control

The dependence of the delay time $T_{d,e}$ on the operating point of the engine requires a feed-forward adaptation of the integral time constant $T_i$. The maximum amplitude of the limit cycle has been constrained by the lambda-window to:

$$\frac{\Delta\lambda_g}{\lambda} \leq 3\,\% \,. \tag{5.30}$$

Equation 5.29 determines the minimum value of the integration time constnant $T_i$. Consequently the lambda control loop reacts relatively slow to dynamic transitions between operating points (see also Figure 5.20). During long transient times, the lambda value leaves the 3 % window. In such situations, noxious emissions are no longer reduced by the catalytic converter.

## 5.1.5   Measurement Results

If we assume the volume of the catalytic converter to be around $V_C \approx 0.016\,m^3$, then it contains an air mass of about $m_a = 0.02\,kg$ (and noxious exhaust gases). At full engine load and speed, a mass air flow of $\dot{m}_a = 600\,kg/h$ shall run through the exhaust pipe. It will stay $t_C = m_a/\dot{m}_a \approx 120\,ms$ in the catalytic converter. At engine idling, mass air flow might be at $6\,kg/s$. This would stay $t_C \approx 12\,s$ in the converter. The frequency of the lambda control limit cycle must therefore be above

- $0.1\,Hz$ at idling,

- $10\,Hz$ at full load and speed.

Figure 5.17 shows measurement results of the lambda control factor $F_\lambda$ at an approximately stationary operating point of the engine. Noxious emissions and $\lambda$ before and after catalytic treatment are shown in one diagram. At stationary engine operation the catalytic converter has a high conversion ratio.

Figures 5.18 and 5.19 show measurement results of the lambda control during dynamic engine transients of a driving test cycle. Rotational speed and load vary a lot during acceleration and deceleration. At dynamic engine transients we observe fast rotational speed variations e.g. caused by gear changes. Since the integration time constant is limited by the lambda control limit cycle amplitude, mismatches occur where the lambda window is left and high peaks of noxious emissions are generated.

## 5.1.6 Adaptive Lambda Control

The dynamic performance of the lambda control is strongly restricted by the following parameters:

- given delay time of the engine $T_{d,e}$ (see Section 5.1.3)

- amplitude of limit cycle $\Delta\lambda_g < 3\%$ (see Equation 5.30)

The integration time constant $T_i$ of the controller is constrained to the lower limit given in Equation 5.29. At engine transients to another operating point, the actual lambda needs up to several seconds for arriving back to the stoichiometric mixture (see Figure 5.20).

During this transition time the lambda window is left. In this section the remaining control errors shall be eliminated by an adaptive feed-forward control. By that the original lambda control is relieved from compensating mismatches in transients.

### Adaptation of a Feed-forward Control Map

The lambda control loop compensates errors of the air-fuel ratio by a multiplicative correction factor $F_\lambda$. These lambda correction factors are stored into a feed-forward control map in all engine operating points. Instead of performing the error compensation by the original lambda control loop, it can now be performed by the right $F_\lambda$ from the feed-forward control map without time delay. The lambda mismatches during transients are thus overcome.

The problem is how to adapt the correction factors $F_\lambda$ in the feed-forward control map, when some engine operating points are only very rarely visited, due to special habits of individual drivers. Eventually an adaptation is even impossible. High noxious exhaust emissions would thus remain during transients into these rarely visited operation points. Therefore a globally valid lambda compensation approach is used rather than a local adaptation of correction factors in all engine operating points.

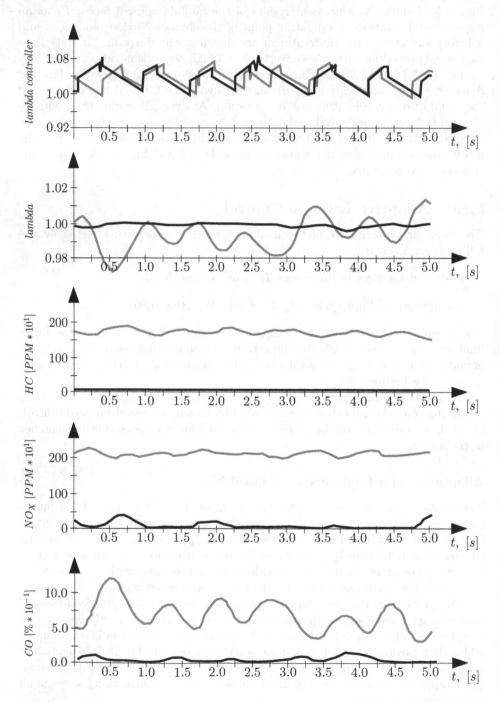

**Figure 5.17**  Emissions of lambda-controlled engine at stationary engine operation with $n = 1800/min$, $T = 65\,Nm$, before and after catalytic converter

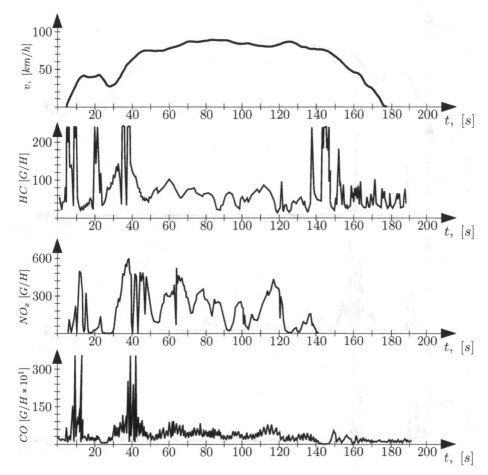

**Figure 5.18** Raw emissions of lambda-controlled engine before catalytic conversion, FTP-HT2 driving-cycle

### Globally Valid Lambda Compensation

The air-fuel ratio errors are assumed to consist of two components (Figure 5.21):

**Additive lambda offset error:** since the absolute value of this offset is identical over the entire engine operating range, its impact is mostly felt at low engine power outputs. At medium or high power output, the relative error from the offset may be neglected. An example is air leakage bypassing the mass air flow meter.

**Multiplicative lambda errors:** since the gradient of the linear lambda function between fuel and air mass flow is affected, its impact is equally felt at any engine operation. An example is the air density error at flap type air flow meters.

This simplified error model is supported by practical experience in engine management systems.

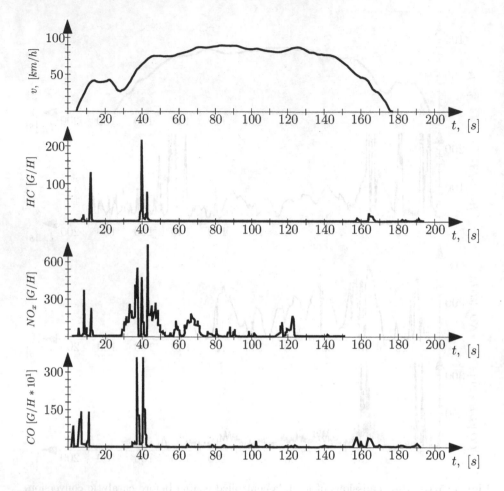

**Figure 5.19** Emissions of lambda-controlled engine after catalytic conversion, FTP-HT2 driving-cycle

The additive and multiplicative errors shall now be compensated. The correct air mass flow would have been

$$\dot{m}_{a,o} = \lambda_0 L_{st} \dot{m}_f \quad . \tag{5.31}$$

The corrupted characteristic is then

$$\begin{aligned}
\dot{m}_a &= \lambda L_{st} \dot{m}_f + \Delta \dot{m}_a \\
&= \frac{\lambda}{\lambda_0} \dot{m}_{a,0} + \Delta \dot{m}_a \quad .
\end{aligned} \tag{5.32}$$

The compensation scheme comprises three steps.

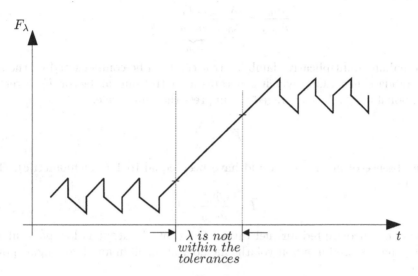

**Figure 5.20** Control action $F_\lambda$ at the transition to a new operating point

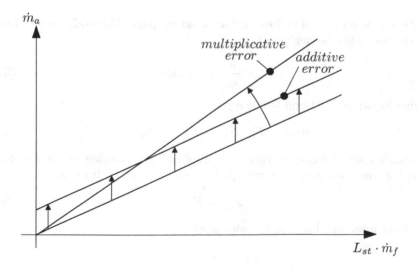

**Figure 5.21** Simplified error model for the lambda characteristic

a.)  At **medium** and **high** engine power outputs, the additive error can be neglected.

$$\frac{\dot{m}_a}{\dot{m}_{a,0}} = \frac{\lambda}{\lambda_0} + \underbrace{\frac{\Delta \dot{m}_a}{\dot{m}_{a,0}}}_{\approx 0} \approx \frac{\lambda}{\lambda_0} \tag{5.33}$$

The remaining multiplicative lambda error can then be compensated by the regular lambda control loop which generates a control output factor $F_\lambda$ inversely proportional to $\frac{\lambda}{\lambda_0}$. It is averaged to suppress the limit cycle.

$$\overline{F}_\lambda = \frac{\lambda_0}{\lambda} \tag{5.34}$$

In the absence of errors, $F_\lambda$ would have been equal to 1 (stoichiometric). The product

$$\overline{F}_\lambda \cdot \frac{\dot{m}_a}{\dot{m}_{a,0}} \approx 1 \tag{5.35}$$

recovers the uncorrupted air-fuel ratio. The control output is low-pass filtered into $\overline{F}_\lambda$ and is stored in a non-volatile memory at medium and high engine power outputs.

$$F_{Hi} = \overline{F}_\lambda \tag{5.36}$$

Taking advantage of the compensation factor $F_{Hi}$, the corrected mass air flow $\dot{m}_{a,0}$ can be calculated from the measured one $\dot{m}_a$.

$$F_{Hi} \cdot \dot{m}_a \approx \dot{m}_{a,0} \tag{5.37}$$

By application of $F_{Hi}$, the gradient of the lambda characteristic is turned back to $\lambda_0$.

b.)  This is now employed at **low** engine power outputs. The additive offset error $\Delta \dot{m}_a$ can no longer be neglected.

$$\dot{m}_a = \frac{\lambda}{\lambda_0} \dot{m}_{a,0} + \Delta \dot{m}_a \tag{5.38}$$

Inserting Equation 5.34 and 5.36 we get

$$\dot{m}_{a,0} = F_{Hi} \cdot \dot{m}_a - F_{Hi} \cdot \Delta \dot{m}_a \quad . \tag{5.39}$$

The lambda control loop generates a multiplicative correction factor $F_\lambda$ also at low engine power output. It is averaged to suppress the limit cycle.

$$F_{Lo} = \overline{F}_\lambda \tag{5.40}$$

The correct mass air flow can be calculated as

$$\dot{m}_{a,0} = F_{Lo} \cdot \dot{m}_a \quad . \tag{5.41}$$

c.)  An additional control loop for compensating the additive error is now installed. Merging Equation 5.39 and 5.41 yields

$$\frac{\Delta \dot{m}_a}{\dot{m}_a} = \frac{F_{Lo} - F_{Hi}}{F_{Hi}} \quad . \tag{5.42}$$

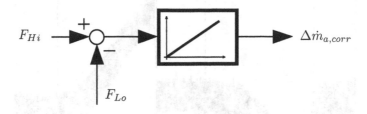

**Figure 5.22** Additional integral controller for offset correction

If the two correction factors

$$F_{Lo} = F_{Hi} \quad , \quad \text{for} \quad \Delta \dot{m}_a = 0 \tag{5.43}$$

would be identical, then the offset error $\Delta \dot{m}_a$ would be eliminated.

This is achieved by an additional integral control loop, which gets the difference $F_{Hi} - F_{Lo}$ at it's input, and which generates the unknown offset $\Delta \dot{m}_{a,corr}$ at it's output.

The lambda characteristic can now be corrected by subtracting $\Delta \dot{m}_{a,corr}$. The corrected lambda characteristic is

$$F_{Hi} \cdot \dot{m}_a = \underbrace{F_{Hi} \cdot \frac{\lambda}{\lambda_0}}_{\approx 1} \cdot \dot{m}_{a,0} + F_{Hi} \underbrace{(\Delta \dot{m}_a - \Delta \dot{m}_{a,corr})}_{\approx 0} \approx \dot{m}_{a,0} \quad . \tag{5.44}$$

Since the original lambda control loop is unloaded from the correction task, there are no more mismatches during engine transients.

Both $F_{Hi}$ and $\Delta \dot{m}_{a,corr}$ are stored in a non-volatile memory, so that they are correcting the lambda characteristic even at open-loop operation.

### Results of the Globally Valid Compensation

In Figure 5.23, a bypass leakage of $0.1\ mm^2$ was introduced into the intake system. At stationary engine operation, the control output factor $F_\lambda$ first corrects the resulting air-fuel ratio. This correction is then slowly shifted from the closed loop control to the adaptive compensation scheme. The control output factor $F_\lambda$ slowly returns to a limit cycle around an average value of 1.

The real gain of global compensation comes when driving through engine transients such as shown in Figure 5.24. The vehicle is following the speed profile of an emission test. The control output factor $F_\lambda$ is limited in its adaptation speed by the minimum value of the integration time constant. When the major portion of all lambda mismatches is globally compensated, the closed loop control is mostly relieved from the correction task. The limited transient speed of $F_\lambda$ is no longer leading to noxious exhaust emission spikes.

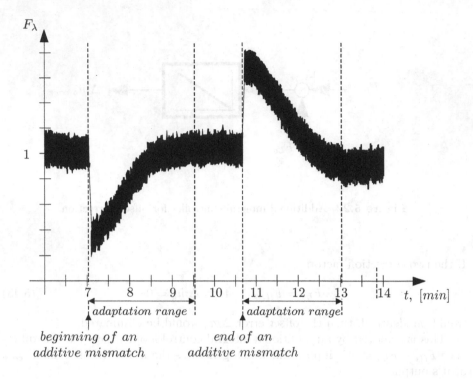

**Figure 5.23** Control output factor $F_\lambda$ with a bypass leakage in the intake system

## 5.2   Idle Speed Control

As a rule of thumb, fuel consumption of internal combustion engines increases proportional to engine speed at idling. Therefore, the idle speed should be made as low as possible. The reduced engine idle speed can be held up with less engine power output. Contrary to that, load torque variations such as the switch-on of the air condition compressor motor stay constant. Engine torque output steps compensating such loads thus increase relative to the basic torque required to keep the engine running. This is a challenge for the idle speed control. The actuation variable at SI engines is the mass air flow into the engine, at Diesel engines the injected fuel amount.

A problem are gas pedal movements of the driver at idling. They modulate the actuation variable in competition to the control actuator, which also varies the same variable. When the driver e.g. slowly increases the mass air flow in SI engines, the controller will reduce its actuator signal in order to regulate the speed to the reference level. If in a next step the driver would release the gas pedal, the control actuation takes some time to adapt to this. With an improper design, the engine might stall in such situations.

The control scheme of SI engines presented in this section measures the engine speed and estimates the intake manifold pressure. The dynamic behavior of the control loop is determined by the intake manifold (see Section 3.2.6), the energy

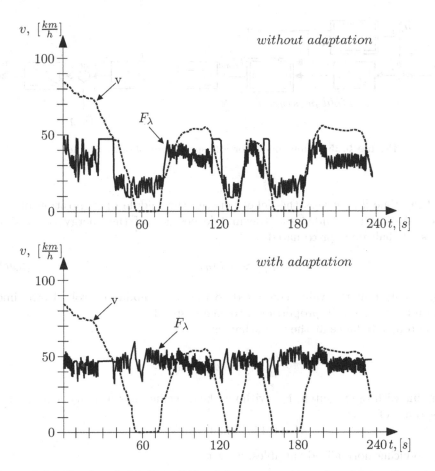

**Figure 5.24** Results of globally valid lambda compensation when driving through test cycle

conversion process and the torque balance at the crankshaft, which are modeled in the following section. At Diesel engines, the intake manifold model and the feedback of the intake manifold pressure are dropped.

## 5.2.1 Energy Conversion Model and Torque Balance

The energy conversion process is extremely complex and highly nonlinear. In a simplified approach, the stationary dependence of the combustion torque $T_{comb}$ from intake manifold pressure and engine speed shall be represented by a nonlinear map $f_2(n, p_m)$, which can be measured at all engine operating points. The dynamic behavior is separately considered by a combination of first order lag time $T_{l,e}$ and a delay time $T_{d,e}$. The lag time approximates the phase-shifted operation of the engine cylinders, as seen in lambda control (see Section 5.1.3).

$$T_{l,e} \approx \frac{2(CYL - 1)}{CYL} \cdot \frac{1}{n} \quad . \tag{5.45}$$

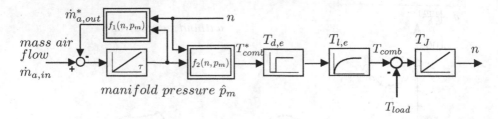

**Figure 5.25**  Plant model for idle speed control of SI engines

The delay time $T_{d,e}$ covers the delay between the middle open position of the intake valves of a cylinder and the middle position of the energy conversion process. It shall be approximated by

$$T_{d,e} \approx 3/(4n)   ,$$ (5.46)

which is only half the value compared to that at lambda control. Both time constants vary inversely proportional to engine speed.

The torque balance at the crankshaft is

$$2\pi J \frac{dn}{dt} = T_{comb} - T_{load}  .$$ (5.47)

An engine with open clutch, i.e. without the driveline, has a moment of inertia in the range of

$$J = 0.15 \ldots 0.30 \, kg \, m^2  .$$

By introducing normalized variables, we get

$$2\pi \cdot \underbrace{\frac{J \cdot n_0}{T_0}}_{T_J} \cdot \frac{d(n/n_0)}{dt} = \frac{T_{comb}}{T_0} - \frac{T_{load}}{T_0}$$ (5.48)

with a time constant

$$T_J = 2\pi \frac{J \cdot n_0}{T_0}   .$$ (5.49)

At maximum torque output and engine speed

$$\begin{aligned} J &= 0.3 \, kg \, m^2, \\ n_0 &= 6000 \, min^{-1}, \\ T_0 &= 300 \, Nm, \end{aligned}$$

the time constant is $T_J = 0.63 \, s$. When accelerating from low engine speed with maximum torque, the time constant $T_J$ is an order of magnitude smaller. Contrary, $T_J$ is an order of magnitude larger at high engine speed and minimum torque output, e.g. when coasting. The load torque comprises friction, auxiliary drives and disturbances. The complete plant model for idle speed control is shown in Figure 5.25.

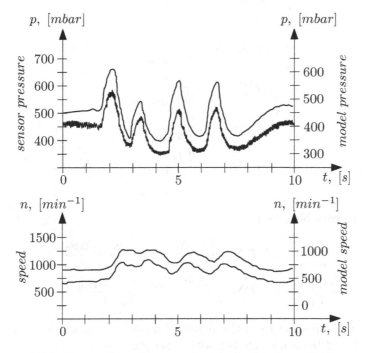

**Figure 5.26** Comparison of measured and calculated manifold pressure

The mass air flow $\dot{m}_a$ into the engine is measured, whereas the average intake manifold pressure $\hat{p}_m$ is calculated by integrating model Equation 3.50. A comparison of measured and calculated manifold pressure is shown in Figure 5.26.

Taking into account the shifted scale for the model variables, there is an excellent tracking of the model to the real engine.

## 5.2.2  State Space Control

In many vehicles, the idle speed is actually controlled with a PID controller. The differentiating D portion is sometimes shifting the ignition angle, due to the smaller delays between angle advance/retard and torque response. In this book, a state space controller shall be used, which feeds back the model pressure $\hat{p}_m$ and the measured engine speed $n$. Any unwanted driver actuation of the mass air flow generates a much faster response of the manifold pressure compared to that of the engine speed. The engine delay time $T_{d,e}$ cannot be compensated by the differentiating D portion of the PID control. For this reason, state space control is superior to PID control of the idle speed at SI engines.

In a first step, the two maps $f_1(n, p_m)$ and $f_2(n, p_m)$ are linearized at the idle speed operation point $\dot{m}_{a,0}, n_0, p_{m0}$. Introducing first order differentials

$$FN_1 = \left.\frac{\partial f_1}{\partial n}\right|_{n=n_0}$$

$$FP_1 = \left.\frac{\partial f_1}{\partial p_m}\right|_{p_m=p_{m0}}$$

$$FN_2 = \left.\frac{\partial f_2}{\partial n}\right|_{n=n_0}$$

$$FP_2 = \left.\frac{\partial f_2}{\partial p_m}\right|_{p_m=p_{m0}} \tag{5.50}$$

and difference variables, we get

$$\frac{\Delta \dot{m}_{ac}^*}{\dot{m}_{a,0}} = FN_1 \frac{n_0}{\dot{m}_{a,0}} \frac{\Delta n}{n_0} + FP_1 \frac{p_{m,0}}{\dot{m}_{a,0}} \frac{\Delta p_m}{p_{m,0}} \quad, \tag{5.51}$$

$$\frac{\Delta T_{comb}^*}{T_0} = FN_2 \frac{n_0}{T_0} \frac{\Delta n}{n_0} + FP_2 \frac{p_{m,0}}{T_0} \frac{\Delta p_m}{p_{m,0}} \quad. \tag{5.52}$$

The differential equation from the manifold model (Equation 3.54) is Laplace-transformed, and becomes together with Equation 5.51

$$s \cdot \tau_n \cdot \frac{\Delta P_m}{p_{m,0}} = -FN_1 \frac{n_0}{\dot{m}_{a,0}} \frac{\Delta N}{n_0} - FP_1 \frac{p_{m,0}}{\dot{m}_{a,0}} \frac{\Delta P_m}{p_{m,0}} + \frac{\Delta \dot{M}_{a,in}}{\dot{m}_{a,0}} \quad. \tag{5.53}$$

The incoming air flow $\Delta \dot{M}_{a,in}$ serves as a control input $\Delta U$. Equation 5.52 is also Laplace-transformed and extended by the engine lag and delay times.

$$\frac{\Delta T_{comb}}{T_0} = FN_2 \frac{n_0}{T_0} \frac{e^{-s\,T_{d,e}}}{1+sT_{l,e}} \frac{\Delta N}{n_0} + FP_2 \frac{p_{m,0}}{T_0} \frac{e^{-s\,T_{d,e}}}{1+sT_{l,e}} \frac{\Delta P_m}{p_{m,0}} \quad. \tag{5.54}$$

This is now inserted into the torque balance (Equation 5.48). Neglecting the disturbance load torque $T_{load}$ for control purposes, we get

$$s\,T_J \cdot \frac{\Delta N}{n_0} = \frac{e^{-s\,T_{d,e}}}{1+sT_{l,e}} \left( FN_2 \frac{n_0}{T_0} \frac{\Delta N}{n_0} + FP_2 \frac{p_{m,0}}{T_0} \frac{\Delta P_m}{p_{m,0}} \right) \quad. \tag{5.55}$$

The stability analysis of the plant model and the controller design shall now be done by neglecting time constants $T_{d,e}$ and $T_{l,e}$. The subsequent approach simplifies to a second order linear state space model

$$s \cdot \begin{bmatrix} \frac{\Delta P_m}{p_{m,0}} \\ \frac{\Delta N}{n_0} \end{bmatrix} = \underbrace{\begin{bmatrix} -\frac{FP_1}{\tau_n} \frac{p_{m,0}}{\dot{m}_{a,0}} & -\frac{FN_1}{\tau_n} \frac{n_0}{\dot{m}_{a,0}} \\ \frac{FP_2}{T_J} \frac{p_{m,0}}{T_0} & \frac{FN_2}{T_J} \frac{n_0}{T_0} \end{bmatrix}}_{A} \cdot \begin{bmatrix} \frac{\Delta P_m}{p_{m,0}} \\ \frac{\Delta N}{n_0} \end{bmatrix} + \underbrace{\begin{bmatrix} \frac{1}{\tau_n} \\ 0 \end{bmatrix}}_{b} \cdot \frac{\Delta U}{\dot{m}_{a,0}} \quad. \tag{5.56}$$

The poles of the open-loop system are obtained from the characteristic equation

$$det(s\underline{I} - \underline{A}) = 0 \tag{5.57}$$

or

$$s^2 + \left( \frac{FP_1}{\tau_n} \frac{p_{m,0}}{\dot{m}_{a,0}} - \frac{FN_2}{T_J} \frac{n_0}{T_0} \right) s + \left( \frac{FN_1 \cdot FP_2 - FP_1 \cdot FN_2}{\tau_n \cdot T_J} \right) \frac{p_{m,0}}{\dot{m}_{a,0}} \cdot \frac{n_0}{T_0} = 0 \quad . \tag{5.58}$$

Inserting Equation 5.49 for $T_J$ and Equation 3.53 for $\tau_n$, this becomes

$$s^2 + \left( \frac{FP_1}{\tau} - \frac{FN_2}{2\pi J} \right) s + \left( \frac{FN_1 \cdot FP_2 - FP_1 \cdot FN_2}{\tau \cdot 2\pi J} \right) = 0 \quad . \tag{5.59}$$

The characteristic equation is independent of a specific normalization of the variables. The two poles are

$$s_{1,2} = -\frac{1}{2} \left( \frac{FP_1}{\tau} - \frac{FN_2}{2\pi J} \right) \pm$$

$$\pm \sqrt{\frac{1}{4} \left( \frac{FP_1}{\tau} - \frac{FN_2}{2\pi J} \right)^2 - \left( \frac{FN_1 \cdot FP_2 - FP_1 \cdot FN_2}{\tau \cdot 2\pi J} \right)} \quad . \tag{5.60}$$

The poles are real for

$$\frac{FP_1}{\tau} - \frac{FN_2}{2\pi J} \geq 2\sqrt{\frac{FN_1 \cdot FP_2 - FP_1 \cdot FN_2}{\tau \cdot 2\pi J}} \quad . \tag{5.61}$$

For

$$\frac{FN_2}{2\pi J} > \frac{FP_1}{\tau} \quad , \tag{5.62}$$

the open-loop idle speed plant becomes unstable.

The controller shall be implemented by a proportional feedback of the manifold pressure and the engine speed.

$$\frac{\Delta U}{\dot{m}_{a,0}} = \begin{bmatrix} -K_P, & -K_N \end{bmatrix} \cdot \begin{bmatrix} \frac{\Delta P_m}{p_{m,0}} \\ \frac{\Delta N}{n_0} \end{bmatrix} \tag{5.63}$$

The second order model of the closed-loop system is then

$$s \cdot \begin{bmatrix} \frac{\Delta P_m}{p_{m,0}} \\ \frac{\Delta N}{n_0} \end{bmatrix} = \underbrace{\begin{bmatrix} -\left( \frac{FP_1}{\tau_n} \frac{p_{m,0}}{\dot{m}_{a,0}} + \frac{K_P}{\tau_n} \right) & -\left( \frac{FN_1}{\tau_n} \frac{n_0}{\dot{m}_{a,0}} + \frac{K_N}{\tau_n} \right) \\ \frac{FP_2}{T_J} \frac{p_{m,0}}{T_0} & \frac{FN_2}{T_J} \frac{n_0}{T_0} \end{bmatrix}}_{\underline{A}_C} \cdot \begin{bmatrix} \frac{\Delta P}{p_{m,0}} \\ \frac{\Delta N}{n_0} \end{bmatrix} \quad . \tag{5.64}$$

The control dynamics shall be determined by pole placement. The characteristic equation of the closed-loop system is

$$det(s\underline{I} - \underline{A}_C) = 0 \quad ,$$

$$s^2 + \left( \frac{FP_1}{\tau_n} \frac{p_{m,0}}{\dot{m}_{a,0}} + \frac{K_p}{\tau_n} - \frac{FN_2}{T_J} \frac{n_0}{T_0} \right) s -$$

$$- \left[ \left( \frac{FP_1}{\tau_n} \frac{p_{m,0}}{\dot{m}_{a,0}} + \frac{K_p}{\tau_n} \right) \frac{FN_2}{T_J} \frac{n_0}{T_0} - \left( \frac{FN_1}{\tau_n} \frac{n_0}{\dot{m}_{a,0}} + \frac{K_N}{\tau_n} \right) \frac{FP_2}{T_J} \frac{p_{m,0}}{T_0} \right] = 0 \quad . \tag{5.65}$$

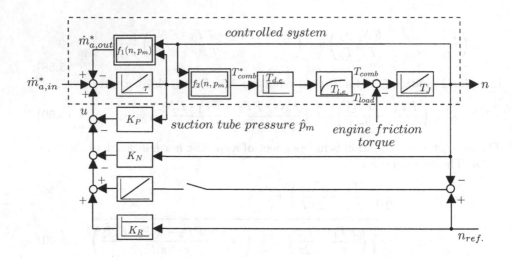

**Figure 5.27** Block diagram of idle speed control

The characteristic equation of a second order system with desired poles $s_1$ and $s_2$ is

$$s^2 - (s_1 + s_2) \cdot s - s_1 \cdot s_2 = 0 \quad . \tag{5.66}$$

A comparison yields the control parameters

$$K_P = -\tau_n(s_1 + s_2) - FP_1 \cdot \frac{p_{m,0}}{\dot{m}_{a,0}} + \frac{\tau_n}{T_J} \cdot FN_2 \cdot \frac{n_0}{T_0} \tag{5.67}$$

and

$$K_N = \frac{T_J \cdot \tau_n}{FP_2 \cdot \frac{p_{m,0}}{T_0}}(s_1 \cdot s_2) - FN_1 \cdot \frac{n_0}{\dot{m}_{a,0}} -$$

$$-\frac{FN_2 \cdot \frac{n_0}{T_0}}{FP_2 \cdot \frac{p_{m,0}}{T_0}} \cdot \tau_n \cdot (s_1 + s_2) + \frac{\tau_n}{T_J} \cdot \frac{FN_2^2 \cdot \frac{n_0^2}{T_0^2}}{FP_2 \cdot \frac{p_{m,0}}{T_0}} \quad . \tag{5.68}$$

In the practical calibration process of the idle speed control to an actual engine, these parameters can be tuned to make up for the neglected lag time $T_{l,e}$ and delay time $T_{d,e}$. The complete block diagram of idle speed control is shown in Figure 5.27.

The multiplication factor $K_R$ for the reference speed $n_{ref}$ is selected, so that the closed-loop system has no offset, i.e. $\Delta n = 0$ and $\Delta p_m = 0$. This is applying for the absence of disturbance load torques. A proportional control does however show still a stationary control offset in the case of disturbance inputs or parameter variations. This is why an additional integral controller is introduced, which reduces stationary offset to zero. The problem with integral control is, that a disturbance input from the driver could result in control actions integrating to its range boundaries. If the driver relinquishes his input, the engine might stall.

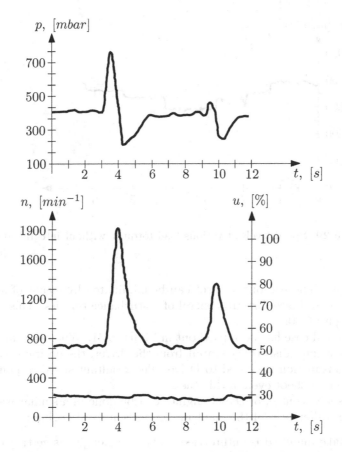

**Figure 5.28** Speed response after driver impulse input, without integral control

There are however a number of ways to overcome such a so-called wind-up effect. A heuristic approach could be used to interrupt integration, when a disturbance input from the driver is detected, e.g. by monitoring the throttle position.

If SI engines would be throttled by a modulation of their intake valves instead of the throttle body butterfly, the intake manifold model could be deleted, simplifying the controller design. The same applies for Diesel engines.

## 5.2.3 Measurement Results

The idle speed controller was applied to a two-liter four cylinder SI-engine with power steering, automatic transmission and air condition. The idle speed $n_0$ in the absence of load disturbance torques was $720\,min^{-1}$. Figure 5.28 shows an aperiodic decay of the speed response after an acceleration impulse from the driver. No additional integral control was applied in this test. There is no undershoot when the speed levels off into its stationary value.

In Figure 5.29 various disturbance loads are applied to the state space control without integral control. The stationary speed level is going down with increas-

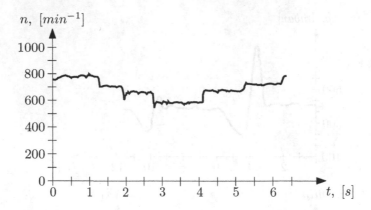

**Figure 5.29** Speed levels at various load torques, without integral control

ing load torque. The stationary offset can be eliminated by means of additional integral control and feed-forward control of disturbance torques. This is demonstrated in Figure 5.30.

A very critical case has been tried out in Figure 5.31. When the engine speed sharply drops after a disturbance input from the driver, the position stick of the automatic transmission is shifted to Drive. The resultant speed response shows only a small undershoot even in this case.

The idle speed control of Diesel engines can be done in a similar way. There are two major differences of the plant in comparison to SI-engines:

1. The intake manifold is unthrottled, so that the engine is getting the maximum possible mass air flow $\dot{m}_a$ in each operation point.

2. With direct fuel injection, the lag time $T_{l,e}$ may be significantly reduced.

These two points simplify the control design. A complication would be turbo charging, which introduces a significant time constant for the response of the mass air flow $\dot{m}_a$ to gas pedal transients.

## 5.3 Knock Control

### 5.3.1 Knocking at SI Engines

During a combustion cycle, a portion of the air-fuel mixture may self-inflame, before it is reached by the flame front coming from the spark plug. The condition for this to happen is, that the self-inflammation time is shorter than the propagation time of the flame front. The self-inflammation delay $\tau_{id}$ is approximated by Woschni [130] as

$$\tau_{id} = 0.44\,ms \cdot \exp(4650\,K/\vartheta) \cdot (p/p_0)^{-1.19} \quad . \tag{5.69}$$

Self-inflammation preferably happens at locations within the combustion chamber, which are distant from the spark plug, and which show high temperature

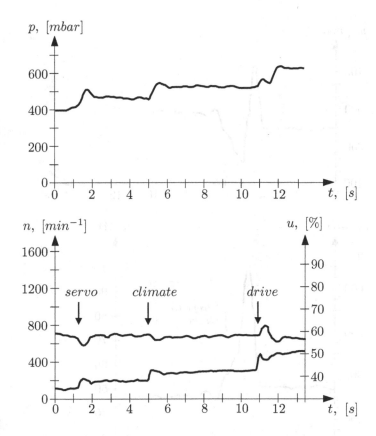

**Figure 5.30** Speed levels at various load torques, with integral control and feed-forward control

levels. In the case of an additional self-inflammation in a remote spot, two flame fronts with opposite directions are generated (Figure 5.32). When colliding, the resulting pressure peak excites acoustical eigen-oscillations, which depend of the geometry of the combustion chamber. These resonances are superimposed to the normal pressure curve (Figure 5.33). Due to very high pressure gradients knock oscillations can lead to significant engine damages by cavitation. In extreme cases, the entire engine may be destroyed in a fraction of a minute.

The sensitivity of SI engines to fuel self-inflammation depends upon several parameters.

- Increased ambient temperature, which leads to higher peak temperatures within the combustion process.

- Increased load pressure, which also increases peak temperatures. This can be caused by higher bariometric pressures, by higher engine load conditions or by turbo charging.

- Bad fuel quality, e.g. low octane number.

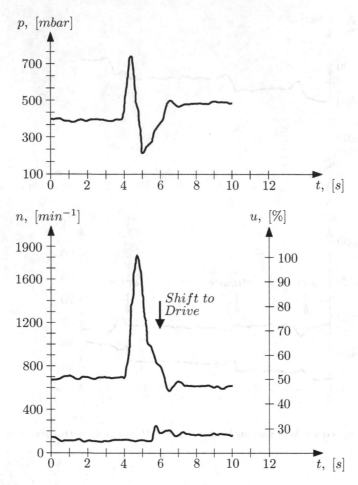

**Figure 5.31**   Disturbance input from driver and simultaneous gear shift to Drive position

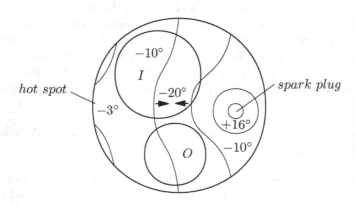

**Figure 5.32**   Self-inflammation with two colliding flame fronts

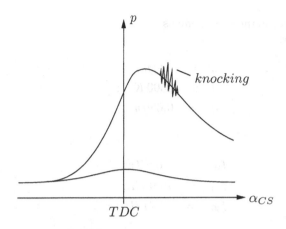

**Figure 5.33** Resonance induced by self-inflammation

The knocking sensitivity of engines can be reduced by a proper design.

- Compact combustion chamber geometry in order to avoid hot spots.

- Central position of the spark plug in order to minimize flame propagation.

- Increased turbulence for faster flame propagation.

- Limitation or regulation of boost pressure at turbo-charged engines.

By retarding the ignition angle, the entire energy conversion process is shifted backwards. Since the combustion pressure is then superimposed to a lower pressure due to adiabatic compression, resulting peak pressures are reduced.

After a short response time gas oscillations lead to resonance waves in the combustion chamber. When the piston is at top dead center (TDC), the radial resonances with frequency modes

$$f_{mn} = c_0 \sqrt{\vartheta/273\,K} \cdot \beta_{mn}/d \qquad (5.70)$$

dominate. The parameters are

$$
\begin{array}{ll}
c_0 & \text{sound propagation velocity at } 273\,K \\
\vartheta & \text{Temperature within combustion chamber} \\
d & \text{cylinder diameter} \\
\beta_{mn} & \text{Bessel function, e.g.} \\
& \quad \beta_{10} = 0.5861 \\
& \quad \beta_{20} = 0.9722 \\
& \quad \beta_{30} = 1.2197
\end{array}
$$

The variable geometry of the combustion chamber due to the piston movement is neglected.

**Example:** Knock resonance frequencies
From the parameters

$$
\begin{aligned}
c_0 &= 330\,m/s \quad, \\
\vartheta &= 2500\,K \quad, \\
d &= 0.089\,m \quad,
\end{aligned}
$$

we can calculate

$$
\begin{aligned}
f_{10} &= 6.6\,kHz \quad, \\
f_{20} &= 10.9\,kHz \quad, \\
f_{30} &= 13.7\,kHz \quad.
\end{aligned}
$$

At the real engine, a resonance frequency of $f_{10} = 6.8\,kHz$ was measured.

## 5.3.2   Knock Sensors

There are several approaches to measure knock oscillations.

a.) Combustion pressure sensor
The direct approach is to measure the combustion pressure. The knock oscillations superimposed to the pressure curve may be filtered out e.g. by a band pass.

|  |  |  |
|---|---|---|
| Advantage : | • | Integral acquisition of all oscillations in the combustion chamber |
| Disadvantages : | • | High costs to harden pressure sensors for the operation in the combustion chamber. |
|  | • | Engine head design may leave no room for a pressure sensor. |

b.) Mechanical Vibration Sensors at engine block
The engine block is transmitting knock oscillations from the different cylinders, which can be sensed by mechanical resonators. An example is shown in Figure 5.34. With a sensor eigen frequency around $25\,kHz$, several knock resonances can be measured. Four cylinder engines need one or two sensors, six and more cylinder engines at least two sensors.

|  |  |  |
|---|---|---|
| Advantage : | • | Low Costs |
|  | • | Straight forward mounting |
| Disadvantages : | • | Strong disturbance noise from closing valves or piston tilting |

c.) Ion Current Measurement
As sensors, the standard spark plugs may be used. During the combustion process of hydro-carbons, electrically charged ions and electrons are generated. The intensity of the chemical reaction and thus the intensity of the ionization depend on the flame temperature, on the air fuel ratio and on the fuel quality. The unmoved mass of a positive ion $H_3O^+$ is approximately 30,000 time larger than that

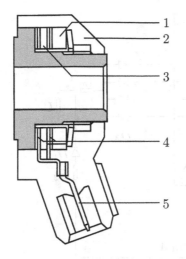

1 — seismic mass
2 — casting resin
3 — piezo ceramic
4 — contact surface
5 — connector

**Figure 5.34** Mechanical knock sensor

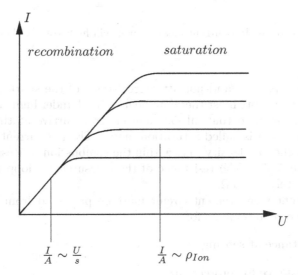

**Figure 5.35** Ion current versus supply voltage

of a negative electron. The voltage polarity at the spark plug gap is therefore selected such that the small area electrode is positive and the large area electrode is negative. The light electrons are accelerated much more than the heavy ions, crossing through a large distance per time. Therefore, the same number of negative electrons can reach the small area electrode as positive ions can reach the large area electrode of the spark plug.

When a low electrical field $U/s$ is applied to the spark plug gap (width s), the ion current density $i/A$ is proportional to the electrical field (A electrode area). The ohm law is applying in this type of operation. Ions and electrons generated

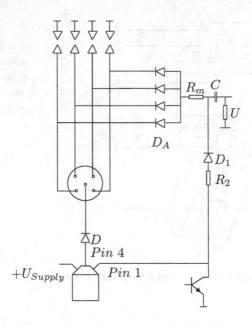

**Figure 5.36** Ion current measurement via high voltage diodes

in the combustion process and not attracted to one of the spark plug gaps are recombining. Ion currents must therefore be measured under high electrical fields at the spark plug gap, so that all ions and electrons arrive at their electrode. This type of operation is called saturation mode. The ion current density then depends only on the ion density $\rho_{Ion}$ within the combustion process and the gap distances (Figure 5.35). The resistance of the measurement loop must be kept below approximately $0.5\,M\Omega$.

The ion current measurement circuit must be protected from high ignition voltages. The two requirements for

- a low resistance at sensing

- a high resistance for protection

can be considered by high voltage diodes (Figure 5.36).

The capacitor $C$ at the primary side of the ignition coil is loaded to e.g. $300 - 400\,V$ during dwell time. After the ignition, the secondary ignition voltage (from Pin 4) decays. After top dead center, the voltage stored in $C$ drives the ionization measurement circuit, in which the current through $R_m$ indicates the ion density and thus the combustion intensity.

A problem with ion current measurement is that it represents only the combustion intensity in a very small volume around the spark plug, not in the entire combustion chamber. Knock detection therefore heavily depends on the position of the spark plug. A central position is advantageous for knock detection, since the first resonance wave has a pressure minimum and a velocity maximum there. The second resonance wave has a pressure maximum and a velocity minimum at

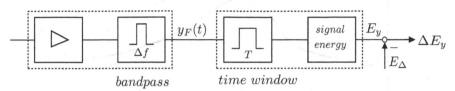

Figure 5.37  Signal processing of knock signal

the center of the combustion chamber. It is not suitable for knock detection by ion current, but rather for an indirect combustion pressure measurement.

Advantages :      • In-cylinder measurement
                  • No mechanical disturbances

Disadvantages :   • Dependence on spark plug position
                  • Measurement in a small portion of the combustion chamber

d.) Light Intensity of Combustion Process
Knock oscillations modulate the intensity of the combustion process. With that comes a modulation of light intensity and color in the combustion chamber. Light measurement is therefore another approach to measure knocking. A cone-formed portion of the combustion chamber is monitored.

A fiber-glass cable is fed through the central electrode of the spark plug, from where the light is forwarded to a remote photo transistor. A severe problem with light measurement in the combustion chamber is that the quartz glass window at the cable end is coated by soot in varying thickness. The measurement sensitivity is thus changing over several orders of magnitude.

Advantages :      • In-cylinder measurement
                  • No mechanical disturbances

Disadvantages :   • Extreme sensitivity variations

## 5.3.3   Signal Processing

If we disregard sensor-specific adaptation circuitry, a uniform methodology for signal processing can be applied (Figure 5.37).

At first the sensor signal amplitude is regulated to a constant level by automatic gain control. The amplifier output signal $y(t)$ is shown in Figure 5.38 for non-knocking and in Figure 5.39 for knocking combustion.

The next step is a bandpass filter $r_{\Delta f}(f)$ which suppresses all spectral information outside the selected knock resonance frequency window $[f_r - \Delta f/2, f_r + \Delta f/2]$. The adiabatic pressure curve is suppressed as well (Figures 5.38 and

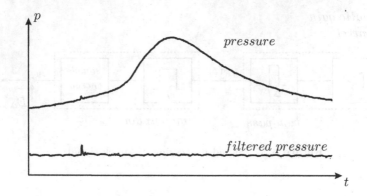

**Figure 5.38** Bandpass output signal for non-knocking combustion

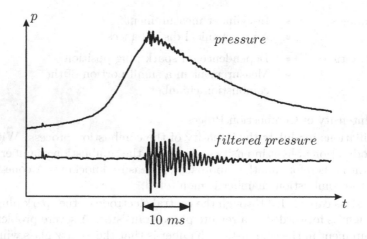

**Figure 5.39** Bandpass output signal for knocking combustion

5.39). The bandpass-filtered signal in the time domain is the convolution

$$y_F(t) = y(t) * r_{\Delta f}(t) \tag{5.71}$$

with the inverse Fourier transform of the rectangular frequency window

$$r_{\Delta f}(t) = \Delta f \frac{\sin(\pi \Delta f t)}{(\pi \Delta f t)} \exp\left(j 2\pi f_r t\right) \quad . \tag{5.72}$$

Knocking can only occur in a limited time interval during combustion. The signal $y_F(t)$ is therefore multiplied by a time window function. In the first place this is a rectangular window $r_T(t)$ with width $T$. If leakage shall be reduced, more sophisticated windows can be applied. Due to windowing, the subsequent integration stretches over a limited time interval. The resulting signal energy is

$$E_y(t) = \int_{t-T/2}^{t+T/2} y_F^2(t)dt = y_F^2(t) * r_T(t) = \left( y(t) * r_{\Delta f}(t) \right)^2 * r_T(t) \quad . \tag{5.73}$$

At each discrete combustion cycle $n$, the signal energy $E_y(n)$ is derived. After substraction of an operation-point depending threshold $E_0$, we get the so-called knock signal

$$\Delta E_y(n) = \begin{cases} E_y(n) - E_0 & , \quad E_y \geq E_0 \\ 0 & , \quad \text{otherwise} \end{cases} \tag{5.74}$$

$\Delta E_y(n)$ may be classified into a few steps in order to simplify knock control.

## 5.3.4 Knock Control

In a classical control circuit, a reference value is given, which must be approached as close as possible by the actual control variable. At knock control, no such reference is available. Because of the high damage potential of only a very few subsequent high-energy knockings in a cylinder, a reaction must be taken immediately after a single knock already.

The usual actuation is a retardation of the ignition angle, shifting the energy conversion process backwards and thus reducing peak pressures and temperatures. An alternative input may be to lower the boost pressure of a turbo charger. The knock control ignition angle is calculated at discrete combustion cycles $n$ as

$$\alpha_k(n) = \alpha_k(n-1) + \Delta\alpha_k - \beta \cdot \Delta E_y(n) \quad , \tag{5.75}$$

where $\Delta\alpha_k$ is a permanent ignition angle advance, and $\beta \cdot \Delta E_y(n)$ the ignition angle retard at knocking. A typical control cycle is shown in Figure 5.40. The knock control ignition angle $\alpha_k(n)$ is added to the ignition angle obtained from the ignition map (Section 3.2.8).

The two parameters $\Delta\alpha_k$ and $\beta$ determine the average knock occurrence rate. For safety reasons, the knock control advance is limited at

$$\alpha_k(n) \leq 0 \quad . \tag{5.76}$$

In case of errors, the ignition angle map determines the most advanced ignition angle. Knock control compensates the influence of parameter variations such as

- ambient temperatures

- bariometric pressures at different altitudes

- octane values at different fuel qualities

- engine manufacturing tolerances and ageing.

The compression ratio of knock controlled engines may be increased by at least 1. Fuel consumption is reduced by around 7 %. At turbo-charged engines, fuel savings are even higher.

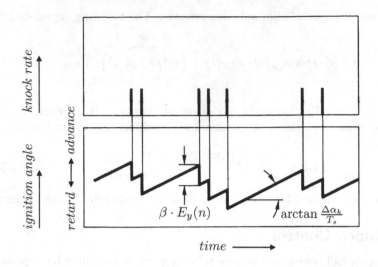

**Figure 5.40** Control of knock occurrence rate by ignition angle shifting

## 5.3.5   Adaptive Knock Control

At dynamic engine transients, mismatches of the ignition angle occur resulting
in increased knock occurrence rates. The response time of knock control can be
reduced by a feed-forward control angle $\alpha_l(n)$ stored in an adaptive ignition angle
map. Contrary to lambda control, a successful global error model has not yet
been found. The values of the ignition angle map must therefore be adapted in
every individual engine operating point for all cylinders (Figure 5.41).

The ignition angle at one cylinder is the sum

$$\alpha_e(n) = \alpha_i(n) + \alpha_k(n) + \alpha_l(n) \quad , \tag{5.77}$$

with

| | | |
|---|---|---|
| $\alpha_e$ | : | effective ignition angle |
| $\alpha_i$ | : | open loop ignition angle from fixed map |
| $\alpha_k$ | : | knock control ignition angle |
| $\alpha_l$ | : | learned ignition angle from adaptive map |

The average knock control ignition angle $\bar\alpha_k(n)$ is the basis to teach the adaptive ignition angle map $\alpha_l(n)$ into the direction of retarding. A fixed advance
angle $\alpha_a$ is superimposed to the teaching process providing a forgetting function
of the thought angles. The learned ignition angle is

$$\alpha_l(n) = (1 - k_l)\alpha_l(n - 1) + k_l(\bar\alpha_k(n - 1) + \alpha_a(n - 1)) \quad . \tag{5.78}$$

The factor $k_l$ determines how fast the learning process is. Z-Transformation of

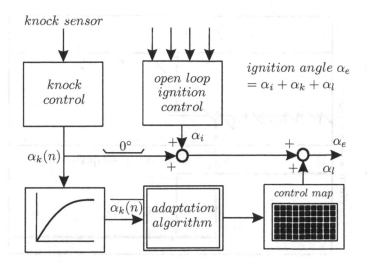

**Figure 5.41** Knock control with feed-forward adaptive ignition angle map

this equation yields

$$\alpha_l(z) = (1 - k_l)z^{-1} \cdot \alpha_l(z) + k_l z^{-1} (\bar{\alpha}_k(z) + \alpha_a(z)) \tag{5.79}$$

$$\alpha_l(z) = \frac{k_l z^{-1}}{1 - (1 - k_l)z^{-1}} (\bar{\alpha}_k(z) + \alpha_a(z)) \tag{5.80}$$

For an evaluation of the learning dynamics, the response of $\alpha_l(n)$ to an input step function

$$\bar{\alpha}_k(z) + \alpha_a(z) = \frac{\alpha_0}{1 - z^{-1}} \tag{5.81}$$

is considered.

$$
\begin{aligned}
\alpha_l(z) &= \frac{k_l \cdot \alpha_0 \, z^{-1}}{(1 - z^{-1})(1 - (1 - k_l)z^{-1})} \\
&= \alpha_0 \left( \frac{1}{1 - z^{-1}} - \frac{1}{1 - (1 - k_l)z^{-1}} \right) \tag{5.82}
\end{aligned}
$$

The discrete response function is then

$$\alpha_l(n) = \alpha_0 \left( 1 - (1 - k_l)^n \right) \quad . \tag{5.83}$$

Reconstructing the continuous-time function with $t = nT_s$, we get

$$\alpha_l(t) \approx \alpha_0 \left( 1 - \exp(-k_l t/T_s) \right) \quad . \tag{5.84}$$

The learning time constant is the sample time $T_s$ divided by the factor $k_l$.

$$T_l \approx T_s/k_l \tag{5.85}$$

Since the sampling is done at each combustion cycle, the sampling time $T_s$ is inversely proportional to engine speed. A compensation may be achieved by letting the factor $k_l$ become also inversely proportional to engine speed.

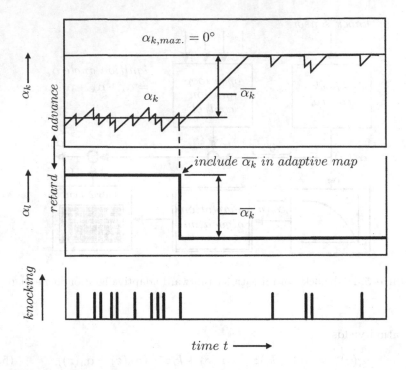

**Figure 5.42** Reduction of knock occurrence rate by transfer from $\alpha_k$ to learned angle $\alpha_l$

In Figure 5.42 the effect of the adaptive feed-forward control on knock control is shown. The knock control angle is oscillating around an average $\bar{\alpha}_k$. In order to prevent the engine from knocking to often, the angle $\bar{\alpha}_k$ is negative, corresponding to a retard from the fixed ignition angle map. Transferring the average knock control angle $\bar{\alpha}_k$ into the adaptive feed-forward control map $\alpha_l$, the knock control angle can return to its maximum advance $\alpha_k = 0$. In Figure 5.42 this transfer has been simplified to happen in one step. In reality the transfer response happens with time constant $T_l$. After the adaptation, the knock control $\alpha_k$ will retard in the event of knocking, however advance only against the maximum limit $\alpha_k = 0$ with steps $\Delta\alpha_k$. Since the advance can not go beyond the limit, ignition angle oscillations are reduced, resulting in a reduced knock occurrence rate in comparison to non-adaptive approaches.

When the engine enters a new operating area (Figure 5.43), the average knock control angle $\bar{\alpha}_k$ follows a transient into a stationary angle. The transient shall be terminated after $n_{max}$ combustion cycles, and the adaptation could start from there on. Since the engine may however stay in the operating area for a shorter time than $n_{max}$ cycles, the adaptation is allowed to start already after $n_{min}$ combustion cycles. In this case, the average knock control angle $\bar{\alpha}_k$ is still somewhat deviating from the stationary value. Therefore, the adaptation factor

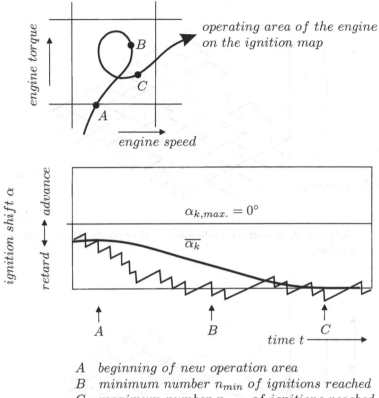

A  beginning of new operation area
B  minimum number $n_{min}$ of ignitions reached
C  maximum number $n_{max}$ of ignitions reached

**Figure 5.43** Operation time in one operating area

$k_l$ must be reduced.

$$k_l = \begin{cases} 0, & n_{Comb} < n_{min} \\ \frac{n_{Comb}}{n_{max}} k_{l0} &, & n_{min} \leq n_{Comb} \leq n_{max} \\ k_{l0} &, & n_{Comb} > n_{max} \end{cases} \qquad (5.86)$$

Typical values are

$$n_{min} = 2 \ldots 10$$
$$n_{max} \leq 500 .$$

In the medium range, the adaptation time constant $T_l$ is increased to compensate for the uncertainty of the average knock control angle $\bar{\alpha}_k$ at $n_{Comb} \leq n_{max}$. The approach allows to adapt the adaptive ignition angle map also in engine operating areas, which are very shortly visited.

Figures 5.44 and 5.45 show the performance of the adaptive knock control. The resulting ignition angle retard $(-\alpha_l)$ is plotted over manifold pressure $p_m$ and engine speed $n$. In Figure 5.44, the engine is operated with a 50 : 50 mixture of 91 octane and 98 octane fuel. The adaptive map must learn a retard in some of the operating points. After returning to an engine operation with 100 % of 98

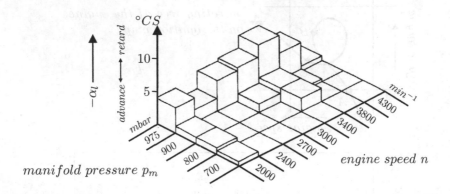

**Figure 5.44** Adaptation of ignition angle map $\alpha_l$ after $50\,km$ ride, $50:50$ mixture of 91 and 98 octane fuel.

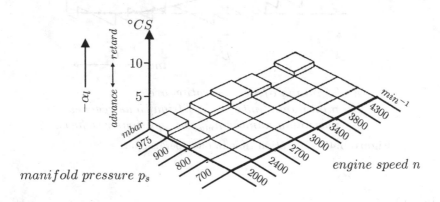

**Figure 5.45** Adaptation of ignition angle map $\alpha_l$ after $50\,km$ ride, $100\,\%$ of 98 octane fuel.

octane fuel, the retard angles are forgotten by integration of the $\alpha_a$-steps.

## 5.4   Cylinder Balancing

In this section an approach to compensate errors of the injected fuel mass

$$\Delta m_{f,i} = m_{f,i} - m_{f,ref} \qquad (5.87)$$

at individual cylinders $i$ is presented. The reference fuel mass $m_{f,ref}$ has been calculated by the engine management system and been transformed into an injection time $t_{inj}$ (see Section 3.2.4). The injection time $t_{inj}$ is decomposed into several very short time portions. Errors due to varying injector rise and fall times thus introduce a relatively large fuel mass error. Another source of errors is the injection pressure difference $\Delta p$ in Equation 3.32. At the first time portion of the injection procedure, the pressure in the fuel rail is decreased at the location of the injector, resulting in pressure oscillations. At subsequent injection time portions, the pressure difference $\Delta p$ is no longer at it's nominal value, contributing to an error of the injected fuel mass $m_{f,i}$. Additional errors are caused by clogging of the injector nozzles. In diesel engines, this error can be as large as

$$\frac{\Delta m_{f,i}}{m_{f,ref}} \leq 25\% \quad . \tag{5.88}$$

### 5.4.1   Residues at Stationary Engine Operation

The non-disturbed fuel mass $m_{f,ref}$ determines the effective work generated by the combustion in one cylinder (Equation 3.12).

$$w_{e,ref} \cdot \frac{V_d}{CYL} = \eta_e \cdot m_{f,ref} \cdot H_f \tag{5.89}$$

The individual fuel mass error $\Delta m_{f,i}$ shall now be indirectly determined by the error of the effective work $\Delta w_{e,i}$ contributed by cylinder $i$. Integration of the torque balance at the crankshaft (Equation 6.39) over one cylinder-related segment of crankshaft angle yields

$$J \int_{\alpha_i - \frac{2\pi}{CYL}}^{\alpha_i + \frac{2\pi}{CYL}} \dot{\alpha}_{CS} \frac{d\dot{\alpha}_{CS}}{d\alpha_{CS}} \, d\alpha_{CS} = \int_{\alpha_i - \frac{2\pi}{CYL}}^{\alpha_i + \frac{2\pi}{CYL}} (T_{comb} - T_{osc} - T_{load}^*) \, d\alpha_{CS} \quad . \tag{5.90}$$

The crankshaft angle $\alpha_i$ represents the center of the combustion process at cylinder $i$.

$$\frac{J}{2} \left( \dot{\alpha}_{CS}^2 (\alpha_i + \frac{2\pi}{CYL}) - \dot{\alpha}_{CS}^2 (\alpha_i - \frac{2\pi}{CYL}) \right) = \frac{V_d}{CYL} (w_{e,i} - w_{osc,i} - w_{load,i}^*) \quad . \tag{5.91}$$

The effective work is partitioned into the reference work $w_{e,ref}$, which is identical for all cylinders in one stationary operating point of the engine, and into the error of the effective work $\Delta w_{e,i}$ at cylinder $i$.

$$w_{e,i} = w_{e,ref} + \Delta w_{e,i} \tag{5.92}$$

Since the work balance at stationary engine operation

$$w_{e,ref} - w_{osc,i} - w_{load,i}^* = 0 \tag{5.93}$$

is zero (the load work does not change as fast), we get the absolute effective work error

$$\Delta w_{e,i} \cdot \frac{V_d}{CYL} = \frac{J}{2} \left( \dot{\alpha}_{CS}^2 (\alpha_i + \frac{2\pi}{CYL}) - \dot{\alpha}_{CS}^2 (\alpha_i - \frac{2\pi}{CYL}) \right) \tag{5.94}$$

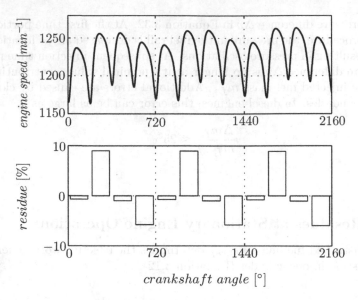

**Figure 5.46** Engine speed and residues at 4-cylinder engine (stationary operation)

and the relative error (residue)

$$R_i = \frac{2\,\eta_e\,m_{f,ref}\,H_f}{J} \cdot \frac{\Delta\omega_{e,i}}{\omega_{e,ref}} = \dot{\alpha}_{CS}^2(\alpha_i + \frac{2\pi}{CYL}) - \dot{\alpha}_{CS}^2(\alpha_i - \frac{2\pi}{CYL}) \quad . \quad (5.95)$$

At stationary engine operation, there is a typical residue $R_i$ for each cylinder $i$ (Figure 5.46). The sum of all residues is zero.

$$\sum_{i=1}^{CYL} R_i = 0 \tag{5.96}$$

This is the basis for the compensation of the fuel mass errors $\Delta m_{f,i}$.

## 5.4.2   Residues at Engine Transients

At engine transients, the work balance in Equation 5.93 is unequal zero. This is due to the fact, that a portion of the effective work $w_{e,ref}$ is now dedicated to the increase of the engine speed, i. e. of the rotational energy $E_{kin}$. Disregarding this would lead to a bias in the residue calculation, violating Equation 5.96. Figure 5.47 shows the uncompensated residues $R_i$ (Equation 5.95) at non-stationary engine operation.

The increase of the rotational kinetic energy due to speed increase over one cylinder-related segment of crankshaft angle is

$$\Delta E_{kin,i} = \frac{J}{2}\left(\bar{\dot{\alpha}}_{CS}^2(\alpha_i + \frac{2\pi}{CYL}) - \bar{\dot{\alpha}}_{CS}^2(\alpha_i - \frac{2\pi}{CYL})\right) \quad . \quad (5.97)$$

This can be easily verified since the average engine speed does not change at stationary operation. Under such conditions, the kinetic energy would remain constant, i. e. $\Delta E_{kin,i} = 0$.

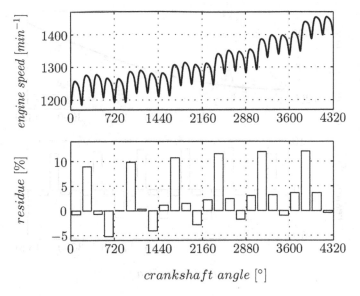

**Figure 5.47** Uncompensated residues at 4-cylinder engine during transient

At engine transients, the work balance becomes

$$\omega_{e,ref} - \Delta E_{kin,i} - \omega_{osc,i} - \omega^*_{load,i} = 0 \quad . \tag{5.98}$$

A compensated residue can therefore be formulated as

$$
R_{comp,i} = \left( \dot{\alpha}^2_{CS}(\alpha_i + \frac{2\pi}{CYL}) - \dot{\alpha}^2_{CS}(\alpha_i - \frac{2\pi}{CYL}) \right) - \\
- \left( \overline{\dot{\alpha}}^2_{CS}(\alpha_i + \frac{2\pi}{CYL}) - \overline{\dot{\alpha}}^2_{CS}(\alpha_i - \frac{2\pi}{CYL}) \right) \quad . \tag{5.99}
$$

Figure 5.48 shows that Equation 5.96 is valid again for such compensated residues.

$$\sum_{i=1}^{CYL} R_{comp,i} = 0 \tag{5.100}$$

The average crankshaft speed $\overline{\dot{\alpha}}$ in Equation 5.97 is calculated by means of an acausal FIR filter. For an even number of cylinders it is

$$\overline{\dot{\alpha}}(\beta_i) = \frac{1}{CYL} \left[ \frac{1}{2}\dot{\alpha}(\beta_{i-\frac{CYL}{2}}) + \sum_{j=i-\frac{CYL}{2}+1}^{i+\frac{CYL}{2}-1} \dot{\alpha}(\beta_j) + \frac{1}{2}\dot{\alpha}(\beta_{i+\frac{CYL}{2}}) \right] \quad . \tag{5.101}$$

The angle $\beta_i$ stands for the two angles

$$\beta_i = \alpha_i \pm \frac{2\pi}{CYL} \tag{5.102}$$

before and after the center of combustion. The computational result of an acausal filter operation for cylinder $i$ in Equation 5.101 is available only at cylinder $i +$

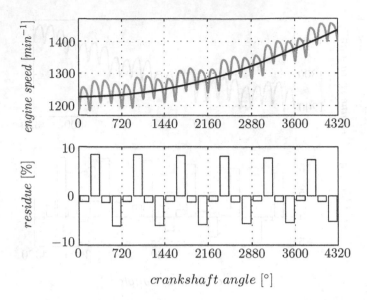

**Figure 5.48** Compensated residues at 4-cylinder engine during transient

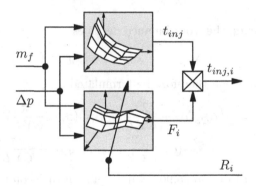

**Figure 5.49** Compensation map for injection timing at cylinder $i$

$\frac{CYL}{2}$. For a four-cylinder engine, the result is delayed by two cylinder segments. Fortunately, this delay does not create any problems, since the residue $R_i$ is required only two crankshaft revolutions later, when the injection for cylinder $i$ is calculated again (see Equation 5.103).

## 5.4.3   Adaptation of Injection Map

The injector map Figure 3.15 in Section 3.2.4 shall now be adapted such that the injection time $t_{inj}$ is compensated by the residues derived in the previous section. Since the original map $t_{inj}(m_f, \Delta p)$ is identical for all cylinders, the compensation is introduced by an additional learning map $F_i(m_f, \Delta p)$ for each cylinder $i$ (Figure 5.49).

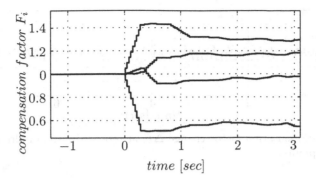

**Figure 5.50** Transient behavior of compensation factors $F_i$ at one operation point

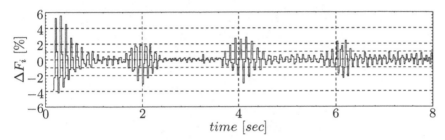

**Figure 5.51** Settling of compensation factor change $\Delta F_i$ after approaching new operation points for the first time

The compensation factor for cylinder $i$ is

$$F_i(n) = (1 - k_l)\, F_i(n - 1) + k_l\big(1 - R_{comp,i}(n - 1)\big) \quad , \qquad (5.103)$$

with $n$ being the discrete iteration time and $k_l < 1$ the weighting factor determining the learning time constant (see Section 5.3.5). After having compensated the injection time at all cylinders, the overall amount of fuel measured to the engine remains unchanged, du to Equation 5.100. For a positive term in the bracket of Equation 5.103, the compensation factor $F_i$ is decreased, since too much fuel has been injected into cylinder $i$. Figure 5.50 shows how the compensation factors $F_i$ are settling at one operation point of the engine. The above adaptation procedure has been verified with an engine running through various operation points where different compensation factors $F_i$ are required for cylinder i. Figure 5.51 shows how the change of $F_i$ is settling rapidly after approaching new engine operation points for the first time every 2 seconds.

When already compensated operation points are revisited, the compensation factors $F_i$ remain at their respective prior values with a tolerance of under 1% (Figure 5.52). This behavior verifies the effectiveness of injector map adaptation for transient engine operation.

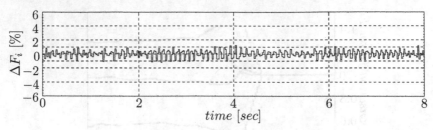

**Figure 5.52** Compensation factor change $\Delta F_i$ at repeated visits of already compensated operation points

# 6 Diagnosis

On-board diagnosis of car engines has become increasingly important because of environmentally based legislative regulations such as OBDII (On-Board Diagnostics-II) [17]. Other reasons for incorporating diagnosis in vehicles are reparability, availability and vehicle protection. Today, due to the legislations, the majority of the code in a modern engine management system is dedicated to diagnosis.

Diverse diagnosis techniques are used in production vehicles [124][57], and the selection of methods depends on the requirements of the more and more restrictive regulations. The simplest forms are based on limit checking of sensors and on active diagnosis. Active diagnosis means that during special operating conditions, the engine is manipulated in such a way that possible faults will be revealed. This is for example used heavily during idle. However, such techniques may be insufficient to fulfill the more restrictive regulations. Further, sometimes the lack of a good diagnosis schemes have forced engineers to choose suboptimal engine control solutions to accommodate the diagnostic requirements. It is therefore desirable to use new diagnosis techniques that perform better and do not rely on special operating conditions and active diagnosis. This leads to the field of model based diagnosis where more physical knowledge about the system is utilized.

The purpose of this chapter is to give an introduction to engine diagnosis. It starts in Section 6.1 with a background in engine diagnosis, and continues in Section 6.2 with a description of one example of an actual regulation, namely the OBDII by the California Air Resource Board [17]. Thereafter the purpose is to demonstrate how the models in previous chapters of this book can be utilized in advanced diagnosis applications, but before doing so some of the general concepts and methods in the field of diagnosis are introduced and discussed in Sections 6.3 to 6.6. Then, in Section 6.7, an important application example, diagnosis of the engine air-intake, is presented as an illustration. Section 6.8 deals with the misfire detection.

# 6.1   Diagnosis of Automotive Engines

Diagnosis of automotive engines has a long history. Since the first automotive engines in the 19:th century, there has been a need for finding faults in engines. For a long time, the diagnosis was performed manually, but diagnostic tools started to appear in the middle of the 20:th century. One example is the stroboscope that is used for determining the ignition time. In the 1960's, exhaust measurements became a common way of diagnosing the fuel system. Until the 1980's, all diagnosis were performed manually and off-board. It was around that time, electronics and gradually microprocessors were introduced in cars. This opened up the possibility to use on-board diagnosis. The objective was to make it easier for the mechanics to find faults.

In 1988, the first legislative regulations regarding On-Board Diagnostics, OBD, were introduced by California Air Resource Board (CARB). In the beginning these regulations applied only to California, but the federal Environmental Protection Agency (EPA) adopted similar regulations that applied for all USA. This enforced manufacturers to include more and more on-board diagnosis capability in the cars. In 1994, the new and more stringent regulations, OBDII, were introduced in California. Today, software for fulfilling OBDII is a major part of engine management systems. Following California and USA, regulations have been introduced in other countries. For example, EU has announced regulations, European On-Board Diagnostics (EOBD).

## 6.1.1   Why On-Board Diagnosis?

There are several reasons for incorporating on-board diagnosis:

- The mechanics can check the stored fault code and immediately replace the faulty component. This implies more efficient and faster repair work.

- If a fault occurs when driving, the diagnosis system can, after detecting the fault, change the operating mode of the engine to *limp home*. This means that the faulty component is excluded from the engine control and a suboptimal control strategy is used until the car can be repaired.

- The engine can potentially be serviced due to the condition of the engine and not due to a service schedule, thus saving service costs.

- The diagnosis system can make the driver aware of faults that can damage the engine, so that the car can be taken to a repair shop in time. This is a way of increasing the *reliability*.

- A fault can often imply increased emission of harmful emission components, dangerous for the environment. As an example, in 1990, the Environmental Protection Agency in USA estimated that 60% of the total tailpipe hydro-carbon emissions from light-duty vehicles, originated from 20% of the vehicles with seriously malfunctioning emission control systems [124]. It is important that such faults are detected so that the car can be repaired as quickly as possible.

The first three items can be summarized as to increase the *availability* of the car. Of all these reasons, the main reason for legislative regulations is the environmental issues. This reason has lead to that, today, automotive engines is one of the major application areas for diagnosis. The number of diagnosis systems in use is larger than for any other application. A characteristic constraint, in contrast to many other applications, is that automotive diagnosis systems are heavily constrained economically since even the slightest extra costs are emphasized because of the large production volumes. All in all, it is thus a challenging field.

## 6.2 OBDII

The requirements of OBDII will be used to illustrate the requirements of modern on-board diagnosis systems. OBDII is an extensive on-board diagnosis requirements that is in actual use. The requirements of the European EOBD are approximately at the same level. OBDII started to apply in 1994, but its requirements are made harder successively. The regulations are valid for all passenger cars, light-duty trucks, and medium-duty vehicles.

Focusing on the SI-engine in its basic configuration there are three main subsystems, each with its special diagnostic requirements:

- air intake system: air mass flow meter, throttle, manifold pressure sensor, engine speed sensor

- fuel/combustion system: fuel injector, spark plug, misfire

- exhaust after-treatment system: lambda sensors, catalyst

Common are also evaporation systems and EGR. All these components need to be diagnosed due to OBDII.

### 6.2.1 Main Characteristics

The main idea is that an instrument panel lamp called Malfunction Indicator Light (MIL) must be illuminated in the case of a fault that can make the emissions exceed the emission limits by more than 50%. The MIL should, when illuminated, display the phrase "Check Engine" or "Service Engine Soon". The OBDII also contains standards for the *scantool*, connectors, communication, and protocols that are used to exchange data between the diagnosis system and the mechanics. Further, it says that the software and data must be encoded to prevent unauthorized changes of the engine management system.

The manufacturer must specify the monitor conditions under which the diagnosis system is able to detect a fault. These monitor conditions must be encountered at least once during the first portion, i.e. Phase I+II, of the FTP75 (Federal Test Procedure) test cycle. FTP75 is a standardized test cycle used in USA and some other countries. It consists of three phases and is defined in vehicle speed as a function of time. Phase I+II is shown in Figure 6.1. The speed data of Phase III equals Phase I, but follows Phase II after 9-11 minutes pause.

Many formulations in the requirements are based on the term *driving cycle*. A driving cycle is defined as engine startup, engine shutoff, and any driving between

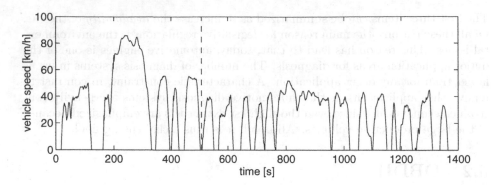

**Figure 6.1** Phase I (to the left of the dashed line) and Phase II (to the right of the dashed line) of the FTP75 test cycle.

these two events. Upon detection of a fault, the MIL must be illuminated and a fault code stored in the computer no later than the end of the next driving cycle during which the monitoring conditions occurs. The information stored in the memory of the computer are a *Diagnostic Trouble Code* (DTC) and *freeze frame data*. Freeze frame data is all information available of the current state of the engine and the control system. After three consecutive fault free driving cycles, the MIL should be turned off. Also, the fault code and freeze frame is erased after 40 fault free driving cycles.

### General

Generally, the components that must be diagnosed in OBDII, are all actuators, sensors, and components connected to the engine management system that can influence emissions in case of a malfunction. Sensors and actuators must be limit checked to be in range. Further, the values must be consistent with each other. Additionally, actuators must be checked using active tests. These general specifications apply therefore to for example mass air flow sensor, manifold pressure sensor, engine speed sensor, and throttle. In addition to these general specifications, OBDII contains specific requirements and technical solutions for many components of the engine. Some of the most important components and their requirements are described next. Except for these, OBDII also contains detailed specifications for heated catalyst, secondary air system, air conditioning refrigerant system, fuel system, and Exhaust Gas Recirculation (EGR) system.

### Misfire

One of the most important parts of OBDII are the requirements regarding misfire. This is because a misfire means that unburned gasoline reach the catalyst, which can be overheated and severely damaged. The diagnosis system must be able to detect a single misfire and also to determine the specific cylinder, in which the misfire occurred. During misfire, the MIL must be blinking.

As described in Section 6.8, the technology used today is primarily signal processing of the RPM-signal [123]. One may note that this is a model based approach. Sometimes an accelerometer is used as a complement. Also, ion current based methods that were mentioned in Section 5.3.2 have been demonstrated to work successfully for misfire detection [3][87].

**Catalyst**

Another central part of OBDII is catalyst monitoring. The catalyst is a critical component for emission regulation. If the efficiency of the catalyst falls below 60%, the diagnosis system must indicate a fault. The technology used today is to use two lambda (oxygen) sensors, one upstream and one downstream the catalyst. For a fully functioning catalyst, the variations, due to the limit cycle enforced by the control system, in the upstream lambda sensor should not be present in the downstream sensor (Figure 5.17).

**Lambda Sensors**

A change in the time constant or an offset of the lambda sensors must be detected. This is done by studying the frequency, comparing the two sensors, and applying steps and studying step responses.

**Purge System**

The purpose of the purge system is to take care of fuel vapor from the fuel tank. It contains a coal canister and some valves to direct the fuel vapor from the tank into the canister and from the canister into the intake manifold. The diagnosis system must be able to detect malfunctioning valves and also a leak in the fuel tank. The technology used here is heavily based on active tests.

**Air Intake System**

The air intake system contains a number of components and sensors that need to diagnosed. Also different types of leakage should be detected. In Section 6.7 it is described how a diagnosis system capable of diagnosing faults in throttle actuator, throttle sensor, air mass flow sensor, and manifold pressure sensor, is constructed.

# 6.3   Introduction to Diagnosis

From a general perspective, including both the medical and technical case, diagnosis can be explained as follows. For a process there are observed variables or behavior for which there are knowledge of what is expected or normal. The task of diagnosis is to, from the observations and the knowledge, generate a fault decision, i. e. to decide whether there is a fault or not and also to identify the fault, as illustrated in Figure 6.2. Thus the basic problems in the area of diagnosis is how the procedure for generating fault decisions should look like, what parameters or behavior that are relevant to study, and how to derive the knowledge of what is expected or normal.

## 6.3.1  Basic Definitions and Concepts

This section presents definitions and concepts that are central for the area of diagnosis. As a step towards a unified terminology, the IFAC Technical Committee SAFEPROCESS has suggested preliminary definitions of some terms in the field [56]. Additional definitions are made in [4]. Following is a list of some common basic terms with explanations. The explanations are partly based on the definitions made by the IFAC Technical Committee SAFEPROCESS.

- **Fault**
  Unpermitted deviation of at least one characteristic property or variable of the system from acceptable/usual/standard/nominal behavior.

- **Failure**
  A fault that implies permanent interruption of a systems ability to perform a required function under specified operating conditions.

- **Disturbance**
  An unknown and uncontrolled input acting on the system.

- **Fault Detection**
  To determine if faults are present in the system and also time of detection.

- **Fault Isolation**
  Determination of the location of the fault, i.e. which component that has failed.

- **Fault Identification**
  Determination of size and time-variant behavior of a fault.

- **Fault Accommodation**
  To reconfigure the system so that the operation can be maintained in spite of a present fault.

- **Fault Diagnosis**
  For the definition of this term, two common views exist in literature. The first view includes fault detection, isolation, and identification, see for example [31]. The second view includes only fault isolation and identification, see for example [54]. Often the word *fault* is omitted so only the word *diagnosis* is used.

- **False Alarm**
  The event that an alarm is generated even though no faults are present.

- **Missed Alarm**
  The event that an alarm is *not* generated in spite of that a fault has occured. This event may also be denoted *missed detection*.

- **Active Diagnosis**
  When the diagnosis is performed by actively exciting the system so that possible faults are revealed.

- **Passive Diagnosis**
  When the diagnosis is performed by passively studying the system without affecting its operation.

### Comment on Terminology

The term *fault diagnosis* is in this text used to denote the whole chain of fault detection, isolation, and identification. This is in accordance with one of the views common in literature. Diagnosis used in this way also serves as a name for the whole area of everything that has to do with diagnosis. If fault detection is excluded from the term *diagnosis*, as in the second view, one gets a problem of finding a word describing the whole area. This can partly be solved by introducing the abbreviation FDI (Fault Detection and Isolation), which is common in papers taking the second view of the definition of the term diagnosis. As noted in some papers, FDI does not strictly contain fault identification. To solve this, also the abbreviation FDII (Fault Detection, Isolation, and Identification) has been used.

### General Structure

The general structure of an application including a diagnosis system is shown in Figure 6.2. Inputs to the diagnosis system are the signals $u(t)$ and $y(t)$, which are equal to or a superset of the control system signals. The plant is affected by faults and disturbances and the task of the diagnosis system is to generate a *fault decision* containing information about at least if a fault has occurred, when it occurred, and the location of the fault, i.e. in what component the fault occurred.

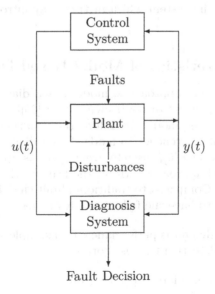

**Figure 6.2** General structure of a diagnosis application.

# 6.4    Model Based Diagnosis

One way to increase the performance of the engine diagnosis system is to increase the use of model based diagnosis in which process knowledge is utilized in the form of a mathematical process model. The diagnosis can to a larger extent be performed passively and over a wider operating range if a proper model is available. It is also possible to handle vehicle to vehicle variations and aging, or variations in environment from day to day, for example because of changes in atmospheric conditions.

An introduction to model based diagnosis of technical systems will be given in the following three sections, Sections 6.3-6.6.

## Traditional Methods

Traditionally diagnosis has been performed by mainly limit checking. When for example a sensor signal level leaves its normal range, an alarm is generated. The normal range is predefined by using thresholds. This normal range can be dependent on the operating conditions. In for example an aircraft, the thresholds, for different operating points defined by altitude and speed, can be stored in a table. This use of thresholds as functions of some other variables, can actually be viewed as a kind of model based diagnosis. In addition to checking signal levels, also trends of signals are often checked against thresholds.

Another traditional approach is duplication (or triplication or more) of hardware. This is called *hardware redundancy*. There are at least three problems associated with the use of hardware redundancy: hardware is expensive, it requires space, and adds weight to the system. In addition, extra components increase the complexity of the system which in turn may introduce extra diagnostic requirements.

## 6.4.1    Some Characteristics of Model Based Diagnosis

As an alternative to traditional approaches, model based diagnosis have shown to be useful either as a complement or on its own [93][31][26][94]. The model can be of any type, from logic based models to differential equations. Depending on the type of model, different approaches to model based diagnosis can be used, for example statistical approach [5], discrete event systems approach [110], AI-based approaches [102], and approaches within the framework of control theory (the focus in this chapter). Compared to traditional limit checking, model based diagnosis has the potential to have the following advantages:

- It can provide higher diagnosis performance, for example smaller faults can be detected and the detection time is shorter.

- It can be performed over a large operating range.

- It can be performed passively.

- Isolation of different faults becomes possible.

- Disturbances can be compensated for which implies that high diagnosis performance can be obtained in spite of the presence of disturbances.

Compared to hardware redundancy, model based diagnosis may be a better solution because of the following reasons:

- It is generally applicable to more kinds of components. Some hardware, such as the plant itself, can not be duplicated.

- No extra hardware is needed, which means for example that it is cheap in production.

It is sometimes believed that model based diagnosis requires much computing power. However, this is often not true which is seen if the diagnosis methods are studied more carefully [90]. Actually for the same level of performance it can be the case that model based diagnosis is *less* computationally intensive than traditional approaches.

The requirement of model based diagnosis is quite naturally the need for a reliable model and possibly a design procedure. Many times a model can be reused, e.g. from control design. Otherwise, it is likely that a major part of the work is spent on building the model. The accuracy of the model is a major factor of the performance of a model based diagnosis system. Compared to the area of model based control, it is more critical that the model is good since model based diagnosis systems operates in open loop. More often than in control, a linear model is not sufficient to provide satisfactory performance.

## 6.5 Faults

The previous chapters of this book have already provided models for the dynamic behavior of an engine, like e.g. the model of the air-intake dynamics in Section 3.2.6. However, in addition to these models of fault free behavior, models of the behavior in malfunction are needed and this leads to the area of fault modeling.

A plant can, as shown in Figure 6.3, be separated into three subsystems: actuators, the process, and sensors. Depending on in what subsystem a fault occurs, a fault is classified to be an *actuator fault*, *process fault*, or *sensor fault*. Process faults are sometimes also called *system faults* or *component faults*.

Typical sensor faults are short-cut or cut-off in connectors and wirings, and drifts, i. e. changes in gain or bias. Also the time response can degrade, i. e.

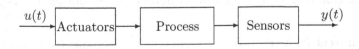

**Figure 6.3** General structure of a plant.

the bandwidth is decreased. Examples of process faults are increased friction, changed mass, leaks, components that get stuck or loose. Examples of faults in an actuator are short-cut or cut-off in connectors and wirings. If the actuator includes an electrical amplifier, there can also be gain and bias faults. Actuators can by them-self be relatively complex systems, containing for example DC-motors, controllers, and sensors. Therefore all examples of sensor and process faults are applicable also to actuators.

In a diagnosis application it may not be sufficient to isolate a faulty (larger) component, e. g. a DC-motor. Often more detailed knowledge is required about the fault, e. g. what part of the DC-motor is faulty. Thus when designing a diagnosis system it is important to have knowledge about what faults that can occur or are most common, and also how different faults affect the system. This is because specific diagnostic solutions are often required for each kind of fault. Such a knowledge can only be obtained from a domain expert and/or through extensive experiments.

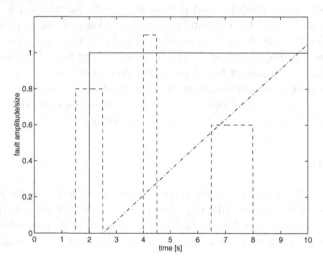

**Figure 6.4**  Different types of fault time-variant behavior.

Faults are often characterized by describing their time-variant behavior. Three basic types are illustrated by examples in Figure 6.4:

- Abrupt, step-faults (solid line) representing for example a component that suddenly brakes.

- Incipient (developing) faults (dash-dotted line) representing for example slow degradation of a component or developing calibration errors of a sensor.

- Intermittent fault (dashed line) caused by for example loose connectors.

These typical time-variant behaviors of faults are often used in test-cases and to describe the performance of diagnosis system. However in a real applications, the time-variant behavior of faults is usually more complex.

## 6.5.1 Fault Modeling

An important concept in model based diagnosis is fault modeling. The fault model is the formal representation of the knowledge of possible faults and how they influence the process. In general, better fault models implies better diagnosis performance, e. g. smaller faults can be detected and more different types of faults can be isolated.

Basically, faults can be modeled as fault signals or changes in parameters. The choice is dependent on what is most natural for a particular fault. When fault signals are used, a specific fault is usually modeled as a *scalar* fault signal. Generally all faults can be modeled as fault signals. Consider for example a fault represented by a change $\Delta G_f(s)$ in the transfer function $G(s)$:

$$y = (G(s) + \Delta G_f(s))u$$

If the fault signal $f$ is defined as $f = \Delta G_f(s)u$ then the fault can also be modeled as

$$y = G(s)u + f$$

The reverse is however not true since for example a bias fault cannot generally be modeled as a change in a parameter.

Two terms that are sometimes used are *additive faults* and *multiplicative faults*, e. g. in Section 5.1.6. Additive faults refers typically to faults that are modeled as fault signals that are added to some signal present in the system, e. g. actuator inputs, sensor outputs or state. Multiplicative faults is just another name for faults modeled as parameter changes.

---

**Example 6.1** *Consider a state-space model of a process:*

$$\dot{x} = Ax + Bu$$
$$y = Cx + Du$$

*Consider also an additive actuator fault $f_a$, a general process fault $f_p$, and an additive sensor fault $f_s$, all modeled as fault signals. Then the influence of the faults to the process can be generally described by the state-space description*

$$\dot{x} = Ax + Bu + Bf_a + Ef_p$$
$$y = Cx + Du + Df_a + Ff_p + f_s$$

---

# 6.6 Principles of Model Based Diagnosis

Having a model of the engine subsystem that is to be diagnosed and appropriate fault models, the next step is to use a design method to obtain the diagnosis system.

Many principles for constructing model based diagnosis systems exists but common to all is the use of *residuals* which are defined as

**Definition 1 [Residual].** *A residual $r(t)$ is a scalar or vector signal that is 0 or small in the fault free case and $\neq 0$ when a fault occurs.*

Model based diagnosis can be separated into two parts: *residual generation* and *residual evaluation*. The internal structure of a diagnosis system consists therefore of two subsystems as is illustrated in Figure 6.5. The purpose of the *residual generator* is to generate the residual signals and the purpose of the *residual evaluator* is to evaluate the residuals and generate a fault decision.

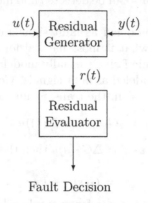

**Figure 6.5** Internal structure of a diagnosis system.

Diagnosis in signal space means that faults are modeled as fault signals. Residual generation is then based on *analytical redundancy* which can be formally defined as

**Definition 2 [Analytical redundancy].** *There exists analytical redundancy if there exists two or more, but not necessarily identical ways to determine a variable where at least one uses a mathematical process model in analytical form.*

A simple example of analytical redundancy is the case when it is possible to both measure an output and, by means of a model, also estimate it. This example is illustrated in Figure 6.6. From the measured and estimated output $y(t)$ and $\hat{y}(t)$, a residual can be formed as

$$r(t) = y(t) - \hat{y}(t)$$

The model used to estimate $\hat{y}(t)$ can be linear or non-linear. If a fault occurs, it will affect the measured output but not the estimated output. In this way the residual will deviate from zero and the fault is detected.

Analytical redundancy exists in two forms:

- *Static Redundancy*
  The instantaneous/static relationship amongst sensors outputs, and actuator inputs of the system. The special case of static redundancy between only outputs, is called *direct redundancy*.

- *Temporal Redundancy*
  The relationship amongst histories or derivatives of sensor outputs and actuator inputs. Equations describing temporal redundancy are generally differential or difference equations.

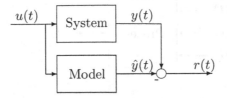

**Figure 6.6** Simple fault detection system

---

**Example 6.2**  *Consider the following system with 3 outputs and one input:*

$$y_1 = g(u)$$
$$y_2 = G(s)u$$
$$y_3 = 2g(u) - G(s)u$$

*For this system, there is for example a static redundancy between the input $u$ and the output $y_1$. In addition temporal redundancy exists between the input $u$ and the output $y_2$. Finally, by substituting the first two equations into the third, arriving at*

$$y_3 = 2y_1 - y_2 \tag{6.1}$$

*it can be seen that also direct redundancy is present.*
*Each of exemplified analytical redundancies can be used to design a residual generator. Residuals based on these redundancies can be formed as*

$$r_1 = y_1 - g(u)$$
$$r_2 = y_2 - G(s)u$$
$$r_3 = y_3 - 2y_1 + y_2$$

---

It should be noted that if the system contains dynamics but detection time is not crucial, analytical redundancy can often be approximated by static redundancy.

## 6.6.1  Residual Generator Design

Residual generation for diagnosis in signal space is treated. First a residual generator is defined:

**Definition 3 [Residual Generator].**  *A residual generator is a system that takes process inputs and outputs as inputs and generates a residual.*

The residual generator can be a static system if it is based on static redundancy, or a dynamic system if it is based on temporal redundancy. The problem of residual generation to achieve structured residuals can be described as follows.

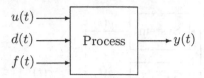

**Figure 6.7** The system with inputs $u$ (known or measurable), $d$ (disturbances and non-monitored faults), $f$ (monitored faults), and output $y$.

Consider the system illustrated in Figure 6.7. The system has three kinds of inputs: known or measurable inputs collected in the vector $u(t)$, disturbances and non-monitored faults in the vector $d(t)$, and the monitored faults in the vector $f(t)$. Now the problem of residual generation can be abstracted to, by only using knowledge of signals $u(t)$ and $y(t)$, generate a signal, i.e. the residual, which is not sensitive to any of the signals $u(t)$, $y(t)$, or $d(t)$, but highly sensitive to the signal $f(t)$.

### General Formulation

Generally a residual generator can be seen as a filter. In the linear case this can be written

$$r = H_y(s)y + H_u(s)u \tag{6.2}$$

where $y$ and $u$ can be scalars or vectors. Let $G(s)$ and $G_f(s)$ be transfer functions from the input $u$ and the faults $f$ respectively. Then the system output is

$$y = G(s)u + G_f(s)f$$

By substituting this expression into (6.2), we see that

$$r = \big(H_y(s)G(s) + H_u(s)\big)u + H_y(s)G_f(s)f = 0 \tag{6.3}$$

In the fault free case, the residual must, according to the definition, be zero and therefore it must hold that

$$\big(H_y(s)G(s) + H_u(s)\big)u = 0 \tag{6.4}$$

If this equation is going to hold for all $u$, then generally for all linear residual generators, the condition

$$H_y(s)G(s) + H_u(s) = 0 \tag{6.5}$$

must be satisfied. Note that $H_u$ can be zero, e.g. in the direct redundancy case, but it must always hold that $H_y \neq 0$.

The expression 6.2 is the residual expressed in inputs and outputs. Because the computation of the residual is based on this expression it is called the *computational form*. The computational form of the residual is sufficient to define the residual generator.

Because of Equation 6.4, it is seen in Equation 6.3 that it is possible to express the residual in only the faults, and when applicable also other signals such as disturbances. Thus, if $d$ is a signal containing all disturbances, the residual can be written as

$$r = G_{rf}(s)f + G_{rd}(s)d$$

This expression, i. e. the residual expressed in only faults and disturbances, is called the *internal form*. In the internal form it is seen directly that the residual is zero if no faults or disturbances are present.

---

**Example 6.3** *Consider again the system described in Example 6.1 (Section 6.5.1). Assume that it is possible to construct an ideal residual generator so that Equation 6.4 is fulfilled. Let the external form of the residual be written as Equation 6.2. Then without considering any other transfer functions than from the different faults $f_a$, $f_p$, and $f_s$, to the system output $y$, we know that an internal form of the residual is*

$$r = H_y(C(sI - A)^{-1}B + D)f_a + H_y(C(sI - A)^{-1}E + F)f_p + H_y f_s$$

---

**Observer formulation**

Above, transfer functions have been used to formulate a residual. An alternative formulation is to use an observer formulation. An example of this is, for example, if a system has two measured outputs $y_1, y_2$. Then an observer, based on only one of the outputs, say $y_2$, can be formulated as

$$\dot{\hat{x}}(t) = g(\hat{x}(t), u(t)) + K(y_2 - \hat{y}_2) \tag{6.6}$$
$$r(t) = y_2(t) - \hat{y}_2(t) \tag{6.7}$$

In this residual, $r(t)$, the first sensor is decoupled. An observer used for diagnosis is called a *diagnostic observer* and does not necessarily estimate any state.

**Additional Design Choices**

The residual represented by the two transfer functions $H_y$ and $H_u$ in Equation (6.2) can alternatively be written as

$$r(\sigma) = \frac{A_1(\sigma)y_1 + \ldots + A_m(\sigma)y_m + B_1(\sigma)u_1 + \ldots + B_k(\sigma)u_k}{C(\sigma)} \tag{6.8}$$

where $m$ is the number of outputs, $k$ the number of inputs, and $A_i(\sigma)$, $B_j(\sigma)$, and $C(\sigma)$, are polynomials in $\sigma$, which represents the differentiation operator $p$ in the continuous case and the time-shift operator $q$ in the discrete case.

The objective of residual generation namely

- The transfer functions from monitored faults to the residual must be non-zero.

- The transfer functions from all other signals to the residual must be zero.

is fulfilled by satisfying Equation 6.5. However, there are additional design choices to be made. The two above requirements, also expressed by Equation 6.5, only introduces a constraint on the numerator polynomial of Equation 6.8. The only constraint on the denominator polynomial $C$ is that the residual generator must be realizable. This means that it must be of an order greater or equal to the largest order of the numerator polynomials $A_i(\sigma)$ and $B_j(\sigma)$. Also it should have its poles placed so that the residual generator becomes stable, and it may further be desirable to place the poles so that not only stability is achieved but also well behaved filtering. It may also be remarked that sometimes, because of properties of the system, the second requirement of perfect decoupling must be relaxed so that only approximate decoupling of disturbances and non-monitored faults are required.

### 6.6.2   Residual Evaluation

The purpose of the residual evaluator is to generate a *fault decision* by processing the residuals. A fault decision is the result of all the tasks fault detection, isolation, and identification. This means that the fault decision contains information about if a fault has occurred and in that case also the following information is contained: when did the fault occur, which is the faulty component, and possibly also the size of the fault. In the presence of disturbances caused by modeling errors and measurement noise, residual evaluation is a non-trivial task.

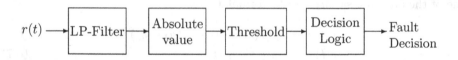

**Figure 6.8**  A simple residual evaluation scheme.

Residual evaluation is essentially to check if the residual is responding to a fault. If it does, we say that the residual *fires*. The residual evaluation can in its simplest form be a thresholding of the residual, i. e. a fault is assumed present if $|r(t)| > J_{th}$ where $J_{th}$ is the threshold. A simple way is to chose the threshold $J_{th}$ a bit above the noise level.

Another straight forward method is used in the application example as depicted in Figure 6.13. The residual is low-pass filtered and then the absolute value is thresholded. For the generation of a fault decision signal containing information about what fault that has occurred, it is sufficient to have a simple logic scheme that generates an alarm when the residuals matches one of the fault columns in the residual structure.

#### Systematic Threshold Selection

Two important properties of a diagnosis system are probability of false alarm and probability of missed detection. If thresholds are selected and the probability distribution of a residual is known, then it is possible to calculate these probabilities.

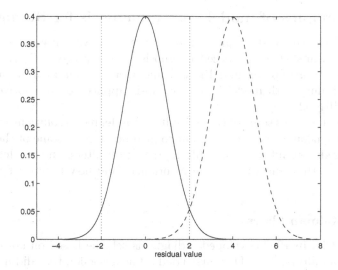

**Figure 6.9** An example of probability density functions for a residual in the fault-free case (solid) and when a constant fault is present (dashed).

Consider for example a residual that is affected by additive Gaussian noise $n(t)$. The internal form becomes

$$r(t) = G_f(s)f(t) + n(t)$$

where $G_f(s)$ is the transfer function from fault to residual. In Figure 6.9, the probability density function of such a residual (after filtering) is shown for the fault free case (solid) and the case when a fault is present (dashed). The fault is assumed to be constant, i.e. $f(t) \equiv c$, and therefore the density function is centered around $G_f(0)c = 4$. The threshold has the value $J_{th} = 2$ and is represented by the dotted lines. There are two lines because it is the absolute value of the residual that is thresholded. By integrating the area below the solid line, to the left and to the right of the threshold, it is possible to calculate the probability of false alarm. Similarly the probability of missed detection can be calculated by integrating the area below the dashed line between the thresholds.

The probabilities of false alarm and missed detection, calculated in this way, are valid for a diagnosis system with only one scalar residual. If the diagnosis system contains several residuals, the calculations become more complicated. The problem quickly becomes intractable as the number of residuals grow. More important, the probability distribution of the residuals are usually not known, which means that they must be estimated.

Instead of simple thresholding it is also possible to use statistical tests for abrupt changes, such as the *Generalized Likelihood Ratio* (GLR) test. For a comprehensive discussion of this topic, see [5]. Another approach is *fuzzy thresholding*, see [27], which is a technique in which the thresholds are made fuzzy and the fault decision is calculated using fuzzy rules. In this way the fault decision will contain information about the certainty that a fault has occurred.

### 6.6.3    Examples of model based diagnosis for SI-engines

In the late 80's, the first results on attempts to solve automotive diagnosis problems by means of model based diagnosis were reported. Early work was published by the groups around Rizzoni and Gertler, but many other groups were active dealing with model based diagnosis applied to automotive engines, [117][83][80][104][52].

This section reviews some of these results in model based solutions to automotive engine diagnosis, with the intention to give pointers to some of the relevant techniques used: extended Kalman filters, detection filters, and residuals. Some of the problems they encountered are mentioned, and how they went around to solve them.

#### Extended Kalman filters

Paolella and Cho in 1990 developed a diagnosis scheme based on non-linear extended Kalman filters, [18]. They first tried a linear version that did not provide satisfactory performance. The non-linear scheme is applied to an automotive powertrain to diagnose faults in engine speed, transmission, and wheel speed sensors, which are also the inputs to the diagnosis system. The scheme is validated in simulations, and in 1991 [92], they test their approach in a real vehicle and they are able to diagnose 100% faults, i. e. the case when sensor outputs are set to zero.

#### Detection Filter

Partly supported by Ford, 1989-1991 Rizzoni [106][107] used an extended version of the detection filter derived from a fourth order linear state-space engine model. The diagnosis system measures throttle angle, manifold pressure and engine speed to produce two residuals. The scheme is shown to be able to diagnose 10% faults in throttle angle sensor and manifold pressure sensor on a real Ford 3.0 liter engine. The use of a linear model restricts the operating range of the model, and data is shown for 56-60 kPa and 1050-1130 rpm. It seems like the residuals are sensitive to engine transients.

In 1993, they used a non-linear model to generate five residuals based on parity equations [105]. Throttle angle, engine speed, manifold pressure and injected fuel is measured to diagnose faults on a real 4 cylinder 1.3 liter engine. The diagnosed components are throttle angle sensor, engine speed sensor, manifold pressure sensor and fuel injectors. The load is decoupled. Plots of the residuals are shown for the case of 10% faults of the diagnosed components.

Later, in 1994 [71][70][77] they used a non-linear discrete NARMAX model being a linear combination of second order polynomials, i. e. linear in the parameters. Inputs to the diagnosis system are demanded throttle angle, injected fuel and measured air mass flow and engine speed. Forward and inverse models are used to generate four corresponding parity equation residuals where the load is decoupled. The scheme is tested on a real Ford 3.0 liter engine over a standardized test schedule and is reported to be able to detect 10% faults in the air mass flow sensor, 20% faults in the engine speed sensor, 15% faults in the throttle actuator and 40% faults in the fuel injector.

**Residuals**

Gertler and his group has been involved in projects with GM. During 1991-1993 [35][34] a simulation study was done. After concluding that a linear model is not sufficient because of its limited operating range, they used a hybrid model with linear core. Five residuals in the form of parity equations are used to diagnose faults in throttle angle sensor, EGR-valve, fuel injectors, manifold pressure sensor, engine speed sensor and lambda sensor. These are also the inputs to the diagnosis system. The residual structure is able to distinguish all faults except fuel injector and lambda sensor faults. A potential problem was that the lambda sensor characteristic was assumed known.

In 1994-1995 they tried this scheme on a real 3.1 liter V6 engine in a production vehicle [32][33]. Both off-line and on-line versions were developed and they stress the importance of simple algorithms because of limited on-board computing power. Instead of using five linear parity equations, they now use six non-linear parity equations as residuals. The same faults as before are diagnosed and the new residual structure provides isolation between all faults. To increase robustness, the residuals are low-pass filtered and threshold crossings are counted. They report that they are able to diagnose faults of 10% size.

# 6.7 Application Example - Air Intake System

As an application example to illustrate the general principles, a model based diagnosis system will be developed for the air intake system of an SI-engine. The air intake system contains a number of components and sensors related by reasonably well understood thermodynamics and physics of gases. The diagnosis system is based on a non-linear semi-physical model and uses a combination of different residual generation methods. Note that many engine variables need not to be considered because of subsystem decoupling. Such variables are for example road load, lubrication, friction, wear, ignition timing, fuel quality, lambda, cooling water and oil temperatures. These variables are present in the other subsystems. Variables that however may affect the air intake system are ambient air temperature, pressure, and humidity. Finally the air intake system does not contain any large disturbances that must be taken into account when designing the diagnosis system.

The undertaking is thus to design a diagnosis system using modeling, residual generation and evaluation, and to investigate overall performance and limiting factors. In Sections 6.7.1 and 6.7.2, a model of the air intake system of a production SAAB 2.3 liter engine is built. From this model, a diagnosis system capable of diagnosing faults in throttle actuator, throttle sensor, air mass flow sensor, and manifold pressure sensor, is constructed in Sections 6.7.3, 6.7.4, 6.7.5, and 6.7.6. The scheme is experimentally validated in Section 6.7.7 on a real production engine.

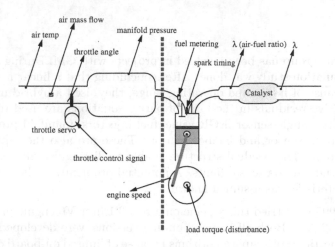

**Figure 6.10** The basic SI engine. The air intake, which is the part left of the dashed line, is to be diagnosed for faults.

## 6.7.1 Modeling the Air Intake System

The engine is a 2.3 liter 4 cylinder SAAB production engine. The measured variables are the same as the ones used for engine control. A schematic picture of the engine is shown in Figure 6.10. The engine has electronic throttle control (drive-by-wire), which is a DC-servo controlled by a PID controller. The air intake system is everything to the left of the dashed line. Also the engine speed must be taken into account because it affects the amount of air that is drawn into the engine. The SI-engine is a non-linear plant and it has been indicated in a pre-study that diagnosis based on a linear model is not sufficient for diagnosing an engine. This has also been concluded by other authors [35][71]. This motivates the choice of a non-linear model.

The model of the air intake system is continuous and consists of two parts, the throttle model and the air dynamics. The model variables and their units are summarized and explained in Table 6.1. The notation is almost the same as in previous chapters but with some minor deviations since that makes it easier to explain the identification actually done. The throttle dynamics is modeled using a standard model for an electrical servo as a second order linear system in which the states are the throttle angular speed and the throttle angle. In addition to that the load torque, $T_{th}$, generated by the air flow on the throttle plate is explicitly modeled using a map, $h_{th}$. The load torque model would typically not be needed for control design, since one would normally instead make a design robust to this load. However, for diagnosis purposes it is needed. The equations describing the throttle are

$$\dot{\omega}_{th} = a\omega_{th} + b(u(t) - T_{th}) \tag{6.9}$$

$$\dot{\alpha}_{th} = \omega \tag{6.10}$$

$$T_{th} = h_{th}(p_m, \dot{m}_{a,in}, \alpha_t) \tag{6.11}$$

**Table 6.1** Model variables and units.

| | |
|---|---|
| $u$ [V] | the output from the throttle controller |
| $p_m$ [kPa] | manifold pressure |
| $R$ [J/(g K)] | the gas constant |
| $\vartheta_m$ [K] | manifold air temperature which is assumed to be equal to the ambient temperature |
| $V_m$ [m³] | manifold volume |
| $\dot{m}_{a,in}$ [kg/s] | air mass flow into the manifold assumed equal to the air flow past the air mass flow meter |
| $\dot{m}_{ac}$ [kg/s] | air mass flow out from the manifold |
| $f$ | static function describing the flow past the throttle |
| $g$ | static function describing the flow into the cylinders (volumetric efficiency) |
| $\alpha_t$ [deg] | throttle angle |
| $T_{th}$ | normalized torque on the throttle plate generated by the air flow past the throttle |
| $h_{th}$ | static function describing the torque on the throttle plate generated by the air flow past the throttle |
| $n$ [rpm] | engine speed |

In the air system application there is no need for extremely fast fault detection, and therefore a mean value model [43] as derived in Section 3.2.6 is chosen, which means that no within cycle variations are covered by the model. Further, even if variations in ambient pressure and temperature do affect the system, they are here assumed to be slowly varying. The fundamental model used is Equation 3.50. The time constant from Equation 3.51, $\tau = \frac{R\vartheta_m}{V_m}$, is written out for identification purposes explained below. The air mass out of the intake into the cylinder, $\dot{m}_{ac}$, is modeled as in Equation 3.49 with a map $f_1 = g(p_m, n)$. The air flow passing the throttle into the intake, $\dot{m}_{a,in}$, is also modeled as a map as in Equation 3.37 but parameterized in pressure and throttle angle as indicated in the paragraph before that equation. The resulting model is

$$\dot{p}_m = \frac{R\vartheta_m}{V_m}(\dot{m}_{a,in} - \dot{m}_{ac}) \tag{6.12}$$

$$\dot{m}_{a,in} = f(p_m, \alpha_t) \tag{6.13}$$

$$\dot{m}_{ac} = g(p_m, n) \tag{6.14}$$

The air dynamics model thus has one state which is the manifold pressure, $\dot{p}_m$. The static funciton $g(p_m, n)$ is equivalent to $f_1(n, p_m)$ in Section 3.2.6.

So far faults have not been considered, so collecting the model equations describes the complete fault free model as

$$\dot{\omega}_{th} = a\omega_{th} + b(u(t) - T_{th}) \tag{6.15}$$

$$\dot{\alpha}_{th} = \omega_{th} \tag{6.16}$$

$$T_{th} = h(p_m, \dot{m}_{a,in}, \alpha_t) \tag{6.17}$$

$$\dot{p}_m = \frac{R\vartheta_m}{V_m}(\dot{m}_{a,in} - \dot{m}_{ac}) \tag{6.18}$$

$$\dot{m}_{a,in} = f(p_m, \alpha_t) \tag{6.19}$$

$$\dot{m}_{ac} = g(p_m, n) \tag{6.20}$$

The final modeling step is now to introduce the faults, and to define available information in terms of inputs and outputs. The process inputs for the diagnosis model are the throttle control signal $u$, and the engine speed $n$. The outputs are throttle angle sensor signal $\alpha_{t,s}$, mass air flow sensor signal $\dot{m}_{a,in,s}$, and manifold pressure sensor signal $p_{m,s}$. The faults are modeled as additive faults. The final model is shown in Figure 6.11.

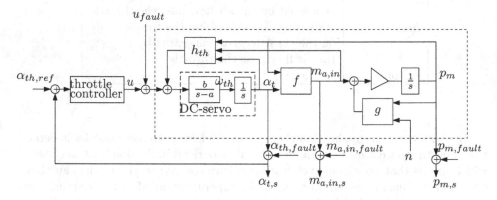

**Figure 6.11** The model of the air intake system including the throttle controller and monitored faults.

## 6.7.2   Model Identification

To find the static functions $f$, $g$ and $h_{th}$, a steady state experiment was performed. The engine is mounted in a test bench together with a Schenk AC dynamometer, and 12 equally spaced engine speeds (1500 to 3000 rpm) and 8 equally spaced manifold pressures (35 to 60 kPa) were used to build a map of $u$, $p_m$, $\dot{m}_{a,in}$, $\alpha_t$ and $n$ measurements in 96 different steady state operating points. To represent the static functions, there is a choice between interpolating in the map or fit for example polynomials to the map. Here interpolation was chosen for $g$ and polynomials for $f$ and $h_{th}$. The choice of terms to be included in the polynomials was guided by studying the correlation coefficient and the final choice was based on a validation against another data set of 96 operating points.

The resulting polynomials used are

$$f(p_m, \alpha_t) = \eta_0 + \eta_1 \alpha_t + \eta_2 \alpha_t^2 + \eta_3 \alpha_t^3 +$$
$$+ \eta_4 \alpha_t p_m + \eta_5 p_m$$
$$h_{th}(p_m, \dot{m}_{a,in}, \alpha_t) = \nu_0 + \sqrt{p_m} + \nu_1 p_m + \nu_2 p_m^2 +$$
$$+ \nu_3 p_m \sqrt{\alpha_t} + \nu_4 p_m^2 \alpha_t +$$
$$+ \nu_5 p_m \alpha_t + \nu_6 p_m \sqrt{\dot{m}_{a,in}}$$

For the air dynamics, the only remaining constant to identify is $V_m$. To identify this constant, a dynamic test was performed. A test-cycle consisting of 4 throttle steps in different operating points was constructed and the already identified functions, $f$ and $g$, were used to find the value of $V_m$ that gave the best fit between measured pressure and estimated pressure. A comparison between measured and simulated pressure for the final air dynamic model is shown in Figure 6.12.

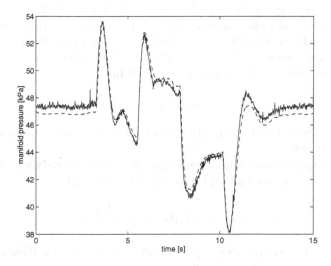

**Figure 6.12** Measured (solid line) and simulated (dashed line) pressure.

The parameters, $a$ and $b$, used in the throttle model, were identified by applying a pseudorandom binary signal to the throttle input and measuring the throttle angle. This was done when the engine was not running so the term $T_{th}$ in Equation 6.15 was zero. The throttle is marginally stable in open loop, so this model is not as easily validated as the air dynamics. Closed loop experiments were performed validating the model, showing that the model is acceptable but less accurate than the air dynamics model. This is later reflected in diagnosis performance where 20 % throttle faults are comparable to 10 % faults in other components as can be seen in Section 6.7.7.

### 6.7.3    The Diagnosis System

The inputs to the diagnosis system are $\dot{m}_{a,in,s}$, $u$, $\alpha_{t,s}$, $p_{m,s}$, and $n$. The diagnosed components are the throttle actuator, throttle angle sensor, air mass flow meter, and manifold pressure sensor. This means that one actuator fault and three sensor faults are considered. It is assumed that only one fault can occur at the same time. The engine speed sensor is not diagnosed because it is diagnosed sufficiently well by existing diagnosing techniques. This is because the speed is measured by counting pulses form a magnetic sensor on a toothed wheel. A magnetic sensor failure is detected directly from its distortion effects on the pulse train. Nevertheless, the residual structure (see Table 6.2) of the diagnosis system has the property that also the engine speed sensor could be included among the diagnosed components.

### 6.7.4    Residual Generation

By using static relationships. i. e. direct redundancy, and diagnostic observers estimating measurable signals, it is possible to construct a large number of residuals. However, many of them are based on the same part of the model that makes them redundant. It is thus desirable to find a small, non-redundant set of residuals with good isolation properties, without loosing robustness. Starting with 18 such residuals only six were selected to be used in the residual generator. The computational form of these six residuals are described next.

The first residual is a direct redundancy residual:

$$r_1 = \dot{m}_{a,in,s} - f(p_{m,s}, \alpha_{t,s})$$

This residual relies on a static relationship in the model and checks the consistency of Equation 6.19.

The other five residuals are derived using temporal redundancy. The second residual checks the consistency of Equations 6.18 and 6.20:

$$r_2 = \dot{m}_{a,in,s} - g(p_{m,s}, n) - \frac{V_m}{R\vartheta_m}\hat{\dot{p}}_m$$

It is assumed that the derivative of $p_m$ can be estimated with sufficient accuracy. If the derivative is computed as a difference, $r_2$ is a parity equation. Note that the static functions (here $g$) do not need analytical expressions, they can be represented by maps as usual.

Next are two dedicated observer residuals. The point of dedicated observer residuals is to make the residuals sensitive to only one sensor fault by measuring only one output ($\alpha_{t,s}$ is not considered to be an output in this case). This can be seen in $r_3$, which measures only $p_m$, and $r_4$, which measures only $\dot{m}_{a,in}$. The residuals are formed as

$$\hat{\dot{p}}_{m,3} = \frac{R\vartheta_m}{V_m}\big(f(\hat{p}_{m,3}, \alpha_{t,s}) - g(\hat{p}_{m,3}, n) +$$
$$K_1(p_{m,s} - \hat{p}_{m,3})\big)$$
$$r_3 = p_{m,s} - \hat{p}_{m,3}$$

$$\dot{\hat{p}}_{m,4} = \frac{R\vartheta_m}{V_m}\left(f(\hat{p}_{m,4}, \alpha_{t,s}) - g(\hat{p}_{m,4}, n) + \right.$$
$$\left. K_2(\dot{m}_{a,in,s} - \dot{\hat{m}}_{a,in,4})\right)$$
$$\dot{\hat{m}}_{a,in,4} = f(\hat{p}_{m,4}, \alpha_{t,s})$$
$$r_4 = \dot{m}_{a,in,s} - \dot{\hat{m}}_{a,in,4}$$

In $r_3$, the manifold pressure, $p_m$, is estimated by means of a non-linear diagnostic observer. The pressure $p_m$ is measured and the estimation error is fed back into the observer. The residual $r_3$ equals the estimation error. The fourth residual, $r_4$, is constructed similarly but the estimation error of $\dot{m}_{a,in}$, that equals $r_4$, is fed back into the observer.

The fifth residual, $r_5$, is also based on an observer but in contrast to $r_3$ and $r_4$, the estimation error does not equal $r_5$. This means that to compute $r_5$, both $p_m$ and $\dot{m}_{a,in}$ are measured so $r_5$ is not a dedicated observer residual and therefore becomes sensitive to both manifold pressure sensor and air mass flow sensor faults. The residual is formed as

$$\dot{\hat{p}}_{m,5} = \frac{R\vartheta_m}{V_m}\left(f(\hat{p}_{mn,5}, \alpha_{t,s}) - g(\hat{p}_{m,5}, n) + \right.$$
$$\left. K(\dot{m}_{a,in,s} - \dot{\hat{m}}_{a,in,5})\right)$$
$$\dot{\hat{m}}_{a,in,5} = f(\hat{p}_{m,5}, \alpha_{t,s})$$
$$r_5 = p_{m,s} - \hat{p}_{m,5}$$

The remaining residual is observer based as well, and its purpose is strictly for diagnosis of the throttle actuator:

$$\dot{\hat{\omega}}_{th} = a\hat{\omega}_{th} + b\left(u(t) - h(p_{m,s}, \dot{m}_{a,in,s}, \alpha_{t,s})\right) + k_1(\alpha_{t,s} - \hat{\alpha}_t)$$
$$\dot{\hat{\alpha}}_t = \hat{\omega}_{th} + k_2(\alpha_{t,s} - \hat{\alpha}_t)$$
$$r_6 = \alpha_{t,s} - \hat{\alpha}_t$$

The choice of the observer gains $k_1$, $k_2$ and also $K_1$ and $K_2$ were done by pole placement. The observer poles are functioning like a low-pass filter so the pole placement is a compromise between fast fault response and sensitivity to disturbances and noise.

The six residuals form a set of structured residuals. Different residuals are sensitive to different faults. This can be seen by studying the equations of the residuals and this is summarized in Table 6.2. Some residuals should be sensitive to specific faults due to the model equations, but it turned out in the experiments that the effect of a fault on some residuals was not significant enough to guarantee a robust diagnosis system. These cases are marked with an X (don't care) in Table 6.2. In one case, marked $X_1$, this has a straightforward physical explanation: the air flow past the throttle is not dependent on manifold pressure for supersonic air speeds [46]. This is the case for manifold pressures below 50 kPa. The throttle model has a larger model uncertainty, compared to the other parts of the model, so that if any of the other residuals were fed by $u$, then it would degrade the performance of these residuals. That is the reason for choosing to have the five zeros in the $u$ column. In practice this can have the effect that a small $\alpha_{t,s}$, $\dot{m}_{a,in,s}$ or $p_{m,s}$-fault is interpreted as a $u$-fault.

**Table 6.2** The residual structure of the diagnosis system. Each column represents a faulty signal.

|       | $u$ | $\alpha_{t,s}$ | $\dot{m}_{air,s}$ | $p_{man,s}$ | $n$ |
|-------|-----|----------------|-------------------|-------------|-----|
| $r_1$ | 0   | 1              | 1                 | $X_1$       | 0   |
| $r_2$ | 0   | 0              | 1                 | 1           | 1   |
| $r_3$ | 0   | 1              | 0                 | 1           | 1   |
| $r_4$ | 0   | 1              | 1                 | 0           | 1   |
| $r_5$ | 0   | 1              | 1                 | X           | 1   |
| $r_6$ | 1   | X              | X                 | X           | 0   |

## 6.7.5   Residual Evaluation

The six residuals should respond in accordance with Table 6.2 if a fault occurs, and take the value zero in the fault free case. The purpose of the residual evaluation is to check this and generate a fault decision. Taking disturbances into account is not trivial. The X:s in Table 6.2 are important. For a specific fault, residuals marked with an X have an uncertain response. This means that the fault decision must not depend on these residuals.

The residual evaluation scheme chosen here is shown in Figure 6.13. First, the residuals are low-pass filtered with 0.8 Hz cut-off frequency and then normalized with their standard deviations in the fault free case. The last step is fuzzy thresholding [27]. The fuzzy thresholding is chosen before regular thresholding to make the output from the diagnosis system contain more information. With fuzzy thresholding the output is not just a fault or not a fault, but a relative degree of how probable it is that a fault has occurred.

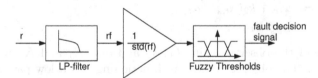

**Figure 6.13** Residual evaluation.

Three fuzzy sets are used and they are shown in Figure 6.14. The way they are chosen, only the constant $\gamma$ has to be determined. The criterion used to determine $\gamma$, is that in the fault free case, there should be no false alarm. Also in the case of a fault, all residuals that should respond due to Table 6.2, must reach the fuzzy *high* value and residuals that should not respond must be fuzzy *low*. Here, a value of $\gamma = 4$ proved experimentally to be good choice. Since the residual is normalized with its standard deviation, $\gamma = 4$ corresponds to a value of the residual that is four times its noise standard deviation.

Table 6.2 is transferred to fuzzy rules and an example of this is the $p_{man,s}$-column that corresponds to the fuzzy rule

IF $r_2$ is *high* AND $r_3$ is *high* AND $r_4$ is *low*

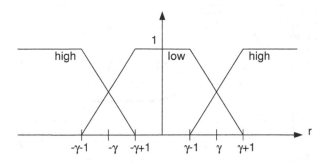

**Figure 6.14** The fuzzy sets used in the fuzzy thresholding.

THEN *pressure sensor fault*

The fault decision is then computed by executing all the rules, one for each fault. The connective AND is implemented as multiplication. Neither fuzzy implication nor defuzzification is needed because the result of each fuzzy rule is a crisp value, not a membership function. Further, note that no aggregation between the rules is needed because each rule is a fuzzy system on its own and therefore executed independently of the other fuzzy rules. The fault decision output of the diagnosis system is then four fault decision signals, one for each fault, representing how probable it is that the particular fault has occurred.

## 6.7.6 Implementation

The diagnosis system was implemented in Matlab Simulink. However all measurements were made on the real engine and sampled at 240 Hz with 40 Hz first order anti-alias filters. The sampling frequency 240 Hz is probably unnecessary high compared to what is needed to satisfy OBDII. The diagnosis was then performed off-line.

An important issue is how much computing power that is needed to compute the residuals. By studying the equations of the residuals and assuming that the static functions are implemented by maps, it can be concluded that there is about one multiplication, some table look-ups, and additions in each residual. This is not a problem in an on-board implementation in production vehicles.

## 6.7.7 Validation of the Diagnosis System

To validate the diagnosis system, experiments were performed on a single production engine in a test cell. A short test-cycle was constructed to represent realistic driving and cover a large part of the range of the model. The test-cycle lasts for 60 seconds and is shown in Figure 6.15.

All faults are simulated by adding a pulse to the corresponding signal in accordance with Figure 6.11. The pulse starts after 10 seconds and lasts for 40 seconds. It has an amplitude of about 10% of the signal mean for all cases except for the throttle actuator fault where 20% is used. The actuator fault is

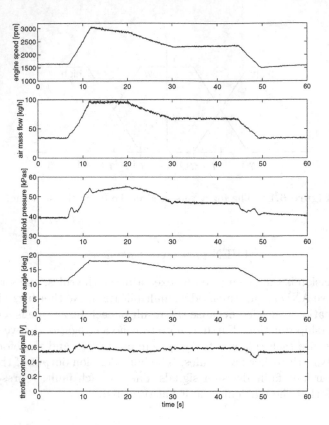

**Figure 6.15** The 60 seconds test-cycle.

larger because of the comparably larger model uncertainty in the throttle model. The throttle actuator fault is simulated by adding a pulse to the output of the throttle controller. Due to the closed loop throttle control, the throttle angle returns to its reference value shortly after the start of the pulse. Although the operation of the throttle is only marginally affected by the fault, the diagnosis is not degraded as can be seen in Figure 6.17. The sensor faults are simulated by adding a constant to the measured value. How the faults are introduced in the simulations is illustrated in Figure 6.11.

**Validation Plots**

In Figure 6.16 to 6.20, the responses (the absolute values) of the six residuals are shown for no fault, $u$-fault, $\alpha_{t,s}$-fault, $\dot{m}_{air,s}$-fault, and $p_{man,s}$-fault respectively. The residuals have been filtered and normalized so it is really the input to the fuzzy thresholding that is shown. Only the range from 0 to 8 is shown in the plots. Therefore residuals values higher then 8 is not visible. It can be seen that the residuals are noisy and non zero even in the case of no fault. The reason for this is the model error, which is the dominating factor that limits the performance of the diagnosis system.

In each plot, the two horizontal solid lines represent the value of $\gamma \pm 1$ in the membership functions (see Figure 6.14). The dashed line is the crossing between the membership functions. Also shown in the figures (in the 7:th plot) is the fault decision signal from the diagnosis system. Only the fault decision signal corresponding to the simulated fault is visible, because the other fault decision signals are constantly zero. The dashed line in this plot is the 0.5 level. This can be used as a criterion when to trigger the fault alarm. If this criterion is used, it can be seen that for these 10% and 20% faults the response of the diagnosis system is distinct.

All faults are detected and correctly isolated. Also there are no false alarms. Even if in several cases, one or two residuals incorrectly indicates false, there is no erroneous diagnosis because the residual structures dictated by Table 6.2 do not coincide. In this way, a kind of robustness against false alarms is achieved by means of the residual structure. In some cases a residual that should react, is not very high. This can be seen in for example Figure 6.17 and 6.19 where a residual reaches the fuzzy threshold area, which results in a weaker fault decision signal.

Going to vehicle implementation one has to consider robustness with respect to vehicle to vehicle variations, environmental changes and aging. Regarding this issue Gertler [32] reports that atmospheric variables, as ambient temperature, pressure and humidity, affects the residuals at a maximum by 10% to 15%.

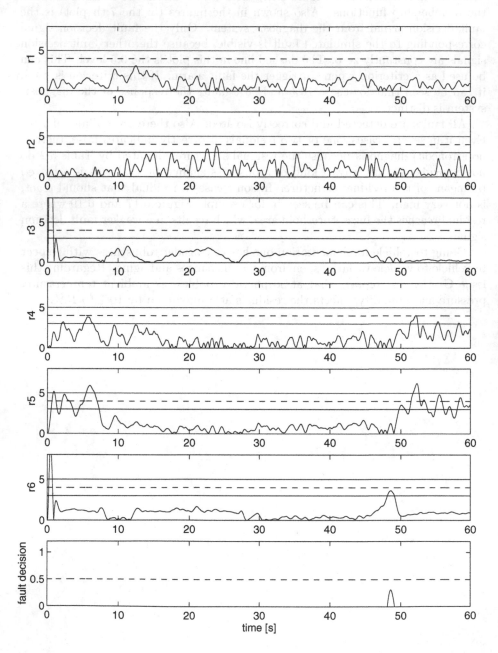

**Figure 6.16** The residuals for no fault.

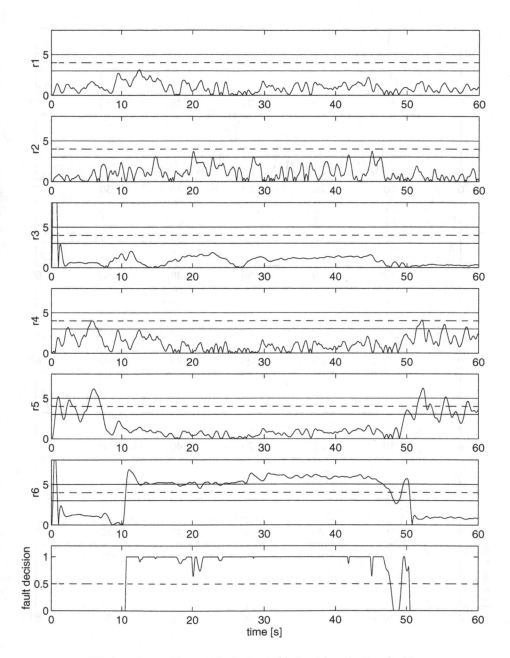

**Figure 6.17** The residuals for 20% throttle actuator fault.

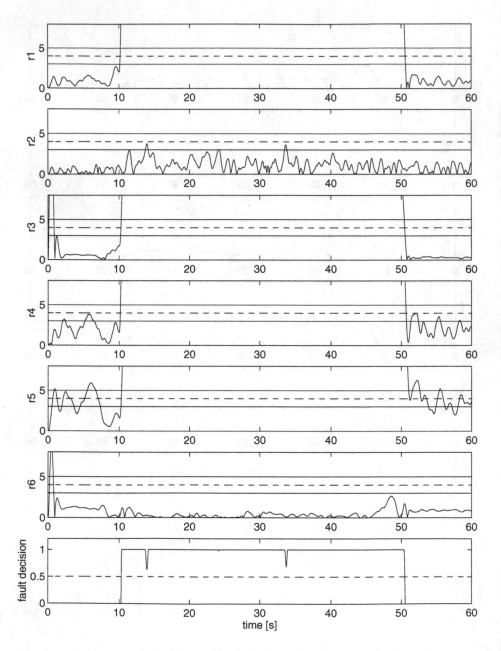

**Figure 6.18** The residuals for 10% throttle sensor fault.

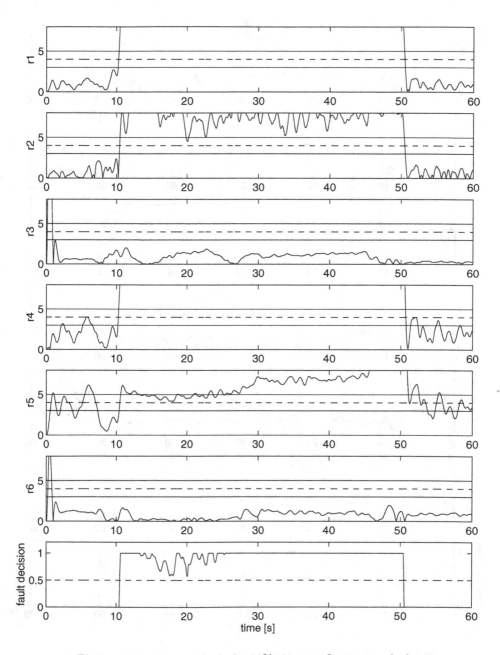

**Figure 6.19** The residuals for 10% air mass flow sensor fault.

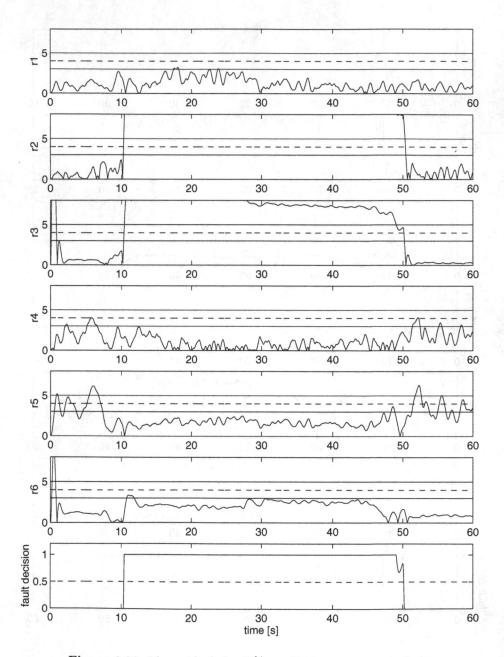

**Figure 6.20** The residuals for 10% manifold pressure sensor fault.

## 6.8 Misfire Detection

### 6.8.1 Crankshaft Moment of Inertia

For diagnostic purposes, the correct combustion must be monitored during engine operation. In case of misfires, unburnt gases would be generated. The engine would then no longer meet legal limits for noxious emissions. Therefore, the combustion torque shall be estimated from the crankshaft speed of the engine, in order to detect such misfires.

The torque balance at the crankshaft is given by

$$T_{comb} - T_{mass} - T_{load} - T_{fric} = 0 \quad . \tag{6.21}$$

In Section 3.1.1 the combustion torque $T_{comb}$ was obtained (see Equation 3.8) as

$$T_{comb} = \sum_{j=1}^{CYL} (p_j(\alpha_{CS}) - p_0) A_p \cdot \frac{ds_j(\alpha_{CS})}{d\alpha_{CS}} \quad .$$

The mass torque shall be derived from the kinetic energy $E_{mass}$ of the engine masses in motion.

$$E_{mass} = \int_0^{2\pi} T_{mass} d\alpha_{CS} = \frac{1}{2} J \dot{\alpha}_{CS}^2 \quad . \tag{6.22}$$

The mass torque $T_{mass}$ is then the derivative

$$
\begin{aligned}
T_{mass} &= \frac{dE_{mass}}{d\alpha_{CS}} = \frac{1}{2} \left( \frac{dJ}{d\alpha_{CS}} \dot{\alpha}_{CS}^2 + J \frac{d}{d\alpha_{CS}} \left( \dot{\alpha}_{CS}^2 \right) \right) \\
&= \frac{1}{2} \left( \frac{dJ}{d\alpha_{CS}} \dot{\alpha}_{CS}^2 + J \frac{d}{dt} \left( \dot{\alpha}_{CS}^2 \right) \cdot \frac{1}{d\alpha_{CS}/dt} \right) \\
&= J \ddot{\alpha}_{CS} + \frac{1}{2} \frac{dJ}{d\alpha_{CS}} \dot{\alpha}_{CS}^2
\end{aligned}
\tag{6.23}
$$

The first term represents the rotational masses, the second term the oscillating ones.

In [23] a two-mass approach is presented as a model of the connecting rod. The overall rod mass $m_{rod}$ is separated into

- an oscillating portion

$$m_{rod,osc} = m_{rod} \cdot \frac{l_{osc}}{l} \quad , \tag{6.24}$$

- and a rotational portion

$$m_{rod,rot} = m_{rod} \cdot \frac{l_{rot}}{l} \quad . \tag{6.25}$$

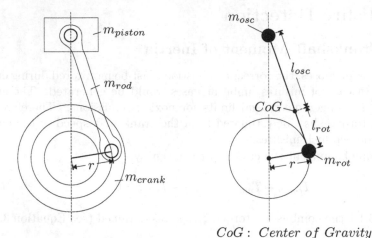

CoG :  Center of Gravity

**Figure 6.21**  Two-mass model for oscillating and rotating masses.

The two lengths $l_{osc}$ and $l_{rot}$ with

$$l_{osc} + l_{rot} = l$$

are defined by the location of the center of gravity of the connecting rod. Thus the oscillating mass at each cylinder is

$$m_{osc} = m_{piston} + m_{rod} \cdot \frac{l_{osc}}{l} \quad , \qquad (6.26)$$

and the rotational mass of the crankshaft portion at one cylinder

$$\frac{m_{rot}}{CYL} = \frac{m_{crank}}{CYL} + m_{rod} \cdot \frac{l_{rot}}{l} \quad . \qquad (6.27)$$

The crankshaft mass is deducted from the moment of inertia

$$m_{crank} = \frac{J_{crank}}{r^2} \quad . \qquad (6.28)$$

The kinetic energy of the engine masses in motion shall now be calculated.

$$E_{mass} = \frac{1}{2} \frac{m_{rot}}{CYL} \sum_{j=1}^{CYL} v_{rot,j}^2 + \frac{1}{2} m_{osc} \sum_{j=1}^{CYL} v_{osc,j}^2 \qquad (6.29)$$

The speed of the oscillating mass $v_{osc,j}$ is the time derivative of the respective piston stroke $s_j$ (see Equation 3.9).

$$v_{osc,j} = \dot{s}_j(\alpha_{CS}) = \dot{s}\left(\alpha_{CS} - (j-1)\frac{4\pi}{CYL}\right) \qquad , \quad j = 1, ..., CYL \qquad (6.30)$$

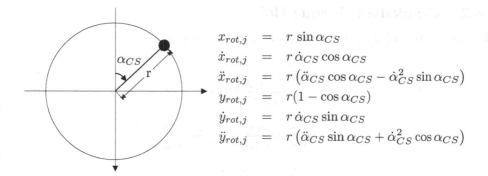

$$x_{rot,j} = r \sin \alpha_{CS}$$
$$\dot{x}_{rot,j} = r \dot{\alpha}_{CS} \cos \alpha_{CS}$$
$$\ddot{x}_{rot,j} = r \left( \ddot{\alpha}_{CS} \cos \alpha_{CS} - \dot{\alpha}_{CS}^2 \sin \alpha_{CS} \right)$$
$$y_{rot,j} = r(1 - \cos \alpha_{CS})$$
$$\dot{y}_{rot,j} = r \dot{\alpha}_{CS} \sin \alpha_{CS}$$
$$\ddot{y}_{rot,j} = r \left( \ddot{\alpha}_{CS} \sin \alpha_{CS} + \dot{\alpha}_{CS}^2 \cos \alpha_{CS} \right)$$

**Figure 6.22** Rotational Motion at the crankshaft

The rotational speed must be composed from the two components

$$\underline{v}_{rot,j} = [\dot{x}_{rot,j}, \dot{y}_{rot,j}]^T \quad , \quad |\underline{v}_{rot,j}|^2 = \dot{x}_{rot,j}^2 + \dot{y}_{rot,j}^2 \quad . \tag{6.31}$$

The time derivative of the kinetic energy $E_{mass}$ is

$$\frac{dE_{mass}}{dt} = \frac{dE_{mass}}{d\alpha_{CS}} \cdot \frac{d\alpha_{CS}}{dt} = T_{mass} \cdot \dot{\alpha}_{CS}$$

$$= \frac{m_{rot}}{CYL} \sum_{j=1}^{CYL} (\dot{x}_{rot,j} \cdot \ddot{x}_{rot,j} + \dot{y}_{rot,j} \cdot \ddot{y}_{rot,j}) + m_{osc} \sum_{j=1}^{CYL} \dot{s}_j \ddot{s}_j$$

$$= \frac{m_{rot}}{CYL} \sum_{j=1}^{CYL} \left( r^2 \cos^2 (\alpha_{CS}) \dot{\alpha}_{CS} \ddot{\alpha}_{CS} + r^2 \sin^2 (\alpha_{CS}) \dot{\alpha}_{CS} \ddot{\alpha}_{CS} \right) +$$

$$+ m_{osc} \sum_{j=1}^{CYL} \frac{ds_j}{d\alpha_{CS}} \cdot \dot{\alpha}_{CS} \left( \frac{d^2 s_j}{d\alpha_{CS}^2} \cdot \dot{\alpha}_{CS}^2 + \frac{ds_j}{d\alpha_{CS}} \cdot \ddot{\alpha}_{CS} \right)$$

$$\frac{dE_{mass}}{dt} =$$

$$\underbrace{\left[ \underbrace{\left( m_{rot} \cdot r^2 + m_{osc} \sum_{j=1}^{CYL} \left( \frac{ds_j}{d\alpha_{CS}} \right)^2 \right)}_{J} \ddot{\alpha}_{CS} + \frac{1}{2} \underbrace{\left( 2 m_{osc} \sum_{j=1}^{CYL} \frac{ds_j}{d\alpha_{CS}} \frac{d^2 s_j}{d\alpha_{CS}^2} \right)}_{\frac{dJ}{d\alpha_{CS}}} \dot{\alpha}_{CS}^2 \right] \dot{\alpha}_{CS}}_{T_{mass}} \tag{6.33}$$

In this, the equations in Figure 6.22 and Equation 3.6 in Section 3.1.1 are used. The moment of inertia is

$$J(\alpha_{CS}) = m_{rot} \cdot r^2 + m_{osc} \sum_{j=1}^{CYL} \left( \frac{ds_j}{d\alpha_{CS}} \right)^2 \quad . \tag{6.34}$$

## 6.8.2   Crankshaft Torque Balance

The friction torque $T_{fric}$ is given by the Coulomb law as

$$
\begin{aligned}
T_{fric} &= \sum_{j=1}^{CYL} \underbrace{F_{fric,j}}_{c_f \cdot \dot{s}_j} \cdot \frac{ds_j}{d\alpha_{CS}} \\
&= \sum_{j=1}^{CYL} c_f \frac{ds_j}{d\alpha_{CS}} \cdot \frac{d\alpha_{CS}}{dt} \cdot \frac{ds_j}{d\alpha_{CS}} \\
&= c_f \sum_{j=1}^{CYL} \left( \frac{ds_j}{d\alpha_{CS}} \right)^2 \cdot \dot{\alpha}_{CS}
\end{aligned}
\tag{6.35}
$$

The torque balance (see Equation 6.21) is then

$$
\sum_{j=1}^{CYL} (p_j(\alpha_{CS}) - p_0) A_p \frac{ds_j(\alpha_{CS})}{d\alpha_{CS}} - \left[ m_{rot} r^2 + m_{osc} \sum_{j=1}^{CYL} \left( \frac{ds_j(\alpha_{CS})}{d\alpha_{CS}} \right)^2 \right] \ddot{\alpha}_{CS}
$$

$$
- \frac{1}{2} \left[ 2 m_{osc} \sum_{j=1}^{CYL} \frac{ds_j(\alpha_{CS})}{d\alpha_{CS}} \cdot \frac{d^2 s_j(\alpha_{CS})}{d\alpha_{CS}^2} \right] \dot{\alpha}_{CS}^2 -
\tag{6.36}
$$

$$
- c_f \sum_{j=1}^{CYL} \left( \frac{ds_j(\alpha_{CS})}{d\alpha_{CS}} \right)^2 \dot{\alpha}_{CS} - T_{load}(\alpha_{CS}) = 0
$$

## 6.8.3   Transformation into Linear System Representation

For further calculations, the torque balance is regrouped into an angle-dependent differential equation with time-derivatives.

$$
J(\alpha_{CS}) \ddot{\alpha}_{CS} = T_{comb}(\alpha_{CS}) - f(\alpha_{CS}) \cdot \dot{\alpha}_{CS}^2 - T_{load}^*(\alpha_{CS})
\tag{6.37}
$$

The angle-dependent function is

$$
f(\alpha_{CS}) = \sum_{j=1}^{CYL} \left( m_{osc} \frac{ds_j(\alpha_{CS})}{d\alpha_{CS}} \cdot \frac{d^2 s_j(\alpha_{CS})}{d\alpha_{CS}^2} \right) \quad .
\tag{6.38}
$$

The extended load torque

$$
T_{load}^*(\alpha_{CS}) = T_{load}(\alpha_{CS}) + \underbrace{c_f \sum_{j=1}^{CYL} \left( \frac{ds_j(\alpha_{CS})}{d\alpha_{CS}} \right)^2 \dot{\alpha}_{CS}}_{T_{fric}}
$$

is an approximation and shall comprise also the friction torque $T_{fric}$. The combustion torque is

$$
T_{comb}(\alpha_{CS}) = \sum_{j=1}^{CYL} (p_j(\alpha_{CS}) - p_0) A_p \cdot \frac{ds_j(\alpha_{CS})}{d\alpha_{CS}} \quad .
$$

The second derivative of the crankshaft angle may be reformulated by substituting

$$\ddot{\alpha}_{CS} = \frac{d^2\alpha_{CS}}{dt^2} = \frac{d}{dt}(\dot{\alpha}_{CS}) = \frac{d\dot{\alpha}_{CS}}{d\alpha_{CS}} \cdot \frac{d\alpha_{CS}}{dt} = \frac{d\dot{\alpha}_{CS}}{d\alpha_{CS}} \cdot \dot{\alpha}_{CS}$$

into Equation 6.37.

$$\dot{\alpha}_{CS} \cdot d\dot{\alpha}_{CS} = \frac{1}{J(\alpha_{CS})}\Big(T_{comb}(\alpha_{CS}) - f(\alpha_{CS}) \cdot \dot{\alpha}_{CS}^2 - T_{load}^*(\alpha_{CS})\Big)d\alpha_{CS} \quad (6.39)$$

Integration results into an equation which depends only on the square of the crankshaft angle speed $\dot{\alpha}_{CS}$ instead on both crankshaft angle and time.

$$\dot{\alpha}_{CS}^2(n+1) - \dot{\alpha}_{CS}^2(n) =$$

$$\frac{2}{J(\alpha_{CS})} \int\limits_{\alpha_{CS}(n)}^{\alpha_{CS}(n+1)} \Big(T_{comb}(\alpha_{CS}) - f(\alpha_{CS}) \cdot \dot{\alpha}_{CS}^2 - T_{load}^*(\alpha_{CS})\Big)d\alpha_{CS} \quad (6.40)$$

Over the discrete angular step $\Delta\alpha_{CS} = \alpha_{CS}(n+1) - \alpha_{CS}(n)$, the integration may be approximated as

$$\dot{\alpha}_{CS}^2(n+1) - \dot{\alpha}_{CS}^2(n) \approx \frac{2\Delta\alpha_{CS}}{J(n)}\Big(T_{comb}(n) - f(n)\dot{\alpha}_{CS}^2(n) - T_{load}^*(n)\Big) \quad . \quad (6.41)$$

With a 60 teeth crankshaft sensor wheel, the angular step $\Delta\alpha_{CS}$ is 6°. Instead of multiples of the sample time $n \cdot T_s$, we are now calculating with multiples of the angular step $n \cdot \Delta\alpha_{CS}$. By regarding the square of the crankshaft rotational speed $\dot{\alpha}^2(n)$ as a state variable $x_1$, we can linearize the calculation. By this we obtain a linear discrete state space model of the crankshaft motion.

$$x_1(n+1) = \left(1 - \frac{2\Delta\alpha_{CS}}{J(n)} \cdot f(n)\right)x_1(n) + \frac{2\Delta\alpha_{CS}}{J(n)}x_2(n) \quad (6.42)$$

with the state variables

$$\begin{aligned}
x_1(n) &= \dot{\alpha}_{CS}^2(n) \quad , \\
x_2(n) &= T_{comb}(n) - T_{load}^*(n) \quad , \\
x_3(n) &= x_2(n+1) \quad .
\end{aligned} \quad (6.43)$$

## 6.8.4 Kalman Filter Design

The combustion torque is the physical cause which generates the crankshaft motion. Kalman filters employ a Markovian system model, i.e. a first order time-discrete linear system which is excited by white noise. Oscillations of the combustion pressure torque and load torque decrease at higher engine order. This behavior is not reflected by a white noise excitation. Therefore the torque state variable $x_2(n)$ is modelled to be the output of a second order low pass system $H(z)$ which is excited by white noise $U(z)$ [44].

$$X_2(z) = H(z) \cdot U(z) \quad (6.44)$$

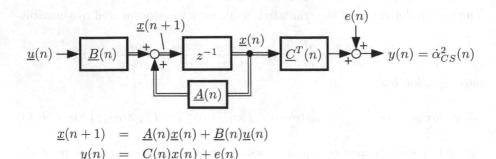

$$\underline{x}(n+1) = \underline{A}(n)\underline{x}(n) + \underline{B}(n)\underline{u}(n)$$
$$\underline{y}(n) = \underline{C}(n)\underline{x}(n) + \underline{e}(n)$$

**Figure 6.23** Linear model for Kalman estimation of squared crankshaft speed.

The filter output $x_2(n)$ is a so-called colored noise, the power density spectrum of which decreases indeed over increasing engine order. $H(z)$ has a double pole on the real axis.

$$H(z) = \frac{(1 - \exp(-\delta \cdot \Delta\alpha_{CS}))^2}{(z - \exp(-\delta \cdot \Delta\alpha_{CS}))^2} \tag{6.45}$$

In the discrete angle domain, the state variable $x_2(n)$ is modelled as

$$x_2(n+2) - 2x_2(n+1) \cdot \exp(-\delta \cdot \Delta\alpha_{CS}) + x_2(n) \cdot \exp(-2\delta \cdot \Delta\alpha_{CS})$$
$$= \left(1 - \exp(-\delta \cdot \Delta\alpha_{CS})\right)^2 \cdot u(n) \quad . \tag{6.46}$$

Introducing another state variable $x_3(n) = x_2(n+1)$, this can be expressed in first order state space form as a basis for Kalman filtering.

$$\underline{x}(n) = \left( \begin{array}{ccc} \dot\alpha_{CS}^2(n) & T_{comb}(n) - T_{load}^*(n) & T_{comb}(n+1) - T_{load}^*(n+1) \end{array} \right)^T \tag{6.47}$$

$$\underline{y}(n) = \dot\alpha_{CS}^2(n) \tag{6.48}$$

$$\underline{A}(n) = \begin{pmatrix} 1 - \frac{2 \cdot f(n) \cdot \Delta\alpha_{CS}}{J(n)} & \frac{2 \cdot \Delta\alpha_{CS}}{J(n)} & 0 \\ 0 & 0 & 1 \\ 0 & -\exp(-2\delta \cdot \Delta\alpha_{CS}) & 2 \cdot \exp(-\delta \cdot \Delta\alpha_{CS}) \end{pmatrix} \tag{6.49}$$

$$\underline{B}(n) = \left( \begin{array}{ccc} 0 & 0 & (1 - \exp(-\delta \cdot \Delta\alpha_{CS}))^2 \end{array} \right)^T \tag{6.50}$$

$$\underline{C}^T(n) = \left( \begin{array}{ccc} 1 & 0 & 0 \end{array} \right) \tag{6.51}$$

Based upon this linearized model, a Kalman estimation can be performed [11]. Appropriate values for $\delta$, $R_{uu}(n) = \sigma_u^2$ and $R_{ee}(n) = \sigma_e^2$ are chosen. By measuring the square of the rotational crankshaft speed $y(n) = \dot\alpha_{CS}^2(n)$, the torque difference $\hat{x}_2(n) = T_{comb}(n) - T_{load}^*(n)$ can be estimated.

$$\underline{\hat{x}}(n+1) = \underline{H}(n)\,\underline{\hat{x}}(n) + \underline{K}(n)\,\underline{y}(n) \tag{6.52}$$

$$\underline{\hat{x}}(n+1) = \underline{A}(n)\,\underline{\hat{x}}(n) + \underline{K}(n)\left(\underline{y}(n) - \underline{\hat{y}}(n)\right) \tag{6.53}$$

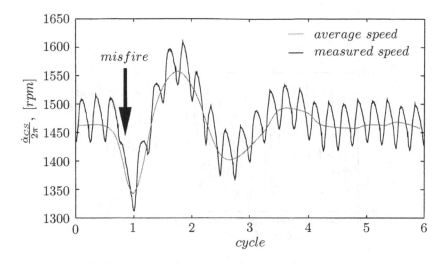

**Figure 6.24** Measured angular speed signal with single misfire

For various applications such as engine misfire detection, the absolute combustion torque must be obtained. At top and bottom dead centers (TDC and BDC) of a four-cylinder engine, the combustion torque is zero, because the piston stroke derivative $ds_j(\alpha_{CS})/d\alpha_{CS}$ is zero in those points. Thus, the load torque can be separately calculated in the TDC and BDC points.

$$\hat{x}_2(n_{TDC,BDC}) \approx -T^*_{load}(n) \tag{6.54}$$

When the estimated torque difference $\hat{x}_2(n)$ is corrected by an interpolated load torque $\hat{x}_2(n_{TDC,BDC})$ between TDC/BDC-Points, the combustion torque $T_{comb}(n)$ can be also calculated separately.

## 6.8.5   Results

In the case of a misfire, driveline oscillations are excited. The estimation process is required to distinguish between crankshaft speed and rotational driveline oscillations. Figure 6.24 shows the measured crankshaft speed signal of a test engine at 1500 *rpm*. At the end of the first cycle a misfire occurs. Figure 6.25 shows the combustion torque estimated by the Kalman filter, and the load torque interpolated between the TDCs. Figure 6.26 contains the corrected combustion torque. The torque fluctuations induced by the driveline are now effectively suppressed.

At higher engine speeds the oscillating mass torques increase. The signal-to-noise ratio decreases rapidly, and the performance of the Kalman filter decreases. The current estimation limit of this approach is between 3000 and 4000 *rpm*.

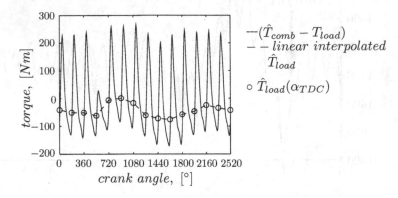

**Figure 6.25** Estimated combustion torque difference by Kalman filter

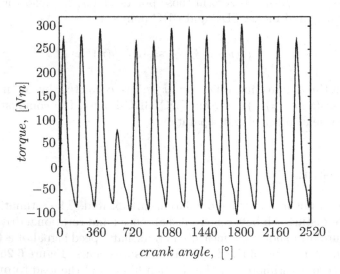

**Figure 6.26** Estimated combustion torque $T_{comb}$ by Kalman filter with load torque correction

# 6.9 Engineering of Diagnosis Systems

This last section of this chapter aims at recalling and discussing the construction of model based diagnosis systems from an engineering point of view. When designing a diagnosis system for a real industrial application, other considerations, than the ones treated in this chapter, must probably be taken into account. These considerations include developing-time constraints, market requirements as well as economical constraints. However, the procedure for model based diagnosis system engineering should approximately look like the following step-by-step procedure:

1. Obtain requirements on what faults that need to be diagnosed, time constraints such as detection time and isolation time, and any requirements regarding fault identification.

2. Study and acquire knowledge about the system and particularly the faults that need to be diagnosed.

3. Build a model of the process and also model how faults and disturbances influence the system. This step consists of three parts: selection of model structure, parameter identification, and model validation.

4. By using the model, design residuals that satisfy the isolation requirements.

5. Evaluate performance of the residuals and select the best ones. If the performance is not satisfactory, the model needs to be refined.

6. Design the residual evaluation scheme. This includes at least selection of thresholds.

7. Test the diagnosis system in simulation and in reality.

8. Do a final implementation of the diagnosis system.

Step 1 and step 2 can only be performed by acquiring information from experiments, long-time experience, and domain experts. Little literature is available. Step 3, the model building, is probably the most well documented of all steps. As noted before in this text, it may be the major part of the work. There is generally available literature on modeling of specific classes of systems and also on general model building, e. g. in system identification literature. Many methods for step 4 have been developed but there still exist many design choices to be made by the engineer. Also the selection of residuals and constructing a residual evaluator, i. e. step 5 and 6, require many design choices. Systematic ways to assist engineers performing in steps 4-6 is an important area of development. Step 7 is again an area in which not much literature is available making it partly an art form. In conclusion, there are several engineering steps involved in the design of diagnosis systems.

# 7 Driveline Control

## 7.1 Driveline Modeling

A vehicular power train consists of engine and driveline. The main parts of the driveline are clutch, transmission, shafts and wheels. The driveline is a fundamental part of a vehicle and its dynamics has been modeled in different ways depending on the purpose. The frequency range important for control is the regime including the lowest resonance modes of the driveline [85, 95]. Vibrations and noise contribute to a higher frequency range [120, 37] which is not treated here.

Section 7.1.1 covers the derivation of basic equations describing a driveline. An illustrative example is presented in Section 7.1.4. Experiments are performed with a Scania heavy truck. The aim of the modeling is to find the most important physical effects explaining the oscillations in the measured engine speed, transmission speed, and wheel speed. The models are combinations of rotating inertias connected by damped shaft flexibilities. The generalized Newton's second law is used to derive the models. The main part of the experiments used for modeling considers low gears. The reason for this is that the lower the gear is, the higher the torque transferred in the drive shaft is. This means that the shaft torsion is higher for lower gears, and hereby also the problems with oscillations.

Furthermore, the amplitudes of the resonances in the wheel speed are higher for lower gears, since the load and vehicle mass appear reduced by the high conversion ratio.

Section 7.2 treats the modeling when the driveline is separated in two parts, which is the case when in neutral gear or when the clutch is disengaged.

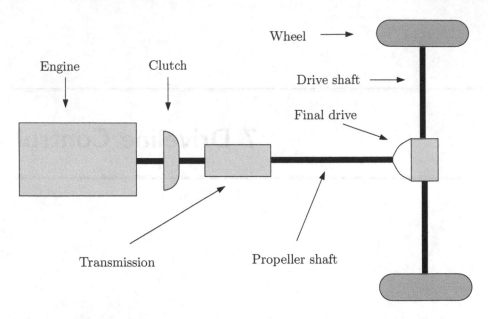

**Figure 7.1** A rear-driven vehicular power train consisting of engine and driveline.

## 7.1.1   Basic Driveline Equations

Figure 7.1 depicts a driveline of a rear-driven vehicle. It consists of clutch, transmission, propeller shaft, final drive, drive shafts, and wheels. (The propeller shaft is sometimes also called a drive shaft, but in our context here it is important to have different names for the shafts since they relate to different torsional forces that enter the modeling differently.) Fundamental equations for the driveline will be derived by using the generalized Newton's second law of motion [81]. Some basic equations regarding the forces acting on the wheel are obtained, influenced by the complete dynamics of the vehicle. This means that effects from, for instance, vehicle mass and trailer will be included in the equations describing the wheels. Figure 7.2 shows the labels, the inputs, and the outputs of each subsystem of the driveline type considered in this work. Relations between inputs and outputs will in the following be described for each part.

**Engine:**   The output torque of the engine is characterized by the driving torque ($T_e$) resulting from the combustion, the internal friction from the engine ($T_{fric,e}$), and the external load from the clutch ($T_c$). Newton's second law of motion gives the following model

$$J_e \ddot{\alpha}_{CS} = T_e - T_{fric,e} - T_c \qquad (7.1)$$

where $J_e$ is the mass moment of inertia of the engine, $\alpha_{CS}$ is the crankshaft angle and $\dot{\alpha}_{CS} = 2\pi n$ is the engine speed.

**Clutch:**   A friction clutch found in vehicles equipped with a manual transmission consists of a clutch disk connecting the flywheel of the engine and the transmission's input shaft. When the clutch is engaged, and no internal

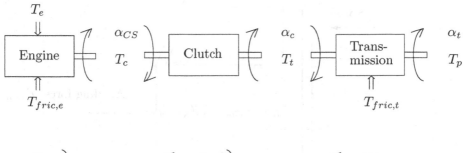

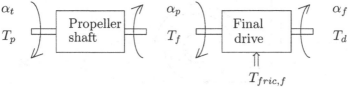

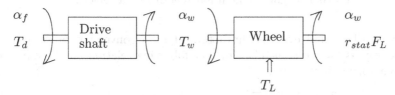

**Figure 7.2** Subsystems of a vehicular driveline with their respective angle and torque labels.

friction is assumed, $T_c = T_t$ is obtained. The transmitted torque is a function of the angular difference $(\alpha_{CS} - \alpha_c)$ and the angular velocity difference $(\dot{\alpha}_{CS} - \dot{\alpha}_c)$ over the clutch

$$T_c = T_t = f_c(\alpha_{CS} - \alpha_c, \ \dot{\alpha}_{CS} - \dot{\alpha}_c) \tag{7.2}$$

**Transmission:** A transmission has a set of gears, each with a conversion ratio $i_t$. This gives the following relation between the input and output torque of the transmission

$$T_p = f_t(T_t, \ T_{fric,t}, \ \alpha_c - \alpha_t i_t, \ \dot{\alpha}_c - \dot{\alpha}_t i_t, \ i_t) \tag{7.3}$$

where the internal friction torque of the transmission is labeled $T_{fric,t}$. The reason for considering the angle difference $\alpha_c - \alpha_t i_t$ in (7.3) is the possibility of having torsional effects in the transmission.

**Propeller shaft:** The propeller shaft connects the transmission's output shaft with the final drive. No friction is assumed ($\Rightarrow T_p = T_f$), giving the following model of the torque input to the final drive

$$T_p = T_f = f_p(\alpha_t - \alpha_p, \ \dot{\alpha}_t - \dot{\alpha}_p) \tag{7.4}$$

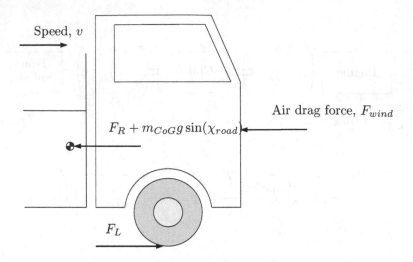

**Figure 7.3** Longitudinal forces acting on a vehicle.

**Final drive:** The final drive is characterized by a conversion ratio $i_f$ in the same way as for the transmission. The following relation for the input and output torque holds

$$T_d = f_f(T_f,\ T_{fric,f},\ \alpha_p - \alpha_f i_f,\ \dot{\alpha}_p - \dot{\alpha}_f i_f, i_f) \tag{7.5}$$

where the internal friction torque of the final drive is labeled $T_{fric,f}$.

**Drive shafts:** The drive shafts connect the wheels to the final drive. Here it is assumed that the rotational wheel speed $\dot{\alpha}_w$ is the same for the two wheels. Neglecting vehicle dynamics, the rotational equivalent wheel speeds shall be equal to the speed of the vehicle body's center of gravity.

$$\dot{\alpha}_w = \frac{v_{Rij}}{r_{stat}} \approx \frac{v_{CoG}}{r_{stat}}$$

Therefore, the drive shafts are modeled as one shaft. When the vehicle is turning and the speed differs between the wheels, both drive shafts have to be modeled. No friction ($\Rightarrow T_w = T_d$) gives the model equation

$$T_w = T_d = f_d(\alpha_f - \alpha_w,\ \dot{\alpha}_f - \dot{\alpha}_w) \tag{7.6}$$

**Wheel:** Figure 7.3 shows the forces acting on a vehicle with mass $m_{CoG}$ and speed $v_{CoG}$. Newton's second law in the longitudinal direction gives

$$F_L = m_{CoG}\dot{v}_{CoG} + F_{wind} + F_R + m_{CoG}g\sin(\chi_{road}) \tag{7.7}$$

The friction force ($F_L$) is described by the sum of the following quantities [37].

- $F_{wind}$, the air drag, is approximated by

$$F_{wind} = \frac{1}{2}c_{air}A_L\rho_a v_{CoG}^2 \tag{7.8}$$

where $c_{air}$ is the drag coefficient, $A_L$ the maximum vehicle cross section area, and $\rho_a$ the air density. However, effects from, for instance, open or closed windows will make the force difficult to model.

- $F_R$, the rolling resistance, is approximated by

$$F_R = m_{CoG}(c_{r1} + c_{r2}v_{CoG}) \tag{7.9}$$

where $c_{r1}$ and $c_{r2}$ depend on, for instance, tires and tire pressure.

- $m_{CoG}g\sin(\chi_{road})$, the gravitational force, where $\chi_{road}$ is the slope of the road.

The resulting torque due to $F_L$ is equal to $F_L r_{stat}$, where $r_{stat}$ is the wheel radius. Newton's second law gives

$$J_W \ddot{\alpha}_w = T_w - F_L r_{stat} - T_L \tag{7.10}$$

where $J_W$ is the mass moment of inertia of the wheel, $T_w$ is given by (7.6), and $T_L$ is the friction torque. Including (7.7) to (7.9) in (7.10) together with $v_R = r_{stat}\dot{\omega}_{ij}$ gives

$$\begin{aligned}(J_W + m_{CoG}r_{stat}^2)\ddot{\alpha}_w &= T_w - T_L - \frac{1}{2}c_{air}A_L\rho_a r_{stat}^3\dot{\alpha}_w^2 \\ -r_{stat}m_{CoG}(c_{r1} &+ c_{r2}r_{stat}\dot{\alpha}_w) - r_{stat}m_{CoG}g\sin(\chi_{road})\end{aligned} \tag{7.11}$$

The dynamical influence from the tire has been neglected in the equation describing the wheel.

A complete model of the driveline with the clutch engaged is described by Equations (7.1) to (7.11). So far the functions $f_c$, $f_t$, $f_p$, $f_f$, $f_d$, and the friction torques $T_{fric,t}$, $T_{fric,f}$, and $T_L$ are unknown. In the following, assumptions will be made about these, resulting in a series of driveline models, with different complexities.

## 7.1.2 A Basic Complete Model

A basic model will be developed. Assumptions about the fundamental equations above are made in order to obtain a model with a lumped inertia. Labels are according to Figure 7.2. The clutch and the shafts are assumed to be stiff. The transmission and the final drive are assumed to multiply the torque by the conversion ratio, without losses.

**Engine:** The engine is modeled as in Equation (7.1)

$$J_e\ddot{\alpha}_{CS} = T_e - T_{fric,e} - T_c \tag{7.12}$$

**Clutch:** The clutch is assumed to be stiff, which gives the following equations for the torque and the angle

$$T_c = T_t, \quad \alpha_{CS} = \alpha_c \tag{7.13}$$

**Transmission:** The transmission is described by one rotating inertia $J_t$. The friction torque is assumed to be described by a viscous damping coefficient $d_t$. The model of the transmission, corresponding to (7.3), is

$$\alpha_c = \alpha_t i_t \tag{7.14}$$
$$J_t \ddot{\alpha}_t = T_t i_t - d_t \dot{\alpha}_t - T_p \tag{7.15}$$

By using (7.13) and (7.14), the model can be rewritten as

$$J_t \ddot{\alpha}_{CS} = T_c i_t^2 - d_t \dot{\alpha}_{CS} - T_p i_t \tag{7.16}$$

The equation above illustrates the general methodology, but from now on a simplified version will be used by neglecting the inertia and the damping losses. This is done by using $J_t = 0$ and $d_t = 0$, which results in the following transmission model

$$\alpha_c = \alpha_t i_t \tag{7.17}$$
$$T_t i_t = T_p \tag{7.18}$$

**Propeller shaft:** The propeller shaft is also assumed to be stiff, which gives the following equations for the torque and the angle

$$T_p = T_f, \quad \alpha_t = \alpha_p \tag{7.19}$$

**Final drive:** In the same way as for the transmission, the final drive can in general be modeled by one rotating inertia $J_f$ and a friction torque that is assumed to be described by a viscous damping coefficient $d_f$. However, also here, for the basic model, the inertia and the damping are neglected by $J_f = 0$ and $d_f = 0$.

The model of the final drive, corresponding to (7.5), is then

$$\alpha_p = \alpha_f i_f \tag{7.20}$$
$$T_f i_f = T_d \tag{7.21}$$

**Drive shaft:** The drive shaft is assumed to be stiff, which gives the following equations for the torque and the angle

$$T_w = T_d, \quad \alpha_f = \alpha_w \tag{7.22}$$

**Wheel:** The force and torque balances on the wheel includes the dynamics of the vehicle and is modeled as before in Equation (7.11). Neglecting the wheel friction, $T_L = 0$, leads to

$$(J_W + m_{CoG} r_{stat}^2)\ddot{\alpha}_w = T_w - \frac{1}{2} c_{air} A_L \rho_a r_{stat}^3 \dot{\alpha}_w^2 \tag{7.23}$$
$$-r_{stat} m_{CoG}(c_{r1} + c_{r2} r_{stat} \dot{\alpha}_w) - r_{stat} m_{CoG} g \sin(\chi_{road})$$

## 7.1.3    Combining the equations

The equations from engine to wheel, Equations (7.12)-(7.23), now constitutes a complete chain. Start the elimination of intermediate variables by using (7.13) and (7.14)-(7.22) to obtain the relationships for angles and torques for a stiff driveline ($\alpha_{CS} = \alpha_c = i_t\alpha_t = i_t\alpha_p = i_t i_f\alpha_f = i_t i_f\alpha_w$, and in the same way for the torques). This results in

$$i_t i_f T_c = T_w, \quad \alpha_{CS} = i_t i_f \alpha_w \tag{7.24}$$

The complete description is now reduced to this equation, (7.24), in combination with (7.12) and (7.23) which are

$$J_e\ddot{\alpha}_{CS} = T_e - T_{fric,e} - T_c/i_t i_f \tag{7.25}$$

$$(J_W + m_{CoG}r_{stat}^2)\ddot{\alpha}_w = T_w - \frac{1}{2}c_{air}A_L\rho_a r_{stat}^3\dot{\alpha}_w^2 \tag{7.26}$$

$$-r_{stat}m_{CoG}(c_{r1} + c_{r2}r_{stat}\dot{\alpha}_w) - r_{stat}mgsin(\chi_{road})$$

There is now an arbitrary choice whether to write the final model in terms of engine speed or wheel speed. Here the wheel speed is chosen. This means that the three variables $T_d, T_f, \alpha_{CS}$ should be eliminated from the four equations (7.24)-(7.26). The result is one single equation constituting the complete model, named the Basic Driveline Model.

---

**Model 7.1 The Basic Driveline Model**

$$(J_W + m_{CoG}r_{stat}^2 + i_t^2 i_f^2 J_e)\ddot{\alpha}_w = i_t i_f(T_e - T_{fric,e}) \tag{7.27}$$

$$-m_{CoG}c_{r2}r_{stat}^2\dot{\alpha}_w - \frac{1}{2}c_{air}A_L\rho_a r_{stat}^3\dot{\alpha}_w^2$$

$$-r_{stat}m_{CoG}(c_{r1} + gsin(\chi_{road}))$$

For low gears and speeds, the influence from the air drag is low and by neglecting $\frac{1}{2}c_{air}A_L\rho_a r_{stat}^3\dot{\alpha}_w^2$ in (7.27), the model is linear in the state $\dot{\alpha}_w$, but nonlinear in the parameters.

---

There is one term in the above Basic Driveline Model that requires additional modeling. It is the term $T_e - T_{fric,e}$ that describes the effective torque from the engine. The torque generated by combustion, $T_e$, is modeled differently for SI or CI engines. The losses in the engine, $T_{fric,e}$, depend, among other things, on pumping losses, and usually requires measurements. Both these aspects will be illustrated in the next section.

## 7.1.4    An Illustrative Modeling Example

A Scania heavy truck is used for experiments. Figure 7.4 shows a Scania 144L 6x2 truck that has the configuration as follows.

- 14 liter V8 turbo-charged diesel engine with maximum power of 530 Hp and maximum torque of 2300 Nm. The fuel metering is governed by an in-line injection pump system [10].

**Figure 7.4** Scania 144L truck.

- The engine is connected to a manual range-splitter transmission GRS900R (Figure 7.6) via a clutch. The transmission has 14 gears and a hydraulic retarder. It is also equipped with the automatic gear shifting system Opti-Cruise [91].

- The weight of the truck is $m = 24\ 000$ kg.

**Engine**

The V8 engine in the 144L truck uses an in-line pump system with one fuel pump supplying all eight cylinders with fuel. Driveline modeling will be influenced by a number of subsystems of the engine that are common for both engine types. These are

**Maximum torque delimiter** The injected fuel amount is restricted by the physical character of the engine (i.e., engine size, number of cylinders, etc.), together with restrictions that the engine control system uses, for utilizing the engine in the best possible way. The maximum torque decreases below 1100 rpm and above 1500 rpm.

**Diesel smoke delimiter** If the turbo pressure is low and a high engine torque is demanded, diesel smoke emissions will increase to an unacceptable level. This is prevented by restricting the fuel amount to a level with acceptable emissions at low turbo pressures.

**Transfer function from fuel amount to engine torque** The engine torque is the torque resulting from the explosions in the cylinders. A static function relating the engine torque to injected fuel can be obtained in a dynamometer

**Figure 7.5** Scania 14 liter V8 DSC14 engine.

**Figure 7.6** Scania GRS900R range-splitter transmission with retarder and OptiCruise automatic gear-shifting system.

**Table 7.1** Measured variables transmitted on the CAN-bus.

| Measured Variables | | | |
|---|---|---|---|
| *Variable* | *Node* | *Resolution* | *Rate* |
| Engine speed, $\dot{\alpha}_{CS}$ | Engine | 0.013 rad/s | 20 ms |
| Engine torque, $T_e$ | Engine | 1% of max torque | 20 ms |
| Engine temp, $\vartheta_e$ | Engine | 1° C | 1 s |
| Wheel speed, $\dot{\alpha}_w$ | ABS | 0.033 rad/s | 50 ms |
| Transmission speed, $\dot{\alpha}_t$ | Transmission | 0.013 rad/s | 50 ms |

test. For a diesel engine this function is fairly static, and no dynamical models are used in this work.

**Engine friction** The engine output torque transferred to the clutch is equal to the engine torque (the torque resulting from the explosions) subtracted by the engine internal friction. Friction modeling is thus fundamental for driveline modeling and control.

### Sensor System

The velocity of a rotating shaft is measured by using an inductive sensor [89], which detects the time when cogs from a rotating cogwheel are passing. This time sequence is then inverted to get the angle velocity. Hence, the bandwidth of the measured signal depends on the speed and the number of cogs the cogwheel is equipped with.

Three speed sensors are used to measure the speed of the flywheel of the engine ($\dot{\alpha}_{CS}$), the speed of the output shaft of the transmission ($\dot{\alpha}_t$), and the speed of the driving wheel ($\dot{\alpha}_w$). The transmission speed sensor has fewer cogs than the other two sensors, indicating that the bandwidth of this signal is lower.

By measuring the injected fuel flow, $\dot{m}_f$, that is fed to the engine, a measure of the driving torque, $T_e(\dot{m}_f)$, is obtained from dynamometer tests, as mentioned before. The output of the engine is the driving torque subtracted by the engine friction, $T_{fric,e}$. This signal, $u = T_e(\dot{m}_f) - T_{fric,e}$, is the torque acting on the driveline, which is a pulsating signal with torque pulses from each cylinder explosion. However, the control signal $u = T_e(\dot{m}_f) - T_{fric,e}$ is treated as a continuous signal, which is reasonable for the frequency range considered for control design. A motivation for this is that an eight-cylinder engine makes 80 strokes/s at an engine speed of 1200 rev/min. This means that a mean-value engine model is assumed (neglects variations during the engine cycle).

The truck is equipped with a set of control units, each connected with a CAN-bus [97]. These CAN nodes are the engine control node, the transmission node, and the ABS brake system node. Each node measures a number of variables and transmits them via the bus.

### Experiments for Driveline Modeling

A number of test roads at Scania were used for testing. They have different known slopes. The variables in Table 7.1 are logged during tests that excite

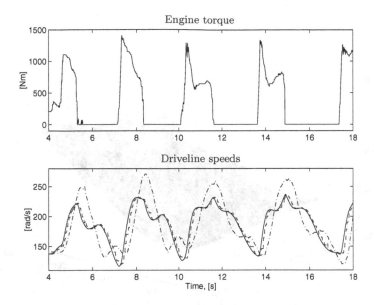

**Figure 7.7** Logged data on the CAN-bus during step inputs in accelerator position with the 144L truck. The transmission speed (dashed) and the wheel speed (dash-dotted) are scaled to engine speed in solid. The main flexibility of the driveline is located between the output shaft of the transmission and the wheel, since the largest difference in speed is between the measured transmission speed and wheel speed.

driveline resonances. Figure 7.7 shows a test with the 144L truck where step inputs in accelerator position excite driveline oscillations. In Figure 7.7 it is seen that the main flexibility of the driveline is located between the output shaft of the transmission and the wheel, since the largest difference in speed is between the measured transmission speed and wheel speed.

**Engine Friction Modeling**

The engine friction $T_{fric,e}$ is modeled as a function of the engine speed and the engine temperature

$$T_{fric,e} = T_{fric,e}(\dot{\alpha}_{CS}, \vartheta_e) \qquad (7.28)$$

The influence from the load is neglected. With neutral gear engaged, the engine speed is controlled to 20 levels between 600 and 2300 RPM, while measuring the engine torque and temperature. The resulting friction map for the 144L truck is shown in Figure 7.8.

The logged engine torque, $T_e(\dot{m}_f)$, as a function of the injected fuel flow, is recalculated to control signal to the driveline by subtracting the engine friction from the engine torque as

$$u = T_e(\dot{m}_f) - T_{fric,e}(\dot{\alpha}_{CS}, \vartheta_e) \qquad (7.29)$$

Engine Friction, [Nm]

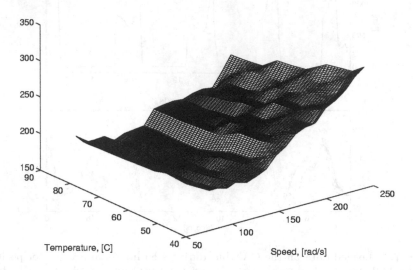

**Figure 7.8** Engine friction for the 144L truck, modeled as a function of engine speed and engine temperature.

## Obtaining a Set of Models

The measured engine speed, transmission speed, and wheel speed for the 144L truck is explained by deriving a set of models of increasing complexity. Figure 7.7 shows that the main difference in speed is between the measured transmission speed and wheel speed, indicating that the important flexibility of the driveline is located between the output shaft of the transmission and the wheel. This leads to a first model with a lumped engine and transmission inertia connected to the wheel inertia by a drive-shaft flexibility. The reason for this is that the drive shaft is subject to the relatively largest torsion. This is mainly due to the high torque difference that results from the amplification of the engine torque by the conversion ratio of the transmission $(i_t)$ and the final drive $(i_f)$. This number $(i_t i_f)$ can be as high as 60 for the lowest gear. A total of three models will be derived for the 144L truck, all based on the basic driveline equations derived in Section 7.1.1.

## Model with Drive-Shaft Flexibility

The simplest model with a drive-shaft flexibility is developed first. Assumptions about the fundamental equations in Section 7.1.1 are made in order to obtain a model with a lumped engine and transmission inertia and a drive-shaft flexibility. Labels are according to Figure 7.2. The clutch and the propeller shafts are assumed to be stiff, and the drive shaft is described as a damped torsional flexibility. The transmission and the final drive are assumed to multiply the torque by the conversion ratio, without losses.

**Clutch:** The clutch is assumed to be stiff, which gives the following equations for the torque and the angle

$$T_c = T_t, \quad \alpha_{CS} = \alpha_c \tag{7.30}$$

**Transmission:** The transmission is described by one rotating inertia $J_t$. The friction torque is assumed to be described by a viscous damping coefficient $d_t$. The model of the transmission, corresponding to (7.3), is

$$\alpha_c = \alpha_t i_t \tag{7.31}$$
$$J_t \ddot{\alpha}_t = T_t i_t - d_t \dot{\alpha}_t - T_p \tag{7.32}$$

By using (7.30) and (7.31), the model can be rewritten as

$$J_t \ddot{\alpha}_{CS} = T_c i_t^2 - d_t \dot{\alpha}_{CS} - T_p i_t \tag{7.33}$$

**Propeller shaft:** The propeller shaft is also assumed to be stiff, which gives the following equations for the torque and the angle

$$T_p = T_f, \quad \alpha_t = \alpha_p \tag{7.34}$$

**Final drive:** In the same way as for the transmission, the final drive is modeled by one rotating inertia $J_f$. The friction torque is assumed to be described by a viscous damping coefficient $d_f$. The model of the final drive, corresponding to (7.5), is

$$\alpha_p = \alpha_f i_f \tag{7.35}$$
$$J_f \ddot{\alpha}_f = T_f i_f - d_f \dot{\alpha}_f - T_d \tag{7.36}$$

Equation (7.36) can be rewritten with (7.34) and (7.35) which gives

$$J_f \ddot{\alpha}_t = T_p i_f^2 - d_f \dot{\alpha}_t - T_d i_f \tag{7.37}$$

Converting (7.37) to a function of engine speed is done by using (7.30) and (7.31) resulting in

$$J_f \ddot{\alpha}_{CS} = T_p i_f^2 i_t - d_f \dot{\alpha}_{CS} - T_d i_f i_t \tag{7.38}$$

By replacing $T_p$ in (7.38) with $T_p$ in (7.33), a model for the lumped transmission, propeller shaft, and final drive is obtained

$$(J_t i_f^2 + J_f)\ddot{\alpha}_{CS} = T_c i_t^2 i_f^2 - d_t \dot{\alpha}_{CS} i_f^2 - d_f \dot{\alpha}_{CS} - T_d i_f i_t \tag{7.39}$$

**Drive shaft:** The drive shaft is modeled as a damped torsional flexibility, having stiffness $k$, and internal damping $d$. Hence, (7.6) becomes

$$T_w = T_d = k(\alpha_f - \alpha_w) + d(\dot{\alpha}_f - \dot{\alpha}_w) = k(\alpha_{CS}/i_t i_f - \alpha_w) \tag{7.40}$$
$$+ \; d(\dot{\alpha}_{CS}/i_t i_f - \dot{\alpha}_w)$$

where (7.30), (7.31), (7.34), and (7.35) are used. By replacing $T_d$ in (7.39) with (7.40) the equation describing the transmission, the propeller shaft, the final drive, and the drive shaft, becomes

$$(J_t i_f^2 + J_f)\ddot{\alpha}_{CS} = T_c i_t^2 i_f^2 - d_t \dot{\alpha}_{CS} i_f^2 - d_f \dot{\alpha}_{CS} \tag{7.41}$$
$$- k(\alpha_{CS} - \alpha_w i_t i_f) - d(\dot{\alpha}_{CS} - \dot{\alpha}_w i_t i_f)$$

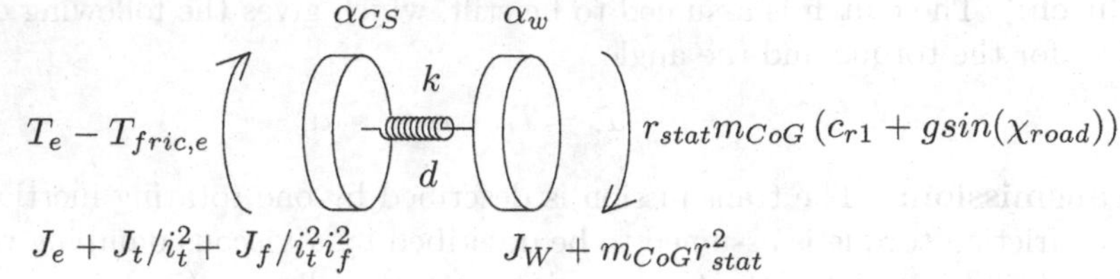

$$\alpha_{CS} \qquad \alpha_w$$

$$T_e - T_{fric,e} \left( \begin{array}{c} k \\ \text{-----} \\ d \end{array} \right) r_{stat} m_{CoG}\,(c_{r1} + gsin(\chi_{road}))$$

$$J_e + J_t/i_t^2 + J_f/i_t^2 i_f^2 \qquad J_W + m_{CoG} r_{stat}^2$$

**Figure 7.9** The Drive-shaft model consists of a lumped engine and transmission inertia connected to the wheel inertia by a damped torsional flexibility.

**Wheel:**  If (7.11) is combined with (7.40), the following equation for the wheel is obtained:

$$(J_W + m_{CoG} r_{stat}^2)\ddot{\alpha}_w = k(\alpha_{CS}/i_t i_f - \alpha_w) + d(\dot{\alpha}_{CS}/i_t i_f - \dot{\alpha}_w) \quad (7.42)$$
$$-d_w\dot{\alpha}_w - \frac{1}{2}c_{air}A_L\rho_a r_{stat}^3\dot{\alpha}_w^2 - mc_{r2}r_{stat}^2\dot{\alpha}_w$$
$$-r_{stat}m_{CoG}\,(c_{r1} + g\sin{(\chi_{road})})$$

where the friction torque is described as viscous damping, with label $d_w$.

The complete model, named the Drive-shaft model, is obtained by inserting $T_c$ from (7.41) into (7.1), together with (7.42), which gives the following equations. An illustration of the model can be seen in Figure 7.9.

---

## Model 7.2 The Drive-Shaft Model

$$
\begin{aligned}
(J_e + J_t/i_t^2 + J_f/i_t^2 i_f^2)\ddot{\alpha}_{CS} &= T_e - T_{fric,e} - (d_t/i_t^2 + d_f/i_t^2 i_f^2)\dot{\alpha}_{CS} \quad (7.43) \\
&\quad -k(\alpha_{CS}/i_t i_f - \alpha_w)/i_t i_f \\
&\quad -d(\dot{\alpha}_{CS}/i_t i_f - \dot{\alpha}_w)/i_t i_f \\
(J_W + m_{CoG} r_{stat}^2)\ddot{\alpha}_w &= k(\alpha_{CS}/i_t i_f - \alpha_w) + d(\dot{\alpha}_{CS}/i_t i_f - \dot{\alpha}_w) \quad (7.44) \\
&\quad -(d_w + m_{CoG}c_{r2}r_{stat}^2)\dot{\alpha}_w - \frac{1}{2}c_{air}A_L\rho_a r_{stat}^3\dot{\alpha}_w^2 \\
&\quad -r_{stat}m_{CoG}\,(c_{r1} + g\sin{(\chi_{road})})
\end{aligned}
$$

The *Drive-shaft model* is the simplest model of three considered. The drive-shaft torsion, the engine speed, and the wheel speed are used as states according to

$$x_1 = \alpha_{CS}/i_t i_f - \alpha_w, \quad x_2 = \dot{\alpha}_{CS}, \quad x_3 = \dot{\alpha}_w \qquad (7.45)$$

More details of state-space descriptions are given in Section 7.3.5. For low gears, the influence from the air drag is low and by neglecting $\frac{1}{2}c_{air}A_L\rho_a r_{stat}^3\dot{\alpha}_w^2$ in (7.44), the model is linear in the states, but nonlinear in the parameters.

---

## Parameter estimation of the **Drive-shaft** model

A data set containing engine torque, engine speed, and wheel speed measurements are used to estimate the parameters and the initial conditions of the **Drive-shaft** model. The estimated parameters are

$$
\begin{aligned}
i &= i_t i_f, \quad l = r_{stat} m_{CoG} \left( c_{r1} + g \sin(\chi_{road}) \right) \\
J_1 &= J_e + J_t/i_t^2 + J_f/i_t^2 i_f^2, \quad J_2 = J_W + m_{CoG} r_{stat}^2 \\
d_1 &= d_t/i_t^2 + d_f/i_t^2 i_f^2, \quad d_2 = d_w + m_{CoG} c_{r2} r_{stat}^2
\end{aligned}
\tag{7.46}
$$

together with the stiffness, $k$, and the internal damping, $d$, of the drive shaft. The estimated initial conditions of the states are labeled $x_{10}$, $x_{20}$, and $x_{30}$, according to (7.45).

Figure 7.10 shows an example of how the model fits the measured data. The measured driveline speed are shown together with the model output, $x_1$, $x_2$, and $x_3$. According to the model, the clutch is stiff, and therefore, the transmission speed is equal to the engine speed scaled with the conversion ratio of the transmission ($i_t$). In the figure, this signal is shown together with the measured transmission speed. The plots are typical examples that show that a major part of the driveline dynamics is captured with a linear mass-spring model with the drive shafts as the main flexibility.

## Results of parameter estimation

- The main contribution to driveline dynamics from driving torque to engine speed and wheel speed is the drive shaft, explaining the first main resonance of the driveline.

- The true drive-shaft torsion ($x_1$) is unknown, but the value estimated by the model has physically reasonable values. These values will be further validated in Section 7.5.6.

- The model output transmission speed ($x_2/i_t$) fits the measured transmission speed data reasonably well, but there is still a systematic dynamics lag between model outputs and measurements.

## Influence from **Propeller-Shaft** Flexibility

The **Drive-shaft** model assumes stiff driveline from the engine to the final drive. The propeller shaft and the drive shaft are separated by the final drive, which has a small inertia compared to other inertias, e.g., the engine inertia. This section covers an investigation of how the model parameters of the **Drive-shaft** model are influenced by a flexible propeller shaft.

A model is derived with a stiff driveline from the engine to the output shaft of the transmission. The propeller shaft and the drive shafts are modeled as damped torsional flexibilities. As in the derivation of the **Drive-shaft** model, the transmission and the final drive are assumed to multiply the torque with the conversion ratio, without losses.

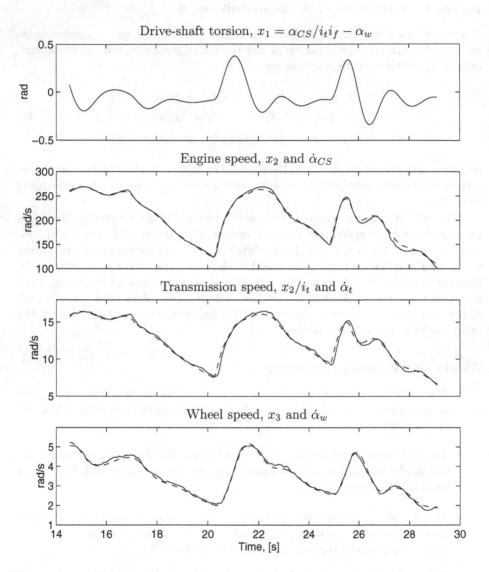

**Figure 7.10** The parameters of the Drive-shaft model estimated on data with step inputs in accelerator position using gear 1. The top figure shows the estimated drive-shaft torsion, and the bottom figures show the model outputs ($x_2$, $x_3$) in dashed lines, together with the measured driveline speeds in solid. The plots are typical examples of that a major part of the dynamics is captured by a linear model with a drive-shaft flexibility.

The derivation of the Drive-shaft model is repeated here with the difference that the model for the propeller shaft (7.34) is replaced by a model of a flexibility with stiffness $k_p$ and internal damping $d_p$

$$T_p = T_f = k_p(\alpha_t - \alpha_p) + d_p(\dot{\alpha}_t - \dot{\alpha}_p) = k_p(\alpha_{CS}/i_t - \alpha_p) + d_p(\dot{\alpha}_{CS}/i_t - \dot{\alpha}_p) \quad (7.47)$$

where (7.30) and (7.31) are used in the last equality. This formulation means that there are two torsional flexibilities, the propeller shaft and the drive shaft. Inserting (7.47) into (7.33) gives

$$J_t \ddot{\alpha}_{CS} = T_c i_t^2 - d_t \dot{\alpha}_{CS} - (k_p(\alpha_{CS}/i_t - \alpha_p) + d_p(\dot{\alpha}_{CS}/i_t - \dot{\alpha}_p)) i_t \quad (7.48)$$

By combining this with (7.1) the following differential equation describing the lumped engine and transmission results

$$(J_e + J_t/i_t^2)\ddot{\alpha}_{CS} = T_e - T_{fric,e} - d_t/i_t^2 \dot{\alpha}_{CS} \quad (7.49)$$
$$-\frac{1}{i_t}(k_p(\alpha_{CS}/i_t - \alpha_p) + d_p(\dot{\alpha}_{CS}/i_t - \dot{\alpha}_p))$$

The final drive is described by inserting (7.47) in (7.36) and using (7.35)

$$\alpha_p = \alpha_f i_f \quad (7.50)$$
$$J_f \ddot{\alpha}_f = i_f(k_p(\alpha_{CS}/i_t - \alpha_p) + d_p(\dot{\alpha}_{CS}/i_t - \dot{\alpha}_p)) - d_f \dot{\alpha}_f - T_d \quad (7.51)$$

Including (7.50) in (7.51) gives

$$J_f \ddot{\alpha}_p = i_f^2(k_p(\alpha_{CS}/i_t - \alpha_p) + d_p(\dot{\alpha}_{CS}/i_t - \dot{\alpha}_p)) - d_f \dot{\alpha}_p - i_f T_d \quad (7.52)$$

The equation for the drive shaft (7.40) is repeated with new labels

$$T_w = T_d = k_d(\alpha_f - \alpha_w) + d_d(\dot{\alpha}_f - \dot{\alpha}_w) = k_d(\alpha_p/i_f - \alpha_w) + d_d(\dot{\alpha}_p/i_f - \dot{\alpha}_w) \quad (7.53)$$

where (7.50) is used in the last equality.

The equation for the final drive (7.52) now becomes

$$J_f \ddot{\alpha}_p = i_f^2(k_p(\alpha_{CS}/i_t - \alpha_p) + d_p(\dot{\alpha}_{CS}/i_t - \dot{\alpha}_p)) - d_f \dot{\alpha}_p \quad (7.54)$$
$$-i_f(k_d(\alpha_p/i_f - \alpha_w) + d_d(\dot{\alpha}_p/i_f - \dot{\alpha}_w))$$

The equation for the wheel is derived by combining (7.11) with (7.53). The equation describing the wheel becomes

$$(J_W + m_{CoG} r_{stat}^2)\ddot{\alpha}_w = k_d(\alpha_p/i_f - \alpha_w) + d_d(\dot{\alpha}_p/i_f - \dot{\alpha}_w) \quad (7.55)$$
$$-d_w \dot{\alpha}_w - \frac{1}{2} c_{air} A_L \rho a r_{stat}^3 \dot{\alpha}_w^2 - m_{CoG} c_{r2} r_{stat}^2 \dot{\alpha}_w$$
$$-r_{stat} m_{CoG}(c_{r1} + g \sin(\chi_{road}))$$

where again the friction torque is assumed to be described by a viscous damping coefficient $d_w$. The complete model with drive shaft and propeller shaft flexibil-

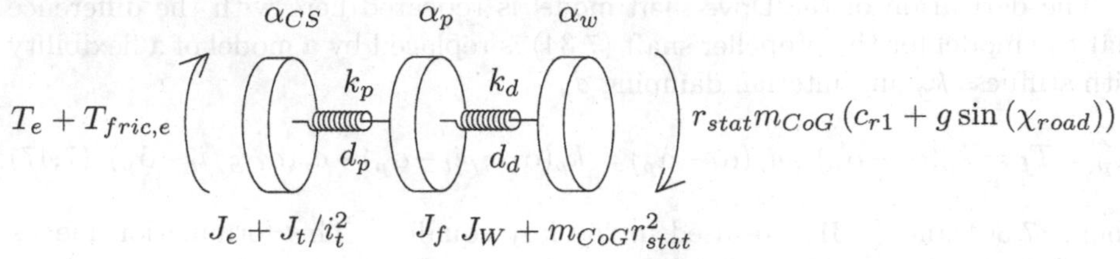

**Figure 7.11** Model with flexible propeller shaft and drive shaft.

ities is the following, which can be seen in Figure 7.11.

$$(J_e + J_t/i_t^2)\ddot{\alpha}_{CS} = T_e - T_{fric,e} - d_t/i_t^2\dot{\alpha}_{CS} \tag{7.56}$$
$$- \frac{1}{i_t}\left(k_p(\alpha_{CS}/i_t - \alpha_p) + d_p(\dot{\alpha}_{CS}/i_t - \dot{\alpha}_p)\right)$$

$$J_f\ddot{\alpha}_p = i_f^2\left(k_p(\alpha_{CS}/i_t - \alpha_p) + d_p(\dot{\alpha}_{CS}/i_t - \dot{\alpha}_p)\right) \tag{7.57}$$
$$- d_f\dot{\alpha}_p - i_f\left(k_d(\alpha_p/i_f - \alpha_w) + d_d(\dot{\alpha}_p/i_f - \dot{\alpha}_w)\right)$$

$$(J_W + m_{CoG}r_{stat}^2)\ddot{\alpha}_w = k_d(\alpha_p/i_f - \alpha_w) + d_d(\dot{\alpha}_p/i_f - \dot{\alpha}_w) \tag{7.58}$$
$$- (d_w + m_{CoG}c_{r2}r_{stat}^2)\dot{\alpha}_w - \frac{1}{2}c_{air}A_L\rho_a r_{stat}^3\dot{\alpha}_w^2$$
$$- r_{stat}m_{CoG}(c_{r1} + g\sin(\chi_{road}))$$

The model equations (7.56) to (7.58) describe the **Drive-shaft model** extended with the propeller shaft with stiffness $k_p$ and damping $d_p$. The three inertias in the model are

$$\begin{aligned} J_1 &= J_e + J_t/i_t^2 \\ J_2 &= J_f \\ J_3 &= J_W + m_{CoG}r_{stat}^2 \end{aligned} \tag{7.59}$$

If the magnitude of the three inertias are compared, the inertia of the final drive ($J_f$) is considerably less than $J_1$ and $J_2$ in (7.59). Therefore, the model will act as if there are two damped springs in series. The total stiffness of two undamped springs in series is

$$k = \frac{k_p i_f^2 k_d}{k_p i_f^2 + k_d} \tag{7.60}$$

whereas the total damping of two dampers in series is

$$d = \frac{d_p i_f^2 d_d}{d_p i_f^2 + d_d}$$

The damping and the stiffness of the drive shaft in the previous section will thus typically be underestimated due to the flexibility of the propeller shaft. This effect will increase with lower conversion ratio in the final drive, $i_f$. The individual stiffness values obtained from parameter estimation are somewhat lower than the values obtained from material data.

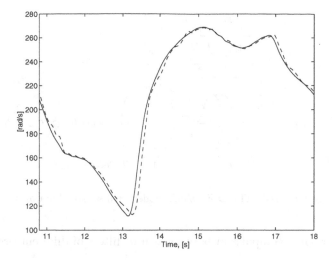

**Figure 7.12** Measured engine speed (solid) and transmission speed (dashed). The transmission speed is multiplied with the conversion ratio of the transmission, $i_t$.

### Deviations between Engine Speed and Transmission Speed

As mentioned above, there is good agreement between model output and experimental data for $u = T_e - T_{fric,e}$, $\dot{\alpha}_{CS}$, and $\dot{\alpha}_w$, but there is a slight deviation between measured and estimated transmission speed. With the Drive-shaft model, stiff dynamics between the engine and the transmission is assumed, and hence the only difference between the model outputs engine speed and transmission speed is the gain $i_t$ (conversion ratio of the transmission). However, a comparison between the measured engine speed and transmission speed shows that there is not only a gain difference according to Figure 7.12. This deviation has a character of a phase shift and some smoothing (signal levels and shapes agree). This indicates that there is some additional dynamics between engine speed, $\dot{\alpha}_{CS}$, and transmission speed, $\dot{\alpha}_t$. Two natural candidates are additional mass-spring dynamics in the driveline, or sensor dynamics. The explanation is that there is a combined effect, with the major difference explained by the sensor dynamics. The motivation for this is that the high stiffness of the clutch flexibility (given from material data) cannot result in a phase shift form of the magnitude shown in Figure 7.12. Neither can backlash in the transmission explain the difference, because then the engine and transmission speeds would be equal when the backlash is at its endpoints.

As mentioned before, the bandwidth of the measured transmission speed is lower than the measured engine and wheel speeds, due to fewer cogs in the sensor. It is assumed that the engine speed and the wheel speed sensor dynamics are not influencing the data for the frequencies considered. The speed dependence of the transmission sensor dynamics is neglected. The following sensor dynamics are

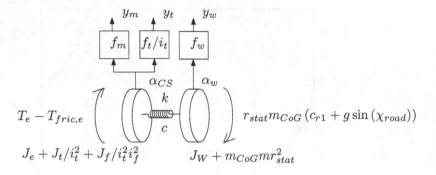

**Figure 7.13** The Drive-shaft model with sensor dynamics.

assumed, after some comparison between sensor filters of different order,

$$
\begin{aligned}
f_m &= 1 \\
f_t &= \frac{1}{1+\gamma s} \\
f_w &= 1
\end{aligned}
\tag{7.61}
$$

where a first order filter with an unknown parameter $\gamma$ models the transmission sensor. Figure 7.13 shows the configuration with the Drive-shaft model and sensor filter $f_m$, $f_t$, and $f_w$. The outputs of the filters are $y_m$, $y_t$, and $y_w$.

Now the parameters, the initial condition, and the unknown filter constant $\gamma$ can be estimated such that the model outputs $(y_m, y_t, y_w)$ fit the measured data. The result of this is seen in Figure 7.14 for gear 1. The conclusion is that the main part of the deviation between engine speed and transmission speed is due to sensor dynamics. Figure 7.15 shows an enlarged plot of the transmission speed, with the model output from the Drive-shaft model with and without sensor filtering.

### Results of parameter estimation

- If the Drive-shaft model is extended with a first order sensor filter for the transmission speed, all three velocities $(\dot{\alpha}_{CS}, \dot{\alpha}_t, \dot{\alpha}_w)$ are estimated by the model. The model outputs fit the data except for some time intervals where there are deviations between model and measured data (see Figure 7.15). However, these deviations will in the following be related to nonlinearities at low clutch torques.

### Model with Flexible Clutch and Drive Shaft

The clutch has so far been assumed stiff and the main contribution to low-frequency oscillations is the drive-shaft flexibility. However, measured data suggests that there is some additional dynamics between the engine and the transmission. The candidate which is most flexible is the clutch. Hence, the model will include two torsional flexibilities, the drive shaft, and the clutch. With this

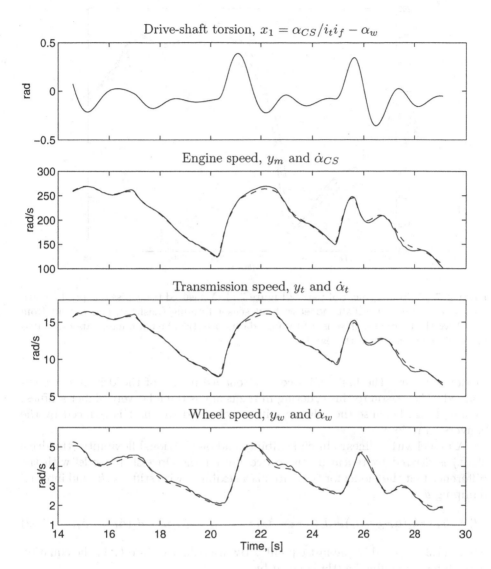

**Figure 7.14** Parameter estimation of the Drive-shaft model as in Figure 7.10, but with sensor dynamics included. The top figure shows the estimated drive-shaft torsion, and the bottom figures show the model outputs $(y_m, y_t, y_w)$ in dashed, together with the measured data in solid. The main part of the deviation between engine speed and transmission speed is due to sensor dynamics. See also Figure 7.15.

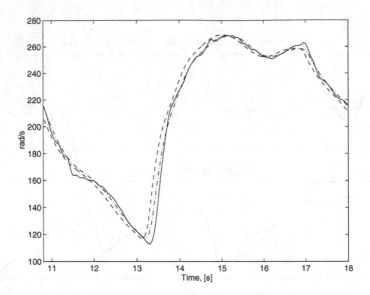

**Figure 7.15** Enlargement of part of Figure 7.14. Measured transmission speed (solid), output from the Drive-shaft model without sensor filtering (dashed), and output from the Drive-shaft model with sensor filtering (dash-dotted). The parameters are estimated based on experiments with gear 1.

model structure, the first and second resonance modes of the driveline are explained. The reason to this ordering in frequency is the relatively higher stiffness in the clutch, because the relative stiffness of the drive shaft is reduced by the conversion ratio.

A model with a linear clutch flexibility and one torsional flexibility (the drive shaft) is derived by repeating the procedure for the Drive-shaft model with the difference that the model for the clutch is a flexibility with stiffness $k_c$ and internal damping $d_c$

$$T_c = T_t = k_c(\alpha_{CS} - \alpha_c) + d_c(\dot{\alpha}_{CS} - \dot{\alpha}_c) = k_c(\alpha_{CS} - \alpha_t i_t) + d_c(\dot{\alpha}_{CS} - \dot{\alpha}_t i_t) \quad (7.62)$$

where (7.31) is used in the last equality. By inserting this into (7.1) the equation describing the engine inertia is given by

$$J_e \ddot{\alpha}_{CS} = T_e - T_{fric,e} - (k_c(\alpha_{CS} - \alpha_t i_t) + d_c(\dot{\alpha}_{CS} - \dot{\alpha}_t i_t)) \quad (7.63)$$

Also by inserting (7.62) into (7.32), the equation describing the transmission is

$$J_t \ddot{\alpha}_t = i_t (k_c(\alpha_{CS} - \alpha_t i_t) + d_c(\dot{\alpha}_{CS} - \dot{\alpha}_t i_t)) - d_t \dot{\alpha}_t - T_p \quad (7.64)$$

$T_p$ is derived from (7.37) giving

$$(J_t + J_f/i_f^2)\ddot{\alpha}_t = i_t (k_c(\alpha_{CS} - \alpha_t i_t) + d_c(\dot{\alpha}_{CS} - \dot{\alpha}_t i_t)) - (d_t + d_f/i_f^2)\dot{\alpha}_t - T_d/i_f$$
$$(7.65)$$

which is the equation describing the lumped transmission, propeller shaft, and final drive inertia.

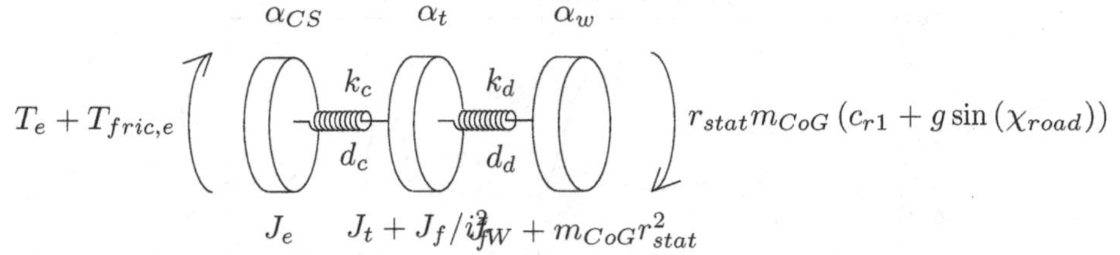

$$\alpha_{CS} \qquad \alpha_t \qquad \alpha_w$$

**Figure 7.16** The Clutch and drive-shaft model: Linear clutch and drive-shaft torsional flexibility.

The drive shaft is modeled according to (7.40) as

$$T_w = T_d = k_d(\alpha_f - \alpha_w) + d_d(\dot{\alpha}_f - \dot{\alpha}_w) = k_d(\alpha_t/i_f - \alpha_w) + d_d(\dot{\alpha}_t/i_f - \dot{\alpha}_w) \quad (7.66)$$

where (7.34) and (7.35) are used in the last equality.

The complete model, named the Clutch and drive-shaft model, is obtained by inserting (7.66) into (7.65) and (7.11). An illustration of the model can be seen in Figure 7.16.

---

## Model 7.3 The Clutch and Drive-Shaft Model

---

$$
\begin{aligned}
J_e\ddot{\alpha}_{CS} &= T_e - T_{fric,e} - (k_c(\alpha_{CS} - \alpha_t i_t) \\
&\quad + d_c(\dot{\alpha}_{CS} - \dot{\alpha}_t i_t)) \quad (7.67) \\
(J_t + J_f/i_f^2)\ddot{\alpha}_t &= i_t(k_c(\alpha_{CS} - \alpha_t i_t) + d_c(\dot{\alpha}_{CS} - \dot{\alpha}_t i_t)) \quad (7.68) \\
&\quad -(d_t + d_f/i_f^2)\dot{\alpha}_t - \frac{1}{i_f}(k_d(\alpha_t/i_f - \alpha_w) \\
&\quad + d_d(\dot{\alpha}_t/i_f - \dot{\alpha}_w)) \\
(J_W + m_{CoG}r_{stat}^2)\ddot{\alpha}_w &= k_d(\alpha_t/i_f - \alpha_w) + d_d(\dot{\alpha}_t/i_f - \dot{\alpha}_w) \quad (7.69) \\
&\quad -(d_w + c_{r2}r_{stat})\dot{\alpha}_w - \frac{1}{2}c_{air}A_L\rho_a r_{stat}^3\dot{\alpha}_w^2 \quad (7.70) \\
&\quad -r_{stat}m_{CoG}(c_{r1} + g\sin(\chi_{road}))
\end{aligned}
$$

The clutch torsion, the drive-shaft torsion, and the driveline speeds are used as states according to

$$x_1 = \alpha_{CS} - \alpha_t i_t, \quad x_2 = \alpha_t/i_f - \alpha_w, \quad x_3 = \dot{\alpha}_{CS}, \quad x_4 = \dot{\alpha}_t, \quad x_5 = \dot{\alpha}_w \quad (7.71)$$

---

More details about state-space representations and parameters are covered in Section 7.3.4. For low gears, the influence from the air drag is low and by neglecting $\frac{1}{2}c_{air}A_L\rho_a r_{stat}^3\dot{\alpha}_w^2$ in (7.69), the model is linear in the states, but nonlinear in the parameters. The model equipped with the sensor filter in (7.61) gives the true sensor outputs ($y_m, y_t, y_w$).

---

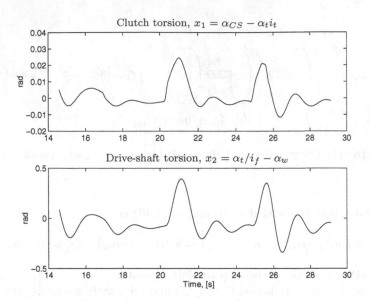

**Figure 7.17** Clutch torsion (top figure) and drive-shaft torsion (bottom figure) resulting from parameter estimation of the Clutch and drive-shaft model with sensor filtering, on data with gear 1. The true values of these torsions are not known, but the plots show that the drive-shaft torsion has realistic values.

## Parameter estimation of the Clutch and drive-shaft model

The parameters and the initial conditions of the Clutch and drive-shaft model are estimated with the sensor dynamics described above, in the same way as the Drive-shaft model in this section. A problem when estimating the parameters of the Clutch and drive-shaft model is that the bandwidth of the measured signals is not enough to estimate the stiffness $k_c$ in the clutch. Therefore, the value of the stiffness given from material data is used and fixed, and the rest of the parameters are estimated.

The resulting clutch torsion ($x_1$) and the drive-shaft torsion ($x_2$) are shown in Figure 7.17. The true values of these torsions are not known, but the figure shows that the amplitude of the drive-shaft torsion has realistic values that agree with material data. However, the clutch torsion does not have realistic values (explained later), which can be seen when comparing with the static nonlinearity in Figure 7.18.

The model output velocities ($\dot{\alpha}_{CS}$, $\dot{\alpha}_t$, $\dot{\alpha}_w$) show no improvement compared to those generated by the Drive-shaft model with sensor dynamics, displayed in Figure 7.14.

## Results of parameter estimation

- The model including a linear clutch does not improve the data fit. The interpretation of this is that the clutch model does not add information for frequencies in the measured data.

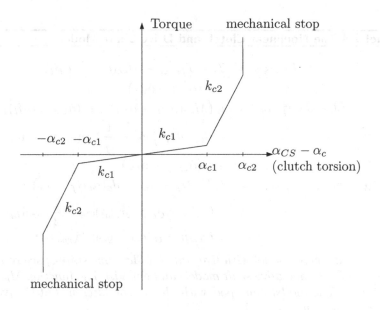

**Figure 7.18** Nonlinear clutch characteristics.

## Nonlinear Clutch and Drive-Shaft Flexibility

When studying a clutch in more detail it is seen that the torsional flexibility is a result of an arrangement with smaller springs in series with springs with much higher stiffness. The reason for this arrangement is vibration insulation. When the angle difference over the clutch starts from zero and increases, the smaller springs, with stiffness $k_{c1}$, are being compressed. This ends when they are fully compressed at $\alpha_{c1}$ radians. If the angle is increased further, the stiffer springs, with stiffness $k_{c2}$, are beginning to be compressed. When $\alpha_{c2}$ is reached, the clutch hits a mechanical stop. This clutch characteristics can be modeled as in Figure 7.18. The resulting stiffness $k_c(\alpha_{CS} - \alpha_c)$ of the clutch is given by

$$k_c(x) = \begin{cases} k_{c1} & \text{if } |x| \leq \alpha_{c1} \\ k_{c2} & \text{if } \alpha_{c1} < |x| \leq \alpha_{c2} \\ \infty & \text{otherwise} \end{cases} \qquad (7.72)$$

The torque $M_{kc}(\alpha_{CS} - \alpha_c)$ from the clutch nonlinearity is

$$M_{kc}(x) = \begin{cases} k_{c1}x & \text{if } |x| \leq \alpha_{c1} \\ k_{c1}\alpha_{c1} + k_{c2}(x - \alpha_{c1}) & \text{if } \alpha_{c1} < x \leq \alpha_{c2} \\ -k_{c1}\alpha_{c1} + k_{c2}(x + \alpha_{c1}) & \text{if } -\alpha_{c2} < x \leq -\alpha_{c1} \\ \infty & \text{otherwise} \end{cases} \qquad (7.73)$$

If the linear clutch in the Clutch and drive-shaft model is replaced by the clutch nonlinearity according to Figure 7.18, the following model, called the Nonlinear clutch and drive-shaft model, is derived.

---

## Model 7.4 The Nonlinear Clutch and Drive-Shaft Model

$$J_e \ddot{\alpha}_{CS} = T_e - T_{fric,e} - M_{kc}(\alpha_{CS} - \alpha_t i_t) \tag{7.74}$$
$$- d_c(\dot{\alpha}_{CS} - \dot{\alpha}_t i_t)$$

$$(J_t + J_f/i_f^2)\ddot{\alpha}_t = i_t \left( M_{kc}(\alpha_{CS} - \alpha_t i_t) + d_c(\dot{\alpha}_{CS} - \dot{\alpha}_t i_t) \right) \tag{7.75}$$
$$- (d_t + d_f/i_f^2)\dot{\alpha}_t - \frac{1}{i_f}\left( k_d(\alpha_t/i_f - \alpha_w) \right.$$
$$\left. + d_d(\dot{\alpha}_t/i_f - \dot{\alpha}_w) \right)$$

$$(J_W + m_{CoG}r_{stat}^2)\ddot{\alpha}_w = k_d(\alpha_t/i_f - \alpha_w) + d_d(\dot{\alpha}_t/i_f - \dot{\alpha}_w) \tag{7.76}$$
$$- (d_w + m_{CoG}c_{r2}r_{stat})\dot{\alpha}_w - \frac{1}{2}c_{air}A_L\rho_a r_{stat}^3 \dot{\alpha}_w^2$$
$$- r_{stat}m_{CoG}\left( c_{r1} + g\sin(\chi_{road}) \right)$$

Nonlinear driveline model with five states. (The same state-space representation as for the Clutch and drive-shaft model can be used.) The function $M_{kc}(\cdot)$ is given by (7.73). The model equipped with the sensor filter in (7.61) gives the true sensor outputs ($y_m$, $y_t$, $y_w$).

---

### Parameter estimation of the Nonlinear clutch and drive-shaft model

When estimating the parameters and the initial conditions of the Nonlinear clutch and drive-shaft model, the clutch static nonlinearity is fixed with known physical values and the rest of the parameters are estimated, except for the sensor filter which is the same as in the previous model estimations.

The resulting clutch torsion ($x_1 = \alpha_{CS} - \alpha_t i_t$) and drive-shaft torsion ($x_2 = \alpha_t/i_f - \alpha_w$) are shown in Figure 7.19. The true values of these torsions are not known as mentioned before. However, the figure shows that both angles have realistic values that agree with other experience. The model output velocities ($\dot{\alpha}_{CS}$, $\dot{\alpha}_t$, $\dot{\alpha}_w$) show no improvement compared to those generated by the Drive-shaft model with sensor dynamics, displayed in Figure 7.14.

In Figure 7.15 it was seen that the model with the sensor filtering fitted the signal except for a number of time intervals with deviations. The question is if this is a result of some nonlinearity. Figure 7.20 shows the transmission speed plotted together with the model output and the clutch torsion. It is clear from this figure that the deviation between model and experiments occurs when the clutch angle passes the area with the low stiffness in the static nonlinearity (see Figure 7.18).

### Results of parameter estimation

- The model including the nonlinear clutch does not improve the overall data fit for frequencies in the measured data.

- The model is able to estimate a clutch torsion with realistic values.

- The estimated clutch torsion shows that when the clutch passes the area with low stiffness in the nonlinearity, the model deviates from the data. The reason is unmodeled dynamics at low clutch torques [8].

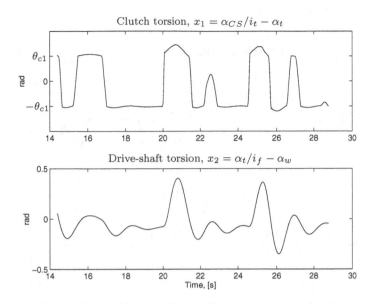

**Figure 7.19** Clutch torsion (top figure) and drive-shaft torsion (bottom figure) resulting from parameter estimation of the Nonlinear clutch and drive-shaft model with sensor filtering, on data with gear 1. The true values of these torsions are not known, but the plots show that they have realistic values.

### Model Validity

In the parameter estimation, the unknown load, $l$, which vary between the trials, is estimated. The load can be recalculated to estimate road slope, and the calculated values agree well with the known values of the road slopes at Scania. Furthermore, the estimation of the states describing the torsion of the clutch and the drive shaft shows realistic values. This gives further support to model structure and parameters.

The assumption about sensor dynamics in the transmission speed influencing the experiments, agrees well with the fact that the engine speed sensor and the wheel speed sensor have considerably higher bandwidth (more cogs) than the transmission speed sensor.

When estimating the parameters of the Drive-shaft model, there is a problem when identifying the viscous friction components $d_1$ and $d_2$. The sensitivity in the model to variations in the friction parameters is low, and the same model fit can be obtained for a range of friction parameters. However the sum $d_1 i^2 + d_2$ is constant during these tests. The problem with estimating viscous parameters will be further discussed later.

### Summary of Modeling Example

Parameter estimation shows that a model with one torsional flexibility and two inertias is able to fit the measured engine speed and wheel speed in a frequency regime including the first main resonance of the driveline. By considering the difference between measured transmission speed and wheel speed it is reasonable to deduce that the main flexibility is the drive shafts.

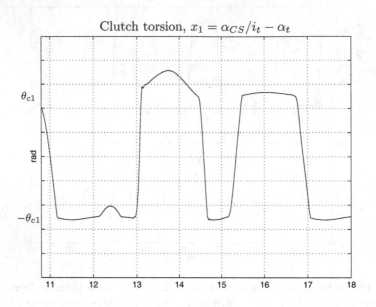

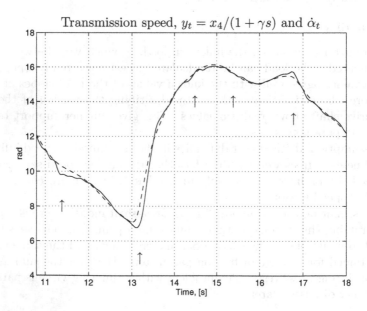

**Figure 7.20** Clutch torsion (top figure) and measured and estimated transmission speeds (bottom figure) from the Nonlinear clutch and drive-shaft model with sensor dynamics with gear 1. The result is that the main differences between model (dashed) and experiments (solid) occur when the clutch torsion passes the area with the low stiffness ($|\theta| < \theta_{c1}$) in the static clutch nonlinearity.

In order for the model to also fit the measured transmission speed, a first order sensor filter is added to the model, in accordance with properties of the sensor system. It is shown that all three velocities are fitted accurately enough. Parameter estimation of a model with a nonlinear clutch explains that the difference between the measured data and the model outputs occurs when the clutch transfers zero torque.

Further supporting facts of the validity of the models are that they give values to the non-measured variables, drive shaft and clutch torsion, that agree with experience from other sources. Furthermore, the known road slopes are well estimated.

The result is a series of models that describe the driveline in increasing detail by, in each extension, adding the effect that seems to be the major cause for the deviation still left.

The result, from a user perspective, is that, within the frequency regime interesting for control design, the Drive-shaft model with some sensor dynamics gives good agreement with experiments. It is thus suitable for control design. The major deviations left are captured by the nonlinear effects in the Nonlinear clutch and drive-shaft model, which makes this model suitable for verifying simulation studies in control design.

# 7.2 Modeling of Neutral Gear

An important basis for design of driveline management is to understand the dynamic behavior of the driveline before and after going from a gear to neutral. This requires additional modeling of the driveline since it is separated in two parts when in neutral. Such a decoupled model is the topic of this section, together with its use for analysis of possible oscillation patterns of the decoupled driveline.

A decoupled model has several applications. It is the basis for a diagnosis system of gear-shift quality. The analysis cast light on the sometimes, at first sight, surprising oscillations that occur in an uncontrolled driveline. It is also an indication for the value of feedback control.

## 7.2.1 Stationary Gear-Shift Experiments

First, a series of gear-shifts with a stationary driveline are performed without using driveline torque control. This means that a speed controller controls the engine speed to a desired level, and when the driveline speeds have reached stationary levels, engagement of neutral gear is commanded. Figure 7.21 shows two of these trials where the engine speed is 1400 RPM and 2100 RPM respectively, on a flat road with gear 1. The behavior of the engine speed, the transmission speed, and the wheel speed is shown in the figure. At $t = 14$ s, a shift to neutral is commanded. A gear shift is performed by using a gear lever actuator driven by air pressure. A delay-time from commanded gear shift to activated gear-lever movement is seen in the experiments. This is a combined effect from a delay in the actuator, and a delay in building up the air pressure needed to overcome friction. This delay is longer the higher the speed is.

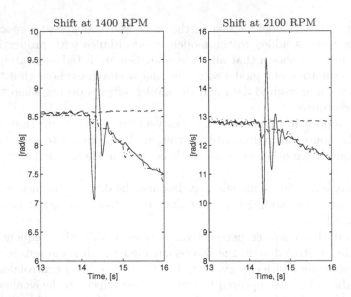

**Figure 7.21** Engagement of neutral gear commanded at 14 s, with stationary driveline at 1400 RPM and 2100 RPM on a flat road with gear 1. Engine speed (dashed) and wheel speed (dash-dotted) are scaled to transmission speed which is seen in solid lines. After a delay time, neutral gear is engaged, causing the driveline speeds to oscillate. The amplitude of the oscillating transmission speed is higher the higher the speed is.

After the shift, the driveline is decoupled into two parts. The movement of the engine speed is independent of the movement of the transmission speed and the wheel speed, which are connected by the propeller shaft and the drive shaft, according to Figure 7.1. The speed controller maintains the desired engine speed also after the gear shift. The transmission speed and the wheel speed, on the other hand, are only affected by the load (rolling resistance, air drag, and road inclination), which explains the decreasing speeds in the figure.

The transient behavior of the transmission speed and the wheel speed differ however, and the energy built up in the shafts is seen to affect the transmission speed more than the wheel speed, giving an oscillating transmission speed. The higher the speed is, the higher amplitude of the oscillations is obtained. The amplitude value of the oscillations for 1400 RPM is 2.5 rad/s, and 5 rad/s for 2100 RPM.

## 7.2.2   Dynamical Gear-Shift Experiments

In the previous trials there was no relative speed difference, since the driveline was in a stationary mode. If a relative speed difference is present prior to the gear-shift, there will be a different type of oscillation. Figures 7.22 and 7.23 describe two trials where neutral gear is engaged with an oscillating driveline without torque control. The oscillations are a result of an engine torque pulse at 11.7 s.

There is only a small difference between the measured engine speed and transmission speed prior to the gear shift. This difference was in Section 7.1 explained

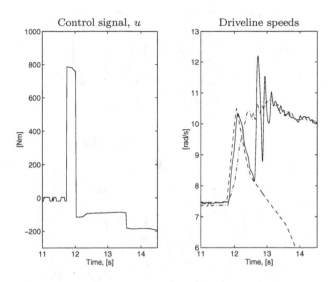

**Figure 7.22** Engaged neutral gear without torque control at 12.5 s in a trial with oscillating driveline as a result of a provoking engine torque pulse at 11.7 s in the left figure. Engine speed (dashed) and wheel speed (dash-dotted) are scaled to transmission speed (solid) in the right figure. After the gear shift the transmission speed oscillates.

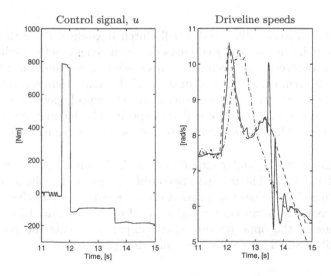

**Figure 7.23** Same field trial as in Figure 7.22, but with engaged neutral gear at 13.2 s. Engine speed (dashed) and wheel speed (dash-dotted) are scaled to transmission speed (solid) in the right figure. After the gear shift the transmission speed oscillates.

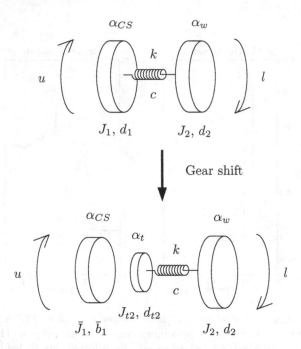

**Figure 7.24** Description of how the driveline model changes after engagement of neutral gear. The first model is the Drive-shaft model, which is then separated into two sub-models when neutral gear is engaged. The left part consists of the engine and one part of the transmission. The right part of the model consists of the rest of the transmission and the drive shaft out to the wheels, called the Decoupled model.

to be a result of a sensor filter and a stiff clutch flexibility. After the gear shift, the energy built up in the shafts is released, which generates the oscillations and minimizes the difference between the transmission speed and the wheel speed. The two speeds then decrease as a function of the load. Hence, a relative speed difference between the transmission speed and the wheel speed at the shift moment gives oscillations in the transmission speed. The larger the relative speed difference is, the higher the amplitude of the oscillating transmission speed will be.

Figure 7.23 shows a similar experiment as in Figure 7.22, but with neutral gear engaged at 13.2 s. The relative speed difference has opposite sign compared to that in Figure 7.22. The transmission speed transfers to the wheel speed, and these two decrease as a function of the load. However, initially the transmission speed deviates in the opposite direction compared to how the relative speed difference indicates, which seems like a surprising behavior.

## 7.2.3   A Decoupled Model

A decoupled model is needed to analyze and explain the three different types of oscillations described by Figures 7.21, 7.22, and 7.23. Engaging neutral gear can be described as in Figure 7.24. Before the gear shift, the driveline dynamics is described by the Drive-shaft model. This model assumes a lumped engine and transmission inertia, as described previously. When neutral gear is engaged, the driveline is separated into two parts as indicated in the figure. The two parts move independent of each other, as mentioned before. The engine side of the model consists of the engine, the clutch, and part of the transmission (characterized by the parameters $J_{t1}$ and $d_{t1}$ according to Section 7.1.4). The parameters describing the lumped engine, clutch, and part of transmission are $\bar{J}_1$ and $\bar{b}_1$ according to the figure. The wheel side of the model consists of the rest of the transmission (characterized by the parameters $J_{t2}$ and $d_{t2}$) and the drive-shaft flexibility out to the wheels, which is named the Decoupled model. The model is described by the following equations.

---

**Model 7.5 The Decoupled Model**

---

$$J_{t2}\ddot{\alpha}_t = -d_{t2}\dot{\alpha}_t - k(\alpha_t/i_f - \alpha_w)/i_f - d(\dot{\alpha}_t/i_f - \dot{\alpha}_w)/i_f \qquad (7.77)$$
$$J_2\ddot{\alpha}_w = k(\alpha_t/i_f - \alpha_w) + d(\dot{\alpha}_t/i_f - \dot{\alpha}_w) - d_2\dot{\alpha}_w - l \qquad (7.78)$$

*The model equipped with the sensor filter in (7.61) gives the true sensor outputs* $(y_t, y_w)$.

---

All these parameters were estimated in Section 7.1.4, except the unknown parameters $J_{t2}$ and $d_{t2}$. The model is written in state-space form by using the states $x_1 =$ drive-shaft torsion, $x_2 =$ transmission speed, and $x_3 =$ wheel speed.

Note that the Decoupled model after the gear shift has the same model structure as the Drive-shaft model, but with the difference that the first inertia is considerable less in the Decoupled model, since the engine and part of the transmission are decoupled from the model.

### Quality of the Decoupled Model

The unknown parameters $J_{t2}$ and $d_{t2}$ can be estimated if the dynamics described by the Decoupled model is excited. This is the case when engaging neutral gear at a transmission torque level different from zero, giving oscillations. One such case is seen in Figure 7.25, where the oscillating transmission speed is seen together with the Decoupled model with estimated parameters $J_{t2}$ and $d_{t2}$, and initial drive-shaft torsion, $x_{10}$. The rest of the parameters are the same as in the Drive-shaft model, which were estimated in Section 7.1.4. The rest of the initial condition of the states (transmission speed and wheel speed) are the measured values at the time for the gear shift. The model output ($y_t$ and $y_w$ with sensor filter) are fitted to the measured transmission speed and wheel speed. The conclusion is that the Decoupled model is able to capture the main resonance in the oscillating transmission speed.

If the initial states (drive-shaft torsion, transmission speed, and wheel speed) of the Decoupled model are known at the time for engaging neutral gear, the behavior of the speeds after the shift can be predicted.

The different characteristic oscillations seen in the experiments after engaged neutral gear are explained by the value of the drive-shaft torsion and the relative speed difference at the time of engagement. The Decoupled model can be used to

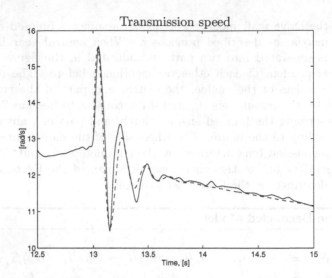

**Figure 7.25** Measured oscillations after a gear shift at 13.0 s in solid line. The outputs of the Decoupled model are fitted to data, shown in dashed line. The Decoupled model is able to capture the main resonance in the oscillating transmission speed after the gear shift.

predict the behavior of the driveline speeds if these initial variables are known. The demonstration of problems with an uncontrolled driveline motivates the need for feedback control in order to minimize the oscillations after a gear shift.

## 7.3 Driveline Control

There are two types of variables that are of special interest in driveline control:

- (rotational) velocities

- torques

Since the parts of a vehicular driveline (engine, clutch, transmission, shafts, and wheels) are elastic, velocities and torques differ along the driveline. It also means that mechanical resonances may occur. The handling of such resonances is basic for functionality and driveability, but is also important for reducing mechanical stress and noise. New driveline-management applications and high-powered engines increase the need for strategies for how to apply engine torque in an optimal way.

Two important applications for driveline control are

- driveline speed control

- driveline control for gear shifting

These two applications will be treated in the following sections. The first application is important to handle wheel-speed oscillations following from a change in accelerator pedal position or from impulses from towed trailers and road rough-ness, known as vehicle shuffle.

The second application, driveline control for gear shifting, is used to imple-ment automatic gear shifting. In todays traffic it is desired to have an automatic gear shifting system on heavy trucks. One approach at the leading edge of tech-nology is gear shifting by engine control [91]. With this approach, disengaging the clutch is replaced by controlling the engine to a state where the transmission transfers zero torque, and by that realizing a virtual clutch. After neutral gear is engaged, the engine speed is controlled to a speed such that the new gear can be engaged. The gear shifting system uses a manual transmission with automated gear lever, and a normal friction clutch that is engaged only at start and stop.

The total time needed for a gear shift is an important quality measure. One reason for this is that the vehicle is free-rolling, since there is no driving torque, which may be serious with heavy loads and large road slopes. The difference in engine torque before a gear shift and at the state where the transmission transfers zero torque is often large. Normally, this torque difference is driven to zero by sliding the clutch. With gear shifting by engine control, the aim is to decrease the time needed for this phase by using engine control. However, a fast step in engine torque may lead to excited driveline resonances. If these resonances are not damped, the time to engage neutral gear increases, since one has to wait for satisfactory gear-shift conditions. Furthermore, engaging neutral gear at a non-zero transmission torque results in oscillations in the transmission speed, which is disturbing for the driver, and increases the time needed to engage the new gear. These problems motivate the need for using feedback control in order to reach zero transmission torque. Two major problems must be addressed to obtain this. First, the transmission torque must be estimated and validated. Then a strategy must be derived that drives this torque to zero with damped driveline resonances.

## 7.3.1 Background

### Fuel-Injection Strategy for Speed Control

As described in the previous section, fuel-injection strategy can be of torque con-trol type or speed control type. For diesel engines, speed control is often referred to as RQV control, and torque control referred to as RQ control [6]. With RQ control, the driver's accelerator pedal position is interpreted as a desired engine torque, and with RQV control the accelerator position is interpreted as a desired engine speed. RQV control is essentially a proportional controller calculating the fuel amount as function of the difference between the desired speed set by the driver and the actual measured engine speed. The reason for this controller structure is the traditionally used mechanical centrifugal governor for diesel pump control [6]. This means that the controller will maintain the speed demanded by the driver, but with a stationary error (velocity lag), which is a function of the controller gain and the load (rolling resistance, air drag, and road inclination). With a cruise controller, the stationary error is compensated for, which means that the vehicle will maintain the same speed independent of load changes. This requires an integral part of the controller which is not used in the RQV control concept.

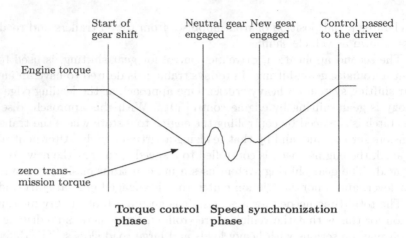

**Figure 7.26** Engine torque during the different phases in automatic gear shifting by engine control. The engine torque is controlled to a state where the transmission transfers zero torque, whereafter neutral gear is engaged without using the clutch. After the speed synchronization phase, the new gear is engaged, and control is transferred back to the driver.

### Automatic Gear Shifting in Heavy Trucks

Traditionally a gear shift is performed by disengaging the clutch, engaging neutral gear, shifting to a new gear, and engaging the clutch again. In todays traffic it is desired to have an automatic gear shifting system on heavy trucks. The following three approaches are used:

**Automatic transmission** This approach is seldom used for the heaviest trucks, due to expensive transmissions and problems with short life time. Another drawback is the efficiency loss compared to manual transmissions.

**Manual transmission and automatic clutch** A quite common approach, which needs an automatic clutch system [88]. This system has to be made robust against clutch wear.

**Manual transmission with gear shifting by engine control** By means of this approach the automatic clutch is replaced by engine control, realizing a virtual clutch. The only addition needed to a standard manual transmission is an actuator to move the gear lever. Lower cost and higher efficiency characterize this solution.

With this last approach a gear shift includes the phases described in Figure 7.26, where the engine torque during the shift event is shown.

## 7.3.2   Field Trials for Problem Demonstration

A number of field trials are performed in order to describe how driveline resonances influence driveline management.

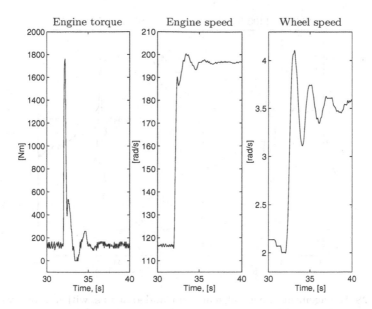

**Figure 7.27** Measured speed response of a step in accelerator position at t=32 s. An RQV speed controller controls the engine speed to 2000 RPM. The engine speed is well damped, but the resonances in the driveline is seen to give oscillating wheel speed, resulting in vehicle shuffle.

### Driveline speed control

A specific example of how the RQV speed controller performs is seen in Figure 7.27. The figure shows how the measured engine speed and wheel speed respond to a step input in accelerator position. It is seen how the engine speed is well behaved with no oscillations. With a stiff driveline this would be equivalent with also having well damped wheel speed. The more flexible the driveline is, the less sufficient a well damped engine speed is, since the flexibility of the driveline will lead to oscillations in the wheel speed. This will be further discussed and demonstrated in later sections.

If it is desired to decrease the response time of the RQV controller (i.e., increase the bandwidth), the controller gain must be increased. Then the amplitude of the oscillations in the wheel speed will be higher.

### Driveline torque control

When using gear shifting by engine control, the phases in Figure 7.26 are accomplished. First, control is transferred from the driver to the control unit, entering the *torque control phase*. The engine is controlled to a torque level corresponding to zero transferred torque in the transmission. After neutral gear is engaged, the *speed synchronization phase* is entered. Then the engine speed is controlled to track the transmission speed (scaled with the conversion ratio of the new gear), whereafter the new gear is engaged. Finally, the torque level is transferred back to the level that the driver demands.

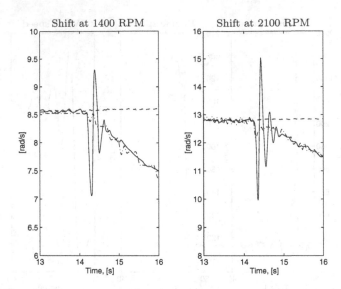

**Figure 7.28** Engagement of neutral gear commanded at 14 s, with stationary driveline at 1400 RPM and 2100 RPM on a flat road with gear 1. The engine speed (dashed) and wheel speed (dash-dotted) are scaled to transmission speed (solid) with the conversion ratio of the driveline. After a short delay time, neutral gear is engaged, causing the driveline speeds to oscillate. The amplitude of the oscillating transmission speed is higher the higher the stationary speed is.

The total time needed for a gear shift is important to minimize, since the vehicle is free-rolling with zero transmission torque. In Figure 7.28, neutral gear is engaged, without a torque control phase, at a constant speed. This means that there is a driving torque transferred in the transmission, which clearly causes the transmission speed to oscillate. The amplitude of the oscillations is increasing the higher the stationary speed is. This indicates that there must be an engine torque step in order to reach zero transmission torque and no oscillations in the transmission speed.

Figure 7.29 shows the transmission speed when the engine torque is decreased to 46 Nm at 12.0 s. Prior to that, the stationary speed 2200 RPM was maintained, which requested an engine torque of about 225 Nm. Four trials are performed with this torque profile with engaged neutral gear at different time delays after the torque step. After 12.4 s there is a small oscillation in the transmission speed, after 13.3 s and 14.8 s there are oscillations with high amplitude, and at 13.8 s there are no oscillations in the transmissions speed. This indicates how driveline resonances influence the transmission torque, which is clearly close to zero for the gear shift at 12.4 s and 13.8 s, but different from zero at 13.3 s and 14.8 s. The amplitude of the oscillating transmission torque will be higher if the stationary speed is increased or if the vehicle is accelerating.

One way this can be handled is to use a ramp in engine torque according to the scheme in Figure 7.26. However, this approach is no good for optimizing shift time, since the ramp must be conservative in order to wait until the transmission torque fluctuations are damped out.

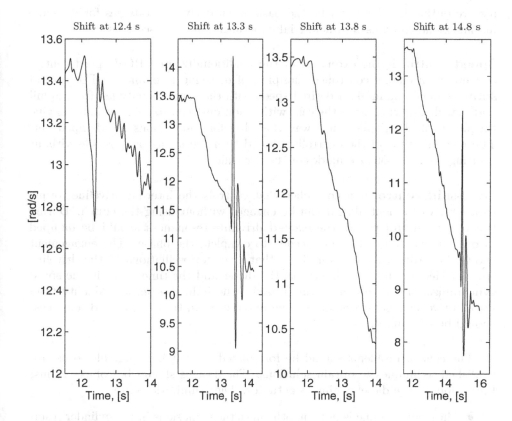

**Figure 7.29** Gear shifts with the engine at the stationary speed 2200 RPM with gear 1. At 12.0 s there is a decrease in engine torque to 46 Nm in order to reach zero transmission torque. The transmission speed is plotted when neutral gear is engaged at 12.4 s, 13.3 s, 13.8 s, and 14.8 s (with the same torque profile). The different amplitudes in the oscillations show how the torque transmitted in the transmission is oscillating after the torque step. Note that the range of the vertical axes differ between the plots.

The gear shift at 13.3 s in Figure 7.29 shows the effect of a gear shift at a transmission torque different from zero. This leads to the following problems:

- Disturbing to the driver, both in terms of noise and speed impulse.

- Increased wear on transmission.

- Increased time for the speed synchronization phase, since the transmission speed, which is the control goal, is oscillating. The oscillations are difficult to track for the engine and therefore one has to wait until they are sufficiently damped.

## 7.3.3    Goals of Driveline Control

Based on the field-trial demonstration of problems with driveline handling, the goals for reducing the influence from oscillations in performance and driveability are outlined. These will be the basis when deriving strategies for driveline management, to be used in field-trial experiments in later sections.

**Speed control** is the extension of the traditionally used RQV speed control concept with engine controlled damping of driveline resonances. The control strategy should maintain a desired speed with the same velocity lag from uphill and downhill driving, as in the case with traditional control. All available engine torque should be applied in a way that driveline oscillations are damped out. The response time of the controller should be made as fast as possible without exciting higher resonance modes of the driveline.

**Gear-shift control** is a controller that controls the internal driveline torque to a level where neutral gear can be engaged without using the clutch. During the torque control phase, the excited driveline resonances should be damped in order to minimize the time needed to complete the phase. The engagement should be realized at a torque level that gives no oscillations in the driveline speeds. Hereby, the disturbances to the driver and the time spent in the speed synchronization phase can be minimized. The influence on shift quality from initial driveline resonances and torque impulses from trailer and road roughness should be minimized.

The control problems should be formulated so that it is possible to use established techniques to obtain solutions. The designs should be robust against limitations in the diesel engine as actuator. These limitations are:

- The engine torque is not smooth, since the explosions in the cylinder result in a pulsating engine torque.

- The output torque of the engine is not exactly known. The only measure of it is a static torque map from dynamometer tests.

- The dynamical behavior of the engine is also characterized by the engine friction, which must be estimated. Many variables influence engine friction and it is necessary to find a simple yet sufficiently detailed model of the friction.

- The engine output torque is limited in different modes of operation. The maximum engine torque is restricted as a function of the engine speed, and the torque level is also restricted at low turbo pressures.

The resulting strategies should be possible to implement on both in-line pump and unit pump injection engines, with standard automotive driveline sensors.

## 7.3.4 Comment on Architectures for Driveline Control

There is one architectural issue in driveline control that should be noted. There are different possible choices in driveline control between using different sensor locations, since the driveline normally is equipped with at least two sensors for rotational speed, but sometimes more. If the driveline was rigid, the choice of sensor would not matter, since the sensor outputs would differ only by a scaling factor. However, it will be demonstrated that the presence of torsional flexibilities implies that sensor choice gives different control problems. The difference can be formulated in control theoretic terms e.g., by saying that the poles are the same, but the zeros differ both in number and values. A principle study should not be understood as a study on where to put a single sensor. Instead, it aims at an understanding of where to invest in increased sensor performance in future driveline management systems. This issue will also be investigated in later design sections.

### 7.3.5 State-Space Formulation

The input to the open-loop driveline system is $u = T_e - T_{fric,e}$, i.e., the difference between the driving torque and the friction torque. Possible physical state variables in the models of Section 7.1 are torques, angle differences, and angle velocity of any inertia. The angle difference of each torsional flexibility and the angle velocity of each inertia are used as state variables. The state space representation is

$$\dot{\underline{x}} = \underline{A}\,\underline{x} + \underline{B}\,u + \underline{H}\,l \tag{7.79}$$

where $\underline{A}$, $\underline{B}$, $\underline{H}$, $\underline{x}$, and $l$ are defined next for the Drive-shaft model and for the Clutch and drive-shaft model defined in Section 7.1.

**State-space formulation of the linear Drive-shaft model:**

$$
\begin{aligned}
x_1 &= \alpha_{CS}/i_t i_f - \alpha_w \\
x_2 &= \dot{\alpha}_{CS} \\
x_3 &= \dot{\alpha}_w \\
l &= r_{stat} m_{CoG}\left(c_{r1} + g\sin\left(\chi_{road}\right)\right)
\end{aligned} \tag{7.80}
$$

giving

$$
\underline{A} = \begin{pmatrix}
0 & 1/i & -1 \\
-k/iJ_1 & -(d_1 + d/i^2)/J_1 & d/iJ_1 \\
k/J_2 & d/iJ_2 & -(d + d_2)/J_2
\end{pmatrix}, \tag{7.81}
$$

$$
\underline{B} = \begin{pmatrix} 0 \\ 1/J_1 \\ 0 \end{pmatrix}, \quad
\underline{H} = \begin{pmatrix} 0 \\ 0 \\ -1/J_2 \end{pmatrix} \tag{7.82}
$$

where

$$
\begin{aligned}
i &= i_t i_f \\
J_1 &= J_e + J_t/i_t^2 + J_f/i_t^2 i_f^2 \\
J_2 &= J_W + m_{CoG}r_{stat}^2 \\
d_1 &= d_t/i_t^2 + d_f/i_t^2 i_f^2 \\
d_2 &= d_w + m_{CoG}c_{r2}r_{stat}^2
\end{aligned} \tag{7.83}
$$

**State-space formulation of the linear Clutch and drive-shaft model:**

$$
\begin{aligned}
x_1 &= \alpha_{CS} - \alpha_t i_t \\
x_2 &= \alpha_t/i_f - \alpha_w \\
x_3 &= \dot\alpha_{CS} \\
x_4 &= \dot\alpha_t \\
x_5 &= \dot\alpha_w
\end{aligned}
\tag{7.84}
$$

$\underline{A}$ is given by the matrix

$$
\begin{pmatrix}
0 & 0 & 1 & -i_t & 0 \\
0 & 0 & 0 & 1/i_f & -1 \\
-k_c/J_1 & 0 & -d_c/J_1 & d_c i_t/J_1 & 0 \\
k_c i_t/J_2 & -k_d/i_f J_2 & d_c i_t/J_2 & -(d_c i_t^2 + d_2 + d_d/i_f^2)/J_2 & d_d/i_f J_2 \\
0 & k_d/J_3 & 0 & d_d/i_f J_3 & -(d_3 + d_d)/J_3
\end{pmatrix}
$$

and

$$
\underline{B} = \begin{pmatrix} 0 \\ 0 \\ 1/J_1 \\ 0 \\ 0 \end{pmatrix}, \quad
\underline{H} = \begin{pmatrix} 0 \\ 0 \\ 0 \\ 0 \\ -1/J_2 \end{pmatrix}
\tag{7.85}
$$

where

$$
\begin{aligned}
J_1 &= J_e \\
J_2 &= J_t + J_f/i_f^2 \\
J_2 &= J_W + m_{CoG} r_{stat}^2 \\
d_2 &= d_t + d_f/i_f^2 \\
d_3 &= d_w + c_{r2} r_{stat}
\end{aligned}
\tag{7.86}
$$

The model equipped with the sensor filter derived in (7.61) gives the true sensor outputs $(y_m, y_t, y_w)$, according to Section 7.1.

## Disturbance Description

The influence from the road is assumed to be described by the slow-varying load $l$ and an additive disturbance $v$. A second disturbance $n$ is a disturbance acting on the input of the system. This disturbance is considered because the firing pulses in the driving torque can be seen as an additive disturbance acting on the input. The state-space description then becomes

$$
\dot{\underline{x}} = \underline{A}\,\underline{x} + \underline{B}\,u + \underline{B}\,n + \underline{H}\,l + \underline{H}\,v
\tag{7.87}
$$

with $\underline{x}$, $\underline{A}$, $\underline{B}$, $\underline{H}$, and $l$ defined in (7.80) to (7.83) or in (7.84) to (7.86), depending on model choice.

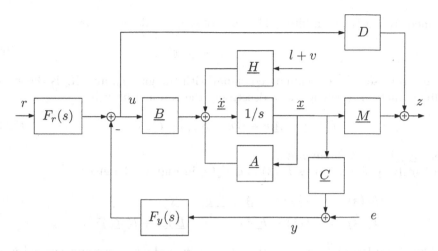

**Figure 7.30** Plant and controllers $F_r$ and $F_y$.

### Measurement Description

For controller synthesis it is of fundamental interest which physical variables of the process that can be measured. In the case of a vehicular driveline the normal sensor alternative is an inductive sensor mounted on a cogwheel measuring the angle, as mentioned before. Sensors that measure torque are expensive, and are seldom used in production vehicular applications.

The output of the process is defined as a combination of the states given by the matrix $\underline{C}$ in

$$y = \underline{C}\,x + e \qquad (7.88)$$

where $e$ is a measurement disturbance.

In this work, only angle velocity sensors are considered, and therefore, the output of the process is one/some of the state variables defining an angle velocity. Especially, the following C-matrices are defined (corresponding to a sensor on $\dot{\alpha}_{CS}$ and $\dot{\alpha}_w$ for the **Drive-shaft model**).

$$\underline{C}_m = (0\ 1\ 0) \qquad (7.89)$$
$$\underline{C}_w = (0\ 0\ 1) \qquad (7.90)$$

### 7.3.6  Controller Formulation

The performance output $z$ is the combination of states that has requirements to behave in a certain way. This combination is described by the matrices $M$ and $D$ in the following way

$$z = \underline{M}\,x + Du \qquad (7.91)$$

The resulting control problem can be seen in Figure 7.30. The unknown controllers $F_r$ and $F_y$ are to be designed so that the performance output (7.91) meets its requirements (defined later).

If state-feedback controllers are used, the control signal $u$ is a linear function of the states (if they are all measured) or else the state estimates, $\hat{\underline{x}}$, which are obtained from a Kalman filter. The control signal is described by

$$u = l_0 r - \underline{K}_c \hat{\underline{x}} \tag{7.92}$$

where $r$ represents the commanded signal with the gain $l_0$, and $K_c$ is the state-feedback matrix. The equations describing the Kalman filter is

$$\dot{\hat{\underline{x}}} = \underline{A}\,\hat{\underline{x}} + \underline{B}\,u + \underline{K}_f(y - \underline{C}\,\hat{\underline{x}}) \tag{7.93}$$

where $\underline{K}_f$ is the Kalman gain.

Identifying the matrices $F_r(s)$ and $F_y(s)$ in Figure 7.30 gives

$$
\begin{aligned}
F_y(s) &= \underline{K}_c(s\underline{I} - \underline{A} + \underline{B}\,\underline{K}_c + \underline{K}_f\underline{C})^{-1}\underline{K}_f \tag{7.94}\\
F_r(s) &= l_0\left(1 - \underline{K}_c(s\underline{I} - \underline{A} + \underline{B}\,\underline{K}_c + \underline{K}_f\underline{C})^{-1}\underline{B}\right)
\end{aligned}
$$

The closed-loop transfer functions from $r$, $v$, and $e$ to the control signal $u$ are given by

$$
\begin{aligned}
G_{ru} &= \left(\underline{I} - \underline{K}_c(s\underline{I} - \underline{A} + \underline{B}\,\underline{K}_c)^{-1}\underline{B}\right) l_0 \tag{7.95}\\
G_{vu} &= \underline{K}_c(s\underline{I} - \underline{A} + \underline{K}_f\underline{C})^{-1}N - \underline{K}_c(s\underline{I} - \underline{A} + \underline{B}\,\underline{K}_c)^{-1}N \tag{7.96}\\
&\quad -\underline{K}_c(s\underline{I} - \underline{A} + \underline{B}\,\underline{K}_c)^{-1}\underline{B}\,\underline{K}_c(s\underline{I} - \underline{A} + \underline{K}_f\underline{C})^{-1}N\\
G_{eu} &= \underline{K}_c\left((s\underline{I} - \underline{A} + \underline{B}\,\underline{K}_c)^{-1}\underline{B}\,\underline{K}_c - I\right)(s\underline{I} - \underline{A} + \underline{K}_f\underline{C})^{-1}\underline{K}_f \tag{7.97}
\end{aligned}
$$

The transfer functions to the performance output $z$ are given by

$$
\begin{aligned}
G_{rz} &= (\underline{M}(s\underline{I} - \underline{A})^{-1}\underline{B} + D)G_{ru} \tag{7.98}\\
G_{vz} &= \underline{M}(s\underline{I} - \underline{A} + \underline{B}\,\underline{K}_c)^{-1}\underline{B}\,\underline{K}_c(s\underline{I} - \underline{A} + \underline{K}_f\underline{C})^{-1}N \tag{7.99}\\
&\quad +\underline{M}(s\underline{I} - \underline{A} + \underline{B}\,\underline{K}_c)^{-1}N + DG_{vu}\\
G_{ez} &= (\underline{M}(s\underline{I} - \underline{A})^{-1}\underline{B} + D)G_{vu} \tag{7.100}
\end{aligned}
$$

Two return ratios (loop gains) result, which characterize the closed-loop behavior at the plant output and input respectively

$$
\begin{aligned}
GF_y &= \underline{C}(s\underline{I} - \underline{A})^{-1}\underline{B}\,F_y \tag{7.101}\\
F_yG &= F_y\underline{C}(s\underline{I} - \underline{A})^{-1}\underline{B} \tag{7.102}
\end{aligned}
$$

When only one sensor is used, these return ratios are scalar and thus equal.

## 7.3.7  Some Feedback Properties

The performance output when controlling the driveline to a certain speed is the velocity of the wheel, defined as

$$z = \dot{\alpha}_w = \underline{C}_w x \tag{7.103}$$

When studying the closed-loop control problem with a sensor on $\dot{\alpha}_{CS}$ or $\dot{\alpha}_w$, two different control problems result. Figure 7.31 shows a root locus with respect to a P-controller gain for two gears using velocity sensor $\dot{\alpha}_{CS}$ and $\dot{\alpha}_w$ respectively.

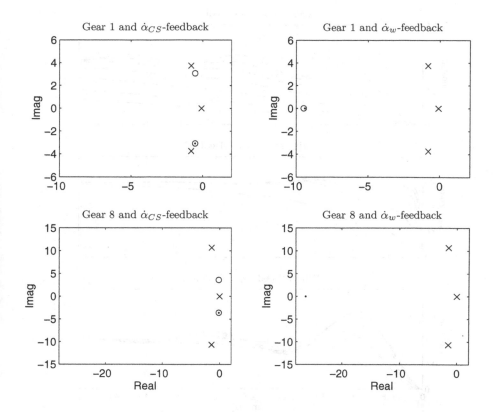

**Figure 7.31** Root locus with respect to a P-controller gain, for gear 1 (top figures) and gear 8 (bottom figures), with sensor on $\dot{\alpha}_{CS}$ (left figures), or $\dot{\alpha}_w$ (right figures). The cross represent the open-loop poles, while the rings represent the open-loop zeros. The system goes unstable when the $\dot{\alpha}_w$-gain is increased, but is stable for all $\dot{\alpha}_{CS}$-gains.

The open-loop transfer functions from control signal to engine speed $G_{um}$ has three poles and two zeros, as can be seen in Figure 7.31. $G_{uw}$ on the other hand has one zero and the same poles. Hence, the relative degree [60] of $G_{um}$ is one and $G_{uw}$ has a relative degree of two. This means that when $\dot{\alpha}_w$-feedback is used, and the gain is increased, two poles must go to infinity which makes the system unstable. When the velocity sensor $\dot{\alpha}_{CS}$ is used, the relative degree is one, and the closed-loop system is stable for all gains. (Remember that $\dot{\alpha}_w$ is the performance output and thus desirable to use.)

The same effect can be seen in step response tests when the P-controller is used. Figure 7.32 demonstrates the problem with resonances that occur with increasing gain for the two cases of feedback. When the engine-speed sensor is used, the engine speed is well damped when the gain is increased, but the resonance in the drive shaft makes the wheel speed oscillate. When using $\dot{\alpha}_w$-feedback it is difficult to increase the bandwidth, since the poles moves closer to the imaginary axis, and give a resonant system.

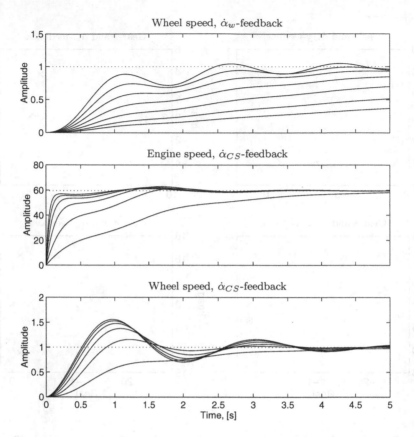

**Figure 7.32** Step responses when using a P-controller with different gains on the Drive-shaft model with gear 1. With $\dot{\alpha}_w$-feedback (top figure), increased gain results in instability. With $\dot{\alpha}_{CS}$-feedback (bottom figures), increased gain results in a well damped engine speed, but an oscillating wheel speed.

The characteristic results in Figures 7.31 and 7.32 only depend on the relative degree, and are thus parameter independent. However, this observation may depend on feedback structure, and therefore a more detailed analysis is performed in the following section.

## 7.3.8  Driveline Control with LQG/LTR

Different sensor locations result in different control problems with different inherent characteristics, as illustrated in the previous section. The topic of this section is to show how this influences control design when using Linear Quadratic design with Loop Transfer Recovery (LQG/LTR) with design of the return ratio at the output of the plant [78].

**Important comment:** LQG/LTR is *one* method to obtain the parameters in a controller structure with state feedback using an observer. Even if the method is unknown to the reader, the presentation should be easy to follow if the reader has a basic course in control and accepts that LQG/LTR is a method to compute $l_0$, $\underline{K}_c$, and $\underline{K}_f$ in Eq. 7.92 and Eq. 7.93.

The reason for using LQG/LTR, in this principle study, is that it offers a control design method resulting in a controller and observer of the same order as the plant model, and it is also an easy method for obtaining robust controllers.

### Transfer Functions

When comparing the control problem of using $\dot{\alpha}_{CS}$ or $\dot{\alpha}_w$ as sensor, the open-loop transfer functions $G_{um}$ and $G_{uw}$ results. These have the same number of poles but different number of zeros, as mentioned before. Two different closed-loop systems are obtained depending on which sensor that is being used.

### Feedback from $\dot{\alpha}_w$

A natural feedback configuration is to use the performance output, $\dot{\alpha}_w$. Then among others the following transfer functions result

$$G_{rz} = \frac{G_{uw}F_yF_r}{1 + G_{uw}F_y} = T_wF_r \qquad (7.104)$$

$$G_{nu} = \frac{1}{1 + G_{uw}F_y} = S_w \qquad (7.105)$$

where (7.95) to (7.100) are used together with the matrix inversion lemma [60], and $n$ is the input disturbance. The transfer functions $S_w$ and $T_w$ are the *sensitivity* function and the *complementary sensitivity* function [78].The relation between these transfer function is, as usual,

$$S_w + T_w = 1 \qquad (7.106)$$

### Feedback from $\dot{\alpha}_{CS}$

The following transfer functions result if the $\dot{\alpha}_{CS}$-sensor is used.

$$G_{rz} = \frac{G_{uw}F_yF_r}{1 + G_{um}F_y} \qquad (7.107)$$

$$G_{nu} = \frac{1}{1 + G_{um}F_y} \qquad (7.108)$$

The difference between the two feedback configurations is that the return difference is $1 + G_{uw}F_y$ or $1 + G_{um}F_y$.

It is desirable to have sensitivity functions that corresponds to $y = \dot{\alpha}_{CS}$ and $z = \dot{\alpha}_w$. The following transfer functions are defined

$$S_m = \frac{1}{1 + G_{um}F_y}, \quad T_m = \frac{G_{um}F_y}{1 + G_{um}F_y} \qquad (7.109)$$

These transfer functions correspond to a configuration where $\dot{\alpha}_{CS}$ is the output (i.e. $y = z = \dot{\alpha}_{CS}$). Using (7.107) it is natural to define $\overline{T}_m$ by

$$\overline{T}_m = \frac{G_{uw}F_y}{1 + G_{um}F_y} = T_m\frac{G_{uw}}{G_{um}} \qquad (7.110)$$

The functions $S_m$ and $\overline{T}_m$ describe the design problem when feedback from $\alpha_{CS}$ is used.

When combining (7.109) and (7.110), the corresponding relation to (7.106) is

$$S_m + \overline{T}_m \frac{G_{um}}{G_{uw}} = 1 \tag{7.111}$$

If $S_m$ is made zero for some frequencies in (7.111), then $\overline{T}_m$ will not be equal to one, as in (7.106). Instead, $\overline{T}_m = G_{uw}/G_{um}$ for these frequency domains.

### Limitations on Performance

The relations (7.106) and (7.111) will be the fundamental relations for discussing design considerations. The impact of the ratio $G_{uw}/G_{um}$ will be analyzed in the following sections.

**Definition 7.1** $\overline{T}_m$ *in (7.110) is the modified complementary sensitivity function.* $G_{w/m} = G_{uw}/G_{um}$ *is the dynamic output ratio.*

### Design Example with a Simple Mass-Spring Model

Linear Quadratic Design with Loop-Transfer Recovery will be treated in four cases, being combinations of two sensor locations, $\dot{\alpha}_{CS}$ or $\dot{\alpha}_w$, and two models with the same structure, but with different parameters. Design without pre-filter ($F_r = 1$) is considered.

The section covers a general plant with $n$ inertias connected by $k-1$ torsional flexibilities, without damping and load, and with unit conversion ratio. There are $(2n - 1)$ poles, and the location of the poles is the same for the different sensor locations. The number of zeros depends on which sensor that is used, and when using $\dot{\alpha}_w$ there are no zeros. When using feedback from $\dot{\alpha}_{CS}$ there are $(2n - 2)$ zeros. Thus, the transfer functions $G_{um}$ and $G_{uw}$, have the same denominators, and a relative degree of 1 and $(2n - 1)$ respectively.

### Structural Properties of Sensor Location

The controller (7.94) has a relative degree of one. The relative degree of $G_{um}F_y$ is thus 2, and the relative degree of $G_{uw}F_y$ is $2n$. When considering design, a good alternative is to have relative degree one in $GF_y$, implying infinite gain margin and high phase margin.

When using $G_{um}F_y$, one pole has to be moved to infinity, and when using $G_{uw}F_y$, $2n-1$ poles have to be moved to infinity, in order for the ratio to resemble a first order system at high frequencies.

When the return ratio behaves like a first order system, also the closed-loop transfer function behaves like a first order system. This conflicts with the design goal of having a steep roll-off rate for the closed-loop system in order to attenuate measurement noise. Hence, there is a trade-off when using $\dot{\alpha}_w$-feedback.

When using $\dot{\alpha}_{CS}$-feedback, there is no trade-off, since the relative degree of $G_{um}$ is one.

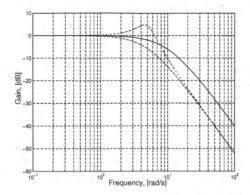

**Figure 7.33** Dynamic output ratio $G_{w/m}$ for Example 7.1a (solid line) and Example 7.1b (dashed line).

## Structure of $G_{w/m}$

We have in the previous simple examples seen that the relative degree and the zeros are important. The dynamic output ratio contains exactly this information and nothing else.

For low frequencies the dynamic output ratio has gain equal to one,

$$|G_{w/m}(0)| = 1$$

(if the conversion ratio is equal to one). Furthermore, $G_{w/m}$ has a relative degree of $2n - 2$ and thus, a high frequency gain roll-off rate of $20(2n - 2)$ dB/decade. Hence, the dynamic output ratio gives the closed-loop transfer function $\overline{T}_m$ a high frequency gain roll-off rate of $q_m + 20(2n - 2)$ dB/decade, where $q_m$ is the roll-off rate of $G_{um}F_y$. When using $\dot{\alpha}_w$-feedback, $T_w$ will have the same roll-off rate as $G_{uw}F_y$.

## Parametric properties of $G_{w/m}$

Typical parametric properties of $G_{w/m}$ can be seen in the following example.

---

**Example 7.1** *Two different plants of the form (7.80) to (7.83) are considered with the following values:*
*a)* $J_1 = 0.0974$, $J_2 = 0.0280$, $k = 2.80$, $c = 0$, $d_1 = 0.0244$, $d_2 = 0.566$, $l = 0$.
*b)* $J_1 = 0.0974$, $J_2 = 0.220$, $k = 5.50$, $c = 0$, $d_1 = 1.70$, $d_2 = 0.660$, $l = 0$.
*with labels according to the state-space formulation in Section 7.3.5. The shape of $G_{w/m}$ can be seen in Figure 7.33. The rest of the section will focus on control design of these two plant models.*

---

## LQG Designs

Integral action is included by augmenting the state to attenuate step disturbances in $v$ [78]. The state-space realization $\underline{A}_a$, $\underline{B}_a$, $\underline{M}_a$, $\underline{C}_{wa}$, and $\underline{C}_{ma}$ results. The

Kalman-filter gain, $\underline{K}_f$, is derived by solving the Riccati equation [78]

$$\underline{P}_f\underline{A}^T + \underline{A}\underline{P}_f - \underline{P}_f\underline{C}^T\underline{V}^{-1}\underline{C}\underline{P}_f + \underline{B}W\underline{B}^T = 0 \qquad (7.112)$$

The covariances $W$ and $\underline{V}$, for disturbances $v$ and $e$ respectively, are adjusted until the return ratio

$$\underline{C}(s\underline{I} - \underline{A})^{-1}\underline{K}_f, \qquad \underline{K}_f = \underline{P}_f\underline{C}^T\underline{V}^{-1} \qquad (7.113)$$

and the closed-loop transfer functions $S$ and $T$ show satisfactory performance. The Nyquist locus remains outside the unit circle centered at $-1$. This means that there is infinite gain margin, and a phase margin of at least $60°$. Furthermore, the relative degree is one, and $|S| \leq 1$.

**Design for $\dot{\alpha}_w$-feedback.** $W$ is adjusted (and thus $F_y(s)$) such that $S_w$ and $T_w$ show satisfactory performance, and that the desired bandwidth is obtained. The design of the driveline models in Example 7.1 is shown in Figure 7.34. Note that the roll-off rate of $T_w$ is 20 dB/decade.

**Design for $\dot{\alpha}_{CS}$-feedback.** $W$ is adjusted (and thus $F_y(s)$) so that $S_m$ and $T_m$ (and thus $\dot{\alpha}_{CS}$) show satisfactory performance. Depending on the shape of $G_{w/m}$ for middle high frequencies, corrections in $W$ must be taken so that $\overline{T}_m$ achieves the desired bandwidth. If there is a resonance peak in $G_{w/m}$, the bandwidth in $\overline{T}_m$ is chosen such that the peak is suppressed. Figure 7.34 shows such an example (the plant in Example 7.1b with $\dot{\alpha}_{CS}$-feedback), where the bandwidth is lower in order to suppress the peak in $G_{w/m}$. Note also the difference between $S_w$ and $S_m$.

The parameters of the dynamic output ratio are thus important in the LQG step of the design.

## Loop-Transfer Recovery, LTR

The next step in the design process is to include $\underline{K}_c$, and recover the satisfactory return ratio obtained previously. When using the combined state feedback and Kalman filter, the return ratio is $GF_y = \underline{C}(s\underline{I} - \underline{A})^{-1}\underline{B}\underline{K}_c(s\underline{I} - \underline{A} + \underline{B}\underline{K}_c + \underline{K}_f\underline{C})^{-1}\underline{K}_f$. A simplistic LTR can be obtained by using $\underline{K}_c = \rho C$ and increasing $\rho$. As $\rho$ is increased, $2n - 1$ poles move towards the open system zeros. The remaining poles move towards infinity (compare to Section 7.3.5). If the Riccati equation

$$\underline{A}^T\underline{P}_c + \underline{P}_c\underline{A} - \underline{P}_c\underline{B}R^{-1}\underline{B}^T\underline{P}_c + \underline{C}^TQ\underline{C} = 0 \qquad (7.114)$$

is solved with $Q = \rho$, and $R = 1$, $\underline{K}_c = \sqrt{\rho}C$ is obtained in the limit, and to guarantee stability, this $\underline{K}_c$ is used for recovery.

Figure 7.35 shows the recovered closed-loop transfer functions for Example 7.1. Nyquist locus and control signal transfer function, $G_{ru} = F_y/(1 + G_{uw}F_y)$, are shown in Figure 7.36.

**Recovery for $\dot{\alpha}_w$-feedback.** There is a trade-off when choosing an appropriate $\rho$. A low $\rho$ gives good attenuation of measurement noise and a low control signal, but in order to have good stability margins, a high $\rho$ must be chosen. This gives

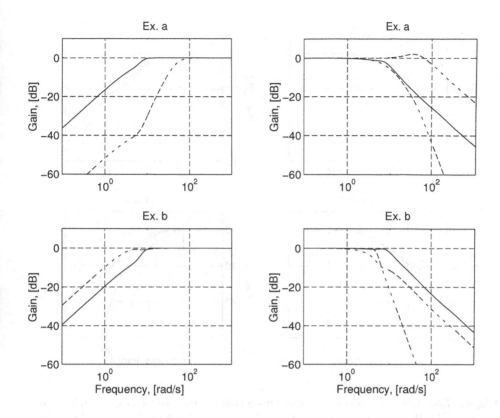

**Figure 7.34** Closed-loop transfer functions $S$ (left figures), and $T$ (right figures). Feedback from $\dot{\alpha}_w$ is seen in solid lines, and feedback from $\dot{\alpha}_{CS}$ in dashed lines. $\overline{T}_m$ is seen in the right figures in dash-dotted lines. For the $\dot{\alpha}_{CS}$-design, $W = 5 \cdot 10^4$ (Ex. 7.1a) and $W = 50$ (Ex. 7.1b) are used, and for the $\dot{\alpha}_w$-design, $W = 15$ (Ex. 7.1a) and $W = 5 \cdot 10^2$ (Ex. 7.1b) are used.

an increased control signal, and a 20 dB/decade roll-off rate in $T_w$ for a wider frequency range.

**Recovery for $\dot{\alpha}_{CS}$-feedback.** There is no trade-off when choosing $\rho$. It is possible to achieve good recovery with reasonable stability margins and control signal, together with a steep roll-off rate.

The structural properties, i.e. the relative degrees are thus dominant in determining the LTR step of the design.

## 7.4 Driveline Speed Control

The background and problems with traditional diesel engine speed control (RQV) were covered in Section 7.3. Driveline speed control is here defined as the extension of RQV control with engine controlled active damping of driveline resonances. Active damping is obtained by using a feedback law that calculates the

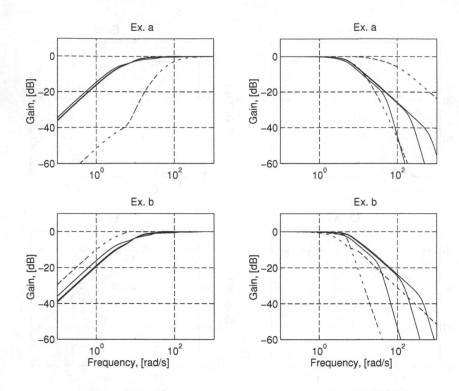

**Figure 7.35** Closed-loop transfer functions $S$ (left figures), and $T$ (right figures) after recovery. Feedback from $\dot{\alpha}_w$ is seen in solid lines, and feedback from $\dot{\alpha}_{CS}$ in dashed lines. $\overline{T}_m$ is seen in the right figures in dash-dotted lines. For the $\dot{\alpha}_{CS}$-design, $\rho = 10^6$ (Ex. 7.1a) and $\rho = 10^5$ (Ex. 7.1b) are used, and for the $\dot{\alpha}_w$-design, $\rho = 10^4$, $10^8$, and $10^{11}$ are used in both Ex. 7.1a and b.

fuel amount so that the engine inertia works in the opposite direction of the oscillations, at the same time as the desired speed is obtained. The calculated fuel amount is a function of the engine speed, the wheel speed, and the drive-shaft torsion, which are states of the Drive-shaft model, derived in Section 7.1. These variables are estimated by a Kalman filter with either the engine speed or the wheel speed as input. The feedback law is designed by deriving a criterion in which the control problem is given a mathematical formulation.

The RQV control scheme gives a specific character to the driving feeling e.g., when going uphill and downhill. This driving character is possible to maintain when extending RQV control with active damping. Traditional RQV control is further explained in Section 7.4.1. Thereafter, the speed control problem keeping RQV characteristics is formulated in Section 7.4.2. The problem formulation is then studied in the following sections. The design based on the Drive-shaft model is simulated together with the more complicated Nonlinear clutch and drive-shaft model as vehicle model. Some important disturbances are simulated that are difficult to generate in systematic ways in real experiments. Finally, some field experiments are shown.

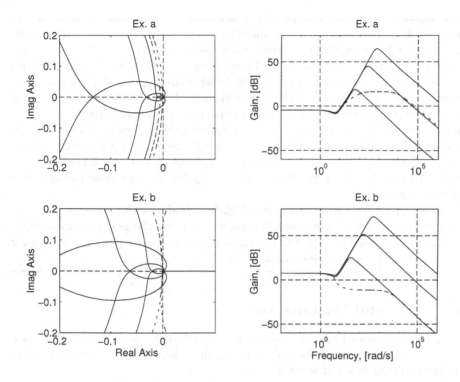

**Figure 7.36** Nyquist plot of return ratio (left figures) and control signal transfer function $F_y/(1 + G_{uw}F_y)$ (right figures). Feedback from $\dot{\alpha}_w$ is seen in solid lines, and feedback from $\dot{\alpha}_{CS}$ in dashed lines. For the $\dot{\alpha}_{CS}$-design, $\rho = 10^6$ (Ex. 7.1a) and $\rho = 10^5$ (Ex. 7.1b) are used, and for the $\dot{\alpha}_w$-design, $\rho = 10^4$, $10^8$, and $10^{11}$ are used in both Ex. 7.1a and b. A dash-dotted circle with radius one, centered at -1, is also shown in the Nyquist plots.

## 7.4.1 RQV Control

RQV control is the traditional diesel engine control scheme covered in Section 7.3. The controller is essentially a proportional controller with the accelerator as reference value and a sensor measuring the engine speed. The RQV controller has no information about the load, and a nonzero load, e.g., when going uphill or downhill, gives a stationary error. The RQV controller is described by

$$u = u_0 + K_p(ri - \dot{\alpha}_{CS}) \qquad (7.115)$$

where $i = i_t i_f$ is the conversion ratio of the driveline, $K_p$ is the controller gain, and $r$ is the reference velocity. The constant $u_0$ is a function of the speed, but not the load since this is not known. The problem with vehicle shuffle when increasing the controller gain, in order to increase the bandwidth, is demonstrated in the following example.

**Example 7.2**  *Consider the truck modeled in Sections 7.1 traveling at a speed of 2 rad/s (3.6 km/h) with gear 1 and a total load of 3000 Nm (≈ 2 % road slope). Let the new desired velocity be r = 2.3 rad/s. Figure 7.37 shows the RQV control law (7.115) applied to the Drive-shaft model with three gains, $K_p$. In the plots, $u_0$ from (7.115) is calculated so that the stationary level is the same for the three gains. (Otherwise there would be a gain dependent stationary error.)*
*When the controller gain is increased, the rise time decreases and the overshoot in the wheel speed increases. Hence, there is a trade-off between short rise time and little overshoot. The engine speed is well damped, but the flexibility of the driveline causes the wheel speed to oscillate with higher amplitude the more the gain is increased.*
*The same behavior is seen in Figure 7.38, which shows the transfer functions from load and measurement disturbances, v and e, to the performance output, when the RQV controller is used. The value of the resonance peak in the transfer functions increases when the controller gain is increased.*

## 7.4.2   Problem Formulation

The goals of the speed control concept were outlined in Section 7.3. These are here given a mathematical formulation, which is solved for a controller using established techniques and software.

The performance output for the speed controller is the wheel speed, $z = \dot{\alpha}_w$, as defined in Section 7.3.4, since the wheel speed rather than the engine speed determines vehicle behavior. Figure 7.39 shows the transfer functions from control signal ($u$) and load ($l$) to the wheel speed ($z$) for both the Drive-shaft model and the Clutch and drive-shaft model. The Clutch and drive-shaft model adds a second resonance peak originating from the clutch. Furthermore, the high frequency roll-off rate is steeper for the Clutch and drive-shaft model than for the Drive-shaft model. Note that the transfer function from the load to the performance output is the same for the two models. This section deals with the development of a controller based on the Drive-shaft model, neglecting the influence from the clutch for higher frequencies.

A first possible attempt for speed control is a scheme of applying the engine torque to the driveline such that the following cost function is minimized

$$\lim_{T \to \infty} \int_0^T (z - r)^2 \tag{7.116}$$

where $r$ is the reference velocity given by the driver. The cost function (7.116) can be made arbitrarily small if there are no restrictions on the control signal $u$, since the plant model is linear. However, a diesel engine can only produce torque in a certain range, and therefore, (7.116) is extended such that a large control signal is penalized in the cost function.

The stationary point $z = r$ is reached if a stationary control signal, $u_0$, is used. This torque is a function of the reference value, $r$, and the load, $l$. For a

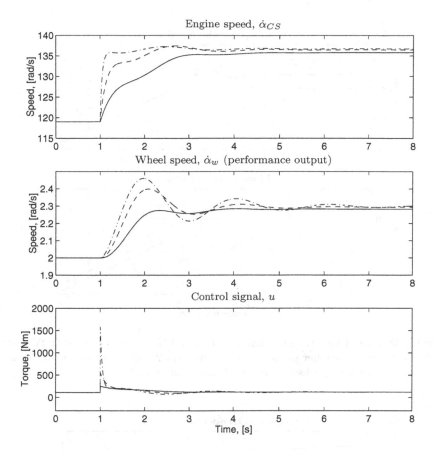

**Figure 7.37** Response of step in accelerator position at t=1 s, with RQV control (7.115) controlling the Drive-shaft model. Controller gains $K_p$=8, $K_p$=25, and $K_p$=85 are shown in solid, dashed, and dash-dotted lines respectively. Increased gain results in a well damped engine speed and an oscillating wheel speed.

given wheel speed, $\dot{\alpha}_w$, and load, the driveline has the following stationary point

$$x_0(\dot{\alpha}_w, l) = \begin{pmatrix} d_2/k & 1/k \\ i & 0 \\ 1 & 0 \end{pmatrix} \begin{pmatrix} \dot{\alpha}_w \\ l \end{pmatrix} = \delta_x \dot{\alpha}_w + \delta_l l \qquad (7.117)$$

$$u_0(\dot{\alpha}_w, l) = \begin{pmatrix} (d_1 i^2 + d_2)/i & 1/i \end{pmatrix} \begin{pmatrix} \dot{\alpha}_w \\ l \end{pmatrix} = \lambda_x \dot{\alpha}_w + \lambda_l l \quad (7.118)$$

The stationary point is obtained by solving

$$\underline{A}\,\underline{x} + \underline{B}\,u + \underline{H}\,l = 0 \qquad (7.119)$$

for $\underline{x}$ and $u$, where $\underline{A}$, $\underline{B}$, and $\underline{H}$ are given by (7.80) to (7.83).

The cost function is modified by using (7.117) and (7.118), such that a control signal that deviates from the stationary value $u_0(r, l)$ adds to the cost function.

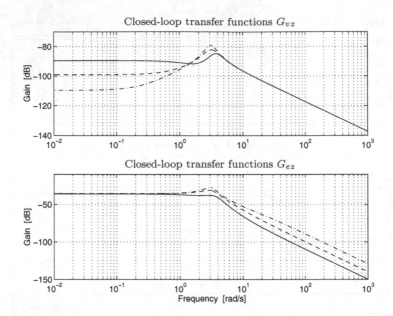

**Figure 7.38** Closed-loop transfer functions $G_{vz}$ and $G_{ez}$ when using the RQV control law (7.115) for the controller gains $K_p=8$ (solid), $K_p=25$ (dashed), and $K_p=85$ (dash-dotted). The resonance peaks increase with increasing gain.

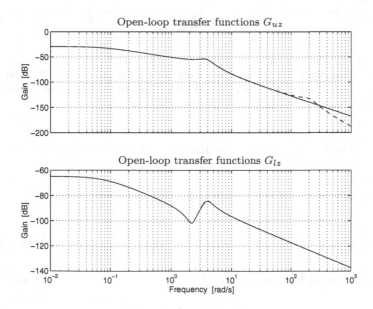

**Figure 7.39** Transfer functions from control signal, $u$, and load, $l$, to performance output, $z$. The Drive-shaft model is shown in solid and the Clutch and drive-shaft model is shown in dashed. The modeled clutch gives a second resonance peak and a steeper roll-off rate.

The extended cost function is given by

$$\lim_{T \to \infty} \int_0^T (z - r)^2 + \eta(u - u_0(r, l))^2 \tag{7.120}$$

where $\eta$ is used to control the trade-off between short rise time and control signal amplitude.

The controller that minimizes (7.120), called the speed controller, has no stationary error, since the load, $l$, is included and thus compensated for. However, it is desirable that the stationary error characteristic for the RQV controller is maintained in the speed controller, as mentioned before. A stationary error comparable with that of the RQV controller can be achieved by using only a part of the load in the criterion (7.120), as will be demonstrated in Section 7.4.3.

## 7.4.3 Speed Control with Active Damping and RQV Behavior

Before continuing, the following is repeated:

**Important comment:** LQG/LTR is *one* method to obtain the parameters in a controller structure with state feedback using an observer. Even if the method is unknown to the reader, the presentation should be easy to follow if the reader has a basic course in control and accepts that LQG/LTR is a method to compute $l_0$, $\underline{K}_c$, and $\underline{K}_f$ in Eq. 7.92 and Eq. 7.93.

The problem formulation (7.120) will be treated in two steps. First without RQV behavior i.e., using the complete load in the criterion, and then extending to RQV behavior. The problem formulation (7.120) is in this section solved with LQG technique. This is done by linearizing the driveline model and rewriting (7.120) in terms of the linearized variables. A state-feedback matrix is derived that minimizes (7.120) by solving a Riccati equation. The derived feedback law is a function of $\eta$ which is chosen such that high bandwidth together with a feasible control signal is obtained.

The model (7.79)

$$\dot{\underline{x}} = \underline{A}\,\underline{x} + \underline{B}\,u + \underline{H}\,l \tag{7.121}$$

is affine since it includes a constant term, $l$. The model is linearized in the neighborhood of the stationary point $(x_0, u_0)$. The linear model is described by

$$\Delta \dot{\underline{x}} = \underline{A}\,\Delta \underline{x} + \underline{B}\,\Delta u \tag{7.122}$$

where

$$
\begin{aligned}
\Delta \underline{x} &= \underline{x} - \underline{x}_0 \\
\Delta u &= u - u_0 \\
\underline{x}_0 &= \underline{x}_0(x_{30}, l) \\
u_0 &= u_0(x_{30}, l)
\end{aligned}
\tag{7.123}
$$

where the stationary point $(x_0, u_0)$ is given by (7.117) and (7.118) ($x_{30}$ is the initial value of $x_3$). Note that the linear model is the same for all stationary points.

The problem is to devise a feedback control law that minimizes the cost function (7.120). The cost function is expressed in terms of $\Delta x$ and $\Delta u$ by using (7.123)

$$\lim_{T \to \infty} \int_0^T (\underline{M}(\underline{x}_0 + \Delta \underline{x}) - r)^2 + \eta(u_0 + \Delta u - u_0(r,l))^2 \quad (7.124)$$

$$= \lim_{T \to \infty} \int_0^T (\underline{M}\Delta \underline{x} + r_1)^2 + \eta(\Delta u + r_2)^2 \quad (7.125)$$

with

$$r_1 = \underline{M}\underline{x}_0 - r \quad (7.126)$$
$$r_2 = u_0 - u_0(r,l)$$

In order to minimize (7.124) a Riccati equation is used. Then the constants $r_1$ and $r_2$ must be expressed in terms of state variables. This can be done by augmenting the plant model $(A, B)$ with models of the constants $r_1$ and $r_2$. Since these models will not be controllable, they must be stable in order to solve the Riccati equation [78]. Therefore the model $\dot{r}_1 = \dot{r}_2 = 0$ is not used because the poles are located on the imaginary axis. Instead the following models are used

$$\dot{r}_1 = -\sigma r_1 \quad (7.127)$$
$$\dot{r}_2 = -\sigma r_2 \quad (7.128)$$

which with a low $\sigma$ indicates that $r$ is a slow-varying constant.

The augmented model is given by

$$\underline{A}_r = \begin{pmatrix} & & & 0 & 0 \\ & \underline{A} & & 0 & 0 \\ & & & 0 & 0 \\ 0 & 0 & 0 & -\sigma & 0 \\ 0 & 0 & 0 & 0 & -\sigma \end{pmatrix}, \quad (7.129)$$

$$\underline{B}_r = \begin{pmatrix} \underline{B} \\ 0 \\ 0 \end{pmatrix}, \quad \underline{x}_r = (\Delta \underline{x}^T \; r_1 \; r_2)^T \quad (7.130)$$

By using these equations, the cost function (7.124) can be written in the form

$$\lim_{T \to \infty} \int_0^T \underline{x}_r^T \underline{Q}\, \underline{x}_r + R\Delta u^2 + 2\underline{x}_r^T \underline{N}\,\Delta u \quad (7.131)$$

with

$$\underline{Q} = (\underline{M}\; 1\; 0)^T(\underline{M}\; 1\; 0) + \eta(0\; 0\; 0\; 0\; 1)^T(0\; 0\; 0\; 0\; 1)$$
$$\underline{N} = \eta(0\; 0\; 0\; 0\; 1)^T \quad (7.132)$$
$$R = \eta$$

The cost function (7.124) is minimized by using

$$\Delta u = -\underline{K}_c \underline{x}_r \quad (7.133)$$

with

$$\underline{K}_c = \underline{Q}^{-1}(\underline{B}_r^T \underline{P}_c + \underline{N}^T) \tag{7.134}$$

where $\underline{P}_c$ is the stabilizing solution to the Riccati equation

$$\underline{A}_r^T \underline{P}_c + \underline{P}_c \underline{A}_r + R - (\underline{P}_c \underline{B}_r + \underline{N})\underline{Q}^{-1}(\underline{P}_c \underline{B}_r + \underline{N})^T = 0 \tag{7.135}$$

The control law (7.133) becomes

$$\Delta u = -\underline{K}_c \underline{x}_r = - \left( \begin{array}{ccc} K_{c1} & K_{c2} & K_{c3} \end{array} \right) \Delta \underline{x} - K_{c4} r_1 - K_{c5} r_2 \tag{7.136}$$

By using (7.123) and (7.126) the control law for the speed controller is written as

$$u = K_0 x_{30} + K_l l + K_r r - \left( \begin{array}{ccc} K_{c1} & K_{c2} & K_{c3} \end{array} \right) \underline{x} \tag{7.137}$$

with

$$\begin{array}{rcl}
K_0 & = & \left( \begin{array}{ccc} K_{c1} & K_{c2} & K_{c3} \end{array} \right) \delta_x - K_{c4} M \delta_x + \lambda_x - K_{c5} \lambda_x \\
K_r & = & K_{c4} + K_{c5} \lambda_x \\
K_l & = & \left( \begin{array}{ccc} K_{c1} & K_{c2} & K_{c3} \end{array} \right) \delta_l - K_{c4} M \delta_l + \lambda_l
\end{array} \tag{7.138}$$

where $\delta_x$, $\delta_l$, $\lambda_x$, and $\lambda_l$ are described in (7.117) and (7.118).

When this control law is applied to Example 7.2 the controller gain becomes

$$u = 0.230 x_{30} + 4470 r + 0.125 l - \left( \begin{array}{ccc} 7620 & 0.0347 & 2.36 \end{array} \right) \underline{x} \tag{7.139}$$

where $\eta = 5 \cdot 10^{-8}$ and $\sigma = 0.0001$ are used. With this controller the phase margin is guaranteed to be at least 60° with infinite amplitude margin [78]. A step-response simulation with the speed controller (7.139) is shown in Figure 7.40.

The rise time of the speed controller is shorter than for the RQV controller. Also the overshoot is less when using speed control. The driving torque is controlled such that the oscillations in the wheel speed are actively damped. This means that the controller applies the engine torque in a way that the engine inertia works in the opposite direction of the oscillation. Then the engine speed oscillates, but the important wheel speed is well behaved as seen in Figure 7.40.

### Extending with RQV Behavior

The RQV controller has no information about the load, $l$, and therefore a stationary error will be present when the load is different from zero. The speed controller (7.137) is a function of the load, and the stationary error is zero if the load is estimated and compensated for. There is however a demand from the driver that the load should give a stationary error, and only when using a cruise controller the stationary error should be zero.

The speed controller can be modified such that a load different from zero gives a stationary error. This is done by using $\beta l$ instead of the complete load $l$ in (7.137). The constant $\beta$ ranges from $\beta = 0$ which means no compensation for the load, to $\beta = 1$ which means fully compensation of the load and no stationary error. The compensated speed control law becomes

$$u = K_0 x_{30} + K_l \beta l + K_r r - \left( \begin{array}{ccc} K_{c1} & K_{c2} & K_{c3} \end{array} \right) \underline{x} \tag{7.140}$$

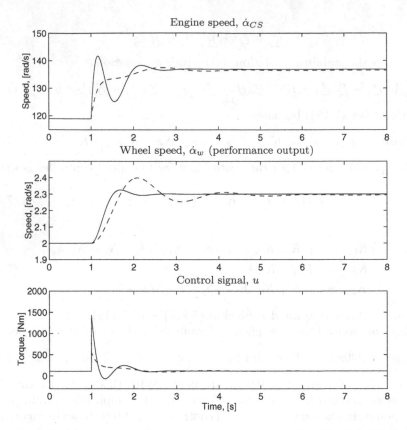

**Figure 7.40** Response of step in accelerator position at t=1 s. The Drive-shaft model is controlled with the speed control law (7.139) in solid lines. RQV control (7.115) with $K_p$=25 is seen in dashed lines. With active damping, the engine speed oscillates, resulting in a well damped wheel speed.

In Figure 7.47, the RQV controller with its stationary error (remember the reference value $r = 2.3$ rad/s) is compared to the compensated speed controller (7.140) applied to Example 7.2 for three values of $\beta$. By adjusting $\beta$, the speed controller with active damping is extended with a stationary error comparable with that of the RQV controller.

## 7.4.4    Influence from Sensor Location

The speed controller investigated in the previous section uses feedback from all states ($x_1 = \alpha_{CS}/i_t i_f - \alpha_w$, $x_2 = \dot{\alpha}_{CS}$, and $x_3 = \dot{\alpha}_w$). A sensor measuring shaft torsion (e.g., $x_1$) is normally not used, and therefore an observer is needed to estimate the unknown states. In this work, either the engine speed or the wheel speed is used as input to the observer. This results in different control problems depending on sensor location. Especially the difference in disturbance rejection is investigated.

The observer gain is calculated using Loop-Transfer Recovery (LTR) [78].

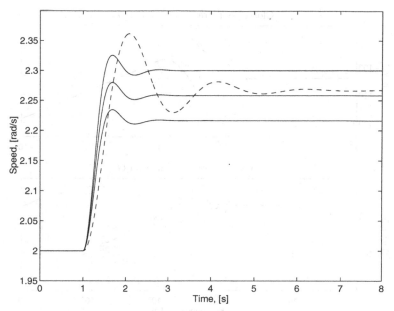

**Figure 7.41** Wheel-speed response of step in accelerator position at t=1 s. The Drive-shaft model is controlled with the RQV controller (7.115) in dashed line, and the speed controller with stationary error (7.140) with $\beta = 0$, 0.5, 1 in solid lines. The speed controller achieves the same stationary level as the RQV controller by tuning $\beta$.

The speed control law (7.137) then becomes

$$u = K_0 x_{30} + K_r r + K_l l - \begin{pmatrix} K_{c1} & K_{c2} & K_{c3} \end{pmatrix} \hat{x} \qquad (7.141)$$

with $K_0$, $K_r$, and $K_l$ given by (7.138). The estimated states $\hat{x}$ are given by the Kalman filter

$$\Delta \dot{\hat{x}} = \underline{A}\,\Delta \hat{x} + \underline{B}\,\Delta u + \underline{K}_f(\Delta y - \underline{C}\,\Delta \hat{x}) \qquad (7.142)$$

$$\underline{K}_f = \underline{P}_f \underline{C}^T \underline{V}^{-1} \qquad (7.143)$$

where $\underline{P}_f$ is derived by solving the Riccati equation

$$\underline{P}_f \underline{A}^T + \underline{A}\underline{P}_f - \underline{P}_f \underline{C}^T \underline{V}^{-1} \underline{C}\,\underline{P}_f + W = 0 \qquad (7.144)$$

The covariance matrices $W$ and $\underline{V}$ correspond to disturbances $v$ and $e$ respectively. The output matrix $\underline{C}$ is either equal to $\underline{C}_m$ (7.89) when measuring the engine speed, or $\underline{C}_w$ (7.90) when measuring the wheel speed.

To recover the properties (phase margin and amplitude margin) achieved in the previous design step when all states are measured, the following values are selected [78]

$$\begin{aligned} V &= 1 \\ W &= \rho B B^T \\ \underline{C} &= \underline{C}_m \text{ or } \underline{C}_w \\ \rho &= \rho_m \text{ or } \rho_w \end{aligned} \qquad (7.145)$$

Equations (7.143) and (7.144) are then solved for $\underline{K}_f$.

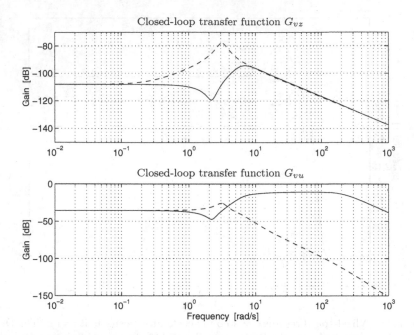

**Figure 7.42** Closed-loop transfer functions from load disturbance, $v$, to performance output, $z$, and to control signal, $u$. Feedback from $\dot{\alpha}_w$ is shown in solid and feedback from $\dot{\alpha}_{CS}$ is shown in dashed lines. With $\dot{\alpha}_{CS}$-feedback the transfer functions have a resonance peak, resulting from the open-loop zeros.

When using LQG with feedback from all states, the phase margin, $\varphi$, is at least $60°$ and the amplitude margin, $a$, is infinity as stated before. This is obtained also when using the observer by increasing $\rho$ towards infinity. For Example 7.2 the following values are used

$$\rho_m = 5 \cdot 10^5 \Rightarrow \varphi_m = 60.5°, \quad a_m = \infty \qquad (7.146)$$
$$\rho_w = 10^{14} \Rightarrow \varphi_w = 59.9°, \quad a_w = 35.0$$

where the aim has been to have at least $60°$ phase margin. The large difference between $\rho_m$ and $\rho_w$ in (7.146) is due to the structural difference between the two sensor locations, according to Section 7.3.4.

The observer dynamics is cancelled in the transfer functions from reference value to performance output ($z = \dot{\alpha}_w$) and to control signal ($u$). Hence, these transfer functions are not affected by sensor location. However, the observer dynamics will be included in the transfer functions from disturbances $v$ and $e$ to both $z$ and $u$.

**Influence from Load Disturbances**

Figure 7.42 shows how the performance output and the control signal are affected by the load disturbance $v$. There is a resonance peak in $G_{vz}$ when using feedback from the engine-speed sensor, which is not present when feedback from the wheel-speed sensor is used. The reason for this can be seen when studying the transfer function $G_{vz}$ in (7.99). By using the matrix inversion lemma [60] (7.99) is rewritten as

$$(G_{vz})_{cl} = \frac{G_{vz} + F_y(G_{uy}G_{vz} - G_{uz}G_{vy})}{1 + G_{uy}F_y} \qquad (7.147)$$

where $G_{ab}$ denotes the transfer function from signal $a$ to $b$, and $cl$ stands for closed loop. The subscript $y$ in (7.147) represents the output of the system, i.e., either $\dot{\alpha}_w$ or $\dot{\alpha}_{CS}$. The controller $F_y$ is given by (7.94) as

$$F_y(s) = \underline{K}_c(s\underline{I} - \underline{A} + \underline{B}\,\underline{K}_c + \underline{K}_f\underline{C})^{-1}\underline{K}_f \qquad (7.148)$$

with $C$ either being $C_m$ for engine-speed feedback, or $C_w$ for wheel-speed feedback. For the speed controller $(z = \dot{\alpha}_w)$, Equation (7.147) becomes

$$(G_{vz})_{cl} = \frac{G_{vw}}{1 + G_{uw}F_y} \qquad (7.149)$$

when the sensor measures the wheel speed. Equation (7.149) is obtained by replacing the subscript $y$ in (7.147) by the subscript $w$. Then the parenthesis in (7.147) equals zero. In the same way, the resulting equation for the $\dot{\alpha}_{CS}$-feedback case is

$$(G_{vz})_{cl} = \frac{G_{vw} + F_y(G_{um}G_{vw} - G_{uw}G_{vm})}{1 + G_{um}F_y} \qquad (7.150)$$

Hence, when using the wheel-speed sensor, the controller is cancelled in the numerator, and when the engine-speed sensor is used, the controller is not cancelled.

The optimal return ratio in the LQG step is

$$\underline{K}_c(s\underline{I} - \underline{A})^{-1}\underline{B} \qquad (7.151)$$

Hence, the poles from $\underline{A}$ is kept, but there are new zeros that are placed such that the relative degree of (7.151) is one, assuring a phase margin of at least 60° $(\varphi > 60°)$, and an infinite gain margin. In the LTR step the return ratio is

$$F_yG_{uy} = \underline{K}_c(s\underline{I} - \underline{A} - \underline{B}\,\underline{K}_c - \underline{K}_f\underline{C})^{-1}\underline{K}_f\underline{C}(s\underline{I} - \underline{A})^{-1}\underline{B} \qquad (7.152)$$

When $\rho$ in (7.145) is increased towards infinity, (7.151) equals (7.152). This means that the zeros in the open-loop system $\underline{C}(s\underline{I} - \underline{A})^{-1}\underline{B}$ are cancelled by the controller. Hence, the open-loop zeros will become poles in the controller $F_y$. This means that the closed-loop system will have the open-loop zeros as poles when using the engine-speed sensor. The closed-loop poles become $-0.5187 \pm 3.0753j$, which causes the resonance peak in Figure 7.42.

### Influence from Measurement Disturbances

The influence from measurement disturbances $e$ is shown in Figure 7.43. The transfer functions from measurement noise to output, (7.100), can be rewritten via the matrix inversion lemma as

$$(G_{ez})_{cl} = -\frac{G_{uz}F_y}{1 + G_{uy}F_y} \qquad (7.153)$$

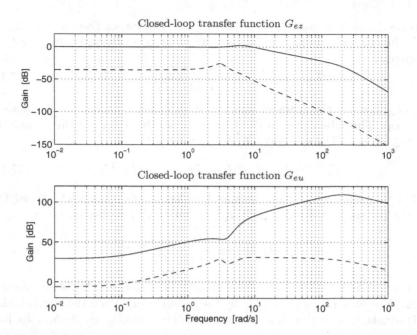

**Figure 7.43** Closed-loop transfer functions from measurement noise, $e$, to performance output, $z$, and to control signal, $u$. Feedback from $\dot{\alpha}_w$ is shown in solid and feedback from $\dot{\alpha}_{CS}$ is shown in dashed lines. The difference between the two feedback principles is described by the dynamic output ratio. The effect increases with lower gears.

The complementary sensitivity function is defined for the two sensor alternatives as

$$T_w = \frac{G_{uw}F_y}{1 + G_{uw}F_y}, \quad T_m = \frac{G_{um}F_y}{1 + G_{um}F_y} \tag{7.154}$$

Then by replacing the subscript $y$ in (7.153) with $m$ or $w$ (for $\dot{\alpha}_{CS}$-feedback or $\dot{\alpha}_w$-feedback), and comparing with (7.154), the following relations hold

$$(G_{ez})_{cl} = -T_w \quad \text{with } \dot{\alpha}_w-\text{feedback} \tag{7.155}$$

$$(G_{ez})_{cl} = -T_m \frac{G_{uw}}{G_{um}} = T_m G_{w/m} \quad \text{with } \dot{\alpha}_{CS}-\text{feedback} \tag{7.156}$$

where the dynamic output ratio $G_{w/m}$ was defined in Definition 7.1. For the Drive-shaft model the dynamic output ratio is

$$G_{w/m} = \frac{ds + k}{i(J_2 s^2 + (d + d_2)s + k)} \tag{7.157}$$

where the state-space description in Section 7.3.4 is used. Especially for low frequencies, $G_{w/m}(0) = 1/i = 1/i_t i_f$. The dynamic output ratio can be seen in Figure 7.44 for three different gears.

When $\rho$ in (7.145) is increased towards infinity, (7.151) equals (7.152), which means that $T_m = T_w$. Then (7.155) and (7.156) gives

$$(G_{ez})_{cl,m} = (G_{ez})_{cl,w} G_{w/m} \tag{7.158}$$

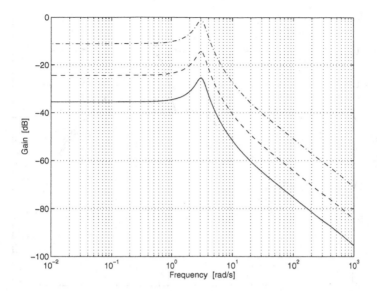

**Figure 7.44** The dynamic output ratio $G_{w/m}$ for gear 1 (solid), gear 7 (dashed), and gear 14 (dash-dotted).

where $cl, m$ and $cl, w$ means closed loop with feedback from $\dot{\alpha}_{CS}$ and $\dot{\alpha}_w$ respectively.

The frequency range in which $T_m = T_w$ is valid depends on how large $\rho$ in (7.145) is made. Figure 7.45 shows the sensitivity functions

$$S_w = \frac{1}{1 + G_{uw}F_y}, \quad S_m = \frac{1}{1 + G_{um}F_y} \tag{7.159}$$

and the complementary sensitivity functions $T_w$ and $T_m$ (7.154) for the two cases of feedback. It is seen that $T_m = T_w$ is valid up to about 100 rad/s ($\approx$ 16 Hz). The roll-off rate at higher frequencies differ between the two feedback principles. This is due to that the open-loop transfer functions $G_{uw}$ and $G_{um}$ have different relative degrees. $G_{uw}$ has a relative degree of two, and $G_{um}$ has a relative degree of one. Therefore, $T_w$ has a steeper roll-off rate than $T_m$.

Hence, the difference in $G_{ez}$ depending on sensor location is described by the dynamic output ratio $G_{w/m}$. The difference in low-frequency level is equal to the conversion ratio of the driveline. Therefore, this effect increases with lower gears.

**Load Estimation**

The feedback law with unknown load is

$$u = K_0 x_{30} + K_r r + K_l \hat{l} - \begin{pmatrix} K_{c1} & K_{c2} & K_{c3} \end{pmatrix} \hat{\underline{x}} \tag{7.160}$$

where $\hat{l}$ is the estimated load. In order to estimate the load, the model used in the Kalman filter is augmented with a model of the load. The load is hard to model correctly since it is a function of road slope. However it can be treated as

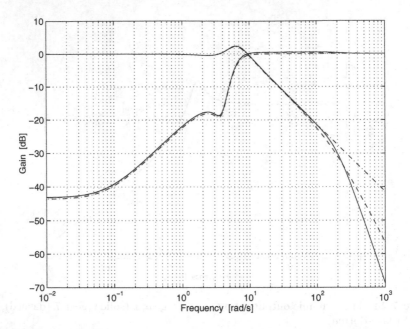

**Figure 7.45** Sensitivity function $S$ and complementary sensitivity function $T$. The dash-dotted lines correspond to the case with all states known. When only one velocity is measured, the solid lines correspond to $\dot{\alpha}_w$-feedback, and the dashed lines correspond to $\dot{\alpha}_{CS}$-feedback.

a slow-varying constant. A reasonable augmented model is

$$x_4 = \hat{l}, \quad \text{with} \quad \dot{x}_4 = 0 \tag{7.161}$$

This gives

$$\dot{\hat{x}} = \underline{A}_l \hat{x}_l + \underline{B}_l u + \underline{K}_f (y - \underline{C}_l \hat{x}_l) \tag{7.162}$$

with

$$\hat{x}_l = \begin{pmatrix} \hat{x} & \hat{l} \end{pmatrix}^T, \tag{7.163}$$

$$\underline{A}_l = \begin{pmatrix} & & & 0 \\ & \underline{A} & & 0 \\ & & & -1/J_2 \\ 0 & 0 & 0 & 0 \end{pmatrix}, \tag{7.164}$$

$$\underline{B}_l = \begin{pmatrix} \underline{B} \\ 0 \end{pmatrix}, \quad \underline{C}_l = \begin{pmatrix} \underline{C} & 0 \end{pmatrix} \tag{7.165}$$

The feedback law is

$$u = K_0 x_{30} + K_r r - \begin{pmatrix} K_{c1} & K_{c2} & K_{c3} & -K_l \end{pmatrix} \hat{x}_l \tag{7.166}$$

## 7.4.5  Simulations

An important step in demonstrating feasibility for real implementation is that a controller behaves well when simulated on a more complicated vehicle model

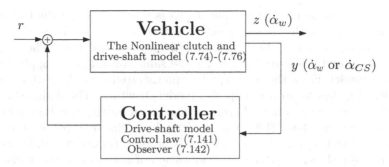

**Figure 7.46** Simulation configuration. As a step for demonstrating feasibility for real implementation, the Nonlinear clutch and drive-shaft model is simulated with the controller based on the Drive-shaft model.

than it was designed for. Even more important in a principle study is that such disturbances can be introduced that hardly can be generated in systematic ways in real experiments. One such example is impulse disturbances from a towed trailer.

The control law based on the reduced driveline model is simulated with a more complete nonlinear model, derived in Section 7.1. The purpose is also to study effects from different sensor locations as discussed in Section 7.4.4. The simulation situation is seen in Figure 7.46. The Nonlinear clutch and drive-shaft model, given by (7.74) to (7.76), is used as vehicle model. The steady-state level for the Nonlinear clutch and drive-shaft model is calculated by solving the model equations for the equilibrium point when the load and speed are known.

The controller used is based on the Drive-shaft model , as was derived in the previous sections. The wheel speed or the engine speed is the input to the observer (7.142), and the control law (7.141) with $\beta = 0$ generates the control signal.

The simulation case presented here is the same as in Example 7.2, i.e., a velocity step response, but a load disturbance is also included. The stationary point is given by

$$\dot{\alpha}_w = 2, \ l = 3000 \ \Rightarrow \ x_0 = \begin{pmatrix} 0.0482 & 119 & 2.00 \end{pmatrix}^T, \ u_0 = 109 \quad (7.167)$$

where (7.117) and (7.118) are used, and the desired new speed is $\dot{\alpha}_w = 2.3$ rad/s. At steady state, the clutch transfers the torque $u_0 = 109$ Nm. This means that the clutch angle is in the area with higher stiffness ($\alpha_{c1} < \alpha_c \leq \alpha_{c2}$) in the clutch nonlinearity, seen in Figure 7.18. This is a typical driving situation when speed control is used. However, at low clutch torques ($\alpha_c < \alpha_{c1}$) the clutch nonlinearity can produce limit cycle oscillations [8]. This situation occurs when the truck is traveling downhill with a load of the same size as the friction in the driveline, resulting in a low clutch torque. This is however not treated here. At $t = 6$ s, a load impulse disturbance is simulated. The disturbance is generated as a square pulse with 0.1 s width and 1200 Nm height, added to the load according to (7.87).

In order to simulate the nonlinear model, the differential equations (7.74) to (7.76) are scaled such that the five differential equations (one for each state) have about the same magnitude. The model is simulated using the Runge Kutta (45) method [118] with a low step size to catch the effect of the nonlinearity.

Figures 7.47 to 7.49 show the result of the simulation. These figures should be compared to Figure 7.40, where the same control law is applied to the Drive-shaft model. From these plots it is demonstrated that the performance does not critically depend on the simplified model structure. The design still works if the extra dynamics is added. Further evidence supporting this is seen in Figure 7.49. The area with low stiffness in the clutch nonlinearity ($\alpha_c < \alpha_{c1}$) is never entered. The load impulse disturbance is better attenuated with feedback from the wheel-speed sensor, which is a verification of the behavior that was discussed in Section 7.4.4.

## 7.4.6   Speed Controller Experiments

Experiments are used to demonstrate that the method is applicable for real implementations in a heavy truck. The goal is further to demonstrate that the simplified treatment of the diesel engine (smooth torque, dynamical behavior, etc, according to Section 7.3) holds in field trials.

The speed control strategy is implemented by discretizing the feedback law and the observer. The controller parameters are tuned for the practical constraints given by the measured signals. Step response tests in engine speed are performed with the strategy and the results are compared to the traditionally used RQV controller for speed control.

The algorithm computed every iteration is as follows.

---

**Model 7.6 Control algorithm**

---

1. *Read engine speed ($\dot{\alpha}_{CS}$) and engine temperature ($\vartheta_e$).*

2. *Calculate engine friction torque, $T_{fric,e}(\dot{\alpha}_{CS}, \vartheta_e)$, as function of the engine speed and the engine temperature. The friction values are obtained from a map, described in Section 7.1.4, by an interpolation routine.*

3. *Read the engine torque ($T_e$) and the variable used as input to the observer (engine speed, $\dot{\alpha}_{CS}$, or wheel speed, $\dot{\alpha}_w$).*

4. *Calculate the control signal $u_k = (T_e - T_{fric,e}(\dot{\alpha}_{CS}, \vartheta_e))$, and update the observer equations.*

5. *Read the reference value ($r_k$), and use the feedback law to calculate the new control signal, $u_{k+1}$.*

6. *The new control signal is transferred to requested engine torque by adding the engine friction torque to the control signal ($u_{k+1} + T_{fric,e}(\dot{\alpha}_{CS}, \vartheta_e)$). The requested engine torque is then sent to the engine control unit.*

---

The repetition-rate of the algorithm is chosen the same as the sampling rate of the input variable to the observer. This means that the sampling-rate is 50 Hz using feedback from the engine-speed sensor. More information about the measured variables are found in Table 7.1. The parameters of the implemented

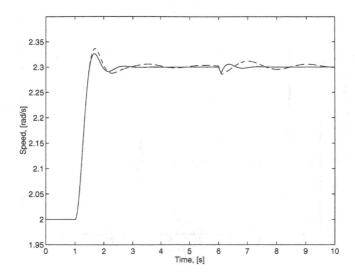

**Figure 7.47** Wheel-speed response of step in accelerator position at t=1 s with the speed controller (7.141) derived from the Drive-shaft model, controlling the Nonlinear clutch and drive-shaft model. The solid line corresponds to $\dot{\alpha}_w$-feedback and feedback from $\dot{\alpha}_{CS}$ is seen in dashed line. At t=6 s, an impulse disturbance $v$ acts on the load. The design still works when simulated with extra clutch dynamics.

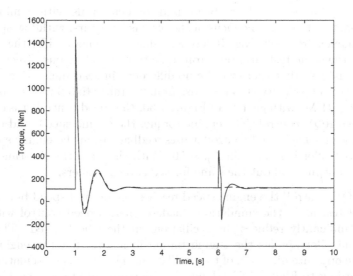

**Figure 7.48** Control signal corresponding to Figure 7.47. There is only little difference between the two sensor alternatives in the step response at t=1 s. However, the load impulse (at t=6 s) generates a control signal that damps the impulse disturbance when feedback from the wheel-speed sensor is used, but not with engine-speed feedback.

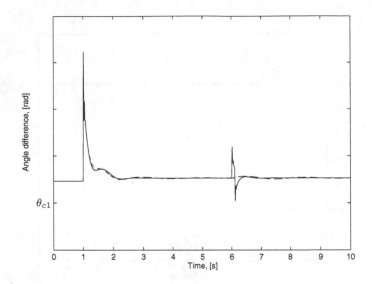

**Figure 7.49** Clutch-angle difference corresponding to Figure 7.47. The influence from the clutch nonlinearity can be neglected, because the area with low stiffness ($\alpha_c < \alpha_{c1}$) is never entered.

algorithm are in the following sections tuned for the practical constraints given by the sensor characteristics.

An almost flat test road has been used for field trials with a minimum of changes from test to test. The focus of the tests is low gears, with low speeds and thus little impact from air drag. Reference values are generated by the computer to generate the same test situation from time to time. Only one direction of the test road is used so that there will be no difference in road inclination. The test presented here is a velocity step response from 2.1 rad/s to 3.6 rad/s (about 1200 RPM to 2000 RPM) with gear 1. In Figure 7.50, the speed controller is compared to traditional RQV control. The engine torque, the engine speed, and the wheel speed are shown. The speed controller uses feedback from the engine speed, and the RQV controller has the gain $K_p = 50$. With this gain the rise-time and the peak torque output is about the same for the two controllers.

With RQV control, the engine speed reaches the desired speed but the wheel speed oscillates, as in the simulations made earlier. Speed control with active damping significantly reduces the oscillations in the wheel speed. This means that the controller applies the engine torque in a way that the engine inertia works in the opposite direction of the oscillation. This gives an oscillating engine speed, according to Figure 7.50. Hence, it is demonstrated that the assumption about the simplified model structure (Drive-shaft model) is sufficient for control design. It is further demonstrated that the design is robust against nonlinear speed dependent torque limitations (maximum torque limitations), and the assumption about static transfer function between engine torque and fuel amount is sufficient.

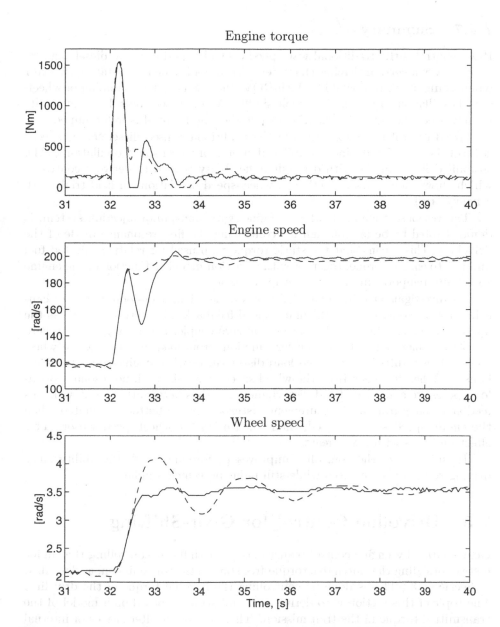

**Figure 7.50** Speed step at t=32 s with active damping and engine-speed feedback (solid) compared to traditional RQV control with $K_p$=50 (dashed). Experiments are performed on a flat road. After 32.5 s, the control signals differ depending on control scheme. With speed control, the engine inertia works in the opposite direction of the oscillations, which are significantly reduced.

### 7.4.7  Summary

RQV control is the traditional way speed control is performed in diesel engines, which gives a certain driving character with a load dependent stationary error when going uphill or downhill. With RQV, there is no active damping of wheel-speed oscillations, resulting in vehicle shuffle. An increased controller gain results in increased wheel-speed oscillations while the engine speed is well damped.

Speed control is the extension of the traditionally used diesel engine speed-control scheme with engine controlled damping of wheel-speed oscillations. The simplified linear model with drive-shaft flexibility is used to derive a controller which shows significant reduction in wheel-speed oscillations in field trials with a heavy truck.

The response time of the diesel engine, with unit-pump injection system, is demonstrated to be fast enough for controlling the first resonance mode of the driveline. This means that the static torque map used for relating injected fuel amount to engine torque, together with a friction model as function of the engine speed and temperature, is sufficient for control.

An investigation using LQG/LTR was done. The open-loop zeros are can-celled by the controller. With engine-speed feedback this is critical, because the open-loop transfer function has a resonant zero couple. It is shown that this zero couple becomes poles of the transfer functions from load disturbances to wheel speed. This results in undamped load disturbances when engine-speed feedback is used. When feedback from the wheel-speed sensor is used, no resonant open-loop poles are cancelled. Load disturbances are thus better attenuated with this feedback configuration. Measurement disturbances are better attenuated when the engine-speed sensor is used, than when using the wheel-speed sensor. This effect increases with lower gears.

To summarize, the controller improves performance and driveability since driving response is increased while still reducing vehicle shuffle.

## 7.5  Driveline Control for Gear-Shifting

Gear shifting by engine control realizes fast gear shifts by controlling the engine instead of sliding the clutch to a torque-free state in the transmission, as described in Section 7.3. This is done by controlling the internal torque of the driveline. The topic of this section is to derive a control strategy based on a model of the transmitted torque in the transmission. There are other alternatives of internal torques that can be used as control objectives, e.g. drive-shaft torque, but this is not used here except in the final subsection. Thus, there is a detailed study of the dynamical behavior of the transmission torque, which should be zero in order to engage neutral gear. A transmission-torque controller is derived that controls the estimated transmission torque to zero while having engine controlled damping of driveline resonances. With this approach, the specific transmission-torque behavior for each gear is described and compensated for.

A model of the transmission is developed in Section 7.5.1, where the torque transmitted in the transmission is modeled as a function of the states and the con-trol signal of the Drive-shaft model. The controller goal was stated in Section 7.3,

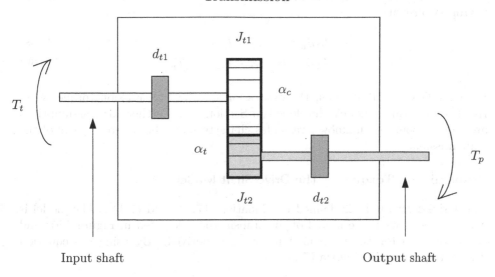

**Figure 7.51** Simplified model of the transmission with two cogwheels with conversion ratio $i_t$. The cogwheels are connected to the input and output shafts respectively. The torque transmitted between the cogwheels is the transmission torque, $z$.

and is formulated in mathematical terms as a gear-shift control criterion in Section 7.5.2. The control law in Section 7.5.3 minimizes the criterion. Influence from sensor location, simulations, and experiments are presented in the sections following.

## 7.5.1 Internal Driveline Torque

There are many possible definitions of internal driveline torque. Since the goal is to engage neutral gear without using the clutch, it is natural to use the minimization of the torque transferred in the transmission as a control goal. The following sections cover the derivation of an expression for this torque, called the *transmission torque*, as function of the state variables and the control signal.

**Transmission Torque**

The performance output, $z$, for the gear-shift controller is the transmission torque transferred between the cogwheels in the transmission. A simplified model of the transmission is depicted in Figure 7.51. The input shaft is connected to bearings with a viscous friction component $d_{t1}$. A cogwheel is mounted at the end of the input shaft which is connected to a cogwheel mounted on the output shaft. The conversion ratio between these are $i_t$, as mentioned in Section 7.1. The output shaft is also connected to bearings with the viscous friction component $d_{t2}$.

By using Newton's second law, the transmission can be modeled by the following two equations

$$J_{t1}\ddot{\alpha}_c = T_t - d_{t1}\dot{\alpha}_c - z \qquad (7.168)$$
$$J_{t2}\ddot{\alpha}_t = i_t z - d_{t2}\dot{\alpha}_t - T_p \qquad (7.169)$$

In the following subsections, the expression for the transmission torque is derived for the three models developed in Section 7.1. Furthermore, assumptions are made about the unknown variables characterizing the different parts of the transmission.

## Transmission Torque for the **Drive-Shaft Model**

The Drive-shaft model is defined by Equations (7.43) and (7.44). The model is here extended with the model of the transmission depicted in Figure 7.51, and the expression for the transmission torque is derived. By using the equation describing the engine inertia (7.1)

$$J_e\ddot{\alpha}_{CS} = T_e - T_{fric,e} - T_c \qquad (7.170)$$

together with (7.30)

$$T_c = T_t, \quad \alpha_{CS} = \alpha_c \qquad (7.171)$$

equation (7.168) is expressed in terms of engine speed

$$(J_e + J_{t1})\ddot{\alpha}_{CS} = T_e - T_{fric,e} - d_{t1}\dot{\alpha}_{CS} - z \qquad (7.172)$$

To describe the performance output in terms of state variables, $\ddot{\alpha}_{CS}$ (which is not a state variable) is replaced by (7.43), which is one of the differential equations describing the Drive-shaft model

$$(J_e + J_t/i_t^2 + J_f/i_t^2 i_f^2)\ddot{\alpha}_{CS} = T_e - T_{fric,e} - (d_t/i_t^2 + d_f/i_t^2 i_f^2)\dot{\alpha}_{CS}$$
$$-k(\alpha_{CS}/i_t i_f - \alpha_w)/i_t i_f$$
$$-d(\dot{\alpha}_{CS}/i_t i_f - \dot{\alpha}_w)/i_t i_f \qquad (7.173)$$

which together with $u = T_e - T_{fric,e}$ gives

$$u - d_{t1}\dot{\alpha}_{CS} - z = \frac{J_e + J_{t1}}{J_e + J_t/i_t^2 + J_f/i_t^2 i_f^2}\left(u - (d_t/i_t^2 + d_f/i_t^2 i_f^2)\dot{\alpha}_{CS}\right.$$
$$\left. -k(\alpha_{CS}/i_t i_f - \alpha_w)/i_t i_f - d(\dot{\alpha}_{CS}/i_t i_f - \dot{\alpha}_w)/i_t i_f\right)$$
$$(7.174)$$

From this equation it is possible to express the performance output, $z$, as a function of the control signal, $u$, and the state variables, $\underline{x}$, according to the state-space description (7.80) to (7.83).

---

## Model 7.7 Transmission Torque for the Drive-Shaft Model

---

$$z = \underline{M}\,\underline{x} + Du \quad \text{with}$$

$$\underline{M}^T = \begin{pmatrix} \frac{(J_e+J_{t1})k}{J_1 i} \\ \frac{J_e+J_{t1}}{J_1}(d_1 + d/i^2) - d_{t1} \\ -\frac{(J_e+J_{t1})d}{J_1 i} \end{pmatrix} \tag{7.175}$$

$$D = 1 - \frac{J_e + J_{t1}}{J_1}$$

*The transmission torque, z, is modeled as a function of the states and the control signal for the Drive-shaft model, where the labels from (7.83) are used.*

---

The unknown parameters in (7.175) are $J_e+J_{t1}$ and $d_{t1}$. The other parameters were estimated in Section 7.1. One way of estimating these unknowns would be to decouple the Drive-shaft model into two models, corresponding to neutral gear. Then a model including the engine, the clutch, and the input shaft of the transmission results, in which the performance output is equal to zero ($z = 0$). Trials with neutral gear would then give a possibility to estimate the unknowns. This will be further investigated in Section 7.2.

In the derivation of the Drive-shaft model in Section 7.1 the performance output, $z$, is eliminated. If $z$ is eliminated in (7.168) and (7.169) and (7.171) is used, the equation for the transmission is

$$(J_{t1}i_t^2 + J_{t2})\ddot{\alpha}_{CS} = i_t^2 T_c - i_t T_p - (d_{t1}i_t^2 + d_{t2})\dot{\alpha}_{CS} \tag{7.176}$$

By comparing this with the equation describing the transmission in Section 7.1, (7.33)

$$J_t\ddot{\alpha}_{CS} = i_t^2 T_c - d_t\dot{\alpha}_{CS} - i_t T_p \tag{7.177}$$

the following equations relating the parameters are obtained

$$J_t = i_t^2 J_{t1} + J_{t2} \tag{7.178}$$

$$d_t = i_t^2 d_{t1} + d_{t2} \tag{7.179}$$

In order to further investigate control and estimation of the transmission torque, the unknowns are given values. It is arbitrarily assumed that the gear shift divides the transmission into two equal inertias and viscous friction components, giving

$$J_{t1} = J_{t2} \tag{7.180}$$

$$d_{t1} = d_{t2}$$

A more detailed discussion of these parameters will be performed in Section 7.5.6. Equations (7.178) and (7.179) then reduce to

$$J_{t1} = \frac{J_t}{1 + i_t^2} \tag{7.181}$$

$$d_{t1} = \frac{d_t}{1 + i_t^2} \tag{7.182}$$

The following combinations of parameters from the Drive-shaft model were estimated in Section 7.1

$$J_1 = J_e + J_t/i_t^2 + J_f/i_t^2 i_f^2 \qquad (7.183)$$

$$d_1 = d_t/i_t^2 + d_f/i_t^2 i_f^2 \qquad (7.184)$$

according to the labels from the state-space formulation in (7.83). From (7.181) and (7.183) $J_e + J_{t1}$ can be derived as

$$
J_e + J_{t1} = J_e + \frac{J_t}{1 + i_t^2} = J_e + \frac{i_t^2}{1 + i_t^2}(J_1 - J_e - J_f/i_t^2 i_f^2)
$$
$$
= J_e \frac{1}{1 + i_t^2} + J_1 \frac{i_t^2}{1 + i_t^2} - J_f \frac{1}{i_f^2(1 + i_t^2)} \qquad (7.185)
$$

A combination of (7.182) and (7.184) gives $d_{t1}$

$$
d_{t1} = \frac{d_t}{1 + i_t^2} = \frac{i_t^2}{1 + i_t^2}(d_1 - d_f/i_t^2 i_f^2) \qquad (7.186)
$$

For low gears $i_t$ has a large value. This together with the fact that $J_f$ and $d_f$ are considerably less than $J_1$ and $d_1$ gives the following approximation about the unknown parameters

$$
J_e + J_{t1} \approx J_1 \frac{i_t^2}{1 + i_t^2} \qquad (7.187)
$$

$$
d_{t1} \approx d_1 \frac{i_t^2}{1 + i_t^2} \qquad (7.188)
$$

## Transmission Torque for the Clutch and Drive-Shaft Model

The performance output expressed for the Clutch and drive-shaft model is given by replacing $T_t$ in (7.168) by equation (7.62)

$$
T_c = T_t = k_c(\alpha_{CS} - \alpha_t i_t) + d_c(\dot{\alpha}_{CS} - \dot{\alpha}_t i_t) \qquad (7.189)
$$

Then the performance output is

$$
z = k_c(\alpha_{CS} - \alpha_t i_t) + d_c(\dot{\alpha}_{CS} - \dot{\alpha}_t i_t) - d_{t1} i_t \dot{\alpha}_t - J_{t1} i_t \ddot{\alpha}_t \qquad (7.190)
$$

This is expressed in terms of state variables by using (7.68)

$$
(J_t + J_f/i_f^2)\ddot{\alpha}_t = i_t \left( k_c(\alpha_{CS} - \alpha_t i_t) + d_c(\dot{\alpha}_{CS} - \dot{\alpha}_t i_t) \right) \qquad (7.191)
$$
$$
- (d_t + d_f/i_f^2)\dot{\alpha}_t - \frac{1}{i_f}\left( k_d(\alpha_t/i_f - \alpha_w) + d_d(\dot{\alpha}_t/i_f - \dot{\alpha}_w) \right)
$$

leading to the following model.

---

**Model 7.8 Transmission Torque for the Clutch and Drive-Shaft Model**

---

$$z = \underline{M}\,\underline{x} \text{ with} \tag{7.192}$$

$$\underline{M}^T = \begin{pmatrix} k_c(1 - \frac{J_{t1}i_t^2}{J_2}) \\ \frac{J_{t1}i_t k_d}{J_2 i_f} \\ d_c(1 - \frac{J_{t1}i_t^2}{J_2}) \\ \frac{J_{t1}i_t^2}{J_2}(i_t^2 d_c + d_2 + d_d/i_f^2) - d_c i_t - d_{t1} i_t \\ -\frac{J_{t1}i_t d_d}{J_2 i_f} \end{pmatrix}$$

*with states and labels according to to the state-space description (7.84) to (7.86).*

---

The following combinations of parameters from the Clutch and drive-shaft model were estimated in Section 7.1

$$J_2 = J_t + J_f/i_f^2 \tag{7.193}$$

$$d_2 = d_t + d_f/i_f^2 \tag{7.194}$$

according to (7.86). From (7.181), (7.182), (7.193), and (7.194), $J_{t1}$ and $d_{t1}$ can be written as

$$J_{t1} = \frac{i_t^2}{1 + i_t^2}(J_2 - J_f/i_f^2) \tag{7.195}$$

$$d_{t1} = \frac{i_t^2}{1 + i_t^2}(d_2 - d_f/i_f^2) \tag{7.196}$$

which are approximated to

$$J_{t1} \approx \frac{i_t^2}{1 + i_t^2}J_2 \tag{7.197}$$

$$d_{t1} \approx \frac{i_t^2}{1 + i_t^2}d_2 \tag{7.198}$$

since $J_f$ and $d_f$ are considerably less than $J_1$ and $d_1$.

**Transmission Torque for the Nonlinear Clutch and Drive-Shaft Model**

The performance output for the Nonlinear clutch and drive-shaft model is derived in the same way as for the Clutch and drive-shaft model, with the difference that (7.189) is replaced by

$$T_c = T_t = T_{kc}(\alpha_{CS} - \alpha_t i_t) + d_c(\dot{\alpha}_{CS} - \dot{\alpha}_t i_t) \tag{7.199}$$

where $T_{kc}$ is the torque transmitted by the clutch nonlinearity, given by (7.73). Then the performance output is defined as

---

### Model 7.9 Transmission Torque for the Nonlinear Clutch and Drive-Shaft Model

$$z = (T_{kc} \quad \dot{\alpha}_t/i_f - \dot{\alpha}_w \quad \dot{\alpha}_{CS} \quad \dot{\alpha}_t \quad \dot{\alpha}_w) \begin{pmatrix} 1 - \frac{J_{t1}i_t^2}{J_2} \\ \frac{J_{t1}i_tk_d}{J_2i_f} \\ d_c(1 - \frac{J_{t1}i_t^2}{J_2}) \\ \frac{J_{t1}i_t^2}{J_2}(i_t^2d_c + d_2 + d_d/i_f^2) - d_ci_t - d_{t1}i_t \\ -\frac{J_{t1}i_td_d}{J_2i_f} \end{pmatrix}$$

$$(7.200)$$

---

The parameters not estimated in the definition above are approximated in the same way as for the performance output for the Clutch and drive-shaft model.

### Model Comparison

Figure 7.52 shows the transmission torque during a test with step inputs in accelerator position with the 144L truck using gear 1. The transmission torque is calculated with (7.175) for the Drive-shaft model, and with (7.192) for the Clutch and drive-shaft model. Figure 7.53 shows the performance output in the frequency domain. The low-frequency level differs between the two models, and the main reason for this is the difficulties to estimate the viscous damping coefficients described in Section 7.1. The difference at higher frequencies is due to the clutch, which gives a second resonance peak for the Clutch and drive-shaft model. Furthermore, the roll-off rate of the Clutch and drive-shaft model is steeper than for the Drive-shaft model.

## 7.5.2   Transmission-Torque Control Criterion

### Problem Formulation

The transmission-torque controller is the controller that drives the transmission torque to zero with engine controlled damping of driveline resonances. Then the time spent in the torque control phase (see Section 7.3) is minimized. The engagement of neutral gear should be at a torque level that gives no oscillations in the driveline speeds. Hereby, the disturbances to the driver and the time spent in the speed synchronization phase can be minimized. The influence on shift quality from initial driveline resonances, and torque impulses from trailer and road roughness should be minimized.

### Control Criterion

The transmission-torque controller is realized as a state-feedback controller, based on the Drive-shaft model. The controller is obtained by deriving a control criterion that describes the control problem of minimizing the transmission torque. The criterion is then minimized by standard software for a controller solving the control problem.

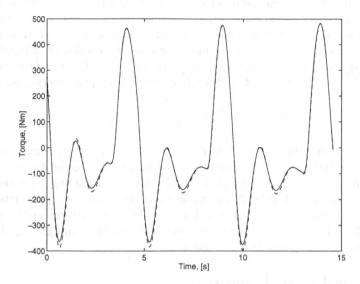

**Figure 7.52** Estimated transmission torque, $z$, in (7.175) and (7.192) for a test with step inputs in accelerator position with the 144L truck. The solid line corresponds to the Drive-shaft model and the dashed line corresponds to the Clutch and drive-shaft model.

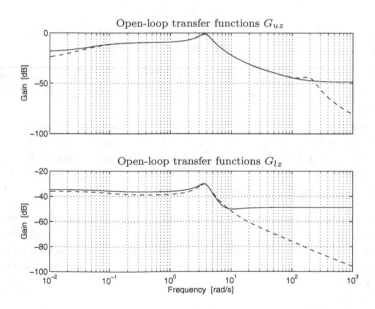

**Figure 7.53** Transfer functions from control signal, $u$, and load, $l$, to transmission torque, $z$. The Drive-shaft model is shown in solid and the Clutch and drive-shaft model is shown in dashed lines. The modeled clutch adds a second resonance peak and a steeper roll-off rate.

The gear-shift problem can be described as minimizing the transmission torque, $z$, but with a control signal, $u$, possible to realize by the diesel engine. Therefore, the criterion consists of two terms. The first term is $z^2$ which describes the deviation from zero transmission torque. The second term describes the deviation in control signal from the level needed to obtain $z = 0$. Let this level be $u_{shift}$, which will be speed-dependent as described later. Then the criterion is described by

$$\lim_{T \to \infty} \int_0^T z^2 + \eta(u - u_{shift})^2 \qquad (7.201)$$

The controller that minimizes this cost function will utilize engine controlled damping of driveline resonances (since $z^2$ is minimized) in order to obtain $z = 0$. At the same time, the control signal is prevented from having large deviations from the level $u_{shift}$. The trade-off is controlled by tuning the parameter $\eta$.

In the following subsections, the influence from each term in the criterion (7.201) will be investigated, and then how these can be balanced together for a feasible solution by tuning the parameter $\eta$.

**Unconstrained Active Damping**

The influence from the first term in the criterion (7.201) is investigated by minimizing $z^2$. The performance output, $z = \underline{M}\,\underline{x} + Du$, is derived in (7.175) for the Drive-shaft model as a function of the states and the control signal. The term $z^2$ can be minimized for a control law, since $z$ includes the control signal and $D$ is scalar. If $u$ is chosen as

$$u = -D^{-1}\underline{M}\,\underline{x} \qquad (7.202)$$

$z = 0$ is guaranteed. This control law is called *unconstrained active damping* and the reason for this is illustrated in the following example.

---

**Example 7.3**   *Consider the 144L truck modeled in Section 7.1 traveling at a speed of 3 rad/s (5.4 km/h) with gear 1 and a total load of 3000 Nm ($\approx$ 2 % road slope).*
*Figure 7.54 shows the resulting transmission torque, the control signal, the engine speed, and the wheel speed, when a gear shift is commanded at t=1 s, with the control signal chosen according to (7.202). Unconstrained active damping is achieved which obtains $z = 0$ instantaneously. The wheel speed decreases linearly, while the engine speed is oscillating.*

---

Unconstrained active damping (7.202) fulfills the control goal, but generates a control signal that is too large for the engine to generate. It can be noted that despite $z = 0$ is achieved this is not a stationary point, since the speed is decreasing. This means that the vehicle is free-rolling which can be critical if lasting too long.

**Gear-Shift Condition**

The influence from the second term in the criterion (7.201) is investigated by minimizing $(u - u_{shift})^2$, resulting in the control law

$$u = u_{shift} \qquad (7.203)$$

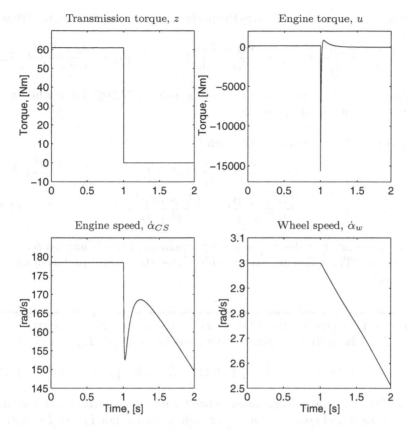

**Figure 7.54** Unconstrained active damping of the Drive-shaft model. At t=1 s, a gear shift is commanded and the control law (7.202) calculates the engine torque such that the transmission torque is driven to zero instantaneously. The oscillations in the transmission torque are damped with an unrealizable large control signal. The wheel speed decreases linearly.

where the torque level $u_{shift}$ is the control signal needed to obtain zero transmission torque, without using active damping of driveline resonances. Hence, $u_{shift}$ can be derived from a stiff driveline model, by solving for $z = 0$.

By using the labels according to Section 7.3.4, the differential equation describing the stiff driveline is

$$(J_1 i + J_2/i)\ddot{\alpha}_w = u - (d_1 i + d_2/i)\dot{\alpha}_w - l/i \tag{7.204}$$

This equation is developed by using the Drive-shaft model in (7.43) and (7.44), and eliminating the torque transmitted by the drive shaft, $k(\alpha_{CS}/i - \alpha_w) + c(\dot{\alpha}_{CS}/i - \dot{\alpha}_w)$. Then, by using $\dot{\alpha}_{CS} = \dot{\alpha}_w i$ (i.e., stiff driveline), (7.204) results.

Equation (7.172) expressed in terms of wheel speed is

$$z = u - d_{t1} i \dot{\alpha}_w - (J_e + J_{t1}) i \ddot{\alpha}_w \tag{7.205}$$

Combining (7.204) and (7.205) gives the performance output for the stiff driveline.

$$z = (1-\frac{(J_e+J_{t1})i^2}{J_1i^2+J_2})u-(d_{t1}i-\frac{(J_e+J_{t1})i}{J_1i^2+J_2}(d_1i^2+d_2))\dot{\alpha}_w+\frac{(J_e+J_{t1})i}{J_1i^2+J_2}l \quad (7.206)$$

The control signal to force $z = 0$ is given by solving (7.206) for $u$ while $z = 0$. Then the torque level $u_{shift}$ becomes

$$u_{shift}(\dot{\alpha}_w, l) \quad = \quad \mu_x\dot{\alpha}_w + \mu_l l \quad \text{with}$$

$$\mu_x \quad = \quad (d_{t1}i - \frac{(J_e+J_{t1})i}{J_1i^2+J_2}(d_1i^2+d_2))(1-\frac{(J_e+J_{t1})i^2}{J_1i^2+J_2})^{-1}$$

$$\mu_l \quad = \quad -\frac{(J_e+J_{t1})i}{J_1i^2+J_2}(1-\frac{(J_e+J_{t1})i^2}{J_1i^2+J_2})^{-1} \quad (7.207)$$

This control law is called the *gear-shift condition*, since it implies zero transmission torque. The following example illustrates the control performance when using (7.207).

---

**Example 7.4**  *Consider the 144L truck in the same driving situation as in Example 7.3. The stationary point is obtained by using (7.117) and (7.118).*

$$x_{30} = 3, \; l = 3000 \; \Rightarrow \; x_0 = (\; 0.0511 \quad 178 \quad 3.00 \;), \quad u_0 = 138 \quad (7.208)$$

*Figure 7.55 shows the resulting transmission torque, the control signal, the engine speed, and the wheel speed when a gear shift is commanded at t=1 s, with the control signal chosen according to (7.207).*
*This control law achieves $z = 0$ with a realizable control signal, but the oscillations introduced are not damped. Therefore, the time needed to obtain zero transmission torque is not optimized. The performance of this approach is worse if the driveline is oscillating at the time for the gear shift, or if there are disturbances present.*

---

**Final Control Criterion**

The final cost criterion for the transmission-torque controller is obtained by including (7.207) in the cost criterion (7.201)

$$\lim_{T\to\infty} \int_0^T z^2 + \eta(u - u_{shift}(\dot{\alpha}_w, l))^2 \quad (7.209)$$

$$= \quad \lim_{T\to\infty} \int_0^T (\underline{M}\underline{x} + Du)^2 + \eta(u - \mu_x\dot{\alpha}_w - \mu_l l)^2$$

If the driveline is stiff, there is no difference between the two terms in the cost function (7.209). Furthermore, the point at which the cost function is zero is no stationary point, since the speed of the vehicle will decrease despite $z = 0$ and $u = u_{shift}$.

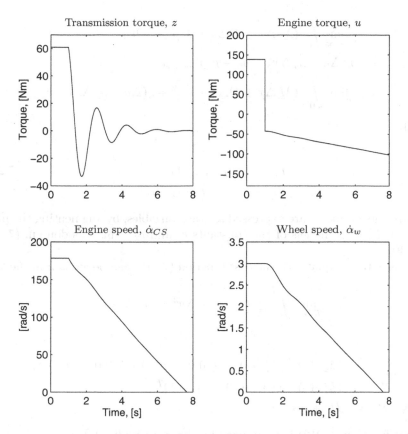

**Figure 7.55** The Drive-shaft model controlled with the gear-shift condition (7.207). At t=1 s, a gear shift is commanded. The speed dependent realizable control signal drives the transmission torque to zero. Undamped oscillations in the transmission torque increase the time needed to fulfill the goal of controlling the transmission torque to zero.

## 7.5.3   Transmission-Torque Control Design

The gear-shift control is in this section given efficient treatment by solving (7.209) for a control law by using LQG technique, and available software. This is done by linearizing the driveline model and rewriting (7.209) in terms of the linearized variables. A state-feedback matrix is derived that minimizes (7.209) by solving a Riccati equation. The derived feedback law is a function of $\eta$, which is chosen such that high bandwidth together with a feasible control signal is obtained.

The linearized driveline model is given by (7.122) and (7.123) in Section 7.4.3. The cost function is expressed in terms of $\Delta \underline{x}$ and $\Delta u$ by using (7.123)

$$\lim_{T \to \infty} \int_0^T (\underline{M}\,\Delta\underline{x} + D\Delta u + \underline{M}\,\underline{x}_0 + Du_0)^2$$

$$+ \quad \eta(\Delta u - \mu_x \Delta x_3 + u_0 - \mu_x x_{30} - \mu_l l)^2$$

$$= \quad \lim_{T \to \infty} \int_0^T (\underline{M}\,\Delta\underline{x} + D\Delta u + r_1)^2 + \eta(\Delta u - \mu_x \Delta x_3 + r_2) \quad (7.210)$$

with

$$r_1 \quad = \quad \underline{M}\,\underline{x}_0 + Du_0 \qquad\qquad\qquad (7.211)$$
$$r_2 \quad = \quad u_0 - \mu_x x_{30} - \mu_l l$$

The constants $r_1$ and $r_2$ are expressed as state variables, by augmenting the plant model $(A, B)$ with models of the constants $r_1$ and $r_2$. This was done in (7.127) to (7.130).

By using these equations, the cost function (7.210) can be written in the form

$$\lim_{T \to \infty} \int_0^T \underline{x}_r^T \underline{Q}\,\underline{x}_r + R\Delta u^2 + 2\underline{x}_r^T N\Delta u \qquad (7.212)$$

with

$$\underline{Q} \quad = \quad (\underline{M}\ 1\ 0)^T(\underline{M}\ 1\ 0) + \eta(0\ 0\ -\mu_x\ 0\ 1)^T(0\ 0\ -\mu_x\ 0\ 1)$$
$$\underline{N} \quad = \quad (\underline{M}\ 1\ 0)^T D + \eta(0\ 0\ -\mu_x\ 0\ 1)^T \qquad\qquad (7.213)$$
$$R \quad = \quad D^2 + \eta$$

The cost function (7.212) is minimized by the state-feedback gain

$$\underline{K}_c = \underline{Q}^{-1}(\underline{B}_r^T \underline{P}_c + \underline{N}^T) \qquad\qquad (7.214)$$

where $\underline{P}_c$ is the stabilizing solution to the Riccati equation (7.135). The resulting control law is

$$\Delta u = -\underline{K}_c \underline{x}_r = -\begin{pmatrix} K_{c1} & K_{c2} & K_{c3} \end{pmatrix} \Delta\underline{x} - K_{c4}r_1 - K_{c5}r_2 \qquad (7.215)$$

which by using (7.211) gives

$$u = K_0 x_{30} + K_l l - \begin{pmatrix} K_{c1} & K_{c2} & K_{c3} \end{pmatrix} \underline{x} \qquad\qquad (7.216)$$

with

$$K_0 \quad = \quad \begin{pmatrix} \lambda_x & \delta_x & \mu_x \end{pmatrix} \underline{Gamma} \qquad\qquad (7.217)$$
$$K_l \quad = \quad \begin{pmatrix} \lambda_l & \delta_l & \mu_l \end{pmatrix} \underline{\Gamma}$$

where $\underline{\Gamma}$ is given by

$$\underline{\Gamma} = \begin{pmatrix} 1 - K_{c4}D - K_{c5} \\ \begin{pmatrix} K_{c1} & K_{c2} & K_{c3} \end{pmatrix} - K_{c4}M \\ K_{c5} \end{pmatrix} \qquad (7.218)$$

with $\lambda$, $\delta$, and $\mu$ given by (7.117), (7.118), and (7.207).

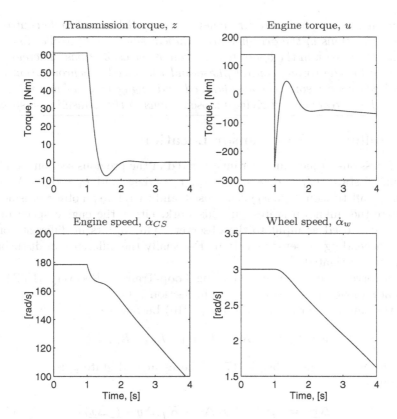

**Figure 7.56** The Drive-shaft model controlled with the transmission-torque controller (7.219), solving the gear-shift criterion (7.209). At t=1 s, a gear shift is commanded. A realizable control signal is used such that the transmission torque is driven to zero, while oscillations are actively damped.

The solution to the gear-shift criterion (7.209) is the transmission-torque controller (7.216), which obtains active damping with a realizable control signal. The parameter $\eta$ is tuned to balance the behavior of the unconstrained active damping solution (7.202) and the gear-shift condition (7.207). The transmission-torque controller with tuned $\eta$ is studied in the following example.

---

**Example 7.5** *Consider the 144L truck in the same driving situation as in Example 7.3. The transmission-torque controller (7.216) then becomes*

$$u = 2.37 \cdot 10^{-4} x_{30} - 0.0327l - \begin{pmatrix} 4.2123 & 0.0207 & -1.2521 \end{pmatrix} \underline{x} \qquad (7.219)$$

*where $\eta = 0.03$ and $\sigma = 0.0001$ are used. With this controller the phase margin is guaranteed to be at least $60°$ and the amplitude margin is infinite [78].*
*Figure 7.56 shows the resulting transmission torque, the control signal, the engine speed, and the wheel speed when a gear shift is commanded at t=1 s, with the control signal chosen according to (7.219).*

*The transmission-torque controller achieves $z = 0$ with a realizable control sig-
nal. The oscillations in the driveline are damped, since the controller forces the
engine inertia to work in the opposite direction of the oscillations. Therefore, the
time needed for the torque control phase and the speed synchronization phase
is minimized, since resonances are damped and engagement of neutral gear is
commanded at a torque level giving no oscillations in the transmission speed.*

## 7.5.4  Influence from Sensor Location

The transmission-torque controller investigated in the previous section uses feed-
back from all states ($x_1 = \alpha_{CS}/i_t i_f - \alpha_w$, $x_2 = \dot{\alpha}_{CS}$, and $x_3 = \dot{\alpha}_w$). A sensor
measuring shaft torsion (e.g., $x_1$) is not used, and therefore an observer is needed
to estimate the unknown states. In this work, either the engine speed or the
wheel speed is used as input to the observer. This results in different control
problems depending on sensor location. Especially the difference in disturbance
rejection is investigated.

The observer gain is calculated using Loop-Transfer Recovery (LTR) [78].
The unknown load can be estimated as in Section 7.4.4.

The transmission-torque control law (7.216) becomes

$$u = K_0 x_{30} + K_l l - \begin{pmatrix} K_{c1} & K_{c2} & K_{c3} \end{pmatrix} \hat{\underline{x}} \qquad (7.220)$$

with $K_0$ and $K_l$ given by (7.217). The estimated state $\hat{\underline{x}}$ is given by the
Kalman filter

$$\Delta \dot{\hat{\underline{x}}} = \underline{A} \Delta \hat{\underline{x}} + \underline{B} \Delta u + \underline{K}_f (\Delta y - \underline{C} \Delta \hat{\underline{x}}) \qquad (7.221)$$

$$\underline{K}_f = \underline{P}_f \underline{C}^T \underline{V}^{-1} \qquad (7.222)$$

where $\underline{P}_f$ is found by solving the Riccati equation (7.144).

When using a LQG-controller with feedback from all states, the phase margin,
$\varphi$, is at least $60°$, and the amplitude margin, $a$, is infinite, as stated before. This
is obtained also when using the observer by increasing $\rho$ towards infinity. For
Example 7.5 the following values are used

$$\rho_m = 10^4 \Rightarrow \varphi_m = 77.3°, \quad a_m = 2.82 \qquad (7.223)$$

$$\rho_w = 10^{11} \Rightarrow \varphi_w = 74.3°, \quad a_w = 2.84 \qquad (7.224)$$

where the aim has been to have at least $60°$ phase margin.

The observer dynamics is canceled in the transfer functions from reference
value, $r$, to performance output, $z$, and to control signal, $u$. Hence, these transfer
functions are not affected by the sensor location. However, the dynamics will be
included in the transfer functions from disturbances to both $z$ and $u$.

### Influence from Load Disturbances

Figure 7.57 shows how the performance output and the control signal are affected by load disturbances, $v$. In Section 7.4.4 it was shown that for the speed
controller, the resonant open-loop zeros become poles of the closed-loop system
when feedback from the engine-speed sensor is used. The same equations are

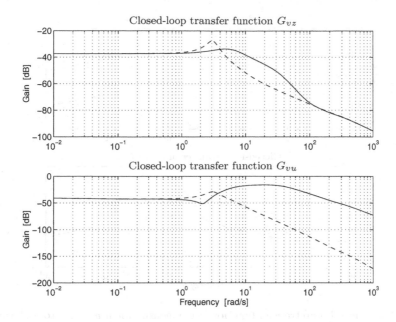

**Figure 7.57** Closed-loop transfer functions from load disturbance, $v$, to performance output, $z$, and to control signal, $u$. Feedback from $\dot{\alpha}_w$ is shown in solid and feedback from $\dot{\alpha}_{CS}$ is shown in dashed lines. With $\dot{\alpha}_{CS}$-feedback the transfer functions have a resonance peak, resulting from the open-loop zeros.

valid for the transmission-torque controller with the minor difference that the $D$ matrix in the performance output, (7.175), is not equal to zero, as for the speed controller. Hence, also the transfer function $DG_{vu}$ should be added to (7.147). The closed-loop transfer function $G_{vu}$ is given by

$$(G_{vu})_{cl} = -\frac{F_y G_{vy}}{1 + F_y G_{uy}} \qquad (7.225)$$

according to (7.96) and the matrix inversion lemma. Thus, the closed-loop transfer function from $v$ to $u$ also has the controller $F_y$ in the numerator. Hence, the closed-loop transfer function from $v$ to $z$ has the open-loop zeros as poles. For $\dot{\alpha}_{CS}$-feedback, this means that a resonance peak is present in the transfer functions from $v$ to performance output and to control signal.

**Influence from Measurement Disturbances**

The influence from measurement disturbances $e$ are shown in Figure 7.58. According to (7.153) the closed-loop transfer function from $e$ to $z$ is

$$(G_{ez})_{cl} = -\frac{G_{uz} F_y}{1 + G_{uy} F_y} \qquad (7.226)$$

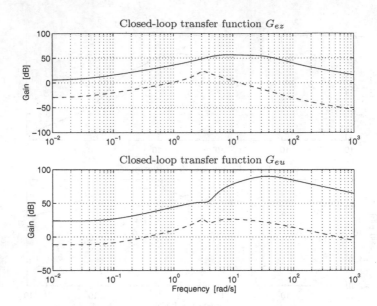

**Figure 7.58** Closed-loop transfer functions from measurement noise, $e$, to performance output, $z$, and control signal, $u$. Feedback from $\dot{\alpha}_w$ is shown in solid and feedback from $\dot{\alpha}_{CS}$ is shown in dashed. The difference between the two feedback principles are described by the dynamic output ratio. The effect increases with lower gears.

Then

$$(G_{ez})_{cl} = -T_w \frac{G_{uz}}{G_{uw}} \quad \text{with} \quad \dot{\alpha}_w\text{--feedback} \qquad (7.227)$$

$$(G_{ez})_{cl} = -T_m \frac{G_{uz}}{G_{um}} \quad \text{with} \quad \dot{\alpha}_{CS}\text{--feedback} \qquad (7.228)$$

with the transfer functions $T_w$ and $T_m$ given by (7.154).

When $\rho$ in (7.145) is increased towards infinity, $T_m = T_w$, as was discussed in Section 7.4.4. Then (7.227) and (7.228) give

$$(G_{ez})_{cl,m} = (G_{ez})_{cl,w} G_{w/m} \qquad (7.229)$$

where $cl, m$ and $cl, w$ denote closed loop with feedback from $\dot{\alpha}_{CS}$ and $\dot{\alpha}_w$ respectively. The dynamic output ratio $G_{w/m}$ was defined in Definition 7.1, and is given by (7.157).

The frequency range in which the relation $T_m = T_w$ is valid depends on how large $\rho$ in (7.145) is made, as discussed in Section 7.4.4. Figure 7.59 shows the sensitivity functions (7.159) and the complementary sensitivity functions $T_w$ and $T_m$ (7.154) for the two cases of feedback. It is seen that $T_m = T_w$ is valid up to about 10 rad/s ($\approx 1.6$ Hz). The roll-off rate at higher frequencies differ between the two feedback principles. This is due to that the open-loop transfer functions $G_{uw}$ and $G_{um}$ have different relative degrees. $T_w$ has a steeper roll-off rate than $T_m$, because that $G_{uw}$ has a relative degree of two, and $G_{um}$ has a relative degree of only one.

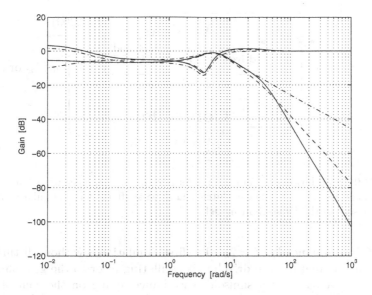

**Figure 7.59** Sensitivity function $S$ and complementary sensitivity function $T$. The dash-dotted lines correspond to the case with all states known. When only one velocity is measured, the solid lines correspond to $\dot{\alpha}_w$-feedback, and the dashed lines correspond to $\dot{\alpha}_{CS}$-feedback.

Hence, the difference in $G_{ez}$ depending on sensor location is described by the dynamic output ratio $G_{w/m}$. The difference in low-frequency level is equal to the conversion ratio of the driveline. Therefore, this effect increases with lower gears.

## 7.5.5 Simulations

As in the case of the speed controller in Section 7.4.5, the feasibility of the gear-shift controller is studied by simulating a more complicated vehicle model than it was designed for. Also here, the disturbances that are difficult to systematically generate in real experiments are treated in the simulations. The control design is simulated with the Nonlinear clutch and drive-shaft model, according to Figure 7.60. The effects from different sensor locations are also studied in accordance with the discussion made in Section 7.5.4.

The Nonlinear clutch and drive-shaft model is given by Equations (7.74) to (7.76). The steady-state level for the Nonlinear clutch and drive-shaft model is calculated by solving the model equations for the equilibrium point when the load and speed are known. By using the parameter relationship (7.180), the equation for the transmission torque is computed by (7.200).

The transmission-torque controller used is based on the Drive-shaft model, and was developed in the previous sections. The wheel speed or the engine speed is input to the observer (7.221), and the control law (7.220) generates the control signal.

Three simulations are performed with the driving situation as in Example 7.5, (i.e., with wheel speed $\dot{\alpha}_w = 3$ rad/s, and load $l = 3000$ Nm). In the simulations,

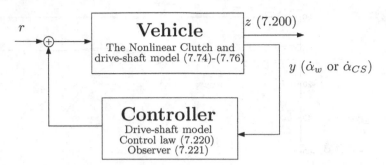

**Figure 7.60** Simulation configuration. As a step for demonstrating feasibility for real implementation, the Nonlinear clutch and drive-shaft model is simulated with the controller based on the Drive-shaft model.

a gear shift is commanded at $t = 2$ s. The first simulation is without disturbances. In the second simulation, the driveline is oscillating prior to the gear shift. The oscillations are a result of a sinusoid disturbance acting on the control signal. The third gear shift is simulated with a load impulse at $t = 3$ s. The disturbance is generated as a square pulse with 0.1 s width and 1200 Nm height.

In order to simulate the nonlinear model, the differential equations (7.74) to (7.76) are scaled such that the five differential equations (one for each state) have about the same magnitude. The model is simulated using the Runge Kutta (45) method [118] with a low step size to catch the effect of the nonlinearity.

Figure 7.61 shows the simulation without any disturbances. This plot should be compared to Figure 7.56 in Example 7.5, where the design is tested on the Drive-shaft model. The result is that the performance does not critically depend on the simplified model structure. The design still works if the extra nonlinear clutch dynamics is added. In the simulation, there are different results depending on which sensor that is used. The model errors between the Drive-shaft model and the Nonlinear clutch and drive-shaft model are better handled when using the wheel-speed sensor. However, neither of the sensor alternatives reaches $z = 0$. This is due to the low-frequency model errors discussed in Section 7.5.1. In Figure 7.62 the simulation with driveline oscillations prior to the gear shift is shown. The result is that the performance of the controller is not affected by the oscillations. Figure 7.63 shows the simulation with a load disturbance. The disturbance is better damped when using feedback from the wheel-speed sensor, than from the engine-speed sensor, which is a verification of the discussion in Section 7.5.4.

## 7.5.6   Gear-Shift Controller Experiments

Some experiments are briefly presented to demonstrate the performance in real field tests. Internal driveline torque control is used to drive the drive-shaft torsion to zero. This is motivated by the fact that the drive shaft is the main flexibility of the driveline, according to Section 7.1. If this torsion is small it is reasonable to believe that the transmission torque also is small, if the dynamical effects in the transmission are neglected.

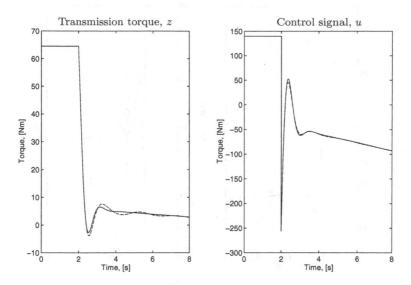

**Figure 7.61** Simulation of the Nonlinear clutch and drive-shaft model with observer and control law based on the Drive-shaft model. A gear shift is commanded at t=2 s. Feedback from the wheel-speed sensor is shown in solid lines, and feedback from the engine-speed sensor is shown in dashed lines. The design still works when simulated with extra clutch dynamics.

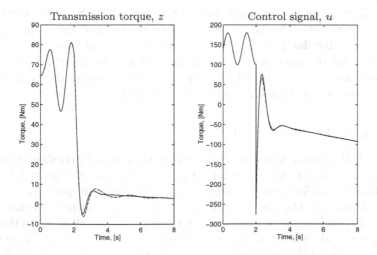

**Figure 7.62** Same simulation case as in Figure 7.61, but with driveline oscillations at the start of the transmission-torque controller. Feedback from the wheel-speed sensor is shown in solid lines, and feedback from the engine-speed sensor is shown in dashed lines. The conclusion is that the control law works well despite initial driveline oscillations.

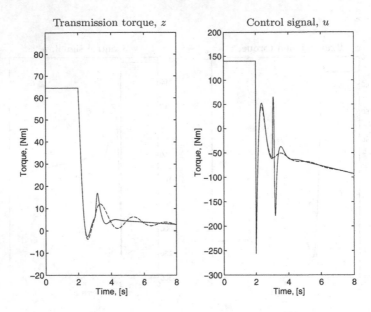

**Figure 7.63** Same simulation case as in Figure 7.61, but with a load disturbance at t=3 s. Feedback from the wheel-speed sensor is shown in solid lines, and feedback from the engine-speed sensor is shown in dashed lines. The conclusion is that the load disturbance is better attenuated when using feedback from the wheel-speed sensor.

## Demonstration of Active Damping in Field Trials

For reason of comparison, Figure 7.64 shows a first trial with a PI controller in dashed lines. The proportional part of the controller gives the speed of the controller, but is not sufficient for damping out the oscillations in the driveline. The result of active damping is seen in Figure 7.64 in solid lines.

Hence, active damping is obtained in field trials with a virtual sensor measuring the drive-shaft torsion. This gives additional support to the Drive-shaft model structure and parameters, derived in Section 7.1.

## Validation of Controller Goal

The drive-shaft torsion is controlled to zero with damped driveline resonances, which was the goal of the controller. However, it is not yet proved that this actually is sufficient for engaging neutral gear with sufficient quality (short delay and no oscillations). The way to prove this is to use the controller demonstrated in Figure 7.64, and engage neutral gear and measure the oscillations in the transmission speed. This is done in Figures 7.65 to 7.67, where the controller is started at 12.0 s and gear shifts are commanded every 0.25 s, starting at 12.25 s.

From these figures, it is clear that controlling the drive-shaft torsion to zero is sufficient for obtaining gear shifts with short delay time. Oscillations in transmission speed are minimized to under 1 rad/s in amplitude with different signs, which is well in the range for giving no disturbance to the driver. Furthermore, the speed synchronization phase, where the engine speed is controlled to match

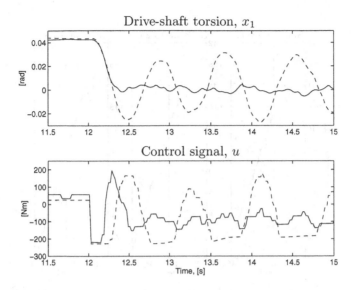

**Figure 7.64** Control signal and drive-shaft torsion when using the gear-shift controller that controls the drive-shaft torsion to zero, started at 12.0 s. Prior to that, the engine has the stationary speed 1900 RPM with gear 1 engaged. In dashed lines, a PI controller is used that gives $x_1 = 0$, but with undamped driveline resonances. The solid lines are extension with the active damping controller.

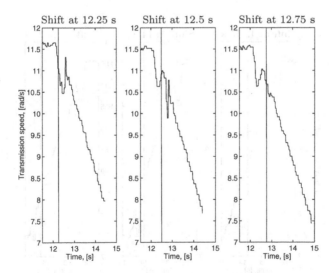

**Figure 7.65** Field trials with start of the gear-shift controller at 12.0 s, all with the same controller controlling the drive-shaft torsion to zero. Engagement of neutral gear is commanded every 0.25 s after the start of the controller, indicated by the vertical lines. The transmission speed is seen when neutral gear is engaged after a delay time. The amplitudes of the transmission speed oscillations after the gear shift are less than 1 rad/s with different signs, which is an acceptable level.

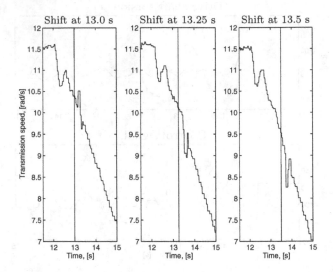

**Figure 7.66** Same type of field experiment as in Figure 7.65, but with commanded engagement of neutral gear at 13.0, 13.25, and 13.5 s. The amplitudes of the transmission speed oscillations after the gear shift are less than 1 rad/s with different signs, which is an acceptable level.

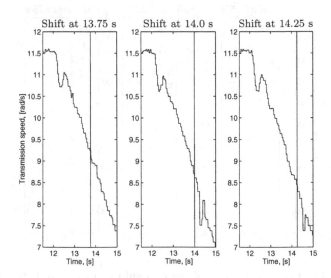

**Figure 7.67** Same type of field experiment as in Figure 7.65, but with commanded engagement of neutral gear at 13.75, 14.0, and 14.25 s. The amplitudes of the transmission speed oscillations after the gear shift are less than 1 rad/s with different signs, which is an acceptable level.

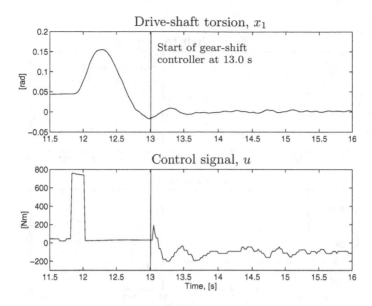

**Figure 7.68** Control signal and drive-shaft torsion during field trials with start of the gear-shift controller at 13.0 s. The driveline is oscillating prior to the gear shift due to an engine torque pulse at 11.7 s. The controller controls the drive-shaft torsion to zero with damped resonances despite initial driveline oscillations.

the propeller shaft speed, can be done fast, since there are only minor oscillations in the transmission speed.

Figure 7.65 also shows that a gear shift can be commanded after only 0.25 s after the controller has started, and an acceptable shift quality is obtained. These results are for gear 1, where the problems with oscillations are largest. The time to a commanded engagement of neutral gear can be decreased further for higher gears.

**Gear Shifts with Initial Driveline Oscillations**

One important problem, necessary to handle, is when a gear shift is commanded at a state where the driveline is oscillating. This was discussed in Section 7.5.5 where the controller was simulated with initial driveline oscillations. To verify that the controller structure can handle this situation and that it also works in real experiments, driveline resonances are excited by an engine torque pulse at 11.7 s, according to Figure 7.68.

Figures 7.68 to 7.70 show the same type of experiments, but the controller is started at different time delays after the engine torque pulse has occured. For all three experiments, the resulting engine torque, calculated by the feedback controller, actively damps the initial driveline oscillations and obtains $x_1 = 0$.

The difference in control signal in Figures 7.68 to 7.70 is a strong evidence that driveline dynamics affects shift performance so much that feedback control is motivated. An open-loop scheme would not be able to handle these initial oscillations, leading to longer time for gear shifts.

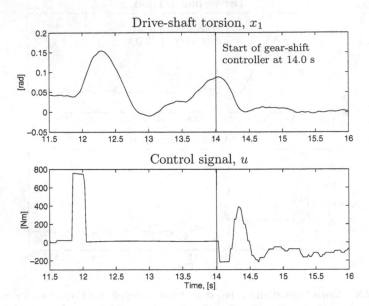

**Figure 7.69** Control signal and drive-shaft torsion during field trials with start of the gear-shift controller at 14.0 s. The controller controls the drive-shaft torsion to zero with damped resonances despite initial driveline oscillations.

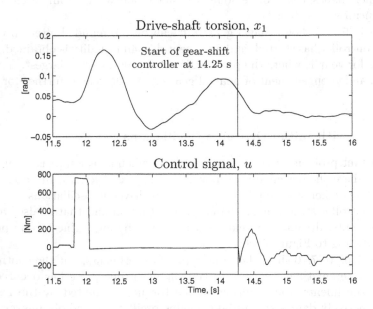

**Figure 7.70** Control signal and drive-shaft torsion during field trials with start of the gear-shift controller at 14.25 s. The controller controls the drive-shaft torsion to zero with damped resonances despite initial driveline oscillations.

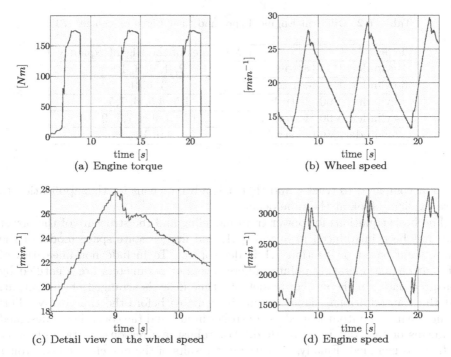

**Figure 7.71** Engine torque and engine speed for *tip in* and *tip out* maneuvers.

# 7.6 Anti-Jerking Control for Passenger Cars

In recent years the engine torque of Diesel engines has been increasing rapidly also for personal cars. Thereby, like for trucks, problems for driveability and comfort arise, and also here the problem with the high engine torque is torsion of the driveline. Due to the transmission ratio of the final drive the torsion primarily occurs at the drive shaft, and it causes power train oscillations combined with jerking of the vehicle. Figure 7.71 shows the behavior of a test car during so-called *tip in*, an abrupt step on the throttle followed by a high torque gradient, and *tip out*, the reverse maneuver and as can be seen the engine speed and wheel speed clearly show oscillations. The oscillations of the engine speed are partly absorbed by the engine mounting whereas the oscillations of the wheel speed correspond to the performance of the car and are therefore directly felt by the passengers.

To avoid these oscillations and to increase the comfort of the passengers, damped driveline control like in Section 7.4 is necessary. This is often called Anti-Jerking Control. Historically, the construction of the power train along with the drive shaft was optimized but the results were not satisfying [39]. Another possible solution would be to avoid high torque gradients, but however, the disadvantage would be restrictions in response.

As an additional illustration of the model-based approach in Section 7.4, Anti-Jerking Control of passenger cars that minimizes driveline oscillations will be introduced. Interesting features are that vehicle parameters of course are quite

**Table 7.2**  Different Engine Types and their Eigenfrequencies [73]

| Type | Multiples of the Eigenfrequency |
|------|--------------------------------|
| 4-cylinder, R4, 4-stroke | 2, 4, 6, 8, 10 |
| 6-cylinder, R4, 4-stroke | 3, 6, 9 |
| 6-cylinder, V60, 4-stroke | 1,5, 4,5, 6, 7,5 |
| 6-cylinder, V90, 4-stroke | 1,5, 3, 4,5, 7,5, 9 |
| 12-cylinder, V60, 4-stroke | 6, 12 |

different compared to trucks, and that it is advantageous for this application to more explicitly look at the time delays.

The basic equations for power train modeling and the structure of the power train has been introduced in Section 7.1, and first, a state space model of the power train based on Section 7.1 is developed. To include non-linearities of different engine operating points, different sets of parameters are identified by measured data. In the system model, the time delay is arranged at the output, and thus, an additional signal output is available before the time delay. This predictive model output is used as controller input, and the controller is designed by means of the root locus method. For robustness reasons, a state observer feedback is included. Finally, measurement results of the model-based controller are discussed. Depending on the power train concept, front or rear wheel drive, the engine torque and the mass of the car, different kind of oscillations can be found in the power train. The following list gives an overview of some oscillations and their causes:

- According to torsion of the shafts jerking and pitching occurs. Because of suspension, damping, mass of the car, the frequency ranges between 2 and 5 $Hz$ [122]. The natural frequency of the car body regarding to pitch motion is between 1 and 2.5 $Hz$ [29, 37].

- The movement of the engine and the characteristic of the engine mounting is responsible for oscillations in the frequency range between 15 and 200 $Hz$ [114].

- Vibrations caused by the gear box housing are in the frequency range between 50 and 80 $Hz$ [7]. Further on high frequency oscillations can occur at high engine load because of gear wheel slackness in the range of $300 - 6000\ Hz$, which in contrast to other oscillations can be heard by passengers [114].

- Combustion process and engine speed are also causes for oscillations which depend on the engine type, number of cylinders and the engine speed. Table 7.2 gives an overview of the engine speed depending oscillations for different engine types.

Other sources for power train oscillations are components like joints and gear wheels. The goal of anti-jerking control is to obtain a well damped behavior of the torsion angle at transients.

## 7.6.1  Model of the Power train of a Passenger Car

For power train modeling a so-called gray-box approach is used. A theoretical model of the power train is introduced in the form of a linear state space model of third order. The parameter values are derived from measurement data. The model input is the requested torque calculated by the engine control unit (ECU). The model output is the difference between engine speed and wheel speed (compensated by transmission ratios).

**Theoretical Model**

The physical structure of the power train of a vehicle is shown in Figure 7.1. It is assumed that the oscillations in the power train are mainly caused by the torsion of the drive shaft, as the drive shaft transmits the relatively largest torque. This assumption has been validated by measurements [96]. For the purpose of jerk damping it is therefore sufficient to model the power train by a two-mass-model connected through a damped torsional element.

For a detailed derivation of this power train model, see Section 7.1. Labels are according to the abstract structure in Figure 7.72. It is assumed that all rotating and oscillating masses inside the engine can be combined to a single mass with the mass moment of inertia $J_e$. The driving torque $T_e$ already includes the internal friction. The mass moment of inertia of the clutch is neglected and the clutch is assumed without friction. Similar to the engine, all mass moments of inertia of the transmission are combined to one rotating inertia $J_t$ which can oscillate viscously damped with the damping coefficient $d_t$. The propeller shaft is assumed to be stiff. The final drive is also modeled by one rotating inertia $J_f$ and the viscous damping coefficient $d_f$. The drive shaft is modeled as a damped torsional element, consisting of stiffness $k$ and damping $d$. The wheel model is strongly simplified. Air drag and rolling resistance are neglected and the road is assumed to be level. Considering all these assumptions yields two differential equations, that are similar to Equations 7.43-7.44,

$$\left(J_e + \frac{J_t}{i_t^2} + \frac{J_f}{i_t^2 i_f^2}\right)\ddot{\alpha}_{CS} = T_e - \left(\frac{d_t}{i_t^2} + \frac{d_f}{i_t^2 i_f^2}\right)\dot{\alpha}_{CS}$$

$$- k\frac{\frac{\alpha_{CS}}{i_t i_f} - \alpha_w}{i_t i_f} - d\frac{\frac{\dot{\alpha}_{CS}}{i_t i_f} - \dot{\alpha}_w}{i_t i_f} \qquad (7.230)$$

and

$$\left(J_w + m_{CoG}r_{stat}^2\right)\ddot{\alpha}_w = k\left(\frac{\alpha_{CS}}{i_t i_f} - \alpha_w\right)$$

$$+ d\left(\frac{\dot{\alpha}_{CS}}{i_t i_f} - \dot{\alpha}_w\right) - d_w\dot{\alpha}_w \ , \qquad (7.231)$$

where $\alpha_w$ is the wheel angle and its derivative $\dot{\alpha}_w$ denotes the rotational wheel speed $v_R/r_{stat}$.

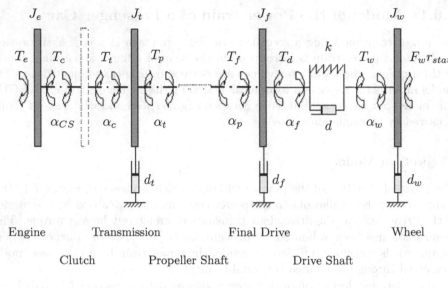

Figure 7.72 Idealized power train Structure

As for Equation 7.46, combining the following parameters

$$i = i_t i_f \ , \quad J_1 = J_e + \frac{J_t}{i_t^2} + \frac{J_f}{i_t^2 i_f^2} \tag{7.232}$$

$$J_2 = J_w + m_{CoG} r_{stat}^2 \ , \quad d_1 = \frac{d_t}{i_t^2} + \frac{d_f}{i_t^2 i_f^2} \ , \quad d_2 = d_w \tag{7.233}$$

results in a two-mass-model. It consists of one inertia $J_1$ with damping instead of the crankshaft, one generalized transmission (instead of transmission and final drive), the flexible drive shaft and one damped inertia $J_2$ instead of the wheel

$$J_1 \ddot{\alpha}_{CS} \ = \ T_e - d_1 \dot{\alpha}_{CS} - k \frac{\frac{\alpha_{CS}}{i} - \alpha_w}{i} \tag{7.234}$$

$$-d \frac{\frac{\dot{\alpha}_{CS}}{i} - \dot{\alpha}_w}{i} \tag{7.235}$$

$$J_2 \ddot{\alpha}_w \ = \ k \left( \frac{\alpha_{CS}}{i} - \alpha_w \right) + d \left( \frac{\dot{\alpha}_{CS}}{i} - \dot{\alpha}_w \right)$$

$$-d_2 \dot{\alpha}_w \ . \tag{7.236}$$

With torsion angle $x_1$, crankshaft speed $\dot{\alpha}_{CS}$, and angular speed of the wheel $\dot{\alpha}_w$ as state variables

$$x_1 = \frac{\alpha_{CS}}{i} - \alpha_w, \ x_2 = \dot{\alpha}_{CS}, \ x_3 = \dot{\alpha}_w \tag{7.237}$$

$$y = \frac{\dot{\alpha}_{CS}}{i} - \dot{\alpha}_w \tag{7.238}$$

Output of the model $y$ is the difference between engine and wheel speed, in the following referred to as speed difference.

The state space model

$$\dot{\underline{x}}(t) = \underline{A}\,\underline{x}(t) + \underline{b}\,u(t) \tag{7.239}$$

$$y(t) = \underline{c}^T\,\underline{x}(t) \tag{7.240}$$

is rewritten on matrix form like Equations 7.81-7.82, which leads to the following system matrices

$$\underline{A} = \begin{pmatrix} 0 & \frac{1}{i} & -1 \\ -\frac{k}{i \cdot J_1} & -\frac{d_1 + \frac{d}{i^2}}{J_1} & \frac{d}{i \cdot J_1} \\ \frac{c}{J_2} & \frac{d}{i \cdot J_2} & -\frac{d_2 + d}{J_2} \end{pmatrix}, \tag{7.241}$$

$$\underline{b} = \begin{pmatrix} 0 \\ \frac{1}{J_1} \\ 0 \end{pmatrix}, \quad \underline{c}^T = \begin{pmatrix} 0 & \frac{1}{i} & -1 \end{pmatrix}. \tag{7.242}$$

and consists of 7 parameters:

$J_1$ : Sum of mass moments of inertia of engine, transmission and final drive referring to the crankshaft

$J_2$ : Mass moment of inertia of the wheel and equivalent mass moment of inertia of the vehicle referring to the drive shaft

$d$ : Damping of the drive shaft

$d_1$ : Damping of transmission and final drive referring to the crankshaft

$d_2$ : Damping of the wheel

$k$ : Stiffness of the drive shaft

$i$ : Transmission ratio between engine speed and wheel speed

## Parameter Identification

The transmission ratio $i$ is usually known or can be calculated from a test drive at constant speed. The other 6 parameters of the state space model can be identified using measurement data. For generation of adequate data sets the power train of a test car can be excited with *tip in* and *tip out* at different engine speeds while measuring the requested torque and the resulting engine speed and wheel speed. To guarantee that the state variables show good compliance with their physical meaning, the quadratic error between simulated and measured engine speed (i.e. $x_2 - \hat{x}_2$), simulated and measured wheel speed (i.e. $x_3 - \hat{x}_3$), and simulated and measured speed difference (i.e. $x_2/i - x_3$, the output error) may be included in the cost function of the identification algorithm. Separate weighting factors $w_j$ for the different quadratic errors yield the desired accuracy for all variables. The cost function $C$ has the following structure

$$C = \sum_j \left[ w_1 \left( x_{2,j} - \hat{x}_{2,j} \right)^2 + w_2 \left( x_{3,j} - \hat{x}_{3,j} \right)^2 \right.$$

$$\left. + w_3 \left( \left( \frac{x_{2,j}}{i} - x_{3,j} \right) - \left( \frac{\hat{x}_{2,j}}{i} - \hat{x}_{3,j} \right) \right)^2 \right]. \tag{7.243}$$

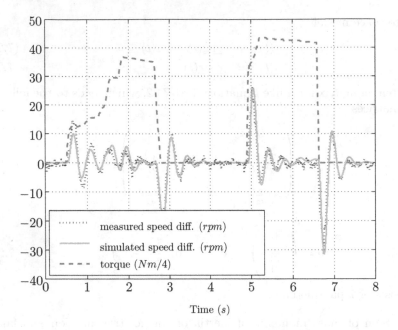

**Figure 7.73**  Measured and Simulated Speed Difference

The first system state is the torsion angle of the overall power train. It is not accessible with standard vehicle sensors and may be neglected during the identification process without significant loss of the overall model accuracy. For good identification results, it is important to utilize a reasonable start vector based on physical knowledge of the parts of the power train. An estimate of the mass moments of inertia can often be calculated from vehicle design data. If no further information is available, the initial values of the stiffness and dampings may be chosen randomly. Due to the nonlinearity of the real process some of the identified values of the parameters vary depending on the engine speed. Storing these parameter values in engine speed-dependent look-up-tables results in a time-varying model. The modeled and the measured speed difference of a front-driven passenger car are shown in Figure 7.73.

Due to the combustion process of the engine, the system contains a delay-time. A requested torque can only be generated with a time delay of at least one cylinder segment, and this time delay depends on the engine speed. The minimum time delay is approximated by

$$\tau_{d,e} = \frac{2}{\text{CYL} \cdot n} \, , \tag{7.244}$$

where CYL is the number of cylinders and $n$ is the engine speed in revolutions per second. During identification this delay-time is treated as an output delay and is separated from the dynamics of the process. By using the signal *before* the time delay as output, the resulting model is predictive and can indicate power train oscillations even before they occur in reality. This behavior is utilized for the design of a controller with a structure similar to a Smith Predictor [119], see Section 7.6.2.

**Transfer to angle-discrete model**

The engine control unit (ECU) calculates in an angular synchronous mode. For a 4 cylinder engine every 180° crankshaft angle, i. e. every cylinder segment, an injection is calculated. As the segment time depends on the engine speed, the sampling time of the ECU and therefore of the power train model is varying roughly between 5 and $50ms$. Thus the model discretization method must contain a varying sampling time and time varying model parameters.

The state equation for discrete time steps $t = t_0 + t_s$ can be written as

$$\underline{x}(t_0 + t_s) = \underline{\Phi}(t_s)\underline{x}(t_0) + \int_0^{t_s}\underline{\Phi}(t_s-\alpha)d\alpha\,\underline{b}\,u_0\,, \qquad (7.245)$$

where $t_s$ is the varying sampling time and $\underline{\Phi}(t_s) = e^{\underline{A}\,(t_s)}$ is the transition matrix. $\underline{\Phi}$ may be approximated with a truncated Taylor polynomial

$$\underline{\Phi}(t_s) = e^{\underline{A}\,t_s} = \sum_{n=0}^{\infty}\frac{1}{n!}(\underline{A}\,t_s)^n \approx \sum_{n=0}^{N}\frac{\underline{A}^n}{n!}t_s^n\,. \qquad (7.246)$$

The integral in Equation 7.245 is approximated accordingly with

$$\int_0^{t_s}\underline{\Phi}(t_s - \alpha)d\alpha\,\underline{b} \approx \sum_{n=0}^{N}\frac{\underline{A}^n}{n!}\,\underline{b}\,\frac{t_s^{n+1}}{n+1}\,. \qquad (7.247)$$

For varying $t_s$ and varying $\underline{A}$ and $\underline{b}$ these calculations are carried out online for every calculation step. The number of elements $N$ of the Taylor expansion affects the accuracy of the approximation. A trade-off between computational effort and accuracy should be found based on simulation results. For $N = 8$ the approximation was found to be sufficiently exact.

## 7.6.2   Controller Design

**Smith Predictor Approach**

Due to the engine delay-time, see Section 7.6.1, the model offers a predictive output: When applying a torque step, the model without delay-time calculates the resulting speed difference before it occurs in reality. The angle-discrete model predicts one cylinder segment which is as much as $50ms$ in a four cylinder engine at low engine speed. Using the predictive model output as controller input gives the possibility to avoid jerking before its very occurrence and therefore damping the oscillations in the power train. This control principle was presented by [119] and is known as the so-called Smith-predictor. The structure is shown in Figure 7.74. As the actuation variable, i.e. the requested torque $T_{req}$, physically differs from the controlled variable, i.e. the speed difference $\Delta n$, the controller is located in the feedback path of the system. This is the difference between the structure employed here and the classical Smith Predictor approach.

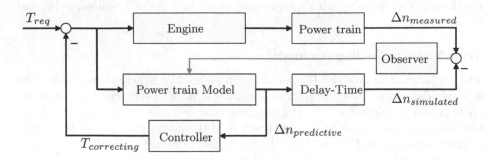

**Figure 7.74** Structure of the controlled System

## State Observer

A significant disadvantage of the classical Smith-predictor approach is its open loop structure. The model output is used in the feedback path. It may drift away from the process output, i.e. the measured speed difference. To overcome such offsets a state observer is introduced. The difference between model and process output is fed back via the observer, see Figure 7.74. Disturbances, e.g. from a bumpy road, and model inaccuracies are thus compensated.

The observer design can be carried out with pole placement. The continuous state equation with observer can be written as

$$\dot{\underline{\hat{x}}} = \underline{A}\,\underline{\hat{x}} + \underline{b}\,u + \underline{L}\,(y - \hat{y}) \;, \tag{7.248}$$

where $y$ is the measured process output and $\hat{y}$ is the calculated model output. The error dynamics $\underline{\tilde{x}} = \underline{x} - \underline{\hat{x}}$ are

$$\dot{\underline{\tilde{x}}} = \left(\underline{A} - \underline{L}\,\underline{c}^T\right)\underline{\tilde{x}} \;. \tag{7.249}$$

The observer dynamics should be slightly faster than that of the process. Then, the error is compensated rapidly. This behavior is obtained by placing the poles of the matrix $\underline{A} - \underline{L}\,\underline{c}^T$ to the left of the poles of the uncontrolled system. In discrete time, the observer poles are placed closer to the origin of the unit circle than the poles of the uncontrolled system.

The observer design is done offline for different engine speeds and the resulting gains are stored in look-up-tables for online operation.

The predictive model output is delayed one calculation step before it is used as observer input in order to compare simulated and measured (i.e. non-predictive) speed difference at the same time instance.

## Controller Design with Root Locus Method

Having the observer described above it would be possible to employ a state regulator. Then the progression of the individual state variables is shaped by pole placement. This approach is justified by the physical understanding of the state variables, where the gradient of engine speed and wheel speed can be specified

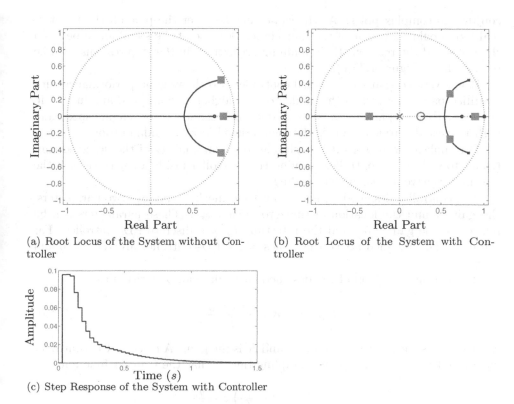

(a) Root Locus of the System without Controller

(b) Root Locus of the System with Controller

(c) Step Response of the System with Controller

**Figure 7.75** Controller Design with Root Locus Method

according to the desired behavior of the Anti-Jerking control. Different system characteristics like comfort or sportive behavior can be obtained depending on the position of the poles.

Here a different approach is taken. It is a reduced order controller, that both give insight into the control problem and into how to tune the control parameters. The controller is located in the feedback path of the system, see Figure 7.74. The root locus method with pole-placement may be applied for designing such controllers. The root locus plot shows the locus of the poles of a closed loop system in dependency of the gain. Information about stability and dynamic behavior of the system can be seen from this [25].

The proceeding of the root locus technique is explained for the discrete model. The root locus curve of the system at low engine speed and high requested torque (tip in case) without controller is shown in Figure 7.75(a). Poles are shown as squares. The conjugated-complex pole pair is responsible for the oscillations of the power train. The desired system behavior is a well damped step response. Therefore, all poles should be moved onto or at least closer to the real axis. Inserting a controller pole in the origin and a controller zero point on the positive real axis near z = 0.25 results in the root locus shown in Figure 7.75(b). The pole positions have changed away from those of the uncontrolled system but real poles are not reached. A still higher gain would reduce the imaginary parts of the conjugated-complex poles. At the same time however the pole on the negative real axis would move out of the unit circle. Because of the small angle between the poles and the real axis the remaining oscillations in the step response can be tolerated, see Figure 7.75(c).

It was verified that the designed controller also shows good performance with modified parameter sets of the model, e.g. at higher engine speed and in the tip out case. High emphasis was laid on good performance at low engine speed and high requested torque as this is the most critical jerking configuration.

For application reasons it is desirable to get an analogy of the designed controller to a classical controller. Tuning the controller inside a car is easier if the parameters have a clear understanding.

The controller designed with the root locus method employs two parameters, the gain $K$ and the location of the zero point $z_{zero}$. These parameters can be related to the gain $K_{PD}$ and the rate time $T_v$ of a discrete PD-controller. For the controller designed above the values of the PD-controller are $K_{PD} = 6$ and $T_v = 0.01$.

The transfer function of the designed controller can be written as

$$G_R(z) = K \cdot \frac{z - z_{zero}}{z} \quad , \tag{7.250}$$

where $z_{zero}$ is the inserted zero point and $K$ is the gain. A discrete PD-controller with rate time $T_v$, gain $K_R$ and sampling time $T_s$ has the transfer function

$$G_{PD} = K_R \frac{\left(1 + \frac{T_v}{T_s}\right) z - \frac{T_v}{T_s}}{z} \quad . \tag{7.251}$$

An analogy between the parameters of the PD-controller and of the controller designed with the root locus method can be derived

$$K_R = K \cdot (1 - z_{zero}) \tag{7.252}$$

$$T_v = \frac{-K \cdot z_{zero} \cdot T_s}{K \cdot (1 - z_{zero})} \quad . \tag{7.253}$$

## 7.6.3  System Performance

Figure 7.73 shows the jerking behavior of a front-driven passenger car without jerk control for tip in and tip out. The plot shows the driver's torque request, the measured speed difference between wheels and engine, the engine speed and the longitudinal acceleration. The performance of the model based control approach is shown in Figure 7.76. For clarity, the corrected torque is plotted.

In Figure 7.73 the typical oscillations of the speed difference can be seen for the uncontrolled case. At acceleration, there is an overshoot for the tip in and an undershoot and oscillations for the tip out occur. The passengers inside the car very much feel the hard overshoot as well as the oscillations of the acceleration.

Figure 7.76 shows the performance of the model-based controller at the lower engine speeds that are the most critical in this application.. The corrected torque shows the influence of the controller. The overshoots at acceleration allow the

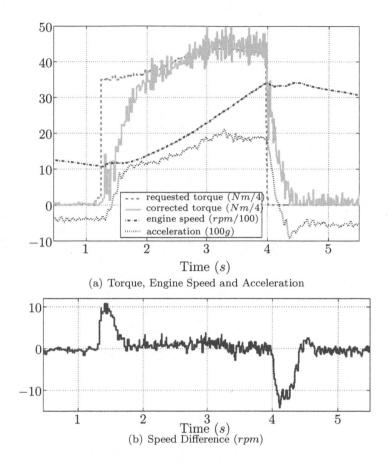

(a) Torque, Engine Speed and Acceleration

(b) Speed Difference $(rpm)$

**Figure 7.76** Performance of the Controlled System

driveline to twist or untwist. The oscillations for tip in and tip out are damped while the fast reaction of the car to the driver's torque request is still preserved.

The system performance for higher engine speeds is similar, but the oscillations are not as large as for lower engine speeds, which means that the results can easily be transferred.

# 8 Vehicle Modelling

## 8.1 Introduction

The last few years have seen scientific and technical competition within the automobile industry become increasingly intense. Because of this it is important for individual manufacturers to ensure that new designs are brought to the market as quickly as possible. In the future it must be possible to carry out the design of a new model, from conception to production, in as little as eighteen months.

These days the equipment within a motor vehicle includes a myriad of electrical and electronic subsystems. Many of these subsystems serve to improve driving comfort, i.e. electric windows, air conditioning systems etc. On top of this there are safety-relevant vehicle subsystems, such as drive dynamics control and anti-lock brake systems.

An ever decreasing design effort for such systems, coupled with increasing amounts of equipment, is only possible due to the far-reaching use of computer simulations in the design of new vehicles. The aim of computer models, such as the one shown schematically in Figure 8.1 is to reveal, as early as possible in the design phase, the effect on the dynamic behavior of the vehicle of new components operating in conjunction with the existing subsystems. With such an approach the effect of a new component can be analyzed in the definition phase, long before the prototype is complete.

To date, modeling efforts have concentrated on reproducing as exactly as possible the behavior of individual components. This approach has yielded, for example, exact descriptions of the wheel dynamics using the finite element method. Simulations of such models are computationally expensive and time consuming. Due to the fact that there is a huge difference between different vehicles in terms of structure and kinematics, such models are very specific, and special know-how is required to alter a model for use with a different vehicle type.

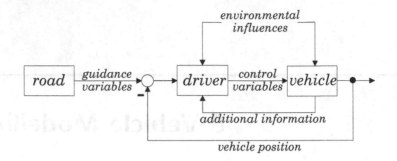

**Figure 8.1** The standard vehicle-driver-road control loop

The modeling approach described in this chapter has the following aims:

- reduction of model complexity to a level sufficient for vehicle dynamics

- implementation on a PC in a common programming language, so that usage
  is widely spread instead of being constrained to a few specialized depart-
  ments.

- interaction between submodels, whereby the design time is concentrated
  upon the subsystem currently under investigation.

- only necessary accuracy, so that time consuming tests with experimental
  vehicles can be reduced.

In order to obtain these aims, the system will be decomposed into its individual
components. When carrying out such a partitioning, it is important to ensure
that meaningful variables are chosen for the interfaces between the different sub-
models. It is thus sensible to choose, insofar as it is possible, torques or forces
and angular or longitudinal velocities.

## 8.2  Co-ordinate Systems

For the theoretical analysis of vehicle dynamics, and for the design of observer-
or control-algorithms, the equations of motion must be known and the physical
interactions between the various subsystems must be written in the form of mathematical equations. For the construction of a vehicle model there are two main
approaches. If the aim is to produce a model which is as exact as possible, the
methods of theoretical physics such as Lagrange or Euler are used. The resultant
models are very precise, however the individual equations lose their reference to
physical quantities as the calculations are carried out for generalized co-ordinate
systems. The alternative approach, as used in many publications, is to attempt
to model the vehicle as simply as possible and with as little computing-time as
possible. To this end emphasis is placed here on the classical *single-track model*,
which gives good results for non-critical driving situations. The first publication
of this method dates back to 1940 [103]. When dealing with simplified models a

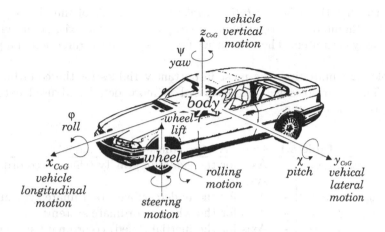

**Figure 8.2** Degrees of freedom of the vehicle

distinction is often made between models for drive dynamics and models for vertical dynamics analysis, and the interactions between the two are ignored. The approach detailed in this chapter follows a middle road between the two modeling methods. The vertical dynamics and the drive dynamics are modeled together and all important non-linearities are included. In contrast to the Lagrange or Euler methods the calculations are limited to 4 co-ordinate systems.

In order to differentiate, indices are introduced for the individual co-ordinate systems:

- "$CoG$" for the chassis (Center of gravity) co-ordinate system,

- "$Un$" for the undercarriage system,

- "$W$" for the wheel co-ordinate system,

- "$In$" for the fixed inertial system.

With the exception of the fixed inertial system, all co-ordinates move during travel. Figure 8.2 shows the 6 degrees of freedom of the vehicle as well as the Center of Gravity (CoG) co-ordinate system.

The *Center of Gravity co-ordinate system*, which has its origin at the vehicle center-of-gravity is of the most importance. All movements of the vehicle body are given with reference to this co-ordinate system. The *undercarriage co-ordinates* differ from this only in the pitch and roll angle. The origin of the undercarriage co-ordinate system lies at road-level in the middle of the perpendicular projection of the rear axle.

Finally, the *wheel co-ordinate system* is required. To be precise there is a co-ordinate system for each individual wheel. In this book, the co-ordinate directions for the rear wheels are the same as for the undercarriage co-ordinate system and for the front wheels only the wheel turn angle (the angle between the longitudinal vehicle axis and the wheel plane) is different.

For the direction of the *rotation angles* there is a rule of thumb:
Orient the thumb of your right hand in the direction of the axis, about which the
vehicle body is rotated. Then the other fingers show in the direction of a positive
rotation angle.

Table 8.2 summarizes the most important variables for the co-ordinate systems. These variables will be discussed in more detail and used extensively
throughout this chapter.

**Table 8.1**  Summary of co-ordinate system variables

| | | |
|---|---|---|
| $x_{CoG}, y_{CoG}, z_{CoG}$ | - | Axis for the center of gravity (chassis) co-ordinate system |
| $x_{Un}, y_{Un}, z_{Un}$ | - | Axis for the undercarriage co-ordinate system |
| $x_W, y_W, z_W$ | - | Axis for the wheel co-ordinate system[1] |
| $x_{In}, y_{In}, z_{In}$ | - | Axis for the inertial (fixed) co-ordinate system |
| $\psi$ | - | Yaw angle (rotation about $z_{CoG}$ ) |
| $\chi$ | - | Pitch angle (rotation about $y_{CoG}$ ) |
| $\varphi$ | - | Roll angle (rotation about $x_{CoG}$ ) |
| $\alpha$ | - | Tire side slip angle (angle between $x_W$ and $v_W$, the wheel ground contact point velocity ) |
| $\delta_W$ | - | Wheel turn angle[2] (angle between $x_{CoG}$ and $x_W$) |
| $\beta$ | - | Vehicle body side slip angle (angle between $x_{CoG}$ and $v_{CoG}$, the vehicle (CoG) velocity) |

## 8.3   Wheel Model

The most important point for the creation of a simulation model for a motor
vehicle is the exact observation of the horizontal forces on the wheel. The task of
the wheel model is to derive these forces. In order to calculate the wheel forces
it is necessary to know wheel slip, tire side slip angle and friction co-efficients,
as these are inputs for the force equations. For the derivation of the wheel slip
it is necessary to determine the individual wheel velocities. For the tire side
slip angle calculation, the wheel caster has to be taken into consideration and
the curve radii of the individual wheels calculated. The order of the necessary
calculations is shown in Figure  8.3.

### 8.3.1   Wheel Ground Contact Point Velocities

This section is concerned with the velocity of the wheels with reference to some
fixed reference point - the so-called *wheel ground contact point velocity* or *wheel
velocity* $v_W$. Here, the rotation of the wheels is not considered. This is in contrast
to the derivation of the *rotational equivalent velocity* $v_R$ of the wheels, which will
be considered later.

---

[1]The suffixes $F$ and $R$ are used to denote front and rear wheels, and $R$ and $L$ to denote
right and left wheels. Thus $x_{WFR}$ refers to the x-direction of the front right wheel.

[2]This angle should not be confused with the steering wheel angle, which is denoted by $\delta_S$.

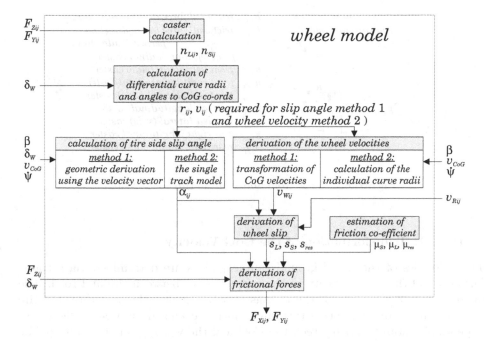

**Figure 8.3** The wheel model

There are two main methods for the derivation of the velocities at the wheel-ground contact points. One approach uses the method of calculating the four wheel velocities via a transformation of the CoG velocity to a two-track (as opposed to the single-track) model. With the two-track model the additional wheel velocity component caused by the yaw rate is no longer orthogonal to the vehicle longitudinal axis and the calculation is significantly more complex. The second approach is to obtain the vehicle movement as an orbit around the instantaneous center with angular velocity $\dot{\psi}$ and $\dot{\beta}$ and then calculate the curve radius to each individual wheel. This is made possible by the orthogonal arrangement between the wheel ground contact point velocity and the line between the wheel-ground contact point and the instantaneous center. Both methods are equivalent and the choice depends on complexity and the opportunity for simplification. For both methods the exact knowledge of the distance from the CoG to the wheel-ground contact point is required. This is dependant upon the dynamic wheel caster. The point of action of the wheel force does not lie in the center of the wheels, but (due to the caster) towards the rear. Caster is the tilting of the steering axis either forward (negative caster) or backwards (positive caster) from the vertical wheel axis. The wheel caster is a measure of the shift of the pressure distribution in the tire contact area. Positive caster provides directional stability and helps the wheels to return to a forward-pointing direction after a turn.

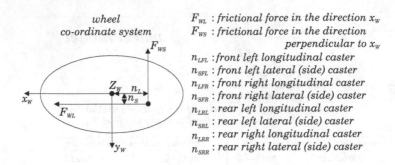

Figure 8.4  View from above of the tire contact area

**Method 1: Transformation of the CoG Velocity**

The velocities of the wheel ground contact points are determined via a transformation of the CoG velocity to the contact point between tire and road. The magnitude and direction of the chassis (CoG) velocity, the yaw rate, the tire side slip angle and the wheel turn angle must be known, as well as the caster dependant distances $r_{ij}$ between the CoG and the wheel ground contact points (see Figure 8.5). To calculate these distances $r_{ij}$ from the CoG to the wheel ground contact points, it is necessary to know the casters $n_L$ and $n_S$. These can be obtained using the approximation formula [12]:

$$n_L = \frac{1}{2}\left(l_0 + l_1\frac{F_Z}{F_{Z0}}\right) \quad \text{and} \quad n_S = 3n_L\tan(\alpha) + \frac{F_Y}{c_{press}} \qquad (8.1)$$

$n_L$ is known as the "dynamic caster", $n_S$ takes into consideration the lateral force influence of the pressure distribution on the CoG. $F_Z$ is the force acting vertically at the wheel ground contact point, and $F_Y$ the lateral wheel force. The parameters of Equation 8.1 are given in Table 8.2.

Table 8.2  Parameters for the wheel model

| $F_{Z0}$ | $5000\,N$ | Nominal vertical force at wheel contact |
|---|---|---|
| $c_{press}$ | $230000\,N/m$ | Parameter to correct for tire pressure distribution |
| $l_0$ | $-0.03\,m$ | caster parameter |
| $l_1$ | $0.12\,m$ | caster parameter |

Figure 8.4 shows an top view of the tire contact area, including the wheel ground contact point.

The middle point of the contact area migrates outwards and creates a torque with the longitudinal force, which during acceleration increases the self aligning torque and during braking decreases it. Whereas the caster is generally assumed to be constant and only the component in the direction of the wheel plane is considered [84], using the above method the caster can be calculated dynamically and in vector form (i.e. direction information included). The frictional forces $F_L$ and $F_S$ act in the direction of the wheel velocity $v_W$ and perpendicular to it. The forces $F_{WL}$ and $F_{WS}$ are obtained by transformation into the wheel co-ordinate system (Eqs 8.25 and 8.26).

The derivation of the wheel ground contact point velocities is carried out in the undercarriage co-ordinate system. One proceeds with the assumption that the vehicle velocity can be described as a superposition of pure translatory motion with magnitude $v_{CoG}$ and angle $\beta$ to the vehicle longitudinal axis, and a purely rotational motion with yaw rate $\dot{\psi}$ around the CoG, as shown in Figure 8.5.

The wheel ground contact point velocities are given by the geometric superposition of the CoG velocity and a part of the magnitude $\left| r_{ij} \cdot \dot{\psi} \right|$ in the direction specified in Figure 8.5. Both the distances $r_{ij}$ and the angles $\vartheta_{ij}$ between the wheel-ground contact points and the undercarriage axis are dependant upon the casters.

Geometric calculations (see Figure 8.6) result in the following equations for the distances:

$$
\begin{aligned}
r_{FL} &= \Big( \big( l_F - n_{LFL} \cos \delta_W + n_{SFL} \sin \delta_W \big)^2 \\
&\quad + \big( \tfrac{b_F}{2} - n_{SFL} \cos \delta_W - n_{LFL} \sin \delta_W \big)^2 \Big)^{1/2} \\
r_{FR} &= \Big( \big( l_F - n_{LFR} \cos \delta_W + n_{SFR} \sin \delta_W \big)^2 \\
&\quad + \big( \tfrac{b_F}{2} + n_{SFR} \cos \delta_W + n_{LFR} \sin \delta_W \big)^2 \Big)^{1/2} \\
r_{RL} &= \Big( \big( l_R + n_{LRL} \big)^2 + \big( \tfrac{b_R}{2} - n_{SRL} \big)^2 \Big)^{1/2} \\
r_{RR} &= \Big( \big( l_R + n_{LRR} \big)^2 + \big( \tfrac{b_R}{2} + n_{SRR} \big)^2 \Big)^{1/2}
\end{aligned}
\tag{8.2}
$$

and for the angles:

$$
\begin{aligned}
\vartheta_{FL} &= \arctan \left( \frac{\tfrac{b_F}{2} - n_{SFL} \cos \delta_W - n_{LFL} \sin \delta_W}{l_F - n_{LFL} \cos \delta_W + n_{SFL} \sin \delta_W} \right) \\
\vartheta_{FR} &= \arctan \left( \frac{l_F - n_{LFR} \cos \delta_W + n_{SFR} \sin \delta_W}{\tfrac{b_F}{2} + n_{SFR} \cos \delta_W + n_{LFR} \sin \delta_W} \right) \\
\vartheta_{RL} &= \arctan \left( \frac{l_R + n_{LRL}}{\tfrac{b_R}{2} - n_{SRL}} \right) \\
\vartheta_{RR} &= \arctan \left( \frac{\tfrac{b_R}{2} + n_{SRR}}{l_R + n_{LRR}} \right)
\end{aligned}
\tag{8.3}
$$

The caster $n_{Lij}$ is always defined positive in the direction $-x_W$, $n_{Sij}$ always in the direction of $-y_W$, hence $n_{Lij}$, $n_{Sij}$ can be negative.

Whereas the course angle changes with the angular velocity ( $\dot{\psi} + \dot{\beta}$ ), the chassis only turns with velocity $\dot{\psi}$ about the vertical axis, as shown in Figure 8.5.

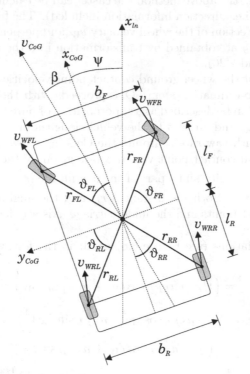

$v_{WFL}$ : *front left wheel velocity*
$v_{WFR}$ : *front right wheel velocity*
$v_{WRL}$ : *rear left wheel velocity*
$v_{WRR}$ : *rear right wheel velocity*
$r_{FL}$  : *distance from CoG to front left wheel ground contact point*
$r_{FR}$  : *distance from CoG to front right wheel ground contact point*
$r_{RL}$  : *distance from CoG to rear left wheel ground contact point*
$r_{RR}$  : *distance from CoG to rear right wheel ground contact point*
$\vartheta_{FL}$ : *angle between chassis (CoG) coordinate system and front left wheel ground contact point*
$\vartheta_{FR}$ : *angle between chassis (CoG) coordinate system and front right wheel ground contact point*
$\vartheta_{RL}$ : *angle between chassis (CoG) coordinate system and rear left wheel ground contact point*
$\vartheta_{RR}$ : *angle between chassis (CoG) coordinate system and rear right wheel ground contact point*
$l_F$  : *distance from CoG to front axle*
$l_R$  : *distance from CoG to rear axle*
$b_F$  : *distance between wheels on front axle*
$b_R$  : *distance between wheels on rear axle*

**Figure 8.5** Velocity components throughout the vehicle

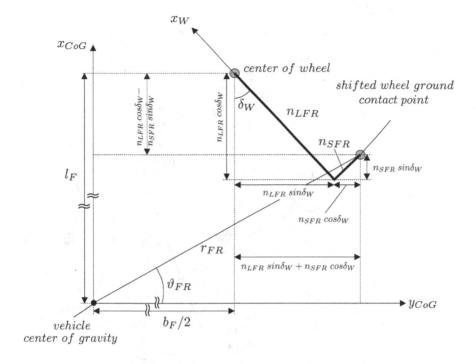

**Figure 8.6** Geometric calculation of the distance $r_{FR}$

The wheel ground contact point velocities consist of two components: the component due to the CoG velocity and the component due to the motion about the vertical vehicle axis, $\dot{\psi}$. A division into longitudinal and lateral vehicle directions gives the following equations, where $\vec{e_X}$ denotes the longitudinal CoG co-ordinate direction and $\vec{e_Y}$ the lateral CoG co-ordinate direction:

$$\underline{v}_{WFL} = \left(v_{CoG}\cos\beta - \dot{\psi}r_{FL}\sin\vartheta_{FL}\right)\vec{e_X} + \left(v_{CoG}\sin\beta + \dot{\psi}r_{FL}\cos\vartheta_{FL}\right)\vec{e_Y}$$

$$\underline{v}_{WFR} = \left(v_{CoG}\cos\beta + \dot{\psi}r_{FR}\cos\vartheta_{FR}\right)\vec{e_X} + \left(v_{CoG}\sin\beta + \dot{\psi}r_{FR}\sin\vartheta_{FR}\right)\vec{e_Y}$$

$$\underline{v}_{WRL} = \left(v_{CoG}\cos\beta - \dot{\psi}r_{RL}\cos\vartheta_{RL}\right)\vec{e_X} + \left(v_{CoG}\sin\beta - \dot{\psi}r_{RL}\sin\vartheta_{RL}\right)\vec{e_Y}$$

$$\underline{v}_{WRR} = \left(v_{CoG}\cos\beta + \dot{\psi}r_{RR}\sin\vartheta_{RR}\right)\vec{e_X} + \left(v_{CoG}\sin\beta - \dot{\psi}r_{RR}\cos\vartheta_{RR}\right)\vec{e_Y}$$

$$(8.4)$$

Equations 8.4 are then the basis for the calculation of the tire side slip angle using the velocity balances (method 1) given in Section 8.3.2. In practice, the calculation variables $\vartheta_{ij}$ and $r_{ij}$, are time dependant because of the time dependant casters. The presented approach is an approximation and enables the direct calculation of both the magnitude and the direction of the wheel ground contact point velocities. The magnitudes of the wheel speeds are calculated by taking the square root of the sum of the squares of the two components $\vec{e_X}$ and $\vec{e_Y}$.

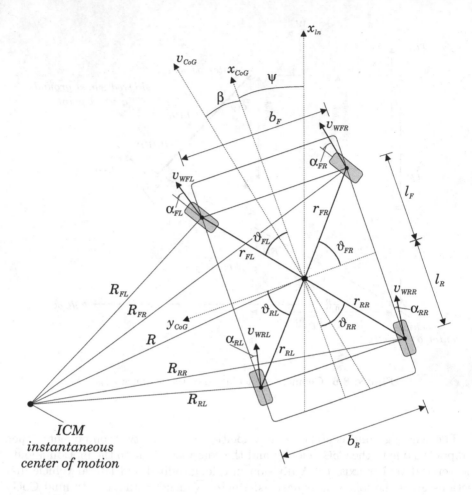

**Figure 8.7**  Individual curve-radii of the wheels when cornering

## Approximation of the Velocities at the Wheel-Ground Contact Points

As mentioned earlier it is assumed that the caster has no dynamics of its own. The caster component perpendicular to the wheel plane is also neglected. This is allowable because the caster effect is very small compared to the distances $l_F$, $l_R$ and $b_F/2$, $b_R/2$. The angles $\vartheta_{ij}$ and the radii $r_{ij}$ can be calculated in advance, so that the evaluation of Eqs. 8.4 requires only a few calculation steps. Moreover, $\sin\beta$ and $\cos\beta$ can be approximated with first order Taylor polynomials; with the assumption that the vehicle body side slip angle $\beta$ is limited to a value which is less than 10°, the error will be less than 0.5 % . To calculate the wheel ground contact point velocities the following approximations are substituted into Eqs. 8.4:

$$\sin\beta = \beta$$
$$\cos\beta = 1 \qquad\qquad (8.5)$$

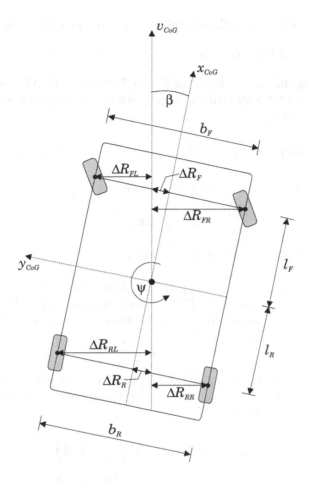

**Figure 8.8** Differential curve radii of the individual wheels

## Method 2: Calculating the Individual Curve Radii

If one considers the vehicle during a turn from a birds eye view, one realizes that each wheel follows an individual curve. The velocities of the CoG and the wheel-ground contact points are perpendicular to the connecting lines to the instantaneous center of motion (ICM), as shown in Figure 8.7.

An exact method for calculating the individual curve radii $R_{ij}$ exists, but has the disadvantage that the distance from the CoG to the instantaneous center (ICM) must be known.

Under the assumption that the distance $R$ from the vehicle CoG to the instantaneous center (ICM) is much larger than the distances $r_{ij}$, one can consider the differential radii $\Delta R_{ij}$ as parallel lines, as shown in Figure 8.8.

With the differential radii approach one has the possibility to determine the velocities of the wheel-ground contact points $v_{Wij}$, without knowing the absolute turn radius $R$ of the CoG.

The values $\Delta R_F$ and $\Delta R_R$ are calculated in the following way:

$$\Delta R_F = l_F \cdot \tan \beta \qquad \text{and} \qquad \Delta R_R = l_R \cdot \tan \beta$$

By disregarding the caster effect, the four differential radii $\Delta R_{ij}$ are found from inspection of Figure 8.8 and substitution of $\Delta R_F$ and $\Delta R_R$, and are given by the following equations:

$$
\begin{aligned}
\Delta R_{FL} &= \left(\frac{b_F}{2} - \Delta R_F\right)\cos\beta = \frac{b_F}{2}\cos\beta - l_F \cdot \sin\beta \\
\Delta R_{FR} &= \frac{b_F}{2}\cos\beta + l_F \cdot \sin\beta \\
\Delta R_{RL} &= \frac{b_R}{2}\cos\beta + l_R \cdot \sin\beta \\
\Delta R_{RR} &= \frac{b_R}{2}\cos\beta - l_R \cdot \sin\beta
\end{aligned}
\tag{8.6}
$$

The velocities of the wheel ground contact points can now be calculated using an additive superposition of the CoG velocity $v_{CoG}$ and the additional angular velocity due to the distance $\Delta R_{ij}$ from the wheel to the center of gravity, i.e.:

$$v_{Wij} = v_{CoG} \mp \dot{\psi}\Delta R_{ij} \tag{8.7}$$

Substituting the approximations (Eq. 8.5) for $cos\beta$ and $sin\beta$ yields the magnitudes of the wheel ground contact point velocities (note that here no information about direction is available):

$$
\begin{aligned}
v_{WFL} &= v_{CoG} - \dot{\psi}\left(\frac{b_F}{2} - l_F\beta\right) \\
v_{WFR} &= v_{CoG} + \dot{\psi}\left(\frac{b_F}{2} + l_F\beta\right) \\
v_{WRL} &= v_{CoG} - \dot{\psi}\left(\frac{b_R}{2} + l_R\beta\right) \\
v_{WRR} &= v_{CoG} + \dot{\psi}\left(\frac{b_R}{2} - l_R\beta\right)
\end{aligned}
\tag{8.8}
$$

## Comparison of the Two Methods for the Calculation of Wheel Velocities

The question now arises whether the methods satisfy the requirements in terms of low computational expense and high accuracy within the area of driving stability. To determine this a simulation run in the limiting range under exact calculation of the wheel-ground contact point velocities is carried out. Figure 8.9 shows the inputs in the form of the wheel turn angles $\delta_W$ and the individual state variables $a_y$, $v_{CoG}$, $\beta$ and $\dot{\psi}$.

The test drive is carried out on dry road conditions with a maximum adhesion coefficient of approximately 1.2. The wheel turn angle $\delta_W$ has a sine wave form with an amplitude of 7° and a frequency of approximately $0.7\,Hz$. The lateral

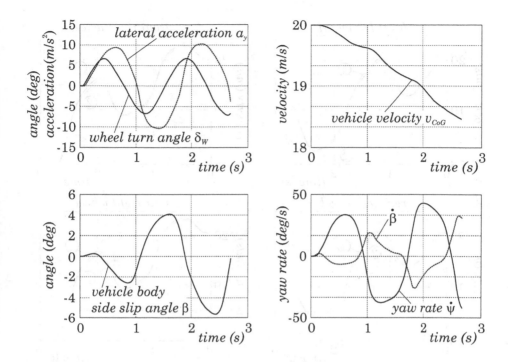

**Figure 8.9** Simulation of stable driving

acceleration $a_Y$ takes values up to $10\,m/s^2$ with which the wheels are well into the nonlinear region. The vehicle velocity $v_{CoG}$ drops because of the turn resistance. The vehicle body side slip angle $\beta$ takes values up to approximately 6°, and the plot shows increasing oscillatory behavior, indicating the onset of an unstable driving situation. Additionally the rate of change of vehicle body side slip angle $\dot{\beta}$ and the yaw rate $\dot{\psi}$ are shown, to give an impression of the turning motion of the chassis. $a_Y$, $\delta_W$ and $\dot{\psi}$ are measured on the experimental vehicle. $v_{CoG}$, $\beta$ and $\dot{\beta}$ are estimated values (see Chapter 9).

In Figure 8.10 the velocities of the wheel ground contact points are calculated from the states from Figure 8.9. The solid lines marked with ①show the calculation from Eq. 8.4 and the corresponding approximation using Eqs. 8.5 (Method 1). Method 2 of calculation using differential radii $\Delta R_{ij}$ and parallelization of the turn radii is marked with ②. The wheel velocity errors are very small.

Method 2, because of its simplicity, is preferable if only the magnitude of the wheel velocities is required. Method 1 must be used when the wheel side slip angle is to be derived individually for all four wheels (See Section 8.3.2, Method 1: Calculation of the tire side slip angle for the two-track model).

## 8.3.2 Wheel Slip and Tire Side Slip Angle

The wheel slip must be very accurately calculated. Due to the extremely high gradient of the cohesion coefficient characteristics, errors in the per-thousand

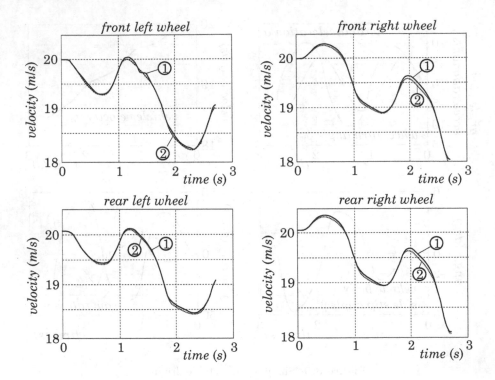

**Figure 8.10** Comparison of the two methods for the derivation of the wheel ground contact point velocities

range can result in force reactions of considerable dimensions. A typical value for the gradient at the origin is 30, this means that a per-thousand wheel slip change causes an adhesion coefficient change of 3 % i.e. a horizontal force of 3 % of the wheel ground contact force.

The calculation of the wheel slip $s_L$, $s_S$ requires the tire side slip angle $\alpha$, as the wheel slip is a vector and thus the velocities in the slip direction must be transformed. Two methods for the calculation of the tire side slip angle $\alpha$ are then compared.

**Wheel Slip Calculation**

If the vehicle drives without tire side slip, the wheel slip is simply the difference between the rotational equivalent wheel velocity $v_{Rij} = \dot{\omega}_{ij} \cdot r_{stat}$ and the CoG velocity $v_{CoG}$. By simultaneous appearance of longitudinal and lateral wheel slip there are various definitions in the literature. Burckhardt calculates the longitudinal wheel slip in the direction of motion of the wheel, as well as the force [14]. In contrast to this Reimpell specifies the longitudinal wheel slip in the direction of the wheel plane [101].

Here, the Burckhardt approach is chosen. The longitudinal slip $s_L$ is defined in the direction of the wheel ground contact point velocity $v_{Wij}$, and the lateral

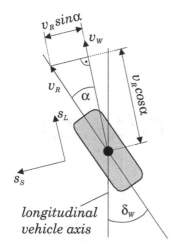

$v_R$ : *rotational equivalent wheel velocity*
$v_W$ : *wheel ground contact point velocity*

**Figure 8.11** Wheel slip calculation

slip $s_s$ at right angles to this (Eq. 8.9). The rotational equivalent wheel velocity $v_{Rij}$ is multiplied with the cosine of the tire side slip angle to obtain the projection in the direction of the wheel velocity $v_W$.

| | **Braking** $v_R \cos \alpha \leq v_W$ | **Driving** $v_R \cos \alpha > v_W$ | |
|---|---|---|---|
| **Longitudinal slip** | $s_L = \frac{v_R \cos \alpha - v_W}{v_W}$ | $s_L = \frac{v_R \cos \alpha - v_W}{v_R \cos \alpha}$ | (8.9) |
| **Side slip** | $s_S = \frac{v_R \cdot \sin \alpha}{v_W}$ | $s_S = \tan \alpha$ | |

The resultant slip $s_L$ must always be between -1 and 1. Therefore, the speed difference is divided by the respective larger speed, i.e. $v_W$ for braking and $v_R$ when accelerating. With these definitions the limiting of $s_L$ is ensured. The resultant wheel slip is the geometrical sum of the longitudinal and side slip:

$$s_{Res} = \sqrt{s_L^2 + s_S^2} \qquad (8.10)$$

**Tire Side Slip Angle Calculation**

**Method 1: Geometric Derivation Using Wheel Velocity Vector**

If the directions of the wheel ground contact point velocities $v_W$ are known (Equation 8.4) the four tire side slip angles $\alpha$ can easily be derived geometrically. The tire side slip angle $\alpha$ is, as shown in Figure8.11, the angle between the wheel plane and the velocity of the wheel ground contact point. For the calculation of the tire-slip angle based on the simplified versions of Eqs 8.4 the following are obtained (see Figure 8.12):

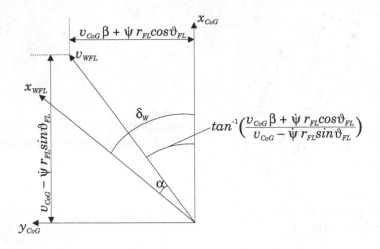

**Figure 8.12** Calculation of the front left tire side slip angle

$$\alpha_{FL} = \delta_W - \arctan\left(\frac{v_{CoG}\beta + \dot{\psi}r_{FL}\cos\vartheta_{FL}}{v_{CoG} - \dot{\psi}r_{FL}\sin\vartheta_{FL}}\right)$$

$$\alpha_{FR} = \delta_W - \arctan\left(\frac{v_{CoG}\beta + \dot{\psi}r_{FR}\sin\vartheta_{FR}}{v_{CoG} + \dot{\psi}r_{FR}\cos\vartheta_{FR}}\right)$$

$$\alpha_{RL} = -\arctan\left(\frac{v_{CoG}\beta - \dot{\psi}r_{RL}\sin\vartheta_{RL}}{v_{CoG} - \dot{\psi}r_{RL}\cos\vartheta_{RL}}\right)$$

$$\alpha_{RR} = -\arctan\left(\frac{v_{CoG}\beta - \dot{\psi}r_{RR}\cos\vartheta_{RR}}{v_{CoG} + \dot{\psi}r_{RR}\sin\vartheta_{RR}}\right) \tag{8.11}$$

### Method 2: The Single-Track Model

In the single-track model, the wheels on each axis are considered as a single unit, as shown in Figure 8.13. Because of this, it is only possible to derive one single tire side slip angle for the left and right wheels on the front axle, and one for the wheels on the rear axle.

The instantaneous center of motion (ICM) is such that the velocity vectors of the CoG and the wheels are perpendicular to the lines connecting these points to the instantaneous center. For a known vehicle body side slip angle $\beta$ the tire side slip angles $\alpha$ can be calculated.

The definition of the tire side slip angles $\alpha_F$ and $\alpha_R$ can be taken from Figure 8.14, and are calculated in accordance with Mitschke [84] by forming the velocity balance equations in the vehicle longitudinal and lateral directions.

The chassis and the wheels have identical velocities at the wheel ground contact points. For the front wheel in the lateral direction (from Figures 8.13 and 8.14):

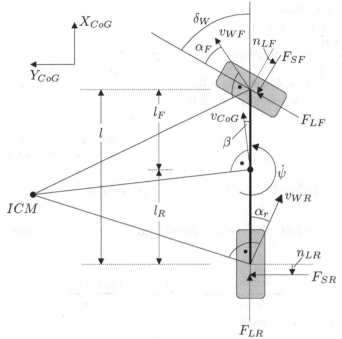

| $F_{LF}$ | : | longitudinal front wheel force |
|---|---|---|
| $F_{LR}$ | : | longitudinal rear wheel force |
| $F_{SF}$ | : | lateral (side) front wheel force |
| $F_{SR}$ | : | lateral (side) rear wheel force |
| $\dot{\psi}$ | : | yaw rate |
| $\delta_W$ | : | wheel turn angle |
| $\alpha_F$ | : | tire side slip angle front |
| $\alpha_R$ | : | tire side slip angle rear |
| $\beta$ | : | vehicle body side slip angle |
| $l$ | : | wheel base |
| $l_F$ | : | distance from CoG to front axle |
| $l_R$ | : | distance from CoG to front axle |
| $n_{LF}$ | : | Caster effect front |
| $n_{LR}$ | : | Caster effect rear |
| $ICM$ | : | instantaneous center(of motion) |
| $v_{WF}$ | : | direction of front wheel velocity |
| $v_{WR}$ | : | direction of rear wheel velocity |
| $v_{CoG}$ | : | direction of CoG velocity |

**Figure 8.13** Variables for the single-track model

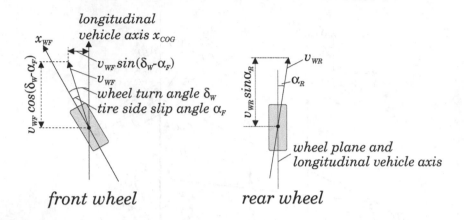

**Figure 8.14** Definition of the tire side slip angle $\alpha$ for the single-track model

$$\underbrace{v_{WF}\sin(\delta_W - \alpha_F)}_{wheel\ velocity} = \underbrace{l_F \cdot \dot\psi + v_{CoG}\sin\beta}_{chassis\ velocity} \qquad (8.12)$$

and in the longitudinal direction:

$$v_{WF}\cos(\delta_W - \alpha_F) = v_{CoG}\cos\beta \qquad (8.13)$$

Dividing the two equations gives the equation for the calculation of the tire side slip angle for the front axle:

$$\tan(\delta_W - \alpha_F) = \frac{l_F \cdot \dot\psi + v_{CoG}\sin\beta}{v_{CoG}\cos\beta} \qquad (8.14)$$

Similarly, for the rear axle:

$$\tan\alpha_R = \frac{l_R \cdot \dot\psi - v_{CoG}\sin\beta}{v_{CoG}\cos\beta} \qquad (8.15)$$

At stable driving conditions, the tire side slip angle $\alpha$ is normally no larger than $5°$ and the above equation can be simplified by substituting $\sin\beta \approx \beta$ and $\cos\beta \approx 1$. The classic equations for the tire side slip angles are then given as:

$$\alpha_F = -\beta + \delta_W - \frac{l_F \cdot \dot\psi}{v_{CoG}} \quad \text{and} \quad \alpha_R = -\beta + \frac{l_R \cdot \dot\psi}{v_{CoG}} \qquad (8.16)$$

Figure 8.15 shows the relationships of the simplified tire side slip angle calculations for the single-track model. To compare the two methods the extreme driving shown in Figure 8.9 is once more used. Figure 8.16 shows the results for the derivation of the tire-slip angle. The tire side slip angles calculated using the transformation of the CoG method (Equation 8.11) are marked with ①. Curve ②, calculated using the single-track model (Eq. 8.16) can not be differentiated.

Important to note is that the extremely simple equations for the calculation of the tire-slip angle based on the single-track model give good results for driving in the limiting range. Thus this simple method can be later used for observer design.

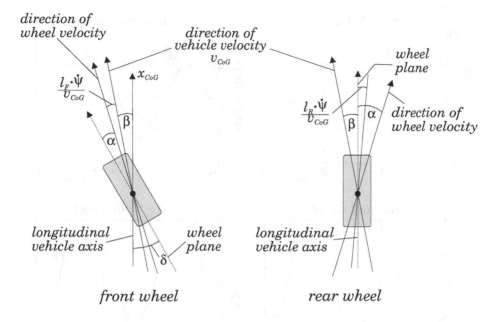

**Figure 8.15** Simplified tire side slip angle calculation for the single-track model

### 8.3.3 Friction Co-efficient Calculation

The friction behavior of the wheels can be approximated with parametric characteristics, as shown in Figure 8.17. The friction, or adhesion co-efficient $\mu$ is defined as the ratio of the frictional force acting in the wheel plane $F_{fric}$ and the wheel ground contact force $F_Z$:

$$\mu = \frac{F_{fric}}{F_Z} \tag{8.17}$$

The calculation of friction forces can be carried out using the method of Burckhardt [14]:

$$\mu(s_{Res}) = c_1 \cdot \left(1 - e^{-c2 \cdot s_{Res}}\right) - c_3 s_{Res} \tag{8.18}$$

The Burckhardt approach, Equation 8.18, can be extended via a pair of factors, where $c_4$ describes the influence of a higher drive velocity and $c_5$ the influence of a higher wheel load. Both factors have a maximum value of 1, i.e. they lead to a reduction of the friction co-efficient. Incorrect tire pressure can also lead to a reduction of the friction co-efficient. This effect however can be disregarded for variations in pressure of less than $0.3\,bar$. The resulting friction co-efficient is then given by:

$$\mu_{Res}(s_{Res}) = \left(c_1 \cdot \left(1 - e^{-c2 \cdot s_{Res}}\right) - c_3 s_{Res}\right) \cdot e^{-c4 \cdot s_{Res} \cdot v_{CoG}} \cdot \left(1 - c_5 F_Z^2\right) \tag{8.19}$$

The parameters $c_1$, $c_2$, and $c_3$ are given for various road surfaces in Table 8.3. Parameter $c_4$ lies in the range $0.002\,s/m$ to $0.004\,s/m$. This results, for a slip of 10\% and velocity of $20\,m/s$, in a friction co-efficient change of almost 8\%. As

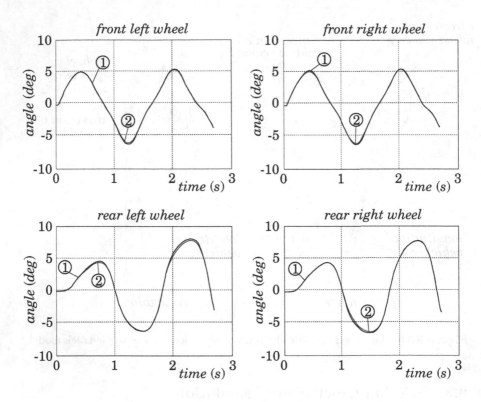

**Figure 8.16**  Results of the two methods of calculating the tire side slip angle

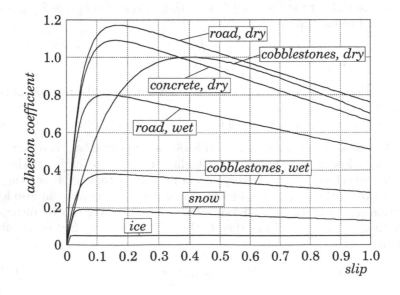

**Figure 8.17**  Typical cohesion coefficient characteristics

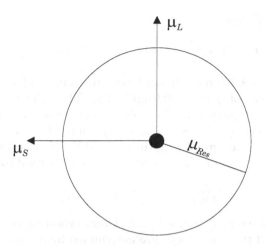

**Figure 8.18** Kamm circle

a first estimate, a homogeneous axle load distribution with static values for the wheel contact force of $F_Z = 4500\,N$ per wheel is assumed. At extreme cornering the wheel contact force can fall almost to zero, or can rise to almost double the normal value. With parameter $c_5 = 0.00015\,(1/kN)^2$ and a dynamic wheel force of $8\,kN$, a reduction in the friction coefficient of $9.6\,\%$ results.

The friction co-efficient values shown in Figure 8.17 are only valid for quasi-static operating points, and not for quickly changing dynamic transition states. For fast changes of the tire side slip angle, an analysis of the time dependency of the cornering forces is given in [1].

The resultant slip $s_{Res}$ is directed in the same direction as the resultant friction co-efficient $\mu_{Res}$. This gives the following equations for the friction co-efficients in the longitudinal and lateral directions:

$$\mu_L = \mu_{Res}\frac{s_L}{s_{Res}} \quad and \quad \mu_S = \mu_{Res}\frac{s_S}{s_{Res}} \tag{8.20}$$

Assuming that the friction behavior of the tire is independent of the direction of the slip, the behavior in Equation 8.20 can be described using a Kamm circle. This gives information about the directional distribution of the friction co-efficients, as shown in Figure 8.18. The contact between the tire and the road, together with the vehicle velocity and the wheel load, determines the level of the maximum resultant friction co-efficient $\mu_{Res,Max}$, which can then be split into longitudinal and lateral friction co-efficients. The sign of the friction coefficients determines the sign of the related friction forces.

In the presence of tread profile the friction behavior can also be dependant on direction. The maximum friction co-efficient in the lateral direction is smaller than in the longitudinal direction. In this case the Kamm circle degenerates to an ellipse. Formula-wise this is expressed using an attenuation factor $k_S$ for the lateral friction co-efficient:

$$\mu_L = \mu_{Res}\frac{s_L}{s_{Res}} \quad and \quad \mu_S = k_s\mu_{Res}\frac{s_S}{s_{Res}} \tag{8.21}$$

For $k_S = 1$ Equation 8.20 and 8.21 are identical. Common low profile tires have an attenuation factor of between 0.9 and 0.95. The behavior can be different during braking and acceleration. This effect is not considered here.

The Kamm circle also describes the maximum force transmission to the road surface. If the geometric sum of longitudinal and lateral wheel forces lies within the Kamm circle, the resultant tire forces can be transmitted to the ground [84]:

$$\sqrt{F_{WLij}^2 + F_{WSij}^2} \leq \mu_{Res} \cdot F_{Zij} \tag{8.22}$$

The $F_{Lij}$ are calculated in section 8.3.4, the lateral wheel forces are approximated in Section 9.5.2 and the wheel loads are determined by means of Equations 9.51 to 9.54. In the case that the friction forces in Equation 8.22 would exceed, the term $\mu_{Res} \cdot F_{Zij}$, a reduction factor $k_{red,ij}$ will be introduced in Section 8.3.4 guarantees the overall limitation.

**Friction Value Characteristics for Various Road Surfaces**

Table 8.3 gives a list of the parameter sets for various road surfaces [14].

With the exception of wet cobblestones the Burckhardt characteristics correspond very precisely to measured characteristics [84]. A measured friction co-efficient characteristic for cobblestones exhibits a higher initial gradient, however this levels out at friction values of about 0.4, and then runs with a smaller gradient to the maximum value, where it can then again be well approximated.

**Table 8.3** Parameter sets for friction co-efficient characteristics (Burckhardt)

|                     | $c_1$  | $c_2$   | $c_3$   |
|---------------------|--------|---------|---------|
| Asphalt, dry        | 1.2801 | 23.99   | 0.52    |
| Asphalt, wet        | 0.857  | 33.822  | 0.347   |
| Concrete, dry       | 1.1973 | 25.168  | 0.5373  |
| Cobblestones, dry   | 1.3713 | 6.4565  | 0.6691  |
| Cobblestones, wet   | 0.4004 | 33.7080 | 0.1204  |
| Snow                | 0.1946 | 94.129  | 0.0646  |
| Ice                 | 0.05   | 306.39  | 0       |

## 8.3.4   Calculation of Friction Forces

The friction co-efficients in the direction of the wheel velocity $v_W$ and orthogonal to it can be calculated using Equations 8.10, 8.19 and 8.21. The frictional forces can then be calculated from the friction co-efficients using Equation 8.17. This provides the frictional forces $F_{WL}$ and $F_{WS}$ in the direction of the wheel ground contact velocity $v_W$, and at right angles to it (see Figure 8.19):

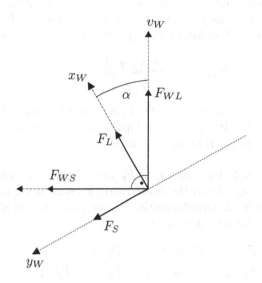

**Figure 8.19** Direction of frictional forces

In the direction $v_W$ :

$$F_{WL} = \mu_L F_Z$$
$$= \mu_{Res} \cdot \frac{s_L}{s_{Res}} \cdot F_Z \qquad (8.23)$$

In the direction at right angles to $v_W$

$$F_{WS} = \mu_S F_Z$$
$$= \mu_{Res} \cdot k_S \cdot \frac{s_S}{s_{Res}} \cdot F_Z \qquad (8.24)$$

Transforming into the wheel co-ordinate system $(x_W, y_W)$:

In the direction $x_W$ : $\quad F_L = F_{WL} \cos\alpha + F_{WS} \sin\alpha \qquad (8.25)$

In the direction $y_W$ : $\quad F_S = F_{WS} \cos\alpha - F_{WL} \sin\alpha \qquad (8.26)$

Substituting Equation 8.23 into Equation 8.25, and Equation 8.24 into Equation 8.26 then yields:

$$F_L = \mu_{Res} \cdot \frac{s_L}{s_{Res}} \cdot F_Z \cdot \cos\alpha + \mu_{Res} \cdot k_S \cdot \frac{s_S}{s_{Res}} F_Z \cdot \sin\alpha$$
$$= \left( \mu_{Res} \cdot \frac{s_L}{s_{Res}} \cdot \cos\alpha + \mu_{Res} \cdot k_S \cdot \frac{s_S}{s_{Res}} \cdot \sin\alpha \right) \cdot F_Z \qquad (8.27)$$

$$F_S = \mu_{Res} \cdot k_S \cdot \frac{s_S}{s_{Res}} \cdot F_Z \cdot \cos\alpha - \mu_{Res} \frac{s_L}{s_{Res}} \cdot F_Z \cdot \sin\alpha$$
$$= \left( \mu_{Res} \cdot k_S \cdot \frac{s_S}{s_{Res}} \cdot \cos\alpha - \mu_{Res} \cdot \frac{s_L}{s_{Res}} \cdot \sin\alpha \right) \cdot F_Z \qquad (8.28)$$

In extreme driving situations the friction forces may be calculated too large. In order to consider that fact, a factor $k_{red}$ is introduced which reduces the longitudinal and lateral wheel forces $F_{WLij}$ and $F_{WSij}$ in Equation 8.22.

$$k_{red,ij} = \frac{\mu_{Res,ij} \cdot F_{Zij}}{\sqrt{F_{WLij}^2 + F_{WSij}^2}} \tag{8.29}$$

The reduction factor $k_{red}$ ensures, that the geometric sum of the forces lies within the Kamm circle. It has to be adapted in every calculation step of the vehicle model in order to guarantee, that the maximum force transmission to the ground is not exceeded.

The longitudinal and side friction forces $F_L$ and $F_S$ are now transformed from the wheel co-ordinate system to the undercarriage co-ordinate system. For the wheels on the rear axle no transformation is necessary as the wheel plane lies parallel to the longitudinal vehicle axis:

$$\begin{aligned}
F_{XRL} &= F_{LRL} \quad , \quad F_{YRL} = F_{SRL} \\
F_{XRR} &= F_{LRR} \quad , \quad F_{YRR} = F_{SRR}
\end{aligned} \tag{8.30}$$

For the wheels on the front axle the forces are transformed by the wheel turn angle $\delta_W$:

$$\begin{aligned}
F_{XFL} &= F_{LFL} \cos \delta_{WL} - F_{SFL} \sin \delta_{WL} \\
F_{YFL} &= F_{SFL} \cos \delta_{WL} + F_{LFL} \sin \delta_{WL} \\
F_{XFR} &= F_{LFR} \cos \delta_{WR} - F_{SFR} \sin \delta_{WR} \\
F_{YFR} &= F_{SFR} \cos \delta_{WR} + F_{LFR} \sin \delta_{WR}
\end{aligned} \tag{8.31}$$

## 8.3.5   Tire Characteristics

In the previous section the frictional forces as well as the wheel slip and tire side slip angle have been derived; this section illustrates the effect of the tire slip. The tire slip is derived geometrically from the vehicle state variables. For control of the tire side slip angle it is necessary to control the rotation around the vehicle vertical axis.

The following three figures show the tire characteristics dependant on tire side slip angle for dry asphalt. The tire characteristics are normally given with the tire side slip angle as a parameter. This does not mean however that the tire side slip angle remains constant during a braking or steering manoeuvre.

Figure 8.20 shows the plot of lateral friction co-efficients against longitudinal friction co-efficients. Without tire side slip angle there is no side force possible, hence a branch of the group of curves lies along the axis. For a 2° tire side slip angle there is already the possibility of lateral friction co-efficients of up to $\mu_S = 0.7$. The cause of this is the fact that the side slip at $\alpha = 2°$ can already lie at $s_S = 3.5\,\%$ and this produces this high friction value for dry road surfaces. If the tangent to the curves is constructed, the Kamm circle of Figure 8.18 is produced. The tendency of the group of curves is easy to understand: the larger the tire side slip angle $\alpha$, the smaller the longitudinal force $F_L$. For a tire side slip angle of 16° a friction value of almost $\mu_L = 1$ is possible in the longitudinal

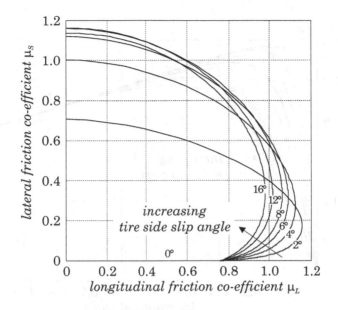

**Figure 8.20** Lateral friction coefficient $\mu_S$ over longitudinal friction coefficient $\mu_L$

direction. Figs. 8.21 and 8.22 show the influence of the tire side slip angle in more detail.

Figure 8.21 shows a plot of longitudinal friction co-efficient $\mu_L$ against longitudinal slip $s_L$. It illustrates the reduction in braking force when braking in a curve. If the driver turns during full braking the result is a tire side slip angle together with a side force. This causes a reduction in the longitudinal force and the braking distance is increased. The maximum wheel turn angle, which is dependant on the vehicle type, is approximately $\delta_W = 30°$. For stable driving situations no tire side slip angle greater than $\alpha = 16°$ can occur, as the vehicle body side slip angle $\beta$ lies in a similar value range. The shift of the maximum of the longitudinal friction co-efficient $\mu_L$ to higher slip values $s_L$ for increasing tire side slip angle $\alpha$ is clear.

Figure 8.22 shows a plot of lateral friction co-efficient $\mu_S$ against longitudinal slip $s_L$. During non-braking turning the lateral friction co-efficient $\mu_S$ makes available the whole traction (adhesion) potential, so that the curves begin with the maximum value. With a tire side slip angle of about $\alpha = 8°$ the lateral friction co-efficient $\mu_S$ assumes its maximum value. If the longitudinal slip $\mu_L$ increases, the side force sinks rapidly. This can be only partly compensated by an increase in the tire side slip angle.

### Effect of the Camber Angle on the Tire Side Slip Angle

Till now it has been assumed that the wheel stands perpendicular to the road. In practice the wheel stands at a camber angle $\gamma$ to the vertical axis. The direction of the camber angle is defined according to [84]: $\gamma$ is negative when the wheel leans

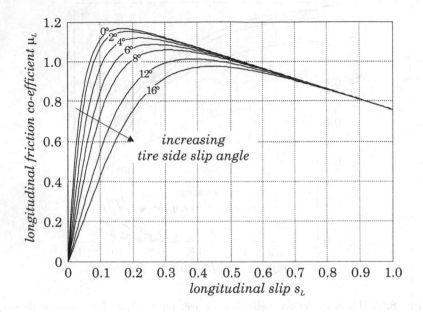

**Figure 8.21** Longitudinal friction coefficient $\mu_L$ over longitudinal slip $s_L$

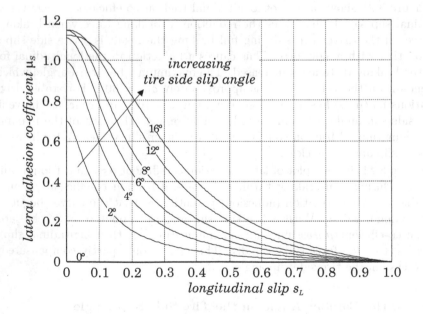

**Figure 8.22** Lateral friction coefficient $\mu_S$ over longitudinal slip $s_L$

towards the center of the turning curve. This corresponds to a *leaning-into-the-turn* of a bike. The tire side slip angle $\alpha$ is made smaller by this. Approximating $s_S = \tan\alpha$ by $s_S \approx \alpha$, the characteristic friction curve $\mu_S(\alpha)$ is shifted upwards with negative $\gamma$. Using the curves from [101] and [84] linear approximations are developed for the calculation of forces in the presence of camber angle. Using the approximately parallel shift of the lateral tire slip curves, the camber angle can be interpreted as a shifting of the tire side slip angle $\alpha$. With this, a negative camber angle $\gamma$ means a shifting to smaller tire side slip angles. The following approximation is used:

$$\alpha^* = \alpha + k_{camb}\gamma \tag{8.32}$$

$k_{camb}$ is chosen as 0.1. With this a $10°$ camber angle gives the same effect as a $1°$ tire side slip angle.

For an exact calculation of the tire camber, one must observe the axle geometry. A camber angle of $\gamma = 0.5°$ is set for the neutral position of the chassis suspension. On top of this a camber angle change of $\gamma = 0.3°$ is generated per $cm$ of spring displacement $\Delta z_W$. In cornering, this results in a stabilizing turning motion of the rear wheels. The maximum camber angle is approximately $\gamma = 3°$.

### 8.3.6 Definition of the Wheel Radius

To calculate the angular wheel velocity $\omega$, a torque balance is formed for each wheel. The accelerating torque is the drive torque $T_{Drive}$ of the driveline. The decelerating effects come from the braking torque $T_{Br}$ and tire friction torque $r_{stat}F_L$. With the moment of inertia of the wheel, $J_W$, one obtains:

$$J_W\dot{\omega} = T_{Drive} - T_{Br} - r_{stat}F_L \tag{8.33}$$

**Static Tire Radius**

The static tire radius $r_{stat}$ relates the stationary wheel ground contact force $F_Z$ to the tire spring stiffness $k_T$, as shown in Figure 8.23:

$$r_{stat} = r_0 - \frac{F_Z}{k_T} \tag{8.34}$$

The static tire radius can also be determined by measuring the length $2 \cdot l$ and afterwards calculating

$$r_{stat} = \sqrt{r_0^2 - l^2} \tag{8.35}$$

## 8.4 The Complete Vehicle Model

If the chassis is considered as a body with 6 degrees of freedom ($x_{CoG}$, $y_{CoG}$, $z_{CoG}$, $\psi$, $\chi$, $\varphi$), with forces from road, gravitation and wind acting upon the chassis, then the structure of Figure 8.24 results. The chassis itself is divided into a rotary part for the calculation of yaw rate $\dot{\psi}$, roll rate $\dot{\varphi}$, pitch rate $\dot{\chi}$ and their integrals, and a translatory part for the calculation of displacement,

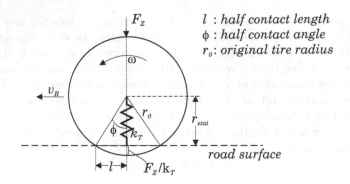

**Figure 8.23** Static wheel radius

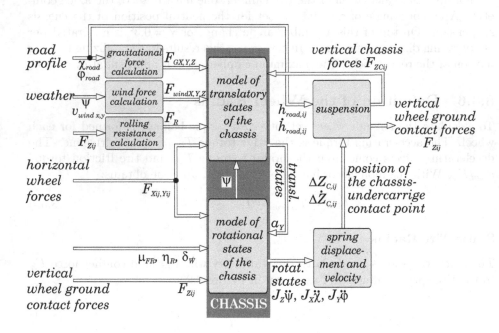

**Figure 8.24** Structure and signal flows for the Complete Vehicle Model

velocity and acceleration in the direction of the three co-ordinate axis. Wind strength is considered in the longitudinal and lateral velocity directions, and the gravitational force in the chassis co-ordinate directions. The wheels together with the chassis are modelled as a vertical spring-damper system. For the vehicle suspension a linear and a nonlinear method are carried out and compared.

## 8.4.1 Chassis Translatory Motion

The translatory variables are calculated in the inertial co-ordinate system and then transformed into the undercarriage co-ordinate system. As well as the horizontally acting wheel forces and the vertically acting chassis forces, the gravitational forces and the wind strength are calculated in the three undercarriage coordinate directions. Because all of the forces relevant for the translatory motion are contact forces to the environment, it is sensible to carry out the integration in the inertial system, as otherwise forces would appear by the transformation between two accelerating co-ordinate systems. The calculation equations are force balances for the three co-ordinate directions of the inertial co-ordinate system:

$$
m_{CoG}
\begin{bmatrix}
\ddot{x}_{In} \\
\ddot{y}_{In} \\
\ddot{z}_{In}
\end{bmatrix}
= \underline{T}_{UIn}
\underbrace{
\begin{bmatrix}
F_{XFL}+F_{XFR}+F_{XRL}+F_{XRR}+F_{windX}+F_{GX}+F_R \\
F_{YFL}+F_{YFR}+F_{YRL}+F_{YRR}+F_{windY}+F_{GY} \\
F_{ZCFL}+F_{ZCFR}+F_{ZCRL}+F_{ZCRR}+F_{windZ}+F_{GZ}
\end{bmatrix}
}_{\text{Forces in the undercarriage co−ordinate system}}
\tag{8.36}
$$

In this, $F_{Xij}$, $F_{Yij}$ are the wheel forces, $F_{ZCij}$ the vertical chassis forces, $F_{wind}$ are the wind forces , $F_G$ the gravitational forces and $F_R$ is the rolling resistance. The wheel forces $F_{Xij}$ and $F_{Yij}$ are calculated using the tire characteristics, as shown in Section 8.3.4. The vertical chassis forces $F_{ZCij}$ are approximated in Section 9.5.1. The derivation of the wind forces, gravitational forces and rolling resistance is given in this section. $\underline{T}_{UIn}$ is a transformation matrix for rotating the vector from the undercarriage to the inertial co-ordinate system.

### Transformation between Co-ordinate Systems

To begin with, the transformation from the CoG co-ordinate system to the inertial co-ordinate system will be considered. Figure 8.25 shows how the transformation is calculated for a rotation of the CoG co-ordinate system by a yaw angle $\psi$ about the vertical axis.

Consider the point $p$, whose co-ordinates in the CoG co-ordinate system are $p_{xCoG}$, $p_{yCoG}$. The $z$-co-ordinate is not affected in the rotation. The projections of these points into the inertial co-ordinate system are shown in Figure 8.25 and yield the following:

$$
\begin{aligned}
p &= (p_{xCoG}\cos\psi - p_{yCoG}\sin\psi)\,\vec{x}_{In} + (p_{xCoG}\sin\psi + p_{yCoG}\cos\psi)\,\vec{y}_{In} \\
&\quad + (p_{zCoG})\,\vec{z}_{In}
\end{aligned}
\tag{8.37}
$$

In matrix form this can be written as

$$
\begin{bmatrix}
x_{In} \\
y_{In} \\
z_{In}
\end{bmatrix}
= \underline{T}_{RotZ}
\begin{bmatrix}
x_{CoG} \\
y_{CoG} \\
z_{CoG}
\end{bmatrix},
\tag{8.38}
$$

where $\underline{T}_{RotZ}$, the matrix which carries out the transformation $T_{CoGIn}$ required due to rotation about the $z$ axis is given by

$$
\underline{T}_{RotZ} =
\begin{bmatrix}
\cos(\psi) & -\sin(\psi) & 0 \\
\sin(\psi) & \cos(\psi) & 0 \\
0 & 0 & 1
\end{bmatrix}.
\tag{8.39}
$$

To rotate in the opposite direction (i.e. from inertial to CoG) $\underline{T}_{RotZ}$ must be inverted. Due to the special structure of this matrix this corresponds to a rotation

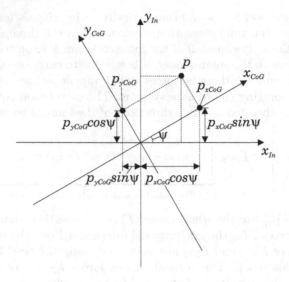

**Figure 8.25**  Rotation of the CoG co-ordinate system relative to the Inertial system

with a negative yaw angle $\psi$ or a transposition of the matrix:

$$\underline{T}_{RotZ}^{-1} = \underline{T}_{RotZ}^{T} = \underline{T}_{RotZ}(-\psi) \tag{8.40}$$

The above then describes a transformation due to a yaw angle (rotation about the vertical $z$ axis). Similar transformation matrices can be derived for the roll and pitch angles:

- **Pitch angle**

$$\underline{T}_{RotY} = \begin{bmatrix} \cos(\chi) & 0 & \sin(\chi) \\ 0 & 1 & 0 \\ -\sin(\chi) & 0 & \cos(\chi) \end{bmatrix} \tag{8.41}$$

- **Roll angle**

$$\underline{T}_{RotX} = \begin{bmatrix} 1 & 0 & 0 \\ 0 & \cos(\varphi) & -\sin(\varphi) \\ 0 & \sin(\varphi) & \cos(\varphi) \end{bmatrix} \tag{8.42}$$

A rotation about several axes corresponds to a multiplication of the rotation matrices. This presents the problem that the sequence of matrix multiplication affects the results. The standard order in the literature is: yaw-pitch-roll.

**Transformation from CoG to Undercarriage Co-ordinate System**

$$
\underline{T}_{CoGU} = \underline{T}_{RotY} \cdot \underline{T}_{RotX} =
\begin{bmatrix}
\cos(\chi) & \sin(\chi)\sin(\varphi) & \sin(\chi)\cos(\varphi) \\
0 & \cos(\varphi) & -\sin(\varphi) \\
-\sin(\chi) & \cos(\chi)\sin(\varphi) & \sin(\chi)\cos(\varphi)
\end{bmatrix}
\quad (8.43)
$$

The inverse transformation can be used to calculate the transformation from undercarriage to CoG co-ordinate systems. This gives:

$$
\underline{T}_{UCoG} = \underline{T}_{CoGU}^{-1} = \underline{T}_{RotX}^{-1} \cdot \underline{T}_{RotY}^{-1} = \underline{T}_{RotX}^{T} \cdot \underline{T}_{RotY}^{T} = \underline{T}_{CoGU}^{T}
\quad (8.44)
$$

Thus for a rotation around all three axis:

$$
\begin{aligned}
\underline{T}_{CoGIn} &= \underline{T}_{RotZYX} = \underline{T}_{RotZ} \cdot \underline{T}_{RotY} \cdot \underline{T}_{RotX} \\
&=
\begin{bmatrix}
\cos\psi\cos\chi & -\sin\psi\cos\varphi-\cos\psi\sin\chi\sin\varphi & \sin\psi\sin\varphi+\cos\psi\sin\chi\cos\varphi \\
\sin\psi\cos\chi & \cos\psi\cos\varphi+\sin\psi\sin\chi\sin\varphi & -\cos\psi\sin\varphi-\sin\psi\sin\chi\cos\varphi \\
\sin\chi & \cos\chi\sin\varphi & \cos\chi\cos\varphi
\end{bmatrix}
\end{aligned}
$$

$$
\begin{aligned}
\underline{T}_{InCoG} &= \underline{T}_{RotXYZ} = \\
&= \underline{T}_{RotZYX}^{-1} = \underline{T}_{RotZYX}^{T} = \underline{T}_{RotZYX}\big|_{\psi=-\psi,\ \chi=-\chi,\ \varphi=-\varphi}
\end{aligned}
\quad (8.45)
$$

Eq. 8.45 can be used to rotate from the CoG to the inertial systems. If however the road is not level, this must be taken into consideration by modifying the angles in $\underline{T}_{RotZYX}$:

$$
\underline{T}_{CoGIn} = \underline{T}_{RotZYX}\big|_{\psi=\psi,\ \chi=\chi-\chi_{road},\ \varphi=\varphi-\varphi_{road}}
\quad (8.46)
$$

Similarly the transformation matrix for a rotation from undercarriage to inertial co-ordinates is:

$$
\underline{T}_{UIn} = \underline{T}_{RotZYX}\big|_{\psi=\psi,\ \chi=-\chi_{road},\ \varphi=-\varphi_{road}}
\quad (8.47)
$$

**Wind Force Calculation**

The wind velocity is directed against the vehicle. In order to calculate the wind force, firstly the external wind velocity is transformed into the undercarriage co-ordinate system, then subtracted from the vehicle velocity, and finally the wind force calculated. The vehicle lift is disregarded.

$$
\begin{bmatrix}
F_{windX} \\
F_{windY} \\
F_{windZ}
\end{bmatrix}
=
\begin{bmatrix}
-c_{airX} A_L \frac{\rho}{2} \left( v_{CoGX} - v_{windX}\cos\psi - v_{windY}\sin\psi \right)^2 \\
-c_{airY} A_S \frac{\rho}{2} \left( v_{CoGY} - v_{wind}^* \right)^2 \cdot \text{sign}\left(-v_{wind}^*\right) \\
0
\end{bmatrix}
$$

$$
(8.48)
$$

with the abbreviation

$$
v_{wind}^* = -v_{windX}\sin\psi + v_{windY}\cos\psi
$$

$c_{airX,Y}$ are the co-efficients of aerodynamic drag, $A_{L,S}$ the front and side vehicle areas, and $v_{windX,Y}$ the wind velocities.

**Gravitational Force Calculation**

The undercarriage co-ordinate system is at angle $\chi_{road}$ (due to road inclination) and $\varphi_{road}$ (due to road camber) to the inertial system. A positive inclination $\chi_{road}$ means an upwards inclined road and a positive camber $\varphi_{road}$ means a road which raises the right hand side of the vehicle. The transformation of the gravitational forces into the undercarriage co-ordinate system corresponds to a multiplication with the rotation matrix from the previous section, with yaw angle $\psi = 0$:

$$
\begin{bmatrix} F_{GX} \\ F_{GY} \\ F_{GZ} \end{bmatrix} =
$$

$$
\begin{bmatrix} \cos(\chi_{road}) & \sin(\chi_{road})\sin(\varphi_{road}) & \sin(\chi_{road})\cos(\varphi_{road}) \\ 0 & \cos(\varphi_{road}) & -\sin(\varphi_{road}) \\ -\sin(\chi_{road}) & \cos(\chi_{road})\sin(\varphi_{road}) & \cos(\chi_{road})\cos(\varphi_{road}) \end{bmatrix} \cdot \begin{bmatrix} 0 \\ 0 \\ mg \end{bmatrix} \tag{8.49}
$$

**Rolling Resistance Calculation**

The tire rolling resistance force $F_R$ is calculated according to [84]:

$$
F_R = -f_{R0}F_Z - f_{R1}F_Z\frac{v}{30} - f_{R2}F_Z\frac{v^4}{30^4} \tag{8.50}
$$

The dependence of the tire rolling resistance on the tire pressure is not regarded here.

**Vehicle Body Side Slip Angle Calculation**

For the direct calculation of the vehicle body side slip angle $\beta$ the CoG (chassis) velocity must be given in inertial co-ordinates $v_{CoG,X}$ and $v_{CoG,Y}$, and the yaw angle subtracted from the chassis direction (Figure 8.7):

$$
\beta = \arctan\left(\frac{v_{CoG,Y}}{v_{CoG,X}}\right) - \psi \tag{8.51}
$$

In Section 9.6.2, a nonlinear observer for the estimation of the vehicle body side slip angle $\beta$ is presented.

## 8.4.2   Chassis Rotational Motion

The rotational variables can be calculated directly in the undercarriage co-ordinate system since the roll and pitch axis are assumed to lie at the road level. For this the torque equations are used:

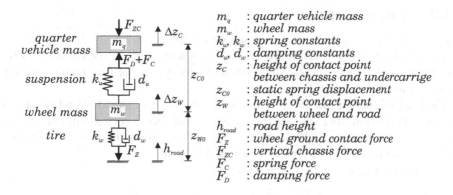

$m_q$ : quarter vehicle mass
$m_w$ : wheel mass
$k_u, k_w$ : spring constants
$d_u, d_w$ : damping constants
$z_C$ : height of contact point
  between chassis and undercarriage
$z_{C0}$ : static spring displacement
$z_W$ : height of contact point
  between wheel and road
$h_{road}$ : road height
$F_Z$ : wheel ground contact force
$F_{ZC}$ : vertical chassis force
$F_C$ : spring force
$F_D$ : damping force

**Figure 8.26** Forces and displacements in the quarter vehicle model

- **Torque balance around the vertical vehicle axis (yaw rate):**

$$
\begin{aligned}
J_Z \ddot{\psi} &= (F_{YFR} + F_{YFL}) \cdot (l_F - n_{LF} \cdot \cos \delta_W) \\
&= -(F_{YRR} + F_{YRL}) \cdot (l_R + n_{LR} \cdot \cos \delta_W) \\
&\quad + (F_{XRR} - F_{XRL}) \cdot \frac{b_R}{2} + F_{XFR} \cdot \left( \frac{b_F}{2} + n_{LFR} \sin \delta_W \right) \\
&\quad - F_{XFL} \cdot \left( \frac{b_F}{2} - n_{LFL} \sin \delta_W \right)
\end{aligned}
\tag{8.52}
$$

In the following equations the vertical chassis forces $F_{ZCij}$ are set equal to the wheel ground contact forces $F_{Zij}$, i.e. neglecting the suspension dynamics. This is justified, since roll and pitch motions are slower than vertical wheel motions on a rough road.

- **Torque balance around the vehicle longitudinal axis (roll rate):**

$$
J_X \ddot{\varphi} = (F_{ZFL} - F_{ZFR}) \cdot \frac{b_F}{2} + (F_{ZRL} - F_{ZRR}) \cdot \frac{b_R}{2} - m_{CoG} a_Y h_{CoG} \tag{8.53}
$$

- **Torque balance around the lateral vehicle axis (pitch rate):**

$$
J_Y \ddot{\chi} = -(F_{ZFL} + F_{ZFR}) \cdot l_F + (F_{ZRL} + F_{ZRR}) \cdot l_R + m_{CoG} a_X h_{CoG} \tag{8.54}
$$

The distances $n_{F,R}$ are due to the longitudinal casters:

$$
n_{LF} = \frac{n_{LFL} + n_{LFR}}{2} \qquad n_{LR} = \frac{n_{LRL} + n_{LRR}}{2} \tag{8.55}
$$

## 8.4.3  Suspension

A quarter-vehicle model is used for each wheel (Figure 8.26).

In contrast to Figure 8.26 however, the mass portion $m_q$ is not constant as the load within the vehicle varies. The normal forces $F_Z$ of the four quarter masses are determined in Section 9.5.1.

Each wheel has an individual mass $m_W$, which is connected via wheel spring $k_W$ and wheel damping $d_W$ to the ground, and via a spring-damper system $(k_U, \ d_U)$ to the chassis. All four wheel suspensions are assumed to be vertically directed. The indices $ij$ are front/rear and left/right.

The forces which act upon the wheel are:

- Wheel ground contact forces

$$
\begin{aligned}
F_{Zij} &= \Delta F_{Zij} + F_{ZCij0} \\
&= k_W \left( h_{road} - \Delta z_{Wij} - z_{Wij0} \right) + d_W \left( \dot{h}_{road} - \dot{z}_{Wij} \right) \quad (8.56)
\end{aligned}
$$

- Suspension spring forces:

$$
F_{Cij} = k_U \left( \Delta z_{Wij} + z_{Wij0} - \Delta z_{Cij} - z_{Cij0} \right) \quad (8.57)
$$

- Suspension damping forces:

$$
F_{Dij} \quad (8.58)
$$

The constants $k_W$, $k_U$, $d_U$ and $d_W$ are equal for wheels on the same axle. The wheel damping constant $d_W$ is set equal to zero. The reference road level (flat road) is $h_{road,0} = 0$. The static spring displacements $z_{Wij0}$ and $z_{Cij0}$ are so chosen that they balance the static forces $F_{ZCij0}$ by exact horizontal positioning of the unloaded vehicle.

$$
F_{ZCij0} = -k_W z_{Wij0} \overset{!}{=} -k_U \left( z_{Wij0} - z_{Cij0} \right) \quad (8.59)
$$

The force balances at the quarter chassis and the wheels are:

$$
\begin{aligned}
m_q \ddot{z}_{Cij} &= k_U \left( \Delta z_{Wij} - \Delta z_{Cij} + z_{Wij0} - z_{Cij0} \right) + F_{ZCij0} + F_{Dij} \\
&= k_U \left( \Delta z_{Wij} - \Delta z_{Cij} \right) + F_{Dij} \quad (8.60)
\end{aligned}
$$

and,

$$
\begin{aligned}
m_W \ddot{z}_{Wij} &= k_W \left( h_{road} - \Delta z_{Wij} - z_{Wij0} \right) + d_W \left( \dot{h}_{road} - \dot{z}_{Wij} \right) - \\
& \quad - k_U \left( \Delta z_{Wij} + z_{Wij0} - \Delta z_{Cij} - z_{Cij0} \right) - F_{Dij} \\
&= k_W \left( h_{road} - \Delta z_{Wij} \right) + d_W \left( \dot{h}_{road} - \dot{z}_{Wij} \right) - \\
& \quad - k_U \left( \Delta z_{Wij} - \Delta z_{Cij} \right) + F_{Dij} \quad (8.61)
\end{aligned}
$$

## Linear Suspension Model

The suspension damping force is proportional to the resulting displacement speed $\dot{z}_W - \dot{z}_C$.

$$F_{Dij} = d_U (\dot{z}_W - \dot{z}_C) \tag{8.62}$$

The damping of the wheel $d_W$ is neglected. The quarter vehicle can then be described by a fourth order state space model of the form,

$$\dot{\underline{x}} = \underline{A}\,\underline{x} + \underline{b}u$$
$$\underline{\dot{y}} = \underline{C}\,\underline{x} + \underline{d}u$$

The state vector is

$$\dot{\underline{x}} = [\Delta z_W, \quad \dot{z}_W, \quad \Delta z_C, \quad \dot{z}_C]^T \quad , \tag{8.63}$$

the input variable is

$$u(t) = h_{road}(t) \quad , \tag{8.64}$$

and the output vector is

$$\underline{y} = [\ddot{z}_W, \quad \ddot{z}_C, \quad (\Delta z_C - \Delta z_W)]^T \quad . \tag{8.65}$$

The road profile $h_{road}$ may be approximated by a randomly distributed Gaussian noise with an assumed covariance

$$c_{hh} = 2\pi A v_{X,CoG} \quad . \tag{8.66}$$

The amplitude factor $A$ is dependent upon the road profile, and $v_{X,CoG}$ is the longitudinal vehicle velocity [42]. For a dynamic analysis, the linear model can be used to define a transfer function between the vertical acceleration $\ddot{z}_C$ and the road profile $h_{road}$:

$$
\begin{aligned}
G(s) &= \frac{s^2 z_C(s)}{h_{road}(s)} \\
&= \frac{k_W d_U s^3 + k_W k_U s^2}{m_W m_q s^4 + d_U (m_W + m_q(k_W + k_U)) s^2 + d_U k_W s + k_W k_U}
\end{aligned}
\tag{8.67}
$$

## Effect of Parameter Variations on the Linear Suspension Model

The influence of the spring stiffness $k_U$, damping $d_U$ and quarter vehicle mass $m_q$ parameters is investigated by simulating an example. For this, the following representative parameter values are substituted into Eq. 8.67:

$$m_q = 350\,kg, \quad m_W = 31\,kg, \quad k_U = 20900\,N/m,$$
$$k_W = 10800\,N/m, \quad d_U = 1140\,Ns/m$$

Simulations are carried out to show how strongly the amplitude and phase of $G(s)$ are affected by changes in the parameters. The following changes are considered: $m_q = 200 - 400\,kg$, $d_U = 500 - 2000\,Ns/m$ and $k_U = 10000 - 30000\,N/m$.

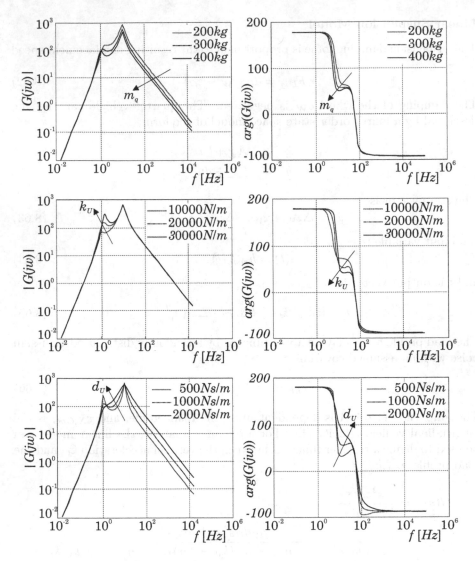

**Figure 8.27** Magnitude and phase characteristics for parameter variations of $G(s)$

Figure 8.27 shows that a change in the quarter vehicle mass $m_q$ only has effects at frequencies under the resonant frequency. The spring constant $k_U$ has a significant influence on the amplitude of $G(s)$ for frequencies around $1\,Hz$, and the resonant frequency also changes slightly.

The variation of the damping constant $d_U$ has a similar effect for frequencies around $6\,Hz$, however the amplitude at the resonant frequency varies much more strongly than with the parameters $m_q$ and $k_U$. This means a strong dependency of driving comfort on shock absorber characteristics.

## Nonlinear Suspension Model

Suspension systems have a nonlinear damping characteristic. When moving upwards the wheel generates a smaller damping force $F_D$ than when moving downwards. The nonlinearity allows an upward bump from the road profile to have a small impact on the chassis, while vertical wheel oscillations are still effectively damped during the downward movement of the wheel. The damping force is approximated by:

$$F_{Dij} \approx d_{U,l}\left(\dot{z}_{Wij} - \dot{z}_{Cij}\right) + d_{U,nl}\sqrt{\left|\dot{z}_{Wij} - \dot{z}_{Cij}\right|}\, sign\left(\dot{z}_{Wij} - \dot{z}_{Cij}\right) \qquad (8.68)$$

Inserting this into Equation 8.60 yields:

$$
\begin{aligned}
m_q \ddot{z}_{Cij} &= k_U\left(\Delta z_{Wij} - \Delta z_{Cij}\right) + d_{U,l}\left(\dot{z}_{Wij} - \dot{z}_{Cij}\right) + \\
&\quad + d_{U,nl}\sqrt{\left|\dot{z}_{Wij} - \dot{z}_{Cij}\right|}\, sign\left(\dot{z}_{Wij} - \dot{z}_{Cij}\right)
\end{aligned}
\qquad (8.69)
$$

Figure 8.28 shows a plot of the damping force $F_D$ over the relative velocity $\dot{z}_W - \dot{z}_C$. It is necessary to measure the required inputs (undercarriage vertical acceleration $\ddot{z}_C$ and relative spring-damper displacement $(\Delta z_W - \Delta z_C)$. The relative spring velocity $(\dot{z}_W - \dot{z}_C)$ is then determined by numerical differentiation from the measured relative spring-damper displacement:

$$\dot{y}(k) \approx \frac{y(k) - y(k-1)}{T_s} \qquad (8.70)$$

Here, $y(k)$ is the measured signal at time $k$ and $T_s$ the sampling time of the process.

This is the classic method of numerical differentiation, which is suitable for linear and nonlinear estimation equations alike. For disturbance and noise-free measurements it gives good results. Noisy measurement signals however, which have high frequency signals superimposed upon them, demand an improvement upon this method [79].

## Simulation Results

To compare the linear and nonlinear suspension models a simulation with stepwise steering angle excitation is carried out. Figure 8.29 shows the results. The top figure shows the assessment of the drive situation, i.e. wheel turn angle, vehicle velocity and lateral acceleration. The test drive took place on dry ground so that the unstable range begins at about $1\,g$ lateral acceleration. In the bottom figure the advantages of the nonlinear damping characteristics can easily be seen. Using the more realistic nonlinear model, a well damped transient behavior results, whilst with the linear model roll angle oscillations appear, which can not be observed in actual driving situations.

## 8.4.4 Reduced Nonlinear Two-track Model

The reduced model should contain only those state variables which are essential for vehicle dynamic control and ABS control. These are the vehicle speed $v_{CoG}$,

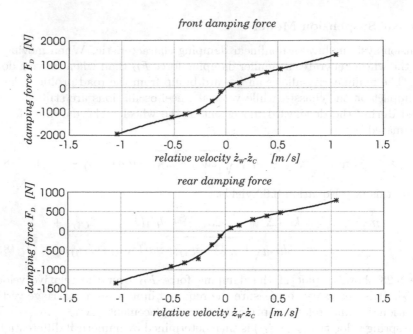

**Figure 8.28** Interpolation of the damping characteristic

the vehicle body side slip angle $\beta$ and the yaw rate $\dot\psi$. Starting from the left portion of Eq. 8.36, only the $x-$ and $y-$ components are considered, i.e. the z-component is disregarded. In this case, the forces $F_{Xij}$ and $F_{Yij}$ are identical in the undercarriage and the CoG co-ordinate systems. From Figure 8.7, the vehicle speed $v_{CoG}$ can be transformed into the fixed inertial co-ordinate system by:

$$\left[\begin{array}{c} \dot{x}_{In} \\ \dot{y}_{In} \end{array}\right] = v_{CoG} \cdot \left[\begin{array}{c} \cos(\beta + \psi) \\ \sin(\beta + \psi) \end{array}\right] \tag{8.71}$$

By differentiation, the accelerations in Equation 8.36 are obtained:

$$\left[\begin{array}{c} \ddot{x}_{In} \\ \ddot{y}_{In} \end{array}\right] = v_{CoG} \cdot \left(\dot\beta + \dot\psi\right) \left[\begin{array}{c} -\sin(\beta + \psi) \\ \cos(\beta + \psi) \end{array}\right] + \dot{v}_{CoG} \left[\begin{array}{c} \cos(\beta + \psi) \\ \sin(\beta + \psi) \end{array}\right] \tag{8.72}$$

These accelerations are now transformed from the inertial into the undercarriage system (which is identical to the CoG system for x- and y- directions). The required transformation matrix $\underline{T}_{RotZ}^{-1}$ (Equation 8.40) is reduced to the order 2, and multiplied into the above equation.

$$\left[\begin{array}{c} \ddot{x}_{CoG} \\ \ddot{y}_{CoG} \end{array}\right] = \left[\begin{array}{cc} \cos\psi & \sin\psi \\ -\sin\psi & \cos\psi \end{array}\right] \cdot \left[\begin{array}{c} \ddot{x}_{In} \\ \ddot{y}_{In} \end{array}\right] =$$
$$= v_{CoG} \cdot \left(\dot\beta + \dot\psi\right) \left[\begin{array}{c} -\sin\beta \\ \cos\beta \end{array}\right] + \dot{v}_{CoG} \left[\begin{array}{c} \cos\beta \\ \sin\beta \end{array}\right] \tag{8.73}$$

If gravitational forces $F_{GX}$ and $F_{GY}$, rolling resistance force $F_R$, lateral wind force $F_{windY}$, and the wind velocity $v_{wind}$ are neglected, the complete Equations 8.36

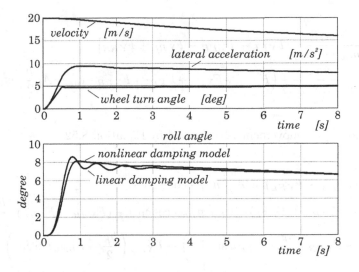

**Figure 8.29** Roll angle from linear and nonlinear suspension models

for horizontal translatory motion are then given in the CoG co-ordinate system by

$$
v_{CoG} \cdot \left( \dot{\beta} + \dot{\psi} \right) \begin{bmatrix} -\sin \beta \\ \cos \beta \end{bmatrix} + \dot{v}_{CoG} \begin{bmatrix} \cos \beta \\ \sin \beta \end{bmatrix} =
$$
$$
= \frac{1}{m_{CoG}} \begin{bmatrix} F_{XFL} + F_{XFR} + F_{XRL} + F_{XRR} + F_{windX} \\ F_{YFL} + F_{YFR} + F_{YRL} + F_{YRR} \end{bmatrix} . \quad (8.74)
$$

The two state equations are resolved for the derivatives of the vehicle speed $\dot{v}_{CoG}$ and the vehicle body side slip angle $\dot{\beta}$:

$$
\dot{v}_{CoG} =
$$
$$
\frac{1}{m_{CoG} \cdot \cos \beta} \left[ F_{XFL} + F_{XFR} + F_{XRL} + F_{XRR} - c_{aerX} A_L \frac{\rho}{2} \cdot v_{CoG}^2 \right] +
$$
$$
+ v_{CoG} \left( \dot{\beta} + \dot{\psi} \right) \tan \beta \qquad (8.75)
$$

$$
\dot{\beta} =
$$
$$
\frac{1}{m_{CoG} \cdot v_{CoG} \cdot \cos \beta} \left[ F_{YFL} + F_{YFR} + F_{YRL} + F_{YRR} - m_{CoG} \dot{v}_{CoG} \sin \beta \right] - \dot{\psi}
$$
$$
(8.76)
$$

In a last step, the mutual interdependence on $\dot{v}_{CoG}$ and $\dot{\beta}$ is eliminated.

(I) $\quad \dot{v}_{CoG} =$
$$
\frac{\cos \beta}{m_{CoG}} \left[ F_{XFL} + F_{XFR} + F_{XRL} + F_{XRR} - c_{aerX} A_L \frac{\rho}{2} \cdot v_{CoG}^2 \right] +
$$
$$
+ \frac{1}{m_{CoG}} \left( F_{YFL} + F_{YFR} + F_{YRL} + F_{YRR} \right) \sin \beta \qquad (8.77)
$$

(II)  $\dot{\beta} =$

$$\frac{\cos\beta}{m_{CoG} \cdot v_{CoG}} [F_{YFL} + F_{YFR} + F_{YRL} + F_{YRR}] -$$

$$-\frac{\sin\beta}{m_{CoG} \cdot v_{CoG}} \left( F_{XFL} + F_{XFR} + F_{XRL} + F_{XRR} - c_{aerX} A_L \frac{\rho}{2} \cdot v_{CoG}^2 \right) - \dot{\psi}$$

(8.78)

The rotational yaw movement is described by Equation 8.52.

(III)  $J_Z \ddot{\psi} =$

$$(F_{YFR} + F_{YFL}) \cdot (l_F - n_{LF} \cdot \cos\delta_W) -$$

$$- (F_{YRR} + F_{YRL}) \cdot (l_R + n_{LR} \cdot \cos\delta_W) + (F_{XRR} - F_{XRL}) \cdot \frac{b_R}{2} +$$

$$+ F_{XFR} \left( \frac{b_F}{2} + n_{LFR} \sin\delta_W \right) - F_{XFL} \left( \frac{b_F}{2} - n_{LFL} \sin\delta_W \right) \quad (8.79)$$

The Equations (I), (II) and (III) represent the nonlinear two-track model. The wheel forces are taken from Equations 8.30 and 8.31.

$$
\begin{aligned}
F_{XFL} &= F_{LFL} \cos\delta_{WL} - F_{SFL} \sin\delta_{WL} \quad , \\
F_{YFL} &= F_{SFL} \cos\delta_{WL} + F_{LFL} \sin\delta_{WL} \quad , \\
F_{XFR} &= F_{LFR} \cos\delta_{WR} - F_{SFR} \sin\delta_{WR} \quad , \\
F_{YFR} &= F_{SFR} \cos\delta_{WR} + F_{LFR} \sin\delta_{WR} \quad , \\
F_{XRL} &= F_{LRL} \quad , \quad F_{YRL} = F_{SRL} \quad , \\
F_{XRR} &= F_{LRR} \quad , \quad F_{YRR} = F_{SRR} \quad .
\end{aligned}
$$
(8.80)

These forces are inserted into the three nonlinear model equations. The wheel side forces $F_S$ are now approximated to be proportional to the tire side slip angles $\alpha$.

$$F_{SFL} = c_{FL} \cdot \alpha_{FL} = c_{FL} \cdot \left( \delta_{WL} - \beta - \frac{l_F \dot{\psi}}{v_{CoG}} \right)$$

$$F_{SFR} = c_{FR} \cdot \alpha_{FR} = c_{FR} \cdot \left( \delta_{WR} - \beta - \frac{l_F \dot{\psi}}{v_{CoG}} \right)$$

$$F_{SRL} = c_{RL} \cdot \alpha_{RL} = c_{RL} \cdot \left( -\beta + \frac{l_R \dot{\psi}}{v_{CoG}} \right)$$

$$F_{SRR} = c_{RR} \cdot \alpha_{RR} = c_{RR} \cdot \left( -\beta + \frac{l_R \dot{\psi}}{v_{CoG}} \right)$$
(8.81)

$c_{ij}$ are the tire side slip constants. They must be adapted (Section 9.5.2). It is assumed that the left and right wheel turn angles are the same, i.e. $\delta_{WL} \approx \delta_{WR} \approx \delta_W$.

The wheel turn angle and the longitudinal wheel forces $F_{Lij}$ are utilized as control inputs for vehicle dynamic control by steering and by applying an appropriate brake pressure. The reduced nonlinear two-track model becomes

$$\text{(I)} \quad f_1 \;=\; \dot{v}_{CoG} =$$

$$= \frac{1}{m_{CoG}} \cdot \Big\{ (F_{LFL} + F_{LFR}) \cdot \cos(\delta_W - \beta)$$

$$+ \Big( F_{LRL} + F_{LRR} - c_{aer} A_L \frac{\rho}{2} \cdot v_{CoG}^2 \Big) \cdot \cos \beta$$

$$- (c_{FL} + c_{FR}) \cdot \Big( \delta_W - \beta - \frac{l_F \cdot \dot{\psi}}{v_{CoG}} \Big) \cdot \sin(\delta_W - \beta)$$

$$+ (c_{RL} + c_{RR}) \cdot \Big( -\beta + \frac{l_R \cdot \dot{\psi}}{v_{CoG}} \Big) \cdot \sin \beta \Big\} \tag{8.82}$$

$$\text{(II)} \quad f_2 \;=\; \dot{\beta} =$$

$$= \frac{1}{m_{CoG} v_{CoG}} \cdot \Big\{ (c_{FL} + c_{FR}) \cdot \Big( \delta_W - \beta - \frac{l_F \cdot \dot{\psi}}{v_{CoG}} \Big) \cdot \cos(\delta_W - \beta)$$

$$+ (F_{LFL} + F_{LFR}) \cdot \sin(\delta_W - \beta)$$

$$- \Big( F_{LRL} + F_{LRR} - c_{aer} A_L \frac{\rho}{2} \cdot v_{CoG}^2 \Big) \cdot \sin \beta$$

$$+ (c_{RL} + c_{RR}) \cdot \Big( -\beta + \frac{l_R \cdot \dot{\psi}}{v_{CoG}} \Big) \cdot \cos \beta \Big\} - \dot{\psi} \tag{8.83}$$

$$\text{(III)} \quad f_3 \;=\; \ddot{\psi} =$$

$$= \frac{1}{J_Z} \cdot \Big\{ (l_F - n_{LF} \cos \delta_W) \cdot (F_{LFL} + F_{LFR}) \cdot \sin \delta_W$$

$$+ (l_F - n_{LF} \cos \delta_W)(c_{FL} + c_{FR}) \Big( \delta_W - \beta - \frac{l_F \dot{\psi}}{v_{CoG}} \Big) \cos \delta_W$$

$$+ \frac{b_F}{2} \cdot (F_{LFR} - F_{LFL}) \cos \delta_W$$

$$- \frac{b_F}{2} \cdot (c_{FR} - c_{FL}) \cdot \Big( \delta_W - \beta - \frac{l_F \cdot \dot{\psi}}{v_{CoG}} \Big) \cdot \sin \delta_W$$

$$- (l_R + n_{LR}) \cdot (c_{RL} + c_{RR}) \cdot \Big( -\beta + \frac{l_R \dot{\psi}}{v_{CoG}} \Big)$$

$$+ \frac{b_R}{2} (F_{LRR} - F_{LRL}) \Big\} \tag{8.84}$$

In state space form, the reduced nonlinear two-track model can be written as:

$$\dot{\underline{x}} \;=\; \underline{A}(\underline{x}, \underline{u})\,\underline{x} + \underline{B}(\underline{x}, \underline{u})\,\underline{u}$$
$$\underline{y} \;=\; \underline{C}(\underline{x}, \underline{u})\,\underline{x}$$

The state vector is:

$$\underline{x} = \begin{bmatrix} v_{CoG}, & \beta, & \dot{\psi} \end{bmatrix}^T \tag{8.85}$$

the control input;

$$\underline{u} = [F_{LFL}, \quad F_{LFR}, \quad F_{LRL}, \quad F_{LRR}, \quad \delta_W]^T \qquad (8.86)$$

and the measurement vector:

$$\underline{y} = \left[v_{CoG}, \quad \dot{\psi}\right]^T \qquad (8.87)$$

The equations are:

$$\underline{\dot{x}} = \begin{bmatrix} \dot{v}_{CoG} \\ \dot{\beta} \\ \ddot{\psi} \end{bmatrix} = \underline{f}(\underline{x}, \underline{u}) = \begin{bmatrix} f_1(\underline{x}, \underline{u}) \\ f_2(\underline{x}, \underline{u}) \\ f_3(\underline{x}, \underline{u}) \end{bmatrix} \qquad (8.88)$$

$$\underline{y} = \underline{C}(\underline{x}, \underline{u})\,\underline{x} = \begin{bmatrix} 1 & 0 & 0 \\ 0 & 0 & 1 \end{bmatrix} \underline{x} \qquad (8.89)$$

The nonlinear model can be linearized around the actual operating point.

$$\underline{f}(\underline{x}, \underline{u}) \approx \underline{f}(\underline{x}_0, \underline{u}_0) + \underbrace{\frac{\partial \underline{f}(\underline{x}, \underline{u})}{\partial \underline{x}}\bigg|_{\substack{\underline{x} = \underline{x}_0 \\ \underline{u} = \underline{u}_0}} \cdot (\underline{x} - \underline{x}_0)}_{\text{Jacobian}} + \underbrace{\frac{\partial \underline{f}(\underline{x}, \underline{u})}{\partial \underline{u}}\bigg|_{\substack{\underline{x} = \underline{x}_0 \\ \underline{u} = \underline{u}_0}} \cdot (\underline{u} - \underline{u}_0)}_{\text{Jacobian}}$$

$$(8.90)$$

The two Jacobians can be found in Appendix A.1.

### 8.4.5   Vehicle Stability Analysis

Analytical stability criteria are not available for nonlinear systems. Therefore, the nonlinear two-track model is further reduced to a linear single-track model of second order.

#### Reduced Linear Single-track Model

No differences are made between the left and right track.

$$\begin{aligned} F_{LF} &= \tfrac{1}{2}\left(F_{LFL} + F_{LFR}\right) & F_{LR} &= \tfrac{1}{2}\left(F_{LRL} + F_{LRR}\right) \\ F_{SF} &= \tfrac{1}{2}\left(F_{SFL} + F_{SFR}\right) & F_{SR} &= \tfrac{1}{2}\left(F_{SRL} + F_{SRR}\right) \end{aligned} \qquad (8.91)$$

$$c_F = \tfrac{1}{2}\left(c_{FL} + c_{FR}\right) \quad c_R = \tfrac{1}{2}\left(c_{RL} + c_{RR}\right) \qquad (8.92)$$

For small vehicle body side slip angles, the trigonometric functions can be linearized using $\sin\beta \approx 0$ and $\cos\beta \approx 1$.

In addition, it is assumed that the vehicle body speed $v_{CoG}$ is constant over a limited time period, i.e. the derivative $\dot{v}_{CoG} = 0$. The wind force $F_{windX}$ is also neglected, as well as force terms $F \cdot \sin \delta_W$. Equation 8.74 for the translatory motion then simplifies to

$$
\begin{aligned}
m_{CoG} \cdot v_{CoG} \cdot \left( \dot{\beta} + \dot{\psi} \right) &= F_{SF} + F_{SR} \\
&= c_F \left( \delta_W - \beta - \frac{l_F \dot{\psi}}{v_{CoG}} \right) + c_R \left( -\beta + \frac{l_R \dot{\psi}}{v_{CoG}} \right)
\end{aligned}
\tag{8.93}
$$

and the rotational yaw movement to

$$
\begin{aligned}
J_Z \ddot{\psi} &= F_{SF} l_F - \underbrace{F_{SF} n_{LF}}_{\approx 0} - F_{SR} l_R - \underbrace{F_{SR} n_{LR}}_{\approx 0} + \\
& \quad \underbrace{(F_{LRR} - F_{LRL}) \frac{b_R}{2} + (F_{LFR} - F_{LFL}) \frac{b_F}{2}}_{\approx 0} \\
&= c_F l_F \left( \delta_W - \beta - \frac{l_F \dot{\psi}}{v_{CoG}} \right) - c_R l_R \left( -\beta + \frac{l_R \dot{\psi}}{v_{CoG}} \right) .
\end{aligned}
\tag{8.94}
$$

The state variables are the vehicle body side slip angle $\beta$ and the yaw rate $\dot{\psi}$. The vehicle speed $v_{CoG}$ is considered as a parameter. As control input, only the wheel turn angle $\delta_W$ remains and as measurement variable the yaw rate $\dot{\psi}$. Thus the linear single-track model is not suited for vehicle dynamic control. The state equations are:

$$
\begin{bmatrix} \dot{\beta} \\ \ddot{\psi} \end{bmatrix} =
\begin{bmatrix}
-\dfrac{c_F + c_R}{m_{CoG} v_{CoG}} & \dfrac{c_R l_R - c_F l_F}{m_{CoG} v_{CoG}^2} - 1 \\[3mm]
\dfrac{c_R l_R - c_F l_F}{J_Z} & -\dfrac{c_R l_R^2 + c_F l_F^2}{J_Z v_{CoG}}
\end{bmatrix}
\cdot
\begin{bmatrix} \beta \\ \dot{\psi} \end{bmatrix}
+
\begin{bmatrix}
\dfrac{c_F}{m_{CoG} v_{CoG}} \\[3mm]
\dfrac{c_F l_F}{J_Z}
\end{bmatrix}
\delta_W
$$

$$
\dot{\psi} = \begin{bmatrix} 0 & 1 \end{bmatrix} \cdot \begin{bmatrix} \beta \\ \dot{\psi} \end{bmatrix}
$$

According to [84], this linear single-track model is only valid for lateral accelerations below $0.4\,g$. By adaptation of the side force constants $c_F$ and $c_R$ (see Section 9.5.2), the validity range may be extended. For stationary operation, $\ddot{\psi} = 0$, $\dot{\beta} = 0$, $\dot{v}_{CoG} = 0$, the vehicle moves on a circular path at constant speed.

## Position of Eigenvalues During Test Drives

The transfer function between yaw rate $\dot{\psi}$ and wheel turn angle $\delta_W$ is

$$
\dot{\psi} = \frac{1}{l} \cdot \frac{v_{CoG}}{1 + v_{CoG}^2 / v_{char}^2} \cdot \delta_W
\tag{8.95}
$$

with characteristic speed

$$
v_{char}^2 = \frac{c_F c_R l^2}{m_{CoG} (c_R l_R - c_F l_F)}
\tag{8.96}
$$

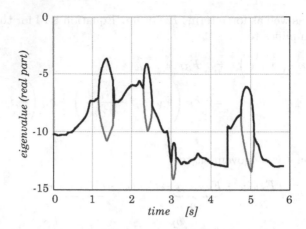

**Figure 8.30**  Real part of the eigenvalues during evasive action manoeuvre

and vehicle length $l = l_F + l_R$. For production cars, the characteristic speed
varies between 68 and $112\,km/h$. A small characteristic speed means that the
car is understeering. The characteristic equation of the linear single-track model
is:

$$s^2 + \frac{\left(J_Z + m_{CoG}l_F^2\right)c_F + \left(J_Z + m_{CoG}l_R^2\right)c_R}{J_Z m_{CoG} v_{CoG}}s +$$

$$+ \frac{c_F c_R l^2 + m_{CoG}v_{CoG}^2\left(c_R l_R - c_F l_F\right)}{J_Z m_{CoG} v_{CoG}^2} = 0 \tag{8.97}$$

Asymptotic stability requires both terms to be positive. The term associated
with s always meets this requirement. The absolute term is only positive if

$$c_R l_R > c_F l_F \quad , \tag{8.98}$$

which is true for all production vehicles.

In Figs 8.30 and 8.31 the position of the real part of the time variant eigenvalues is shown for two driving manoeuvres. Figure 8.30 belongs to an evasive
action manoeuvre, and Figure 8.31 to an ABS braking in a curve.

For the evasive action manoeuvre, both eigenvalues show stable behavior,
with a satisfactory distance from the instability limit. When the two real parts
coincide, there exists a complex- conjugate pair of eigenvalues.

During ABS braking in a curve, stability can become an issue (Figure 8.31).
During ABS braking on stretches with low maximum friction values, the steering
response of the vehicle is poor.

## 8.5   Validation of the Vehicle Model

Once the model is constructed it must be verified with as much information from
the real system as possible. This process is known as *validation*. The necessary

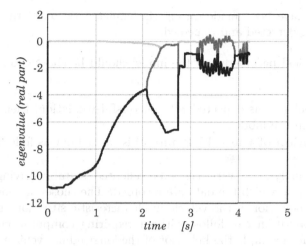

**Figure 8.31** Real part of the eigenvalues during braking manoeuvre in a curve

information consists of *a-priori* knowledge, measured data and experience of the user with the model [76] [55].

The most common method of model validation is to evaluate the reaction of the model to measured data and compare it with actual values. The data used for this should be different from that used for training of the model.

The construction of a model often involves many simplifications through which the outputs of the model deviate to a greater or lesser extent from the real values. By model construction the following question arises: Where are such simplifications allowable in the model, such that the model does not deviate too much from the actual system.

## 8.5.1 Validation Procedure

From [76] the following questions arise with the model validation:

1. Do the model outputs correspond well enough to the measured data?

2. Is the model suitable for the purpose for which it was constructed?

The more data are available from the real system, the better the above questions can be answered. A model can in general be considered validated when, following evaluation with suitable validation data, it satisfies the requirements for which it was constructed. Before the validation, one must clearly know which purpose the model is to be put to, which outputs must be precisely modelled, and where certain errors can be accepted.

The following can be stated as tasks of the validation:

- If simplifications have been carried out during the modelling process which are too large, or not suitable, these must be found out and suitable corrections made.

- If faults have been made during the implementation of the model, these should be detected and remedied.

- The physical parameters of the model should be checked and if necessary adjusted.

If the model behaves as expected with successful validation, then a lot of faith can be placed in the model.

The **validation of the vehicle model** is carried out as follows:

According to the dynamic variables to be validated, suitable driving manoeuvres are defined, during which the data are recorded. The chosen manoeuvres should reproduce the behavior of the vehicle in characteristic situations in such a way that an interpretation can follow without requiring computationally intensive processing. If for example the behavior of the longitudinal vehicle dynamics are to be validated, then,

- Straight ahead braking and accelerating driving

would be suitable, as this would stimulate pitch oscillations, whilst at the same time coupled roll and yaw movements would scarcely appear. If on the other hand the lateral dynamics vehicle variables are to be verified, the following driving manoeuvre would especially stimulate the corresponding dynamics:

- Step changes in the steering angle

- Sine-wave form steering input

The recorded measured variables are compared with the corresponding outputs from the model in order to determine where the model is sufficiently accurate and where tuning of the model structure or its parameters is necessary. Via repeated tuning and comparison of measured and simulated data, one arrives at a version of the model which reproduces the desired drive dynamics with sufficient accuracy.

In order to carry out validation, test drives were carried out with an experimental vehicle, and the following variables recorded:

| Model input variables | | |
|---|---|---|
| | $\delta_S$ | steering wheel angle |
| | $\omega_{ij}$ | wheel angular velocities |
| | | |
| Model output variables | $a_x$ | acceleration in x-direction (at CoG) |
| | $a_y$ | acceleration in y-direction (at CoG) |
| | $v_{CoG}$ | velocity of the CoG |
| | $\dot{\psi}$ | yaw rate |
| | $\dot{\chi}$ | pitch rate |
| | $\dot{\varphi}$ | roll rate |

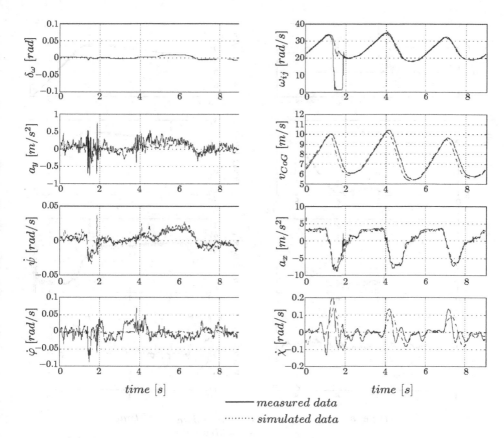

**Figure 8.32** Model validation: straight ahead braking and acceleration

The data recording is carried out using a real driver to provide the steering angle and the desired acceleration, hence the driver model, and the motor/drive/brake model are not applicable here. The outputs produced by driver models, $\delta_S$ and $\omega_{ij}$, are replaced by the measured values. These then act as the inputs for the validation model. Also, the road model is replaced by actual road data for the validation.

The validation model calculates, from the inputs $\delta_S$ and $\omega_{ij}$, among other things the outputs $a_x$, $a_y$, $a_z$, $v_{CoG}$, $\dot{\psi}$, $\dot{\chi}$ and $\dot{\varphi}$, which according to the above list are also available as measured outputs, so that a direct comparison can be carried out between the model and actual outputs.

## 8.5.2 Validation Results

From the multitude of driving situations used to validate the model, the above mentioned three driving manoeuvres will be used as examples in each case.

Figure 8.32 shows the results of a straight ahead braking and acceleration manoeuvre, in order to analyze the estimation quality of the longitudinal variables.

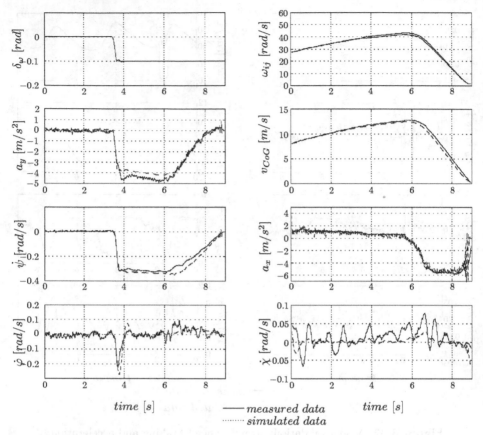

**Figure 8.33**  Model validation: step change in steering angle

The simulated longitudinal dynamics correspond very well to the measured data. Even with the scarcely stimulated lateral dynamics the simulation results show only small errors for the roll motion. The short-term locking of the front wheels at $1.8 - 2\,s$ leads to small disturbances in this range, however these do not negatively influence the simulation.

Similarly good results are given for the test drive with step changes in steering angle, as shown in Figure 8.33. The change in steering angle with a velocity of $50\,km/h$ amounts to a steering wheel angle of almost $100°$, which corresponds to wheel turn angles of $0.1\,rad$. Again the calculated values follow the measured data very well. The roll rate is also well reproduced. The pitch rate, which according to the measured data is to some extent dynamic, is simulated in the model as semi-constant at zero. This can result in differences if the measured data are affected by coupling, which has been disregarded in the model.

As the last example, the sine-wave form driving is considered (Figure 8.34). For this test drive the same can be said as with the step change in steering angle: the simulated values track the measured variables well. Whilst the yaw rate is almost exactly modeled, the modeled roll rate shows too large a peak value

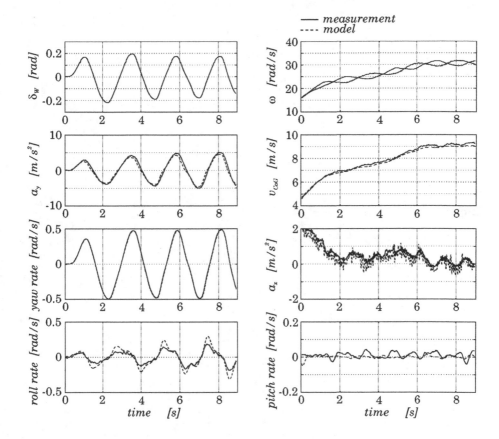

**Figure 8.34** Model validation: sine wave form steering

compared to the actual data. This can be an indication that the moment of inertia $J_x$ around the roll axis has been set too small.

In summary one can say that **with sufficient excitation of the respective dynamics, the longitudinal and lateral dynamics were very well reproduced.** Problems appeared in the simulation of the roll rate during straight ahead driving and the pitch rate during a step change in steering angle and sinewave form steering, as these dynamic variables are scarcely excited. The yaw rate, which is required for many dynamic control operations, was modeled very well for all drive situations investigated.

The simulation model has proved itself suitable for calculation of the relevant drive dynamics variables given steering angle and angular wheel velocity. The same can also be said about the vehicle body side slip angle $\beta$ and the wheel forces; because the measurements were so well simulated, one can also assume that the non-measurable variables will give good results. Hence the wheel forces calculated by the model are used for the identification of the mass moments of inertia (see Section 9.4.2).

# 9 Vehicle Parameters and States

This Chapter describes various approaches for the estimation and observation of variables which are not directly measurable. Section 9.1 presents two methods of obtaining the vehicle velocity in the inertial co-ordinate system, a Kalman filter approach and a fuzzy estimator. In Section 9.2 these methods are also employed for the estimation of the yaw rate. In Section 9.4, various approaches for estimating the friction characteristics, and the mass moments of inertia. In Section 9.5, approximation formulas are given for the wheel ground contact forces. The tire side slip constants are adapted with a simple nonlinear approximation equation. Based on the wheel ground contact forces, the roll and pitch angles are approximated. In Section 9.6 the vehicle body side slip angle is estimated using a nonlinear observer. Section 9.7 presents two methods for road gradient estimation.

## 9.1   Vehicle Velocity Estimation

The vehicle velocity $v_{CoG}$ is obtained via a fusion of the data from all rotational wheel velocities $v_{Rij}$ and the longitudinal acceleration sensor. Via integration of the acceleration a fifth estimate for the vehicle velocity is made available. The estimation must be very accurate, as a basis for the wheel slip calculation (see Section 8.3.2). Some systems only select the maximum rotational wheel speed as the estimate for the vehicle velocity. When all four wheels happen to lock simultaneously, this approach is very inaccurate.

Two alternative estimation methods for the vehicle velocity are regarded, the Kalman filter and the fuzzy estimator.

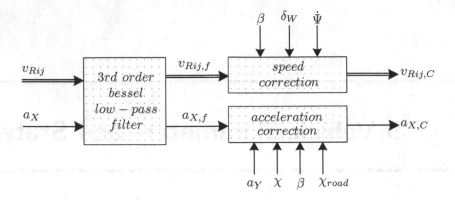

**Figure 9.1** Structure of the sensor data preprocessing

## 9.1.1 Sensor Data Preprocessing

All sensors contain systematic errors which must be corrected (Figure 9.1). The signals are first low-pass filtered. During normal driving conditions, the offset of the longitudinal acceleration $a_x$ sensor signal is compensated by the rotational wheel speed information. Wheel speed differences are due to driving in curves. Wheel speed measurements are therefore corrected by transformation to the CoG. Only for extreme curves must the vehicle body side slip angle $\beta$ be known. Otherwise the yaw rate $\dot{\psi}$ is sufficient. The wheel ground contact point velocities $v_{Wij}$ are approximated by the measurable wheel rotational equivalent velocities $v_{Rij}$. The transformation of Equation 8.8 considers the different directions of the wheel and the vehicle velocities. The wheel speeds are inversely corrected from the wheel ground contact points to the center of gravity.

$$v_{RFL,C} = \left[ v_{RFL} + \dot{\psi} \left( \frac{b_F}{2} - l_F \beta \right) \right] \cdot \cos\left( \delta_W - \beta \right)$$

$$v_{RFR,C} = \left[ v_{RFR} - \dot{\psi} \left( \frac{b_F}{2} + l_F \beta \right) \right] \cdot \cos\left( \delta_W - \beta \right)$$

$$v_{RRL,C} = \left[ v_{RRL} + \dot{\psi} \left( \frac{b_R}{2} + l_R \beta \right) \right] \cdot \cos \beta$$

$$v_{RRR,C} = \left[ v_{RRR} - \dot{\psi} \left( \frac{b_R}{2} - l_R \beta \right) \right] \cdot \cos \beta$$

$$\tag{9.1}$$

Figure 9.2 shows the individual wheel velocities of a slalom drive before and after the transformation. All four corrected rotational equivalent wheel speeds $v_{Rij,C}$ are now effective estimates for the vehicle velocity $\hat{v}_{CoG}$. In order to obtain the desired accuracy an equalization of the radii of the wheels must be carried out. For this, scaling factors for the radii are determined during driving with small steering angles and low accelerations.

The signals from the acceleration sensor are corrupted by several systematic errors, which all arise from the fixed installation on the vehicle body. The sensor is

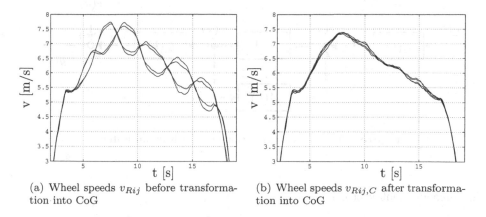

(a) Wheel speeds $v_{Rij}$ before transformation into CoG

(b) Wheel speeds $v_{Rij,C}$ after transformation into CoG

**Figure 9.2** Wheel speeds during a sinusodial drive

adjusted in the direction of the vehicle longitudinal axis, whilst the acceleration affects the vehicle in the direction of travel. The appearing angular offset is however sufficiently small, so that the cosine of it can be set to 1. If the vehicle drives on an incline, the sensor direction is correct with respect to the vehicle body. Because of the tilted position of the sensor in relation to the inertial co-ordinate system however, a component of the gravitational acceleration $g$ is measured, which must be corrected by $(\chi - \chi_{road})$. Pitch angle changes result in a cosine component of the angle offset which can be neglected. Another error source is caused by the vehicle body side slip angle $\beta$. The centripetal force which acts orthogonal to the vehicle velocity yields a lateral acceleration component in the sensor signal. Altogether, the preprocessing of the acceleration sensor is:

$$a_{X,C} = \hat{\ddot{x}} = a_X - g\sin(\chi - \chi_{road}) + a_Y \sin\beta \tag{9.2}$$

## 9.1.2 Kalman Filter Approach

The Kalman filter can determine state variables in a similar way as the Luenberger observer. The difference is that the Kalman Filter is designed for stochastic or time varying processes. An important pre-requisite is that the system input and the measurement noise must be white. The Kalman filter has advantages especially in cases where the stochastic properties of the noise processes are known.

If a Kalman filter is used for the data fusion then it is necessary to formulate the system model in discrete state-space form (see Section 6.8.4) . The state vector (longitudinal acceleration $a_X$, vehicle speed $v_{CoG}$) is:

$$\begin{aligned}
\underline{x}(k+1) &= \underline{A}(k)\underline{x}(k) + \underline{B}(k)\underline{u}(k) \\
&= \begin{bmatrix} 1 & 0 \\ T_s & 1 \end{bmatrix} \begin{bmatrix} a_X(k) \\ v_{CoG}(k) \end{bmatrix} + \begin{bmatrix} 1 & 0 \\ 0 & 1 \end{bmatrix} \begin{bmatrix} u_1(k) \\ u_2(k) \end{bmatrix}
\end{aligned} \tag{9.3}$$

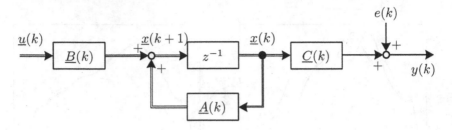

**Figure 9.3** Velocity estimation using a Kalman Filter

The measurement vector is:

$$
\underline{y}(k) = \underline{C}(k)\underline{x}(k) + \underline{e}(k) =
\begin{bmatrix}
a_{X,C}(k) \\
v_{RFL,C}(k) \\
v_{RFR,C}(k) \\
v_{RRL,C}(k) \\
v_{RRR,C}(k)
\end{bmatrix}
$$

$$
=
\begin{bmatrix}
1 & 0 \\
0 & 1 \\
0 & 1 \\
0 & 1 \\
0 & 1
\end{bmatrix}
\begin{bmatrix}
a_X(k) \\
v_{CoG}(k)
\end{bmatrix}
+
\begin{bmatrix}
e_a(k) \\
e_1(k) \\
e_2(k) \\
e_3(k) \\
e_4(k)
\end{bmatrix}
\tag{9.4}
$$

The input noise $\underline{u}$ serves as an excitation to the system. $\underline{e}$ is the vector of measurement noise. Figure 9.3 shows the structure of the Kalman Filter.

The Kalman filter approach assumes the measurement noise to be white. The wheel ground contact point velocities $v_{Wij}$ were approximated by the rotational equivalent wheel speeds $v_{Rij}$; these signals contain however systematic, slip-dependant offsets (Section 8.3.2), which contradicts the assumption of the measurement noise to be white. For the corrected acceleration sensor signals, one can assume a white noise error with constant power density. This is achieved by the preprocessing in Section 9.1.1.

In order to apply a Kalman Filter in spite of systematic errors, four different covariance matrices could be defined. The proper matrix is then selected according to the current driving situation.

- **Large positive acceleration:** the covariance of the non-driven wheel noise is small and that of the driven wheel and acceleration noise is large.

- **Small positive acceleration:** the covariance of all wheels is small and that of the acceleration sensor is large.

- **Small negative acceleration:** the covariance of all wheels and of the acceleration is small.

- **Large negative acceleration:** the covariance of all wheels is large and that of the acceleration sensor is small.

With such an approach, the Kalman filter produces good results. It requires however intensive calculations for the gain matrix. The idea is now to replace the Kalman Filter by a fuzzy logic estimator which also classifies the driving situations in order to determine suitable weighting factors for the sensor signals. The available heuristic knowledge about the vehicle behavior can be included in the fuzzy estimator by the formulation of the rules.

### 9.1.3 Short Introduction to Fuzzy Logic

A fuzzy system [1] can be divided into three parts, as shown in Figure 9.4.

Crisp, continuous inputs are transformed into linguistic variables with membership grades between 0 and 1. This process is known as fuzzification. The linguistic inputs are then evaluated using fuzzy rules and formed into fuzzy outputs. Continuous crisp outputs are then obtained via the process of defuzzification.

A MIN-MAX inference scheme is chosen in this book; AND-operations are carried out using the minimum operator and OR-operations using the maximum operator. The defuzzification is carried out using the centroid method:

$$y_{Def} = \frac{\int\limits_{-\infty}^{+\infty} y \cdot \mu_{res}(y) \cdot dy}{\int\limits_{-\infty}^{+\infty} \mu_{res}(y) \cdot dy} \qquad (9.5)$$

The rule base contains only premises which are combined using the AND-operator. The individual processing levels will now be further explained by an example.

**Example**
The rule base shall consists of two rules:
IF    $T=low$      AND    $P=large$    THEN    $y=middle$
IF    $T=middle$   AND    $P=large$    THEN    $y=small$

The two crisp inputs, $T_o$ and $P_o$ are first fuzzified. One then determines the truth value of the premises $w_1$ ($T = \ldots$) and $w_2$ ($P = \ldots$), from which the activations $\mu_{B1}$ and $\mu_{B2}$ of the rules can be determined, i.e. the membership function of the conclusion can be determined. If more rules exist then the resulting membership function $\mu_{RES}$ of the conclusion is formed from the aggregation

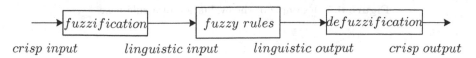

crisp input      linguistic input      linguistic output      crisp output

**Figure 9.4** Processing steps of a fuzzy system

---

[1]Here, only a brief introduction to Fuzzy logic is given. For a more detailed explanation see [132]

of the outputs of all rules. From the area under the final resulting membership function, the crisp output value is determined using Equation 9.5.
The complete process is shown schematically in Figure 9.5.

*evaluation of the first rule*

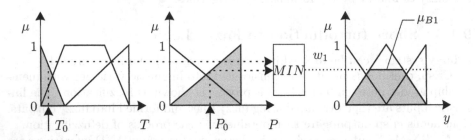

*evaluation of the second rule*

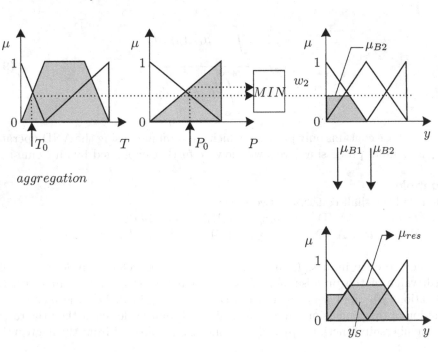

*aggregation*

**Figure 9.5** Example of the evaluation of a rule base

## 9.1.4  Fuzzy Estimator

In this section, the data fusion of the four wheel speed signals and the longitudinal acceleration sensor signal is implemented with a fuzzy estimator. The rule base of the fuzzy estimator contains the heuristic knowledge about the individual sensor

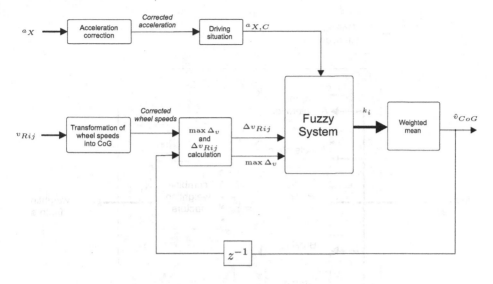

**Figure 9.6** Structure of the fuzzy estimator with data preprocessing

signal errors during different driving situations. Based on the sensor errors, weighting factors are generated by the fuzzy estimator (see Figure 9.6). The four weighting factors $k_1...k_4$ for the wheel speed sensor signals and that for the acceleration signal $k_5$ are employed to determine the estimation value for the center of gravity velocity

$$\hat{v}_{CoG}(k) = \frac{\sum\limits_{i=1}^{4} k_i \cdot v_{Ri,C}(k) + k_5 \left[\hat{v}_{CoG}(k-1) + T_S \cdot a_{X,C}(k)\right]}{\sum\limits_{i=1}^{5} k_i} \quad . \tag{9.6}$$

Equation 9.6 is a weighted average of all sensor signals. The goal of the fuzzy estimator is to determine the weighting factors $k_i$ in Equation 9.6 appropriately.

## Sub-Models

In order to reduce the number of active rules in the rule base of the fuzzy estimator, the fuzzy system is partitioned into five sub-models (see Figure 9.7). The corrected longitudinal acceleration signal $a_{X,C}$ is taken to distinguish between the five driving conditions "strong Acceleration", "Acceleration", "Rolling", "Braking" and "strong Braking". Each of these sub-systems contains a reduced rule base suited for the respective driving situation. The input signals of all five sub-systems are identical and will be presented next.

## Input Signals

Apart from the corrected longitudinal acceleration $a_{X,C}$ only signals containing information about the sensor signal reliability are suitable as inputs into the fuzzy

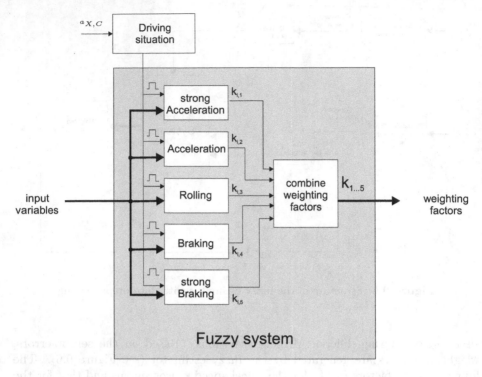

**Figure 9.7** Partitioned fuzzy estimator

estimator.

In addition to the above five signals, the difference between the corrected wheel speeds $v_{Rij,C}$ and the last estimated vehicle velocity value $\hat{v}_{CoG}$ is considered:

$$\Delta v_{Rij} = v_{Rij,C}(k) - \hat{v}_{CoG}(k-1) \tag{9.7}$$

Large deviations $\Delta v_R$ of a specific wheel speed indicate slip at this wheel. Under such conditions, the wheel speed signal of the respective wheel will be inaccurate. Therefore, the respective weighting factor is reduced by the fuzzy estimator. The difference $\Delta v_R$ is correlated to the slip. However, the absolute value of $\Delta v_R$ is usually larger than the slip and less sensitive to noise and errors. Therefore the maximum deviation $\max \Delta_v$ of the corrected wheel speeds $v_{Rij,C}$ is taken to assess the quality of a signal

$$\max \Delta_v = \left| \hat{v}_{CoG} - \max_{ij}\{v_{Rij,C}\} \right| + \left| \hat{v}_{CoG} - \min_{ij}\{v_{Rij,C}\} \right| \quad . \tag{9.8}$$

If $\max \Delta_v$ is very small, then two conditions are fulfilled: firstly, the measured velocity signal from the wheel speed sensors is close to the previously estimated value $\hat{v}_{CoG}$. If the estimated value $\hat{v}_{CoG}$ drifts away, this is detected by $\max \Delta_v$ increasing above a certain threshold, see Figure 9.9(a). Secondly, if both the maximum and the minimum wheel speed are very close to $\hat{v}_{CoG}$, the individual

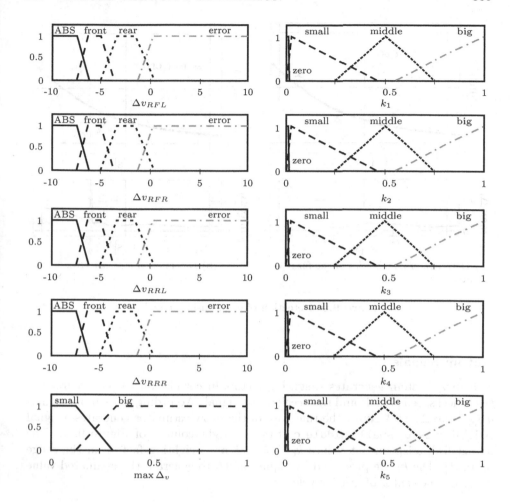

**Figure 9.8** Membership functions of sub-system "strong Braking"

wheel speeds do not deviate significantly from each other. Then, the wheel speed signals are considered reliable and the fuzzy estimator generates a high weight for them.

For values of max $\Delta_v$ around zero, the vehicle velocity can be determined by just averaging the four corrected wheel speeds $v_{Rij}$. In this case, the fuzzy estimator is not used at all.

If max $\Delta_v$ is "small" (see lower left corner of Figure 9.8) then the wheel speed signal deviations may no longer be neglected. In this case, the fuzzy estimator generates individual weighting factors for the wheel speeds and the acceleration signal.

"Big" values of max $\Delta_v$ indicate sensor errors or disturbances (see Figure 9.9(b)), for instance spinning wheels or ABS-braking. The fuzzy estimator then reduces the wheel speed signal weights accordingly.

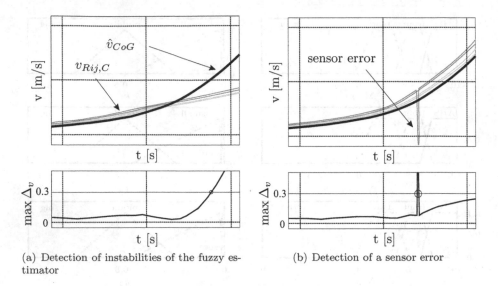

(a) Detection of instabilities of the fuzzy es-     (b) Detection of a sensor error
timator

**Figure 9.9**  Detection of errors using $\max \Delta_v$

## Output Signals

The fuzzy system generates weighting factors in correlation to the accuracy of
the wheel speed signals and the acceleration signal. According to its input signals
$a_{X,C}$, $\max \Delta_v$ and $\Delta v_{Rij}$ the rule base of the fuzzy estimator generates a signal
reliability **zero**, **small**, **middle** or **big** (see right column of Figure 9.8).   The
defuzzified, crisp output values $k_i$, $i = 1...5$ are weighting factors in the range
of $[0...1]$. The $k_i$ are processed by Equation 9.6 to generate the estimated value
$\hat{v}_{CoG}$ for the center of gravity velocity.

## Rule Base

All rules in the rule base contain the AND operator only.   The Mamdani-
implication ([59]) is employed and all the membership functions used in the system are trapezoid to reduce processing complexity.
Due to a sensor-specific drift, the acceleration sensor signal is unreliable. Therefore, its weight should be kept small whenever possible. The time periods, during
which the acceleration signal is integrated to yield the vehicle speed should be
as short as possible. To meet this constraint, the weight of at least one wheel
speed signal generated from the rule base is high. This will normally prevent the
estimated vehicle velocity from drifting away when solely using the integrated
acceleration. When the vehicle velocity still drifts away in some cases, $\max \Delta_v$
and $\Delta v_{Rij}$ are analysed to detect this effect (see Figure 9.9(a)).
For the sub-system "strong Braking" ($a_{X,C} < -3m/s^2$) the membership functions displayed in Figure 9.8, are presented in order to provide an idea of the rule
base structure.
Braking with a deceleration below $-3m/s^2$ causes large slip values on the wheels.
Therefore, a small weighting factor is assigned to the wheel speed signals here.

Generally, the braking force on the front wheels is higher than that on the rear wheels. This increases the probability of ABS cycles on the front axle. Accordingly, the front wheel speed signals are used only if the rear wheel speed signals are erroneous. In such situations, the acceleration sensor provides the best signal. The integrated acceleration signal is then weighted highest. The complete rule base of the subsystem "strong braking" can be seen in Table 9.1.

For the other four sub-systems, the rule base is composed of similar rules. However, these rules are adapted specifically to the respective driving situation.

**Table 9.1**  Example rule base for the Fuzzy-Subsystem "strong braking"

| $\Delta v_{FL}$ | $\Delta v_{FR}$ | $\Delta v_{RL}$ | $\Delta v_{RR}$ | max $\Delta_v$ | k FL | k FR | k RL | k RR | k v(a) |
|---|---|---|---|---|---|---|---|---|---|
| - | - | - | - | small | small | small | small | small | big |
| - | - | front | front | big | zero | zero | middle | middle | big |
| - | - | not front | front | big | zero | zero | zero | big | big |
| - | - | front | not front | big | zero | zero | big | zero | big |
| ABS | ABS | rear | rear | big | zero | zero | small | small | big |
| rear | - | not rear | not front | big | small | zero | zero | zero | big |
| - | rear | not rear | not front | big | zero | small | zero | zero | big |

## 9.1.5  Results of Vehicle Velocity Estimator

A test drive with hard braking is adopted to validate the vehicle velocity estimation.

Figure 9.10 shows how the corrected acceleration signal is used to distinguish between the different fuzzy sub-systems (see Figure 9.7). According to the driving condition, a specific reduced fuzzy estimator is selected with rules fitting to the respective situation.

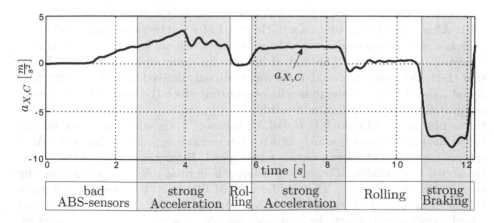

**Figure 9.10**  Corrected longitudinal acceleration $a_{x,C}$ categorizing the driving situation

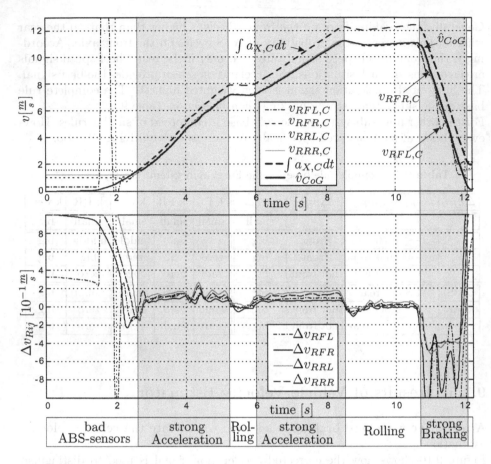

**Figure 9.11**  Velocities (top) and $\Delta v_{Rij}$ (bottom) during an ABS test drive

Figure 9.11 shows a test drive belonging to the acceleration signal in Figure 9.10. After accelerating up to a vehicle speed of $v = 11m/s$, a panic braking to vehicle standstill was conducted. At the bottom of Figure 9.11, the actual driving conditions can be seen. The dashed line for the integrated acceleration shows, that the vehicle velocity would drift away when only derived from the acceleration signal $a_{X,C}$. Therefore, the time windows during which the acceleration signal is integrated are kept as short as possible.

In the first phase of Figure 9.11 ("bad ABS-sensors"), the wheel speed sensors are below their activation threshold. Due to their measuring principle, inductive ABS sensors are activated above a certain wheel speed. Below the activation threshold, the signal is unreliable. The vehicle velocity is derived solely by integrating the acceleration signal $a_{X,C}$. The middle part of Figure 9.11 shows the difference velocities $\Delta v_{Rij}$. In the second phase ("strong Acceleration"), the corrected wheel speeds are all above the estimated vehicle velocity $\hat{v}_{CoG}$ due to drive slip. Between $t = 5.2s$ and $t = 5.6s$, the vehicle is in "Rolling" condition. Almost no slip occurs and the velocity differences are close to zero. In the last phase ("strong

Braking"), the velocity differences $\Delta v_{Rij}$ are significantly below 0 due to a large braking slip. ABS cycles at the front wheels cause velocity drops of $v_{RFL,C}$ and $v_{RFR,C}$. The front wheel speed signals are rated as "ABS" or "erroneous". Thus, the weights for these signals are zero. At the very end of the measurement, the wheel speed signals fall below the activation threshold of the ABS sensors. The velocity is again determined only by integration of $a_{X,C}$.

Figure 9.12 zooms into the start of the strong braking phase of the test drive described above. At the beginning, after approximately $t = 10.6s$, the fuzzy estimator detects large deviations $\Delta v_{Rij}$ of the front wheel speeds and rates the front wheel speed sensor signals as erroneous. The estimated vehicle velocity $\hat{v}_{CoG}$ is then approximately equal to the velocity of the rear wheels. After approximately $t = 10.75s$, the driving condition changes from "Rolling" to "Braking/strong Braking". Now, all the wheel speed signals are rated as unreliable due to slip. In this driving situation the integrated acceleration signal is the main signal for estimating the vehicle velocity. The estimated vehicle velocity in Figure 9.12 is almost always slightly above the highest wheel velocity. ABS control cycles from the wheel speeds are completely suppressed.

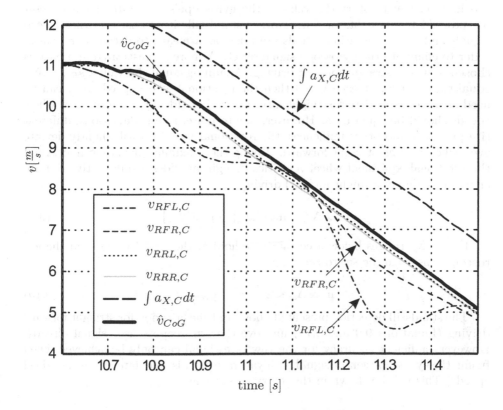

**Figure 9.12** Initial phase of an ABS braking test drive

## 9.2    Vehicle Yaw Rate Estimation

Accurate yaw rate signals $\dot{\psi}$ are crucial for vehicle dynamics control systems. Usually the yaw rate is measured with a gyroscope sensor. One main disadvantage of available gyroscopes is their offset drift caused by temperature changes. In order to increase the accuracy of the yaw rate signal, signals from different sensors are fused for yaw rate calculation. Their weights are determined according to the driving situation. For this procedure, a fuzzy estimator similar to the one presented in Section 9.1.4 is used.

After describing the setup of the fuzzy yaw rate estimator, its quality and robustness shall be validated in Section 9.3 by means of trajectory reconstruction.

### 9.2.1    Data Preprocessing

The gyroscope signal is preprocessed before using it for the yaw rate calculation. The idea is to eliminate the time varying gyroscope offset according to the driving situation. The gyroscope signal value is certainly zero if the vehicle is standing still or when driving exactly straight forward. The goal therefore is to determine these driving situations: standstill and straight forward driving. After turning the ignition key to start the vehicle, the gyroscope's yaw rate signal is reset to zero. At traffic lights or during other standstill situations, the wheel speed signals and the acceleration signals are taken as a means to detect standstill in order to eliminate the gyroscope signal offset. A more sophisticated approach is chosen to detect straight forward driving. Assuming equal slip values $s_{Res,ij}$ and equal tire radii $r_{ij}$ at each wheel then the rotational wheel speeds $\omega_{Rij}$ can be used as a means to detect straight forward driving. Ideally, all rotational wheel speeds should be equal then. However, due to noise, radius deviations, different tire pressures and other influences, the rotational velocities will slightly deviate even when driving straight forward. Taking the maximum deviation max $\Delta_\omega$ of the rotational equivalent wheel velocities though, provides a sufficiently accurate criterion to detect straight forward driving:

$$\max \Delta_\omega = \max_{ij}\{\omega_{ij}\} - \min_{ij}\{\omega_{ij}\} \tag{9.9}$$

If max $\Delta_\omega$ ranges below a certain threshold $\varepsilon$, the signal value from the gyroscope sensor can be set to zero:

$$\max \Delta_\omega < \varepsilon \quad \Rightarrow \quad \dot{\psi}_S \overset{!}{=} 0 \tag{9.10}$$

The standstill detection presented above and the criterion for straight forward driving (Equations 9.9 and 9.10) improve the gyroscope sensor signal already. However, sufficient accuracy for the yaw rate signal can only be achieved when fusing the gyroscope sensor signal with yaw rate signals calculated from the wheel speeds. This will be shown in the following sections.

### 9.2.2    Yaw Rate Calculation using the Wheel Speeds

In curves, the wheels of outer and inner vehicle track run with different velocities. The outer wheels travel a larger distance than the inner wheels. By using a simple triangular approximation, the velocity difference can be used to calculate the yaw rates of front and rear axle:

$$\dot{\psi}_F = \frac{(\omega_{FR} - \omega_{FL}) \cdot r_{stat,F}}{b_F \cdot \cos \delta_W} \tag{9.11}$$

$$\dot{\psi}_R = \frac{(\omega_{RR} - \omega_{RL}) \cdot r_{stat,R}}{b_R} \tag{9.12}$$

Considering the third yaw rate signal $\dot{\psi}_S$ coming from the gyroscope sensor, the fuzzy system generates weighting factors $h_i$, $i = 1...3$, before merging the respective individual sensor signals.

### 9.2.3 Inputs

The estimator inputs should allow to consider the actual driving situation. The rules generate weighting factors corresponding to the individual sensors' measurement errors.

**Wheel Turn Angle $\delta_W$**

The wheel turn angle indicates, whether the curve radius is large or small. The wheel turn angle disturbs the velocity calculation at the front axle. As a consequence, the calculated yaw rate increases with growing steering angle, [115]. Furthermore, for small curve radii the yaw rate calculated from the wheels speeds of the rear track is weighted less. The membership functions of $\delta_W$ are displayed in Figure 9.13(a).

**Longitudinal Acceleration $a_X$**

The longitudinal acceleration signal's membership functions are illustrated in Figure 9.13(b). Accelerations $a_X$ other than little indicate large braking or driving slip, where the wheel speed signals are inaccurate. Therefore, in braking situations, the front axle's wheel speeds are weighted small, whereas those for the rear axle are weighted medium. This is due to the braking force distribution. The braking force and the resulting braking slip are larger at the front axle.

**Lateral Acceleration $a_Y$**

Along with the wheel turn angle $\delta_W$, the lateral acceleration assesses the degree of curve driving. At very high lateral accelerations, wheel loads shifts to the outer wheels and cause large slip values at the inner wheels. Therefore, the inner wheel speeds are weighted less. The membership functions for $a_Y$ are almost equal to the ones of $a_X$.

**Wheel Speed Differences at Front and Rear Axle $\Delta v_F$ and $\Delta v_R$**

$\Delta v_F$ and $\Delta v_R$ describe the two axle's wheel speed differences. For the front axle, this yields

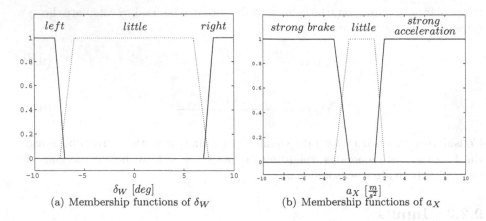

(a) Membership functions of $\delta_W$    (b) Membership functions of $a_X$

**Figure 9.13** Membership functions of system inputs

$$\Delta v_F = |v_{CoG} - v_{RFR,C}| + |v_{RFR,C} - v_{RFL,C}| \qquad . \qquad (9.13)$$

For the rear axle, respectively

$$\Delta v_R = \underbrace{|v_{CoG} - v_{RRR,C}|}_{\text{"condition 1"}} + \underbrace{|v_{RRR,C} - v_{RRL,C}|}_{\text{"condition 2"}} \qquad . \qquad (9.14)$$

$\Delta v_R$ in Equation 9.14 is small, if two conditions are fulfilled: firstly, if the corrected velocity $v_{RRR,C}$ is close to the vehicle velocity ("condition 1"). Secondly, both corrected velocities $v_{RRL,C}$ and $v_{RRR,C}$ must be almost equal ("condition 2"). If both ABS sensors failed, then $v_{RRL,C}$ and $v_{RRR,C}$ would be equal. Without condition 1, the weight for the failing ABS sensors would be high. The estimation results would then be completely wrong. By means of condition 1 though, a large deviation from the vehicle velocity is detected and the sensors are not weighted at all. Condition 1 therefore ensures the stability of the fuzzy estimator. That means, if $\Delta v_R$ is small, then the calculated yaw rate signal is reliable and the weight for the respective sensors is high.

Figure 9.14 shows the membership functions for $\Delta v_F$. The ones for $\Delta v_R$ are similar.

## 9.2.4   Outputs

The output variables of the fuzzy estimator are the three weighting factors, $h_1$ for the gyroscope, $h_2$ for the yaw rate at the front axle and $h_3$ for the yaw rate at the rear axle. The weight ranges between 0...1 (Figure 9.15). If a signal is not reliable at all, then it is weighted "zero". If the reliability increases, the corresponding weight rises and the membership functions are "small", "average" and "large". The weighting factors are used to calculate a weighted mean of the sensor signals according to their reliability:

$$\dot{\psi}_{Fuz} = \frac{h_1 \cdot \dot{\psi}_S + h_2 \cdot \dot{\psi}_F + h_3 \cdot \dot{\psi}_R}{\sum\limits_{i=1}^{3} h_i} \tag{9.15}$$

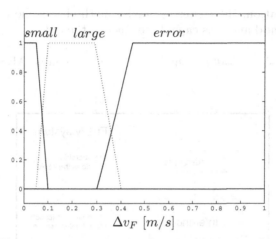

**Figure 9.14** Membership functions of system $\Delta v_F$

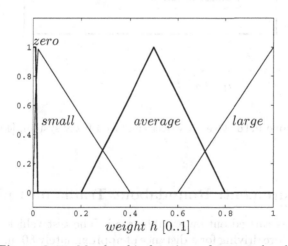

**Figure 9.15** Membership functions of outputs $h_1...h_3$

## 9.2.5   The Fuzzy System

Figure 9.16 shows the setup of the fuzzy system. The inputs are as explained in Section 9.2.3 with an additional status flag "enable ABS". This flag indicates whether the ABS sensor signals can be used for yaw rate calculation or not. Due to the inductive working principle of today's ABS wheel speed sensors, the velocity is not reliable below a certain velocity threshold. In this case, only the yaw rate sensor's signal $\dot{\psi}_S$ is used for the yaw rate estimate $\dot{\psi}_{Fuz}$.

The inference method used in the fuzzy system is the Mamdani-implication. This means, that the logical value of the conclusion is always smaller than the one of the assumption. The linguistic inputs are logically connected with the AND operator.

For the defuzzification, the center of gravity method was chosen. It represents the standard method and it is capable to smooth the output.

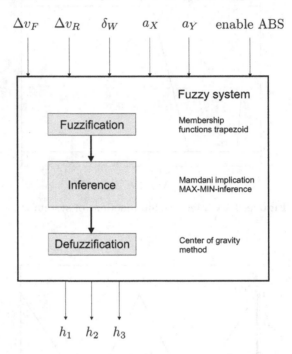

**Figure 9.16** Overview over employed fuzzy system, [49]

## 9.2.6    Measurement: Roundabout Traffic on Public Road

The test drive was carried out on a public road. The test vehicle initially parks along the road before driving for a distance of approximately 80 m. Then it enters a roundabout traffic. The vehicle drives three times through the roundabout traffic and finally leaves it in the opposite direction back to its initial location. Figure 9.17 shows the results of the test drive during which the vehicle turns by a yaw angle of 1260 degrees. The straight horizontal line represents this final value. To validate the yaw rate estimation, the yaw rates from the sensors and from the fuzzy system were integrated to get the yaw angle. As expected, the fuzzy yaw angle $\psi_{Fuz}$ approximates the real value best. The yaw angles from the gyroscope ($\psi_S$) and from the rear axle $\psi_R$ are too large, whereas the one from the front axle $\psi_F$ is too small in this test.

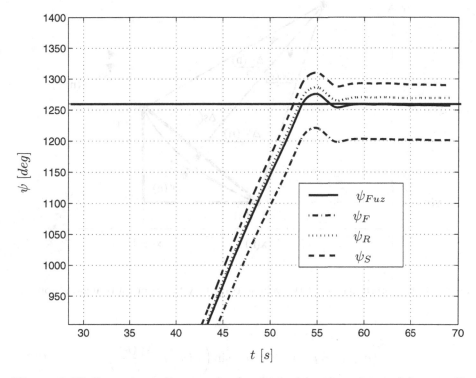

**Figure 9.17** Reconstructed yaw angle of multiple drive through roundabout traffic

# 9.3 Trajectory Reconstruction

The reconstruction of the vehicle trajectory can be used as a means to validate the fuzzy velocity and the fuzzy yaw rate estimator. In this section, the equations for the vehicle location are derived. Then the fuzzy systems are assessed regarding their robustness. An accurate vehicle position is e.g. desired in vehicle navigation systems, should the satellite-based positioning not be available in specific situations.

## 9.3.1 Vehicle Location

The vehicle location is calculated recursively. Based on the old location and heading, the new location is calculated by processing the distance increment $\Delta s(n)$ and the yaw angle increment $\Delta \psi(n)$. Figure 9.18 shows a circular motion increment during one sampling period between $t = n \cdot T_S$ and $t = (n+1) \cdot T_S$. Based on the vehicle location $\underline{x}(n) = [x(n), \quad y(n)]^T$ the new location at time instant $(n+1) \cdot T_S$ is calculated. Using the triangular approximation in Figure 9.18, the location equations for trajectory calculation are

$$x(n+1) = x(n) + \Delta s(n) \cdot \cos \left( \psi(n) + \frac{\Delta \psi(n)}{2} \right) \qquad (9.16)$$

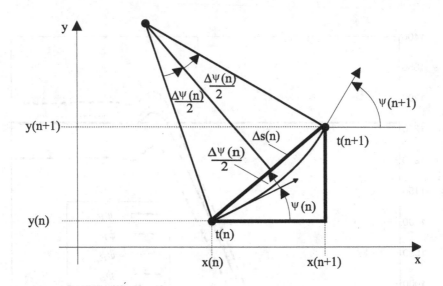

**Figure 9.18** Determination of the vehicle trajectory by means of triangular approximation

$$y(n+1) = y(n) + \Delta s(n) \cdot \sin\left(\psi(n) + \frac{\Delta\psi(n)}{2}\right) \qquad (9.17)$$

The two variables $\Delta s$ and $\Delta\psi$ are calculated using the vehicle velocity $v_{CoG}$ and the yaw rate $\dot{\psi}$

$$\Delta s(n) = v_{CoG}(n-1) \cdot T_S \qquad (9.18)$$

$$\Delta\psi(n) = \dot{\psi}(n-1) \cdot T_S \qquad (9.19)$$

In order to get the absolute distance $s(n)$ and the absolute yaw angle $\psi(n)$, the distance and the time increments of Equations 9.18 and 9.19 are added,

$$s(n) = s(n-1) + \Delta s(n) \qquad (9.20)$$

$$\psi(n) = \psi(n-1) + \Delta\psi(n) \qquad . \qquad (9.21)$$

Equations 9.20 and 9.21 describe a time-discrete integration process. That means, that errors made calculating the time increments $\Delta s$ and $\Delta\psi$ accumulate over time. Therefore, it is crucial for trajectory reconstruction to determine $\Delta s$ and $\Delta\psi$ and due to Equations 9.18 and 9.19 also $v_{CoG}$ and $\dot{\psi}$ very accurately. Therefore, trajectory reconstruction is a good application to validate the fuzzy estimators presented in Sections 9.1 and 9.2.

## 9.3.2   Reconstructed Trajectories

During the test drive through the roundabout traffic, the vehicle trajectory was calculated (see Section 9.2.6). The angular differences between the measured and estimated yaw angles in Figure 9.17 appear to be reasonably small. Regarding

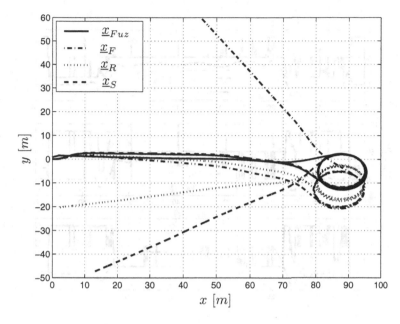

**Figure 9.19** Vehicle trajectory reconstruction of multiple drive through a roundabout traffic

the trajectories of the roundabout traffic drive in Figure 9.19, though, shows that even such small deviations from the real vehicle heading result in a poor reconstruction quality. If only one of the three measured yaw angles is taken for reconstruction the resulting vehicle course is not sufficiently accurate. The gyroscope yaw angle $\psi_S$ drifts away at the end of the measurement. The yaw angles from the wheel speeds $\psi_F$ and $\psi_R$ also yield large deviations from the real vehicle course. Only fuzzy estimation describes the vehicle motion accurately from the beginning to the end of the course. Only there, the vehicle returns to its initial location.

## 9.3.3 Robustness Analysis

Sensor errors are inevitable in real-world measurements. During the test drives, for instance, low battery load caused ABS-sensor failures with significant drops in the wheel speed signal. In order to assess its robustness, the fuzzy system was tested with artificially induced sensor failures. The plots of Figures 9.20 and 9.23 represent the results from the test drive.

In Figure 9.21, velocity drops were artificially inserted every 10 seconds at the front left wheel. That means that the calculated yaw rate from the front axle $\dot{\psi}_F$ was corrupted then. Figure 9.21 also shows, that the wheel speeds fail below $1m/s$ at the very beginning and at the end of the measurement.

Figure 9.22 displays the system inputs into the fuzzy system. During the circular motion in the roundabout traffic (between 17s and 53s), the wheel turn angle is

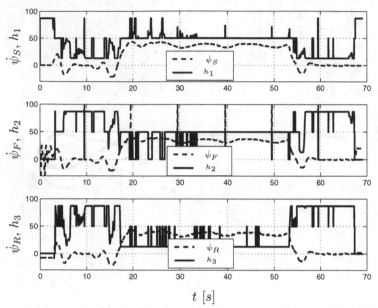

**Figure 9.20** Yaw rates and weighting factors ($\dot{\psi}_F$ with errors)

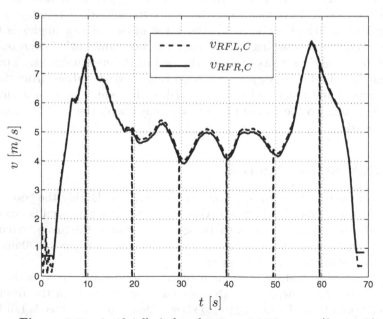

**Figure 9.21** Artificially induced errors of ABS-sensor (front left)

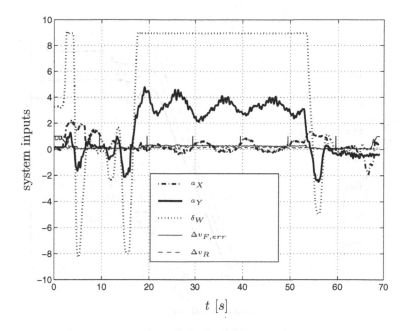

**Figure 9.22** System inputs

high resulting in a high lateral acceleration. Close to the zero-axis, peaks in the $\Delta v_{F,err}$-signal can be recognized resulting from the sensor signals drops.

Figure 9.20 depicts the various yaw rate signals and their weighting factors. At the beginning and at the end of the measurement, the weighting factors $h_2$ and $h_3$ for the wheel speed signals are zero and $h_1$ for the gyroscope signal is high. In such conditions, the wheel speed sensors are below their activation threshold. In the middle of the measurement at high lateral accelerations $a_Y$ and for large wheel turn angles $\delta_W$ the rear axle's yaw rate signal $\dot{\psi}_R$ is only little weighted, whereas the other two yaw rates ($\dot{\psi}_S$ and $\dot{\psi}_F$) are rated as "medium reliable". During the wheel speed signal drops at $t \approx 9s, 19s, ...$ the weighting factor $h_2$ for the front axle drops to zero as well, whereas $h_1$ for the gyroscope signal increases. The sensor signal errors cause sharp yaw rate peaks (dashed) in the middle plot of Figure 9.20. In curves the yaw rate from the rear axle is generally weighted small, so that the sensor signal drops do not influence $h_3$. In the phases before the circular motion ($3s < t < 17s$) and after the circular motion ($53s < t < 68s$) the car is driving almost straight forward. Here, the wheel speeds are generally preferred to the gyroscope. In these phases, the weights $h_2$ and $h_3$ are higher than $h_1$.

As mentioned above, the edges of the membership functions were chosen relatively steep. This causes fast switches between different driving conditions. Figure 9.20 shows, that the sharp velocity signal drops are very quickly detected and the estimation can therefore recover very fast.

Figure 9.23 shows the results of the robustness test for the yaw angle estimation and for the trajectory calculation. In Figure 9.23(a), steps in the yaw angle signal

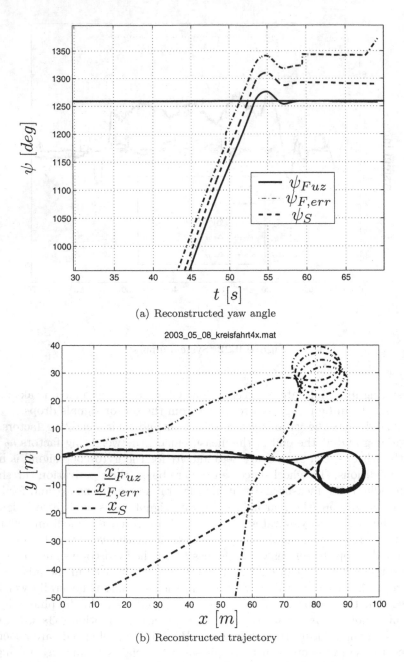

(a) Reconstructed yaw angle

2003_05_08_kreisfahrt4x.mat

(b) Reconstructed trajectory

**Figure 9.23**  Results of a test drive with artificial errors

can be recognized. Driving counter-clockwise through a roundabout traffic results in positive yaw angles. Sensor signal drops at the left wheel cause erroneous positive yaw rates in the $\psi_F$-signal at $t \approx 49s$ and $t \approx 59s$. This is because the velocity difference increases according to equation 9.11. Figure 9.23(a) shows, that the fuzzy estimation completely ignores the false yaw rate signal. Comparing Figures 9.17 and 9.23(a) , the results for $\psi_{Fuz}$ are almost equal with or without sensor signal drops.

The effect of sensor failures on the accuracy of the reconstructed trajectory is illustrated in Figure 9.23(b). The dash-dotted course represents the trajectory reconstructed using the front axle yaw rate $\psi_F$ only. The figure points out that the course is corrupted significantly. Even the circles of the roundabout traffic are no longer centrical. The signal therefore is absolutely inapplicable for trajectory reconstruction. Comparing Figures 9.23(b) and 9.19 shows, that the fuzzy system's trajectory is not at all affected by the sensor signal failures.

## 9.4 Identification of Vehicle Parameters

Using theoretical modeling methods such as those of Chapter 8 a so-called *parametric model* can be obtained, which gives the relationship between the physical data of the system and its parameters. Static parameters can be obtained via measurements or from data sheets. Some parameters are however subject to change over time, and so to improve model quality they must be periodically estimated and updated within the model. The derivation of these parameters is carried out using measured input and output signals together with suitable parameter estimation methods.

Depending upon the model structure, a-priori knowledge, knowledge about disturbances, available computing time (on-line / off-line ) and required estimation quality, an estimation method is chosen from a multitude of available methods [13]. Because of the relatively low computational requirements and its real-time capability, the least squares method is considered here in more detail.

Parameter estimation is applied to the estimation of the friction co-efficients, the mass moments of inertia, the road gradient and the shock absorber characteristics.

In Appendix A.5 the basic theory of least squares estimation is presented, followed by a description of the recursive least squares algorithm and the discrete root filter method in covariant form.

### 9.4.1 Friction Characteristics

The identification of the friction characteristics consists of two processing steps. Firstly, the current friction values $\mu_L$ are obtained and then the complete friction characteristic is estimated using pairs of $(\mu_{Res}, s_{Res})$ values. In order to calculate the friction values the forces at the wheel ground contact point must be known. It would be ideal to know the momentary property of the road surface. Its friction potential is characterized by the maximum friction coefficient. In absence of force braking or strong acceleration, this maximum friction coefficient is however never reached. Therefore, only the currently valid friction can be determined.

Both the estimation of the friction co-efficients and that of the friction characteristic are carried out with a recursive least squares (RLS) algorithm. Outliers must be eliminated via data processing, as they can lead to large inaccuracies in RLS-estimation.

### Estimation of the current Friction Co-efficients During Braking

In order to estimate the friction value $\mu$, the wheel load $F_Z$ and the horizontal force $F_{fric} = \sqrt{F_{WL}^2 + F_{WS}^2}$ are necessary [66], [30]. The longitudinal friction value is defined as the ratio:

$$\mu_L = \frac{F_{WL}}{F_Z} \qquad (9.22)$$

During straight ahead driving, a friction co-efficient estimation can be carried out using a torque balance about the wheel axis (single wheel model). The tire side slip angle $\alpha$ is assumed to be zero under such conditions. This gives the longitudinal frictional force $F_{WL} \approx F_L$, from which the friction co-efficient $\mu_L$ can be calculated:

$$\mu_L(s_L) = \frac{J_W \dot{\omega} + T_{Br} - T_{Drive}}{F_Z \cdot r_{stat}} \qquad (9.23)$$

The different directions of the wheel ground contact point velocity $v_W$ and the rotational equivalent wheel velocity $v_R$ are neglected here. To obtain the friction co-efficient from Equation 9.23, the wheel angular acceleration $\dot{\omega}$, the brake and drive torques $T_{Br}$, $T_{Drive}$ and the wheel ground contact force $F_Z$ must be known.

- The wheel angular acceleration is derived from the difference between two consecutive wheel angular velocity measurements:

$$\dot{\omega}(n) = \frac{\omega(n) - \omega(n-1)}{T_s} \qquad (9.24)$$

  In this $T_s$ is the sampling time. At very low wheel speed differences, the sampling time may be increased in order to get a suitable resolution for the acceleration.

- The wheel ground contact force $F_Z$ is calculated according to Section 9.5.1.

- The drive torque $T_{Drive}$ is derived from the engine torque signal $T_e$, which is provided by today's engine management systems. The effects of the driveline are neglected here.

- The brake torque $T_{Br}$ is either calculated from the measured brake pressure or derived from the motor current in electrical brakes. Such features may not yet be available in production cars.

Using Equation 9.23 the influence of the two torques on the friction characteristics can be estimated. For dry road surfaces, the brake torque dominates by a factor of at least 10. At the other extreme, on an icy road where large rotational wheel accelerations can appear for small brake torques, the values are of the same order.

The friction co-efficient shall be estimated by a recursive least squares estimator. The estimation only yields acceptable results if the excitation torque is sufficiently high. In order to capture time-varying parameters, a forgetting factor $\lambda_{RLS}$ and an additive term $\alpha_{RLS}$ are introduced, as the friction co-efficients may vary during the estimation process (Section A.5.2).

The normalized model equation is:

$$y(n) = u(n) \cdot a(n) \tag{9.25}$$

Comparing Equation 9.25 with 9.23 gives:

$$y(n) = \frac{J_W \dot{\omega} + T_{Br} - T_{Drive}}{F_Z \cdot r_{stat}} \quad , \qquad u(n) = 1 \quad , \qquad a(n) = \mu(n) \quad . \tag{9.26}$$

The estimated parameter $\hat{a}$ is obtained by [55]:

$$\hat{a}(n) = \hat{a}(n-1) + k(n) \cdot [y(n) - u(n) \cdot \hat{a}(n-1)] \tag{9.27}$$

$$k(n) = [\lambda_{RLS} + u_n \cdot p(n-1) \cdot u(n)]^{-1} \cdot p(n-1) \cdot u(n) \tag{9.28}$$

$$p(n) = \frac{1}{\lambda_{RLS}} \cdot [1 - k(n) \cdot u(n)] \cdot p(n-1) + \alpha_{RLS} \tag{9.29}$$

The following values are suitable for the estimation constants $\lambda_{RLS}$ and $\alpha_{RLS}$:

$$\lambda_{RLS} = 0.9$$
$$\alpha_{RLS} = 1000 \cdot [y(n) - u(n) \cdot \hat{a}(n)]^2$$

In an example, the brake torques $T_{BrF,R}$ were calculated from the measured brake pressures $p_{BrF,R}$ using the brake transmission factors $k_{BrF}$ and $k_{BrR}$ (see Equation 10.1):

$$T_{BrF} = r_{stat} k_{BrF} p_{BrF} \qquad T_{BrR} = r_{stat} k_{BrR} p_{BrR} \tag{9.30}$$

The brake transmission factors vary between $\frac{k_{Br}}{A_{Br}} = 0.3$ and $0.4$. Figure 9.24 shows the estimated friction co-efficients for a test drive. The front left wheel runs on wet grass and the front right one on dry asphalt. The anti-locking control begins to reduce the brake torque at about $200\,ms$. The left wheel shows the typical slip control cycles of an ABS-controlled braking. The right wheel displays only a very small brake slip.

## Estimation of the Friction Characteristic Curve over Slip

The above methods allow the estimation of the current friction co-efficients at any instant. It would however desirable to know the entire friction characteristic over slip. At braking, the slip and with it the friction co-efficient are constantly changing with the ABS control cycles. Individually estimated friction values would always lag behind by the estimator dynamics. Assuming a constant road surface, instantaneous values of $\mu$ may be taken from the estimated friction characteristic without any time delays. Of particular interest is the point of maximum friction.

The friction characteristic over slip is estimated, based on the individual friction co-efficient - slip pairs $(\mu_{Res}, s_{Res})$ estimated before. The parameter estimation is again carried out using an RLS algorithm. The following formulation

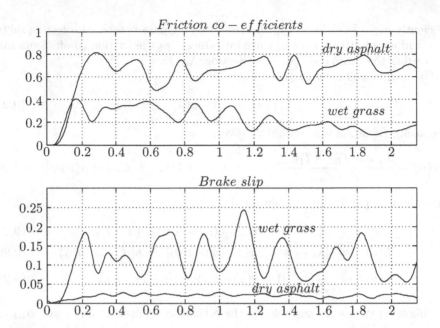

**Figure 9.24** Estimated friction co-efficients and brake slips during braking

[64] of the friction characteristic results in a linear estimation problem:

$$\mu_{Res}(s_{Res}) = \frac{\mu'(0) \cdot s_{Res}}{1 + c_1 s_{Res} + c_2 s_{Res}^2} \tag{9.31}$$

The above approach is therefore utilized as a model, rather than the Burckhardt approach of Equation 8.19. The initial gradient $\mu'(0)$ has a value around 30 for almost all road surface conditions and is fixed beforehand. Thus the number of parameters to be estimated is two.

The model must now be brought into a suitable form for linear estimation. The linearized model is:

$$\begin{aligned} y(n) &= \mu'(0) \cdot s_{Res}(n) - \mu_{Res}(n) \\ &= \left[\mu_{Res}(n) \cdot s_{Res}(n) \, , \; \mu_{Res}(n) \cdot s_{Res}^2(n)\right] \cdot \begin{bmatrix} c_1 \\ c_2 \end{bmatrix} \end{aligned} \tag{9.32}$$

The parameters $c_1$ and $c_2$ can now be derived using a recursive least squares algorithm (Equations A.86 to A.88).

The parameters were estimated for the example of Figure 9.24. The transient behavior of the estimator can be seen in Figures 9.25 and 9.26. For both wheels, the initial characteristic curve used was that for wet asphalt. Figure 9.25 shows the plot of the friction characteristic for the front right wheel. During a braking manoeuvre the friction characteristic is reliably estimated.

Figure 9.26 shows the estimation results for the front left wheel. After approximately $200 \, ms$ a friction characteristic with low maximum is estimated,

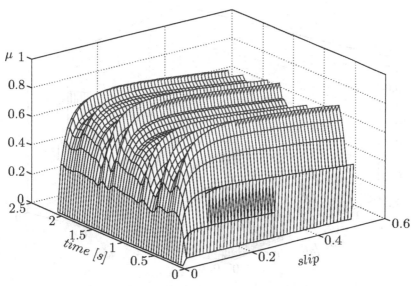

**Figure 9.25** Estimated friction characteristic for the front right wheel (dry asphalt)

which is plausible for wet grass. Such results can only be achieved if there is a sufficient excitation torque and if the friction coefficient reaches levels close to their maximum. Otherwise, the complete friction characteristic curve cannot be estimated. The detection of ABS braking cycles may serve as an indication that the maximum friction coefficient is indeed reached (see Section 10.1.3).

## 9.4.2 Mass Moments of Inertia

For the calculation of the translatory vehicle motions, good results can normally be obtained by assuming a single mass at the CoG. The rotational motion on the other hand depends upon the mass distribution around the relevant axis.

The following mass moments of inertia occur on the vehicle:

- $J_X$ moment of inertia with respect to the roll (longitudinal) axis

- $J_Y$ moment of inertia with respect to the pitch (lateral) axis

- $J_Z$ moment of inertia with respect to the yaw (vertical) axis

For the pitch and roll only the vehicle body moves, and not the so-called "unsprung" vehicle parts (wheels, axles). For yaw motion on the other hand the whole vehicle turns around the axis.

An analytical derivation of these mass moments of inertia is very complex. According to [100] there are at present neither simple measurement techniques nor universally applicable calculation methods available. Part of the identification task is thus the derivation of the mass moments of inertia.

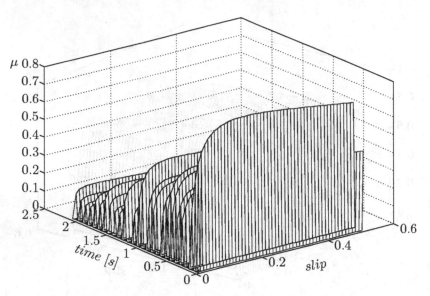

**Figure 9.26** Estimated friction characteristic for the front left wheel (wet grass)

## Method 1 : Approximate Calculation of Mass Moments of Inertia

[100] describes an approximate method using the so-called radii of gyration $i_{X,Y,Z}$, which are recorded in tables for particular vehicle types and loading conditions. Table 9.2 lists values for a limousine in the middle class [100]. The radii of gyration $i_X$ and $i_Y$ relate to the vehicle body, and $i_Z$ to the whole vehicle:

$$
\begin{aligned}
J_X &= (m_{CoG} - m_{Un}) \cdot i_X^2 \\
J_Y &= (m_{CoG} - m_{Un}) \cdot i_Y^2 \\
J_Z &= m_{CoG} \cdot i_Z^2
\end{aligned}
\tag{9.33}
$$

$m_{CoG}$ is the complete vehicle mass and $m_{Un}$ the unsprung vehicle mass (Axles and wheels). For the experimental vehicle the values $80\,kg$ (front) and $130\,kg$ (back) are used for the unsprung masses.

**Table 9.2** Radii of gyration (in m) for a middle-class limousine

| Loading | $i_X$ | $i_Y$ | $i_Z$ |
|---|---|---|---|
| Empty | 0.65 | 1.21 | 1.20 |
| 2 People | 0.64 | 1.13 | 1.15 |
| 4 People | 0.60 | 1.10 | 1.14 |
| 4 People + luggage | 0.56 | 1.13 | 1.18 |

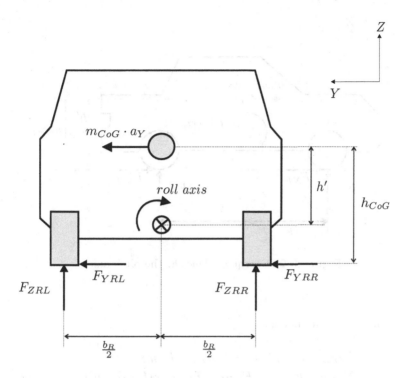

**Figure 9.27** Torque balance at the roll axis

In this method a whole vehicle class is treated the same way. Therefore, the results which are obtained are not precise. The results should be considered as an indication of the variation of the mass moments of inertia dependant upon the loading conditions.

### Method 2: RLS-estimation of the Mass Moments of Inertia

The starting point for the derivation of the estimation equations for the mass moments of inertia comes from the vehicle model of Chapter 8. The roll and pitch axis were assumed to lie at the road level. The torque equations about the roll and pitch axis must now be extended to cover the real torques due to longitudinal and lateral tire forces.

The forces for the roll motion are shown in Figure 9.27.

Constructing the torque balance equations according to Figure 9.27 yields:

$$
\begin{aligned}
J_X \cdot \ddot{\varphi} \;=\; & (F_{ZFL} - F_{ZFR}) \cdot \frac{b_F}{2} + (F_{ZRL} - F_{ZRR}) \cdot \frac{b_R}{2} \\
& + (F_{YFL} + F_{YFR} + F_{YRL} + F_{YRR}) \cdot (h_{CoG} - h') + m_{CoG} \cdot a_Y \cdot h'
\end{aligned}
\tag{9.34}
$$

Similarly, Figure 9.28 shows the situation for the pitch axis, where the longitudinal tire forces are also included.

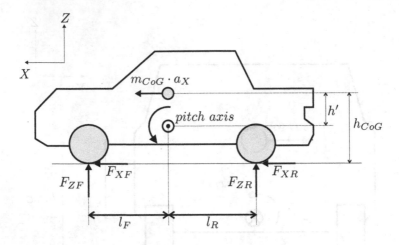

**Figure 9.28** Torque balance at the pitch axis

The pitch torque balance is:

$$J_Y \cdot \ddot{\chi} = -(F_{ZFL} + F_{ZFR}) \cdot l_F + (F_{ZRL} + F_{ZRR}) \cdot l_R \\ + (F_{XFL} + F_{XFR} + F_{XRL} + F_{XRR}) \cdot (h_{CoG} - h') - m_{CoG} \cdot a_X \cdot h' \tag{9.35}$$

As already mentioned, the height of the CoG has no effect on the torque balance equations for yaw, as long as the yaw axis passes through the CoG of the vehicle. Thus the equations can be taken directly from the vehicle model (Equation 8.52):

$$J_Z \cdot \ddot{\psi} = (F_{YFR} + F_{YFL}) \cdot (l_F - n_{LF} \cos \delta_W) \\ -(F_{YRR} + F_{YRL}) \cdot (l_R + n_{LR} \cos \delta_W) \\ -F_{XFL} \cdot \left( \frac{b_F}{2} - n_{LFL} \sin \delta_W \right) \\ +F_{XFR} \cdot \left( \frac{b_F}{2} + n_{LFR} \sin \delta_W \right) \\ + (F_{XRR} - F_{XRL}) \cdot \frac{b_R}{2} \tag{9.36}$$

Equations 9.34 to 9.36 form the basis for the estimation of the mass moments of inertia. They are re-written with measured terms $y(k)$ on the left hand side and parameter terms $\underline{\Theta}(k)$ on the right hand side of the model equation:

$$y(k) = \underline{\Psi}^T(k) \cdot \underline{\Theta}(k) \tag{9.37}$$

The following parameters are assumed to be constant and known:

| | |
|---|---|
| $l_F$ | Front wheel base |
| $l_R$ | Rear wheel base |
| $b_F$ | Front track width |
| $b_R$ | Rear track width |
| $h_{CoG}$ | height of CoG above the road |
| $h_{CoG} - h'$ | height of the pitch and roll axis above the road |

The following variables are measured:

| | |
|---|---|
| $\omega_{ij}$ | Rotational wheel speeds |
| $a_X, a_Y$ | longitudinal and lateral acceleration |
| $\delta_W$ | wheel turn angle |
| $\dot{\psi}$ | yaw rate |
| $\Delta z_{cij}$ | relative displacement of contact points between chassis and undercarriage |

The following variables and parameters are calculated or estimated:

| | |
|---|---|
| $m_{CoG}$ | vehicle mass (Section 9.5.4) |
| $F_{Zij}$ | vertical wheel ground contact forces (Section 9.5.1) |
| $n_{LF}, n_{LR}$ | front and rear casters (Equation 8.1) |
| $c_{ij}$ | tire side slip constants (Section 9.5.2) |
| $\chi, \varphi$ | pitch and roll angles (Section 9.5.3) |
| $v_{CoG}$ | vehicle speed (Section 9.1) |
| $\beta$ | vehicle body side slip angle (Section 9.6) |
| $\alpha_F, \alpha_R$ | front and rear side slip angle (Equation 8.16) |
| $v_{Rij}$ | rotational equivalent wheel speeds (Section 8.3.6) |
| $v_{Wij}$ | wheel ground contact point velocities (Equation 8.8) |
| $s_{Lij}, s_{Sij}$ | longitudinal and lateral slip values (Equation 8.9) |
| $\mu_{Res}$ | friction coefficient (Section 9.4.1) |
| $F_{Xij}, F_{Yij}$ | longitudinal and lateral wheel forces (Section 8.3.4) |

The angular accelerations are obtained via numerical differentiation. As an alternative to the calculation of $F_{Xij}, F_{Yij}$ as in Section 8.3.4, the sum of forces could also be derived from the respective acceleration.

From the above, the following three estimation equations result:

$$\underbrace{(F_{ZFL} - F_{ZFR}) \cdot \frac{b_F}{2} + (F_{ZRL} - F_{ZRR}) \cdot \frac{b_R}{2} + \left( \sum_{i=FL}^{RR} F_{Yi} \right) \cdot (h_{CoG} - h')}_{y(k)}$$

$$= \underbrace{[\ddot{\varphi}, -a_Y]}_{\underline{\Psi}^T(k)} \cdot \underbrace{\left[ \begin{array}{c} J_X \\ m_{CoG} \cdot h' \end{array} \right]}_{\Theta(k)} \qquad (9.38)$$

$$\underbrace{-(F_{ZFL} + F_{ZFR}) \cdot l_F + (F_{ZRL} + F_{ZRR}) \cdot l_R + \left( \sum_{i=FL}^{RR} F_{Xi} \right) \cdot (h_{CoG} - h')}_{y(k)}$$

$$= \underbrace{[\ddot{x}, a_X]}_{\underline{\Psi}^T(k)} \cdot \underbrace{\left[ \begin{array}{c} J_Y \\ m_{CoG} \cdot h' \end{array} \right]}_{\underline{\Theta}(k)} \quad (9.39)$$

$$\underbrace{(F_{YFR} + F_{YFL})(l_F - n_{LF}\cos\delta_W) - (F_{YRR} + F_{YRL})(l_R + n_{LR}\cos\delta_W) + F_{XFR}\left(\frac{b_F}{2} + n_{LFR}\sin\delta_W\right) - F_{XFL}\left(\frac{b_F}{2} - n_{LFL}\sin\delta_W\right)}_{y(k)}$$

$$= \underbrace{\left[ \begin{array}{cc} \ddot{\psi} & -\frac{F_{XRR} - F_{XRL}}{2} \end{array} \right]}_{\underline{\Psi}^T(k)} \cdot \underbrace{\left[ \begin{array}{c} J_Z \\ b_R \end{array} \right]}_{\underline{\Theta}(k)} \quad (9.40)$$

The estimation approach for the yaw moment of inertia $J_Z$ (Equation 9.40) is formulated to include the estimation of the rear track width $b_R$. As $b_R$ is already known, a comparison of the estimated and actual $b_R$ will give an indication as to the quality of the estimation.

### Results for the estimated Mass Moments of Inertia

Good identification results, according to [76], can only be obtained if there is sufficient excitation of the system by the inputs. Thus, for the estimation of the three mass moments of inertia $J_X$, $J_Y$ and $J_Z$, suitable driving manoeuvres must be chosen.

For each of the three mass moments of inertia, test drives with loading conditions of 2 and 5 persons are compared. By comparing the results for the different loading conditions, the extent to which the estimator detects the variations in the moments of inertia can be seen. The results are given in Figs 9.29 and 9.30. The reference values are in each case indicated with dashed lines and each figure contains the results of two test drives.

As seen from Equation 9.38 and 9.39, the product $m_{CoG} \cdot h'$ is estimated as well as the moments of inertia $J_X$ and $J_Y$. If constant and known $h'$ is assumed, the vehicle mass $m_{CoG}$ can also be estimated.

While the estimator very closely approximates the reference values for $J_X$, the estimated values for $J_Y$ have significantly larger oscillations around the expected final value, as shown in the example of Figure 9.30.

The results show that the moment of inertia $J_X$ can be determined quite accurately for suitably chosen test drives. The vehicle mass lies in the expected range, as does the moment of inertia $J_Y$. A reliable detection of the loading conditions, e.g. in this case whether there are 2 or 5 people in the vehicle, is however not always possible. The value of the moment of inertia itself only varies from approximately $2300\,kgm^2$ to $2500\,kgm^2$, i.e. less than $10\,\%$. Complications arise because the vehicle pitch is stimulated to a lesser degree than the roll and yaw, hence the RLS - estimator is excited to a lesser extent and degraded transient oscillations result.

Figure 9.31 shows the results for the moment of inertia $J_Z$ around the vertical vehicle axis.

Here, in accordance with Equation 9.40, the track width $b_R$ is identified as well as the moment of inertia $J_Z$. The reference value for the track width is

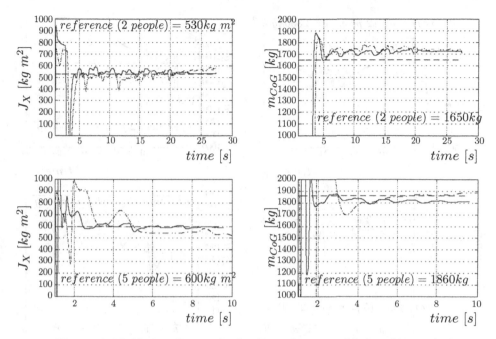

**Figure 9.29** Estimation results for $J_X$ and $m_{CoG}$ with 2 and 5 people

$1.4\,m$. Essentially the same results as for the other two moments of inertia are obtained, namely that the estimated values are of the same order of magnitude as the reference values.

To summarize the results of all three mass moments of inertia, $J_X$, $J_Y$ and $J_Z$:

- The wheel forces required for the estimation can be calculated from the wheel ground contact forces and the wheel model.

- The mass moments of inertia depend upon the loading conditions. They vary less than 10 %. On one hand, this makes the detection of the loading conditions from the mass moments of inertia very difficult. On the other hand, the mass moments of inertia can be applied with reasonable accuracy for different loading conditions.

- The vehicle mass $m_{CoG}$ is derived from the product $m_{CoG} \cdot h'$, whereby $h'$ (the distance of the CoG from the rotation center) is assumed constant and known. The disregarding of the dependence of $h'$ on the loading conditions leads to inaccuracies in the mass calculations. If however the mass is measured separately before the test drive, then one can infer $h'$.

## 9.4.3   Shock Absorber Characteristics

The design of suspension systems for automobiles has gained importance in the last few years, due to their influence on the vehicle characteristics [37]. Also,

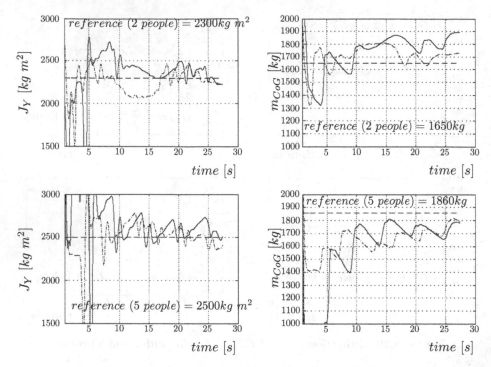

**Figure 9.30** Estimation results for $J_Y$ and $m_{CoG}$ with 2 and 5 people

much effort has been directed towards the design of active and semi-active control algorithms with the purpose of reducing the dynamic wheel forces [16], [64], [14]. Many of these studies however have been carried out in the laboratory, where the shock absorber characteristics are already known.

Basic to the design of a new suspension controller is the reliable estimation of the shock absorber characteristics. In this section the nonlinear suspension model of Section 8.4.3 is taken to estimate its parameters $d_{Ul}$ and $d_{Unl}$.

Re-arranging Equation 8.69 yields:

$$(\Delta z_{Wij} - \Delta z_{Cij}) = -\frac{d_{Ul}}{k_U}(\dot{z}_{Wij} - \dot{z}_{Cij})$$
$$-\frac{d_{Unl}}{k_U}\sqrt{|\dot{z}_{Wij} - \dot{z}_{Cij}|}\,\text{sign}\,(\dot{z}_{Wij} - \dot{z}_{Cij}) + \frac{m_q}{k_U}\ddot{z}_{Cij}$$

$$(9.41)$$

With the measured variable $y(t) = \Delta z_{Wij}(t) - \Delta z_{Cij}(t)$, the general form of the linear RLS estimation problem is:

$$y(k) = a_1 \dot{y}(k) + a_2\sqrt{|\dot{y}(k)|}\,\text{sign}(\dot{y}(k)) + b_0 u(k)$$
$$= \underline{\psi}^T(k) \cdot \underline{\Theta} \tag{9.42}$$

The observation vector is:

$$\underline{\Psi}(k) = \begin{bmatrix} -\dot{y}(k) & \sqrt{|\dot{y}(k)|}\,\text{sign}(\dot{y}(k)) & u(k) \end{bmatrix}^T \tag{9.43}$$

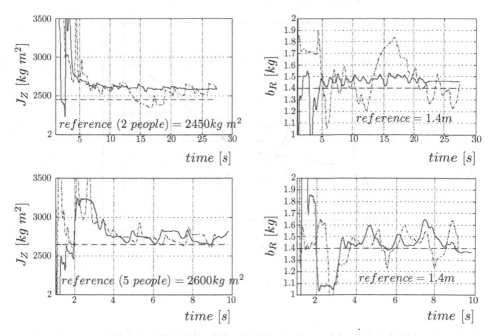

**Figure 9.31** Estimation results for $J_Z$ and $b_R$ with 2 and 5 people

and the estimated parameter vector:

$$\underline{\Theta} = \begin{bmatrix} a_1 & a_2 & b_0 \end{bmatrix}^T$$
$$= \begin{bmatrix} -\dfrac{d_{ul}}{k_U} & \dfrac{d_{Unl}}{k_U} & \dfrac{m_q}{k_U} \end{bmatrix}^T \quad (9.44)$$

Figure 9.32 shows the results for the shock absorber characteristics estimation.

For upward bumps $\dot{z}_C - \dot{z}_W < 0$, the damping force is estimated too high. During downward movements $\dot{z}_C - \dot{z}_W > 0$, the estimate is very precise. Alternative approaches for improving the numerical differentiation are presented in [79].

## 9.5  Approximation of Vehicle Parameters

One of the most important tasks of the identification is to obtain the forces acting upon the wheels, which are very difficult to obtain in terms of measurement techniques. The longitudinal and lateral wheel forces, $F_X$ and $F_Y$ can be determined from the friction co-efficients $\mu_S, \mu_L$ and the wheel ground contact forces $F_{Zij}$ (Section 8.3.4).

The vehicle body side slip angle observer in Section 9.6 uses the wheel forces either implicitly or explicitly as inputs, and observes additional variables which are also available as measurements. If these measured variables are sufficiently well reproduced, then it can be assumed that also the input wheel forces to the observer correspond sufficiently well to reality. In Section 9.6, the variables $v_{CoG}$,

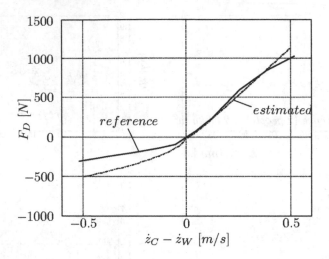

**Figure 9.32** Damping force $F_D$ as a function of velocity $\dot{z}_C - \dot{z}_W$

$\beta$ and $\dot{\psi}$ in the nonlinear two-track model were observed. Of these, $v_{CoG}$ and $\dot{\psi}$ are also measured, and hence can be used for comparison.

### 9.5.1 Calculation of Wheel Ground Contact Forces

In this section, two new acceleration signals $a_{X,Ch}$ and $a_{Y,Ch}$ are introduced. They represent the inertia of the vehicle chassis: if the vehicle accelerates, its acceleration vector directs to the front. However, the chassis will be accelerated backward, i.e. the wheel load will shift to the rear axle. These inertia signals have directions opposite to the vehicle accelerations $a_X$ and $a_Y$

$$a_{X,Ch} = -a_X \tag{9.45}$$

$$a_{Y,Ch} = -a_Y \tag{9.46}$$

If the coupling between roll and pitch is neglected as in [21], the dependencies of the quarter vehicle forces $F_{ZCij}$ on the longitudinal and lateral accelerations $a_{X,Ch}$ and $a_{Y,Ch}$ can be determined separately. By disregarding suspension dynamics, the quarter vehicle forces $F_{ZC}$ are identical to the wheel ground contact forces $F_Z$. According to this approach the common front wheel ground contact force $F_{ZF}$ is formed as shown in Figure 9.33.

The force due to longitudinal acceleration ($m_{CoG} \cdot a_{X,Ch}$) at the CoG causes a pitch torque which reduces the front axle load and increases the rear axle load. The vehicle mass is estimated in Section (9.5.4).

Constructing the torque balance at the rear axis contact point yields:

$$l \cdot F_{ZF} = l_R \cdot m_{CoG} \cdot g - h_{CoG} \cdot m_{CoG} \cdot a_{X,Ch} \tag{9.47}$$

Thus:

$$F_{ZF} = m_{CoG} \cdot \left( \frac{l_R}{l} g - \frac{h_{CoG}}{l} a_{X,Ch} \right) \tag{9.48}$$

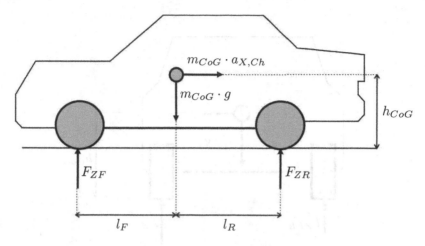

**Figure 9.33** Axle load shifting during acceleration

During cornering the lateral acceleration causes a roll torque as shown in Figure 9.34, whose distribution over the front and rear axle depends on the axle load.

The two axles are considered to be decoupled from one another. In the case of the front axle load a virtual mass $m^*$ is used:

$$m^* = \frac{F_{ZF}}{g} \tag{9.49}$$

From the torque balance equation at the ground contact point of the front left wheel:

$$F_{ZFR} \cdot b_F = F_{ZF} \cdot \frac{b_F}{2} + m^* \cdot a_{Y,Ch} \cdot h_{CoG} \tag{9.50}$$

Substituting the virtual mass of Equation 9.49, and $F_{ZF}$ from Equation 9.48 and solving for $F_{ZFR}$, gives the front right dynamic wheel force (Equation 9.52). By analogy the wheel forces for the other three wheels can then be derived:

$$F_{ZFL} = m_{CoG} \cdot \left( \frac{l_R}{l} g - \frac{h_{CoG}}{l} a_{X,Ch} \right) \left[ \frac{1}{2} - \frac{h_{CoG} \cdot a_{Y,Ch}}{b_F \cdot g} \right] \tag{9.51}$$

$$F_{ZFR} = m_{CoG} \cdot \left( \frac{l_R}{l} g - \frac{h_{CoG}}{l} a_{X,Ch} \right) \left[ \frac{1}{2} + \frac{h_{CoG} \cdot a_{Y,Ch}}{b_F \cdot g} \right] \tag{9.52}$$

$$F_{ZRL} = m_{CoG} \cdot \left( \frac{l_F}{l} g + \frac{h_{CoG}}{l} a_{X,Ch} \right) \left[ \frac{1}{2} - \frac{h_{CoG} \cdot a_{Y,Ch}}{b_R \cdot g} \right] \tag{9.53}$$

$$F_{ZRR} = m_{CoG} \cdot \left( \frac{l_F}{l} g + \frac{h_{CoG}}{l} a_{X,Ch} \right) \left[ \frac{1}{2} + \frac{h_{CoG} \cdot a_{Y,Ch}}{b_R \cdot g} \right] \tag{9.54}$$

Figure 9.35 shows the approximated wheel load with a measured reference for a severe cornering situation. In spite of the simplifications, the approximation is very accurate.

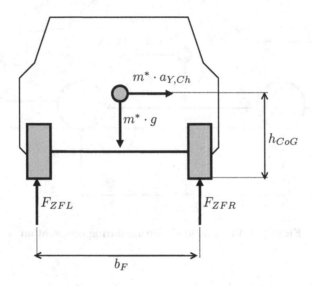

**Figure 9.34** Wheel load shifting during cornering

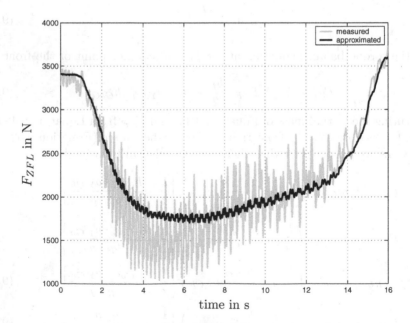

**Figure 9.35** Approximated and measured wheel ground contact forces of a severe cornering maneuver

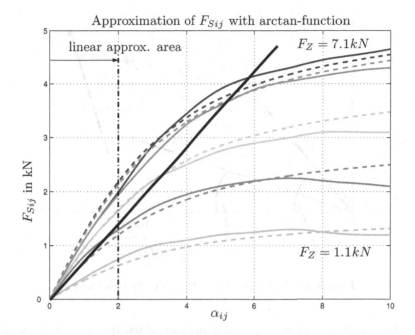

**Figure 9.36** Lateral wheel force characteristic line

## 9.5.2 Adaptation of the Tire Side Slip Constants

In Section 8.4.4, a linear relationship between the tire side slip angles $\alpha_{ij}$ and the lateral wheel forces $F_{Sij}$ was assumed. The lateral tire characteristics was modeled in Equation 8.81 as a spring, where the lateral deflection is represented by the tire side slip angle and the spring constant is the tire side slip constant $c_{ij}$:

$$F_{Sij} = c_{ij} \cdot \alpha_{ij} \qquad (9.55)$$

For lateral accelerations above $4m/s^2$ and large tire side slip angles, though, the linear relationship assumed in Equation 8.81 is not sufficiently accurate any more. In this case, Hooke's Law is invalid, and the tire side slip constant becomes time-variant:

$$F_{Sij}(t) = c_{ij}(t) \cdot \alpha_{ij}(t) \qquad (9.56)$$

The measured characteristic of the lateral force can be seen in Figure 9.36 (solid lines) [84]. It depends on the normal force $F_Z$. The tire side slip constant is represented by the gradient of the straight line. Figure 9.36 shows, that the linear relationship is sufficiently accurate only for small $\alpha$. For larger side slip angles, however, the linear approximation is insufficient. In order to improve the accuracy of the linear relationship in Equation 9.55, the tire side slip constants $c_{ij}$ are adapted. We approximate the nonlinear relationship $F_{Sij}(\alpha_{ij}, F_{Zij})$ of Figure 9.36 with the heuristic approach

$$F_{Sij} = k_{red,ij} \left( k_1 - \frac{F_{Zij}}{k_2} \right) \cdot F_{Zij} \cdot \arctan\left( k_3 \cdot \alpha_{ij} \right) \qquad . \qquad (9.57)$$

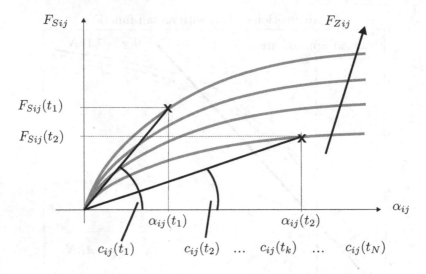

**Figure 9.37** Adaptation of tire side slip constants

In this the reduction factor $k_{red,ij}$ contains the actual friction coefficient from Equation 8.31.

Equation 9.57 considers the influence of varying wheel loads $F_{Zij}$ (see Section 9.5.1) on the lateral wheel forces. The tire side slip angle is calculated by means of Equation 8.16. The errors of this approximation can be seen in Figure 9.36 (dashed lines) for the parameters $k_1 = 1$, $k_2 = 14000$ and $k_3 = 0.36$. For other tires we get different parameters [48]. Based on the lateral wheel forces in Equation 9.57 the tire side slip constants are recalculated in every time step.

$$c_{ij}(t_k) = \frac{F_{Sij}(t_k)}{\alpha_{ij}(t_k)} \tag{9.58}$$

The adaptation process of the tire side slip constants is illustrated in Figure 9.37. The results shall now be verified.
A key variable to determine the lateral vehicle dynamics is the vehicle body side slip angle $\beta$. Thus, the accuracy of vehicle body side slip angle observation (Section 9.6) is taken as a criterion to assess the adaptation of the tire side slip constants.
Figure 9.38 shows the simulation of $c_{ij}$ during a Clothoide driving maneuver. At the beginning, the vehicle is driving straight at a velocity of 50 m/s. From $t = 0s$ on, the wheel turn angle is linearly increased by the driver from 0 to 1 deg. The velocity remains unchanged. The vehicle then moves into a stationary circle. Figure 9.38 illustrates, that the tire side slip constants of inner and outer track deviate significantly due to different wheel loads. In Figure 9.43, the vehicle body side slip angle is shown with and without the above adaptation. Only with adaptation, the transient of the vehicle body side slip angle signal is described well enough. Furthermore, the stationary value of $\beta$ is more accurate. The lateral acceleration during the simulated driving maneuver in Figure 9.38 is between 5 and $9m/s^2$.

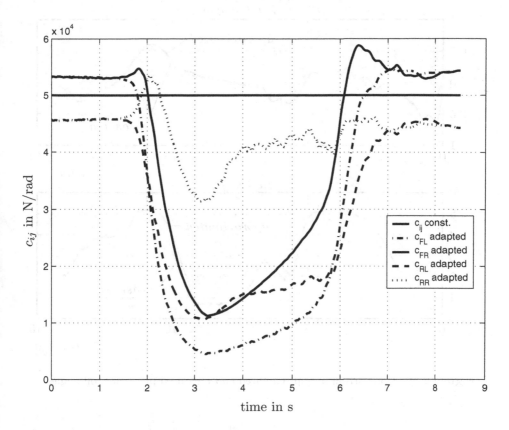

**Figure 9.38** Time varying tire side slip constants during a fast clothoide driving maneuver

### 9.5.3  Approximation of Pitch and Roll Angles

The pitch and roll angles $\chi$, $\varphi$ are calculated from the relative displacement $\Delta z_C$ of the quarter vehicle on top of the suspension. In this simplified approach, the impact of a rough road surface is disregarded.

$$\hat{\chi} = \frac{1}{l} \cdot \left( \frac{\Delta z_{CRL} + \Delta z_{CRR}}{2} - \frac{\Delta z_{CFL} + \Delta z_{CFR}}{2} \right) \qquad (9.59)$$

A positive pitch angle means that the vehicle is diving down at the front.

$$\hat{\varphi} = \frac{1}{b_F + b_R} \cdot \left( \frac{\Delta z_{CFL} + \Delta z_{CRL}}{2} - \frac{\Delta z_{CFR} + \Delta z_{CRR}}{2} \right) \qquad (9.60)$$

A positive roll angle means that the vehicle is leaning towards the right hand side relative to the forward direction. Figure 9.39 shows a comparison between

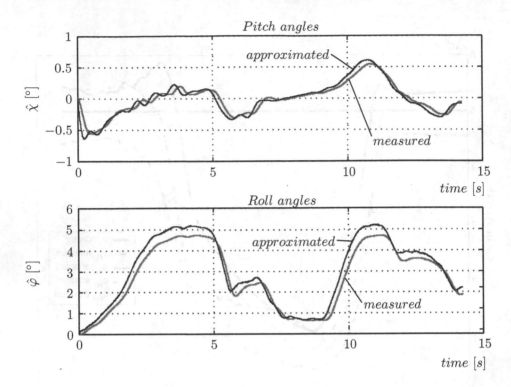

**Figure 9.39** Comparison of approximated and measured pitch and roll angles

measured and approximated pitch and roll angles during a test drive. The approximated angles are ahead of the measured angles in time, and are also not accurate in amplitude. This is mainly due to the fact that the input values $\Delta z_C$ are corrupted. Considering the small computational expense however, this approximation gives still good results.

## 9.5.4  Approximation of Vehicle Mass

The vehicle mass $m_{CoG}$ is a temporally slowly variable parameter. Therefore it is not necessary to estimate the vehicle mass constantly but only for certain driving conditions. The vehicle mass can be estimated from a force-balance. The occurred errors will be analyzed and the approximation will be executed only during suitable driving conditions.

When applying the force-balance all power consumers, e.g. air-condition and even radio reduce the available torque for the acceleration of the vehicle. To avoid all these measurements the estimation is made only in driving conditions with a high engine torque. The engine output will then be reduced only insignificantly by such parasitic consumers. The result is the following force-balance:

$$F_{Acc} = f_{mass} \cdot m_{CoG} \cdot \dot{v}_{CoG} = F_{Drive} - F_{windX} - F_R - F_G \qquad (9.61)$$

The rolling resistance force $F_R$ between tires and road is calculated from Equation 8.50, the wind force will be approximated to:

$$F_{windX} = c_{air} \, A_L \, \frac{\rho_0}{2} \, v_{CoG} \qquad (9.62)$$

$F_G$ is the gravitational force and arises:

$$F_G = m_{CoG} \, g \, \sin \chi_{road} \qquad (9.63)$$

The engine torque is converted with the transmission ratio of the gear and the differential, as well as with the static rolling radius of the wheel into the drive force:

$$F_{Drive} = \frac{T_e}{r_{Stat}} \, i_{gear} \, i_{diff} \qquad (9.64)$$

The dynamic behavior of the drivetrain may be neglected here. The acceleration force $F_{Acc}$ does not correspond to the vehicle mass $m_{CoG}$ multiplied by the vehicle acceleration $\dot{v}_{CoG}$. This is because of the rotatory parts of the engine drivetrain and the wheels which must also be accelerated. Hence, it is necessary to introduce a mass factor $f_{mass}$ to compensate the rotation effect. According to [84] arises:

$$f_{mass} = 1 + \frac{4J_W + i_{diff}^2 (J_{DT} + i_{gear}^2 J_{crank})}{m_{CoG} \, r_{Stat}^2} = 1 + \frac{f_{comp}}{m_{CoG}} \qquad (9.65)$$

Thus results in the following estimation equation for the vehicle mass:

$$m_{CoG} = \frac{F_{Drive} - F_{windX}(v_{CoG}) - F_R(v_{CoG}) - f_{comp} \cdot \dot{v}_{CoG}}{f_{mass}\dot{v}_{CoG} + g \cdot \chi_{road}} \qquad (9.66)$$

Input values into Equation 9.66 are the vehicle velocity $v_{CoG}$, the road gradient $\chi_{road}$, gear and differential transmission ratios $i_{gear}$ and $i_{diff}$ and the engine torque $T_e$. Mass estimation is carried out, when the engine torque is exactly known, and when driving straight. An additional energy consuming part would be the torque converter of the automatic transmission. The mass estimation is executed only if the converter bypass clutch is closed, so that the losses of the torque converter are eliminated. Furthermore, driving conditions with small variations of the throttle angle are selected, since engine map errors occur in dynamic transitions. Modern engine management systems provide an approximated engine torque output signal.

Figure 9.40 shows the estimated mass during a drive on a public road. The estimated mass (dashed line) approaches the real mass after approximately one minute. At the beginning of the test drive, the estimated value still significantly deviates from the real value. This initial deviation is caused by systematic errors of the acceleration sensor signal during acceleration and braking. The corrected acceleration $a_{X,C}$ approximates $v_{CoG}$ here. From about $t = 50s$ on, the velocity changes only slightly. In such driving situations the estimated vehicle mass value fits very well to the real vehicle mass. The estimated value $\hat{m}_{CoG}$ is 1400 kg compared to the real value $m_{CoG}$ of 1420 kg. The relative error of the mass estimation in this example was below 2%.

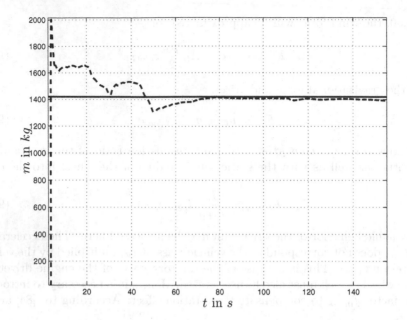

**Figure 9.40** Approximation results of the vehicle mass

## 9.6 Vehicle Body Side Slip Angle Observer

### 9.6.1 Basic Theory of a Nonlinear Observer

In general, observers are implemented when certain state variables cannot be measured with acceptable expense. Since the vehicle model of Section 8.4.4 is nonlinear, a Luenberger observer is not feasible. To overcome this problem, Zeitz [131] proposes an observer design, where the system is linearized and then modeled on the Luenberger observer. Figure 9.41 shows the structure of the nonlinear observer from Zeitz.

The nonlinear system is the starting point for the design:

$$\begin{aligned} \dot{\underline{x}} &= \underline{f}(\underline{x},\underline{u}) \\ \underline{y} &= \underline{c}(\underline{x}) \end{aligned} \tag{9.67}$$

As in the linear case the state observation follows:

$$\begin{aligned} \dot{\hat{\underline{x}}} &= \underline{f}(\hat{\underline{x}},\underline{u}) + \underline{L}(\hat{\underline{x}},\underline{u}) \cdot (\underline{y} - \hat{\underline{y}}) \\ \hat{\underline{y}} &= \underline{c}(\hat{\underline{x}}) \end{aligned} \tag{9.68}$$

The observer gain matrix $\underline{L}$ must now be specified such that the estimation error $\tilde{\underline{x}}(t)$ converges to zero for $t \to \infty$:

$$\lim_{t \to \infty} \tilde{\underline{x}}(t) = \lim_{t \to \infty} \underline{x}(t) - \hat{\underline{x}}(t) = 0 \tag{9.69}$$

The error differential equation is formed, the solution of which can be used to

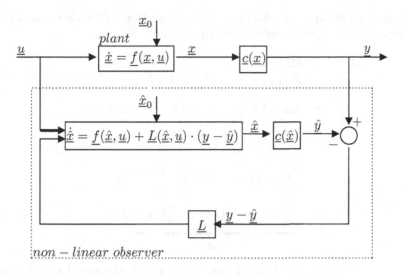

**Figure 9.41** Structure of the nonlinear observer

determine whether or not the condition of Equation 9.69 is satisfied:

$$\dot{\tilde{x}}(t) = \dot{x}(t) - \dot{\hat{x}}(t) = \underline{f}(\underline{x}, \underline{u}) - \underline{f}(\underline{\hat{x}}, \underline{u}) - \underline{L}(\underline{\hat{x}}, \underline{u}) \cdot \underbrace{(\underline{c}(\underline{x}) - \underline{c}(\underline{\hat{x}}))}_{\underline{y} - \underline{\hat{y}}} \qquad (9.70)$$

$\underline{f}(\underline{x}, \underline{u})$ and $\underline{c}(\underline{x})$ are then expanded around the actual estimated state $\underline{\hat{x}}$ using the Taylor series expansion, which is curtailed after the linear term:

$$\underline{f}(\underline{x}, \underline{u}) \approx \underline{f}(\underline{\hat{x}}, \underline{u}) + \frac{\partial \underline{f}}{\partial \underline{x}}(\underline{\hat{x}}, \underline{u}) \cdot (\underline{x} - \underline{\hat{x}}) \qquad (9.71)$$

$$\underline{c}(\underline{x}) \approx \underline{c}(\underline{\hat{x}}) + \frac{\partial \underline{c}}{\partial \underline{x}}(\underline{\hat{x}}) \cdot (\underline{x} - \underline{\hat{x}}) \qquad (9.72)$$

The Jacobian matrix for the reduced nonlinear two-track model (Section 8.4.4)

$$\frac{\partial \underline{f}}{\partial \underline{x}} = \begin{bmatrix} \dfrac{\partial f1}{\partial x1} & \dfrac{\partial f1}{\partial x2} & \cdots & \dfrac{\partial f1}{\partial xn} \\ \vdots & \vdots & \ddots & \vdots \\ \dfrac{\partial fn}{\partial x1} & \dfrac{\partial fn}{\partial x2} & \cdots & \dfrac{\partial fn}{\partial xn} \end{bmatrix} \qquad (9.73)$$

is given in Appendix A.1. Using this linearization approach, one arrives at a linear differential equation for the error $\underline{\tilde{x}}(t)$ which converges to zero for a suitable choice of the observer gain matrix $\underline{L}$.

Substitution of Equations 9.71 and 9.72 in Equation 9.70 leads to the following linearized estimation error differential equation:

$$
\begin{aligned}
\dot{\tilde{x}}(t) &= \dot{x}(t) - \dot{\hat{x}}(t) \\
&= \underline{f}(\underline{x}, \underline{u}) - \underline{f}(\hat{\underline{x}}, \underline{u}) - \underline{L}(\hat{\underline{x}}, \underline{u}) \cdot (\underline{c}(\underline{x}) - \underline{c}(\hat{\underline{x}})) \\
&= \underline{f}(\hat{\underline{x}}, \underline{u}) + \frac{\partial \underline{f}}{\partial \underline{x}}(\hat{\underline{x}}, \underline{u}) \cdot (\underline{x} - \hat{\underline{x}}) - \underline{f}(\hat{\underline{x}}, \underline{u}) \\
&\quad - \underline{L}(\hat{\underline{x}}, \underline{u}) \cdot \left( \underline{c}(\hat{\underline{x}}) + \frac{\partial \underline{c}}{\partial \underline{x}}(\hat{\underline{x}}) \cdot (\underline{x} - \hat{\underline{x}}) - \underline{c}(\hat{\underline{x}}) \right) \\
&= \frac{\partial \underline{f}}{\partial \underline{x}}(\hat{\underline{x}}, \underline{u}) \cdot \tilde{\underline{x}} - \underline{L}(\hat{\underline{x}}, \underline{u}) \cdot \frac{\partial \underline{c}}{\partial \underline{x}}(\hat{\underline{x}}) \cdot \tilde{\underline{x}} \\
&= \underbrace{\left[ \frac{\partial \underline{f}}{\partial \underline{x}}(\hat{\underline{x}}, \underline{u}) - \underline{L}(\hat{\underline{x}}, \underline{u}) \cdot \frac{\partial \underline{c}}{\partial \underline{x}}(\hat{\underline{x}}) \right]}_{\underline{F}(\hat{\underline{x}}, \underline{u})} \cdot \tilde{\underline{x}}
\end{aligned}
\tag{9.74}
$$

The observer gain matrix $\underline{L}$ must now be determined such that the observer dynamic matrix $\underline{F}(\hat{\underline{x}}, \underline{u})$ is constant and its eigenvalues lie to the left of the $j$ - axis, so that the solution $\tilde{\underline{x}}(t)$ converges to zero for $t \to \infty$ indepentent of the initial conditions. A suitable choice of $\underline{L}(\hat{\underline{x}}, \underline{u})$ is given by equating $\underline{F}(\hat{\underline{x}}, \underline{u})$ with a constant matrix $\underline{G}$, whose eigenvalues are predefined according to desired dynamics (pole placement).

$$
\underline{F}(\hat{\underline{x}}, \underline{u}) = \frac{\partial \underline{f}}{\partial \underline{x}}(\hat{\underline{x}}, \underline{u}) - \underline{L}(\hat{\underline{x}}, \underline{u}) \cdot \frac{\partial \underline{c}}{\partial \underline{x}}(\hat{\underline{x}}) \overset{!}{=} \underline{G}
\tag{9.75}
$$

From this, the observer gain matrix $\underline{L}(\hat{\underline{x}}, \underline{u})$ can be calculated:

$$
\underline{L}(\hat{\underline{x}}, \underline{u}) = \left[ \frac{\partial \underline{f}}{\partial \underline{x}}(\hat{\underline{x}}, \underline{u}) - \underline{G} \right] \cdot \left[ \frac{\partial \underline{c}}{\partial \underline{x}}(\hat{\underline{x}}) \right]^{+}
\tag{9.76}
$$

The matrix $\frac{\partial \underline{c}}{\partial \underline{x}}(\hat{\underline{x}})$ is in general non-square. Therefore, the Moore-Penrose pseudo-inverse is used for inversion:

$$
\left[ \frac{\partial \underline{c}}{\partial \underline{x}}(\hat{\underline{x}}) \right]^{+} = \left[ \frac{\partial \underline{c}}{\partial \underline{x}}(\hat{\underline{x}}) \right]^{T} \cdot \left( \left[ \frac{\partial \underline{c}}{\partial \underline{x}}(\hat{\underline{x}}) \right] \cdot \left[ \frac{\partial \underline{c}}{\partial \underline{x}}(\hat{\underline{x}}) \right]^{T} \right)^{-1}
\tag{9.77}
$$

## 9.6.2   Observer Design

The nonlinear observer for the vehicle body side slip angle $\beta$ is based on the reduced nonlinear two-track model (Section 8.4.4). The proof of observability is complicated and therefore done in Appendix A.2.

Equations 8.82 to 8.84 give the nonlinear state space description required for the design of the nonlinear observer according to Equation 9.67. According to these equations, $F_{LFL}$, $F_{LFR}$, $F_{LRL}$, $F_{LRR}$ and $\delta_W$ are the inputs $\underline{u}$ for the observer. The wheel forces in the L-direction are thus explicitly defined as inputs, and the wheel forces in the S-direction appear implicitly in the tire side slip constants $c_F$ and $c_R$ (Equation 8.81).

The Equations 8.82 to 8.84 are brought into the form:

$$\dot{\underline{x}} = \underline{f}(\underline{x}, \underline{u}) \Leftrightarrow \begin{bmatrix} \dot{v}_{CoG} \\ \dot{\beta} \\ \ddot{\psi} \end{bmatrix} = \begin{bmatrix} \underline{f}_1 \left( v_{CoG}, \beta, \dot{\psi}, F_{LFL}, F_{LFR}, F_{LRL}, F_{LRR}, \delta_W \right) \\ \underline{f}_2 \left( v_{CoG}, \beta, \dot{\psi}, F_{LFL}, F_{LFR}, F_{LRL}, F_{LRR}, \delta_W \right) \\ \underline{f}_3 \left( v_{CoG}, \beta, \dot{\psi}, F_{LFL}, F_{LFR}, F_{LRL}, F_{LRR}, \delta_W \right) \end{bmatrix}$$

$$(9.78)$$

The states, inputs and measurements are defined as follows:

State Vector:

$$\underline{x} = \begin{bmatrix} v_{CoG} \\ \beta \\ \dot{\psi} \end{bmatrix}$$

Input Vector:

$$\underline{u} = \begin{bmatrix} F_{LFL}, F_{LFR}, F_{LRL}, F_{LRR}, \delta_W \end{bmatrix}^T$$

Measurement:

$$\underline{y} = \begin{bmatrix} v_{CoG} \\ \dot{\psi} \end{bmatrix}$$

In the input vector $\underline{u}$, the wheel turn angle $\delta_W$ is handled by the driver. By steering, the vehicle body side slip angle $\beta$ is controlled. In dangerous driving situations, the driver may be supported by a vehicle dynamic control system, which uses the longitudinal wheel forces $F_{LFL}, F_{LFR}, F_{LRL}, F_{LRR}$ as additional input variables.

To determine the observer gain matrix $\underline{L}(\hat{\underline{x}}, \underline{u})$ a suitable desired observer dynamic matrix $\underline{G}$ must be chosen in Equation 9.75. The simplest way of determining the eigenvalues of the observer dynamic matrix $\underline{F}$ is to select the matrix $\underline{G}$ in diagonal form. The eigenvalues are the desired eigenvalues for the matrix $\underline{F}(\hat{\underline{x}}, \underline{u})$ (Equation 9.75) and can be read directly from the diagonal elements.

$$\underline{G} = \begin{bmatrix} \lambda_1 & 0 & 0 \\ 0 & \lambda_2 & 0 \\ 0 & 0 & \lambda_3 \end{bmatrix} \qquad (9.79)$$

The three eigenvalues $\lambda_1$, $\lambda_2$ and $\lambda_3$ must be carefully chosen.

With $\underline{y} = \underline{c}(\underline{x}) = \begin{bmatrix} v_{CoG} \\ \dot{\psi} \end{bmatrix}$, $\underline{c}$ is a constant $2 \times 3$ matrix. The pseudo-inverse can be obtained using Equation 9.77 as:

$$\left[ \frac{d\underline{c}}{d\underline{x}}(\hat{\underline{x}}) \right]^+ = \begin{bmatrix} 1 & 0 & 0 \\ 0 & 0 & 1 \end{bmatrix} \qquad (9.80)$$

The Jacobian matrix $\partial \underline{f} / \partial \underline{x}$, the matrix $\underline{G}$ and the pseudo inverse are substituted into Equation 9.76. The elements of the observer gain matrix $\underline{L}(\hat{\underline{x}}, \underline{u})$ are obtained

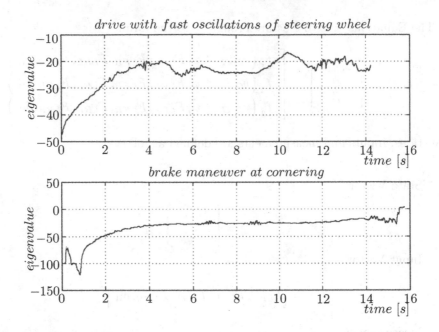

**Figure 9.42** Position of time-variant eigenvalue $\lambda_2$ during test drives

as functions of the Jacobian matrix and the desired eigenvalues:

$$\underline{L}\left(\underline{\hat{x}}, \underline{u}, \lambda_1, \lambda_3\right) = \begin{bmatrix} \dfrac{\partial f_1}{\partial x_1} - \lambda_1 & \dfrac{\partial f_1}{\partial x_3} \\[2ex] \dfrac{\partial f_2}{\partial x_1} & \dfrac{\partial f_2}{\partial x_3} \\[2ex] \dfrac{\partial f_3}{\partial x_1} & \dfrac{\partial f_3}{\partial x_3} - \lambda_3 \end{bmatrix} \tag{9.81}$$

Note that only two of the chosen eigenvalues appear in the above calculation, thus only two of the desired eigenvalues of the matrix $\underline{F}(\underline{\hat{x}}, \underline{u})$ can be placed in the chosen positions. There remains a time-variant eigenvalue $\lambda_2$, which is shown during test drives in Figure 9.42. It can be seen, that also $\lambda_2$ remains in the stable region.

### 9.6.3 Validation of Vehicle Body Side Slip Angle Observer

The vehicle body side slip angle observer was validated with several test drives. Figure 9.43 shows a side slip angle estimation result for a fast clothoide drive. The nonlinear vehicle model is only accurate, when the cornering stiffnesses are adapted according to Section 9.5.2. Otherwise, the model is false and the observer outputs erroneous side slip angle estimation values (dotted). Only the observer based which is based on an adapted nonlinear two track model (dashed) is capable to describe the side slip angle appropriately. The dashed observer value is very

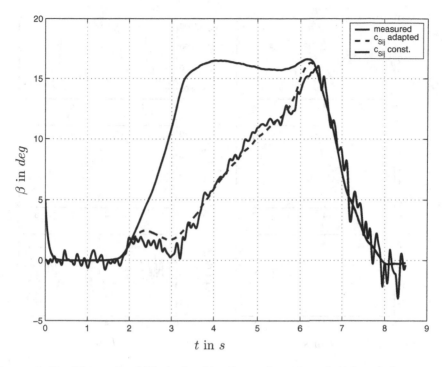

**Figure 9.43** Observed vehicle body side slip angle with and without adaptation of the tire side slip constants

close to the measured reference. Figure 9.44 shows the modeled and observed values (dashed) and the measured references for a sinusoidal test drive. In the upper left corner the vertical wheel force $F_{ZFL}$ is displayed. The dashed line results from the approximation Equations 9.51 - 9.54 in Section 9.5.1. The wheel loads are approximated very accurately. In the upper right plot the lateral wheel forces $F_{Sij}$ are shown. They were calculated by means of Equation 9.57. The lateral wheel forces fit the measured reference very well. The middle plots show the wheel turn angle $\delta_W$ (left) and the yaw rate $\dot{\psi}$ (right): the modeled and the measured yaw rate are identical. That means, that the observer adapts the modeled yaw rate to the measured one. In the lower left corner, the velocity $v_{CoG}$ of the model is compared with the reference. The modeled and the measured reference velocity are almost equal in this test drive. Finally, the plot in the lower right corner displays the observed (dashed) and the measured vehicle body side slip angle. The observed values are very accurate. The results of the second test drive are displayed in Figure 9.45. It was a clothoide drive at the stability limit of the vehicle. The vertical wheel forces are again approximated very well, the lateral wheel forces are too large, though. Apparently, the lateral wheel force cannot be transmitted to the ground any more. The wheel turn angle $\delta_W$ is very large causing enormous yaw rates of almost 70 deg/s. In the lower left corner, the modeled and the measured velocity $v_{CoG}$ deviate significantly. The employed model of the longitudinal wheel forces is not applicable any more. This causes

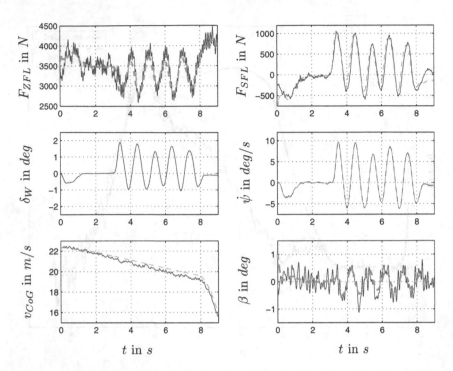

**Figure 9.44**  Validation of vehicle body side slip angle observer with a sinusodial test drive

significant velocity deviations. A state space observer adapts the modeled values to the measured ones. However, if the vehicle body side slip angle observer adapts the inaccurate model velocity to the measured velocity, the observed values for the side slip angle become false. That is the reason, why the velocity may not be used as an output any more, in case the yaw rate exceeds a certain threshold. In such driving situations the velocity is rather calculated from the model. The system switches to a structure, where the observer uses the yaw rate as a measurement variable only. That is the reason, why the observer does not adapt the modeled to the measured velocity in the lower left plot of Figure 9.45. Using only one output variable, the presented side slip angle observer approximates the measured reference very well. The estimation value is very accurate up to a value of $\beta = 15°$, which represents a very critical driving situation. This test drive proves, that the nonlinear observer is capable to describe the vehicle body side slip angle up to the stability limit of the vehicle.

## 9.7  Determination of the Road Gradient

The actual road gradient angle $\chi_{road}$ (positive or negative) has a significant impact on the model accuracy. It is also required for transmission control. Here, two methods of estimating the road gradient are given, and their accuracy investigated.

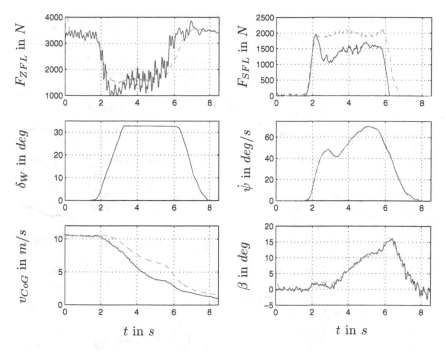

**Figure 9.45** Validation of vehicle body side slip angle observer with a highly dynamical clothoide test drive

## 9.7.1   Method 1: Acceleration and Wheel Speed Method

The starting point of the first method [21] is an offset error in the longitudinal acceleration signal $a_X$. This offset value is dependant upon the gravitational acceleration $g$ and the road gradient $\chi_{road}$:

$$a_X = \ddot{x} - g \cdot \sin \chi_{road} \tag{9.82}$$

For upward driving, i.e. a positive pitch angle $\chi_{road}$, the measured longitudinal acceleration $a_X$ is reduced by the component $g \sin\chi_{road}$ which has the opposite direction as $\ddot{x}$. The acceleration of the vehicle can be calculated using the angular wheel velocity $\omega$:

$$\ddot{x} = r_{stat} \cdot \dot{\omega} \tag{9.83}$$

To avoid errors due to excessive wheel slip, only the angular velocity of the non-driven wheels should be used.

Solving Equation 9.82 for the road gradient angle $\chi_{road}$, and substituting $\sin(\chi) \approx \chi$ for small angles, the equation for the calculation of the road gradient is obtained:

$$\chi_{road} = \frac{r_{stat} \cdot \dot{\omega} - a_X}{g} \tag{9.84}$$

With this equation the road gradient can be quickly obtained from the measured variables $a_X$ and $\omega$. The results are shown in Figure 9.46. The vehicle drives through a "valley", first a negative and then a positive road gradient.

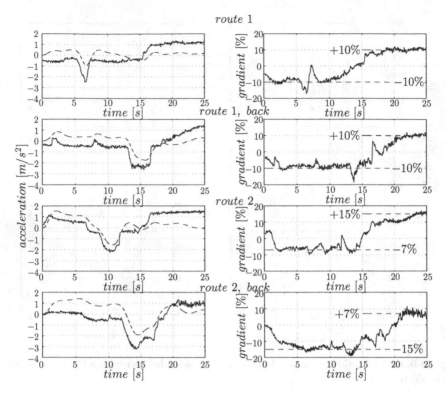

**Figure 9.46** Acceleration signals (left) and estimated road gradients (right)

When the bottom of the valley is reached in the left hand plots of Figure 9.46, the sign of the expression $(\ddot{x} - a_X)$ changes. The right hand plots show the estimated road gradients.

The road gradient can be determined with a resolution of under 5 % of absolute road gradient.

## 9.7.2    Method 2: Model based Road Gradient Observation

The second method is based on a linear Luenberger-observer. Instead of the longitudinal acceleration $a_X$, it employs the vehicle velocity $v_{CoG}$ and its time derivative.

### Force balance for road gradient observation

The effect of the lateral dynamics on road gradient estimation are neglected here. The vehicle body side slip angle is assumed to be zero. This assumption is true for straight on driving situations. Setting up the force balance of the forces displayed in Figure 9.47 yields the nonlinear equation

$$\underbrace{m \cdot \dot{v}_{CoG}}_{F_A} = F_X - \underbrace{m \cdot g \cdot \sin \chi_{road}}_{F_H} - \underbrace{c_{air} \cdot v_{CoG}^2}_{F_{wind}} \quad . \tag{9.85}$$

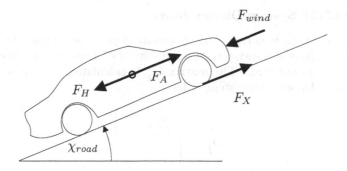

**Figure 9.47** Sketch of longitudinal forces of an ascending car

## Linearization of Equation 9.85

To reduce the computational complexity, a *linear* observer is employed here [47]. Therefore, Equation 9.85 has to be linearized. For the linearization, the following assumptions are made:

- the road gradient angle of public roads is limited to approx. $\pm 12°$ [41], this yields $\sin \chi_{road} \approx \chi_{road}$

- the forces $F_X$ and $F_{wind}$ are merged into a resultant force $F_{res} = F_X - F_{wind}$. This is advantageous because the nonlinear term $F_{wind}$ becomes part of the input. The remaining state space model therefore is linear.

As a consequence of these assumptions Equation 9.85 is simplified

$$m \cdot \dot{v}_{CoG} = F_{res} - m \cdot g \cdot \chi_{road} \quad , \tag{9.86}$$

where $F_X$ is gained by a specific sensor. However, it can be determined from the engine torque. $v_{CoG}$ is obtained from the wheel speed sensors.

## Linear State Space Equations

Equation 9.86 is now transformed to a state space model. The state vector $\underline{x}$ of the linear model contains the velocity $v_{CoG}$ and the road gradient angle $\chi_{road}$. The input $\underline{u}$ is the resultant force $F_{res}$:

$$\underline{x} = \begin{bmatrix} v_{CoG} \\ \chi_{road} \end{bmatrix}, \underline{u} = \begin{bmatrix} F_{res} \\ 0 \end{bmatrix} \tag{9.87}$$

The road gradient model can then be written as

$$\underbrace{\begin{bmatrix} \dot{v}_{CoG} \\ \dot{\chi}_{road} \end{bmatrix}}_{\underline{\dot{x}}} = \underbrace{\begin{bmatrix} 0 & -g \\ 0 & 0 \end{bmatrix}}_{A} \underbrace{\begin{bmatrix} v_{CoG} \\ \chi_{road} \end{bmatrix}}_{\underline{x}} + \underbrace{\begin{bmatrix} \frac{1}{m} & 0 \\ 0 & 0 \end{bmatrix}}_{B} \underbrace{\begin{bmatrix} F_{res} \\ 0 \end{bmatrix}}_{\underline{u}} \tag{9.88}$$

$$\underbrace{\hat{v}_{CoG}}_{y} = \underbrace{\begin{bmatrix} 1 \\ 0 \end{bmatrix}^T}_{C} \underbrace{\begin{bmatrix} v_{CoG} \\ \chi_{road} \end{bmatrix}}_{\underline{x}} \tag{9.89}$$

## Analysis of the System Observability

The observer design is carried out by means of *pole placement*. This, however, requires an analysis of the system observability [25]. The observability matrix $\underline{Q}_B$ must have maximum rank. The matrix $\underline{Q}_B$ is calculated by means of Equation 9.90, where $n$ denotes the system order.

$$
\underline{Q}_B = \begin{bmatrix} \underline{C} \\ \underline{C}\,\underline{A} \\ \vdots \\ \underline{C}\,\underline{A}^{n-1} \end{bmatrix}
\tag{9.90}
$$

For the system of Equations 9.88 and 9.89 $\underline{Q}_B$ becomes

$$
\underline{Q}_B = \begin{bmatrix} \underline{C} \\ \underline{C}\,\underline{A} \end{bmatrix} = \begin{bmatrix} 1 & 0 \\ 0 & -g \end{bmatrix} .
\tag{9.91}
$$

As $\underline{Q}_B$ is square, the maximum rank can be checked by means of the determinant of $\underline{Q}_B$:

$$
\det\left(\underline{Q}_B\right) \overset{!}{\neq} 0
\tag{9.92}
$$

$$
\det\left(\underline{Q}_B\right) = \det \begin{bmatrix} 1 & 0 \\ 0 & -g \end{bmatrix} = -g \neq 0 \quad .
\tag{9.93}
$$

According to Equation 9.93 the determinant of $\underline{Q}_B$ is $-g$ which is the earth's gravitational acceleration. Therefore, the rows of matrix $\underline{Q}_B$ are linearly independent and the linear road gradient model (Equation 9.88) is observable. Thus, the observer design by means of pole placement is feasible.

## Observer design

Since the system order is $n = 2$, the observer gain matrix $\underline{L}$ consists of two elements $l_1$ and $l_2$. In order to calculate these elements, the poles of the observed system must be placed appropriately. Its characteristic polynomial of the closed-loop system is

$$
\begin{aligned}
\det(s\underline{I} - \underline{A} + \underline{L}\underline{C}) &= \det\left( \begin{bmatrix} s & 0 \\ 0 & s \end{bmatrix} - \begin{bmatrix} 0 & -g \\ 0 & 0 \end{bmatrix} + \begin{bmatrix} l_1 & 0 \\ l_2 & 0 \end{bmatrix} \right) \\
&= \det\begin{pmatrix} s + l_1 & g \\ l_2 & s \end{pmatrix} \\
&= s^2 + s \cdot l_1 - g \cdot l_2 \quad .
\end{aligned}
\tag{9.94}
$$

The eigenvalues are denoted as $\lambda_1$ and $\lambda_2$ and are chosen according to the following equation

$$
s^2 + s \cdot l_1 - g \cdot l_2 \overset{!}{=} (s - \lambda_1)(s - \lambda_2) = s^2 - s \cdot (\lambda_1 + \lambda_2) + \lambda_1 \cdot \lambda_2 \quad .
\tag{9.95}
$$

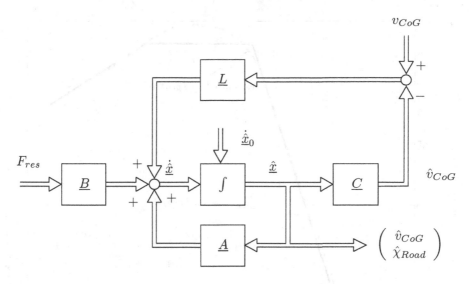

**Figure 9.48** Structure of the developed LUENBERGER-observer

For pole-placement, the coefficients of Equation 9.95 are compared. This yields the elements of the observer gain matrix $\underline{L}$

$$l_1 = -\lambda_1 - \lambda_2 , \quad l_2 = \frac{-\lambda_1 \cdot \lambda_2}{g} \qquad . \tag{9.96}$$

Next, the eigenvalues $\lambda_1$ and $\lambda_2$ are determined. For this, a simulation model with the structure shown in Figure 9.48 was implemented.

The following strategy for pole placement was employed to achieve suitable values of $\lambda_1$ and $\lambda_2$:

- the eigenvalues must be negative, otherwise the observer system becomes unstable.

- if the eigenvalues are too far left in the s-plane, the observer becomes sensitive to noise.

- if the eigenvalues are too close to the imaginary axis, the observer becomes too slow. It then would not be able to follow the driving state of the vehicle properly (e.g. uphill and downhill driving).

Considering these constraints, and running a variety of simulations, the eigenvalues were fixed to

$$\begin{aligned} \lambda_1 &= -2 \\ \lambda_2 &= -3 . \end{aligned} \tag{9.97}$$

The height profile of a road calculated on basis of the estimated road gradient is displayed in Figure 9.49. The test drive was carried out on a test course with a defined road gradient. Starting on a flat road, after $t = 2.5\,s$ the car enters an inclined plane with a gradient of 33 % ($\chi_{road} \approx 18°$). The car moves on this inclined plane for approximately $10\,s$, returns to a flat road again and

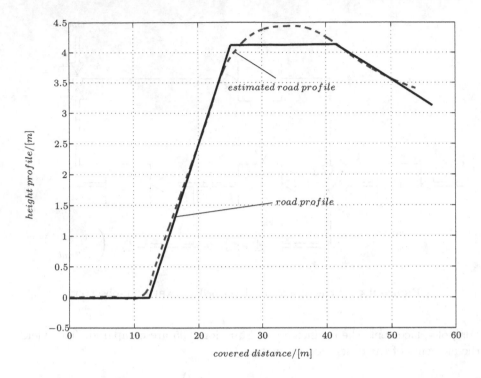

**Figure 9.49** Road gradient of a test course with a defined gradient of 33%

moves downhill at the end. The estimated road gradient of the first ramp is approximately 30%. The deviation to the real road gradient is caused by the linear approximation which is not valid any more. At the end of the inclined plane the vehicle suddenly returns to a flat road. The pitch angle during this transition is responsible for the deviation of the maximum height ($\approx 10\%$). All in all, even for large road gradients, the linear observer provides very good results. The second method has advantages compared to the first one. Due to the observer structure it is less sensitive to errors of the wheel speed and the acceleration signals. The second method requires the longitudinal wheel force $F_X$.

# 10 Vehicle Control Systems

## 10.1 ABS Control Systems

The ABS system aims at minimizing the braking distance while retaining steerability during braking. The shortest braking distance can be reached when the wheels operate at the slip of maximum adhesion co-efficient $\mu_L$ (Figure 8.17). The tire side slip angle is disregarded here, approximating $\cos \alpha \approx 1$.

### 10.1.1 Torque Balance at wheel-road contact

By modeling the torque balance at the wheel-road contact, a better understanding can be obtained of how ABS systems are able to operate around maximum friction without sophisticated estimation algorithms. Figure 10.1 shows the forces acting on the wheel.

In hydraulic brakes, the brake torque at the wheel base depends on the applied braking pressure $p_{Br}$:

$$T_{Br} = F_{Br} \cdot r_{stat} = r_{Br} \cdot \mu_{Br} \cdot A_{Br} \cdot p_{Br} = r_{stat} \cdot k_{Br} \cdot p_{Br} \tag{10.1}$$

Disregarding the drive torque (at braking), the torque balance is (Equation 9.23):

$$J_W \dot{\omega} = r_{stat} \cdot \mu_L(s_L) \cdot F_Z - r_{stat} \cdot k_{Br} \cdot p_{Br} \tag{10.2}$$

The respective block diagram is shown in Figure 10.2.

On applying the braking pressure $p_{Br}$, the brake torque $T_{Br}$ increases. The difference between friction torque $T_{WL}$ and brake torque $T_{Br}$ is negative, resulting in a wheel deceleration. The wheel rotational equivalent velocity $v_R$ (after the integrator in Figure 10.2) starts to decrease and yields an increasing slip $s_L$. At first the friction co-efficient $\mu_L(s_L)$ increases as well, building up the friction torque $T_{WL}$ which narrows the torque difference.

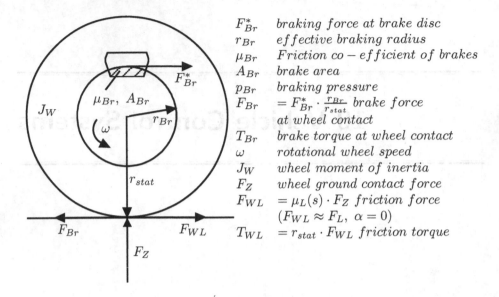

$F_{Br}^*$    braking force at brake disc
$r_{Br}$    effective braking radius
$\mu_{Br}$    Friction co-efficient of brakes
$A_{Br}$    brake area
$p_{Br}$    braking pressure
$F_{Br}$    $= F_{Br}^* \cdot \frac{r_{Br}}{r_{stat}}$ brake force
          at wheel contact
$T_{Br}$    brake torque at wheel contact
$\omega$    rotational wheel speed
$J_W$    wheel moment of inertia
$F_Z$    wheel ground contact force
$F_{WL}$    $= \mu_L(s) \cdot F_Z$ friction force
          $(F_{WL} \approx F_L, \ \alpha = 0)$
$T_{WL}$    $= r_{stat} \cdot F_{WL}$ friction torque

**Figure 10.1**  Torque balance at the wheel-road surface contact

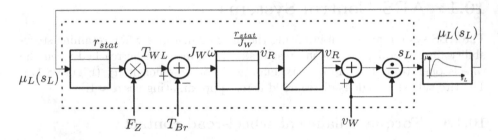

**Figure 10.2**  Block diagram of torque balance at wheel base

After passing the maximum friction co-efficient, the friction curve changes the sign of its gradient. Thus the loop becomes *unstable*, resulting, in the absence of control, in extremely high rotational wheel decelerations: blocking of wheel.

## 10.1.2   Control Cycles of the ABS System

The control cycles are shown in Figure 10.3 [6]. At braking the driver increases the braking pressure (Phase 1).

The rotational equivalent wheel speeds $v_{Rij}$ are measured and differentiated to give the wheel accelerations $\dot{v}_{Rij}$. The point of maximum friction is passed when the wheel speed derivative is below a given threshold $a_1$:

$$\dot{v}_R < -a_1 \tag{10.3}$$

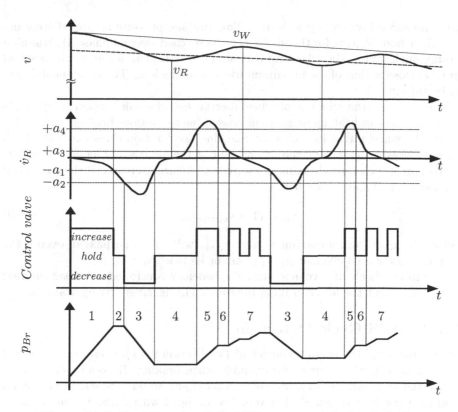

**Figure 10.3** Control cycles of the ABS system with hydraulic brakes

In the very first control cycle, an even lower threshold $a_2$ is applied. Between $a_1$ and $a_2$, braking pressure is held (Phase 2). The introduction of the additional threshold $a_2$ serves to suppress eventual noise influences.

For

$$\dot{v}_R < -a_2 \tag{10.4}$$

braking pressure is decreased (Phase 3). The wheel gains speed again. When threshold $a_1$ is reached again, the pressure drop is stopped (Phase 4). When the wheel speed derivative passes beyond

$$\dot{v}_R > a_4 \tag{10.5}$$

braking pressure is increased in order to prevent the wheel returning to too small slip values (Phase 5). Between

$$a_4 > \dot{v}_R > a_3 \tag{10.6}$$

the pressure is held constant (Phase 6), and below

$$\dot{v}_R < a_3 \tag{10.7}$$

it is slowly raised (Phase 7). When the wheel speed derivative goes below

$$\dot{v}_R < -a_1 \tag{10.8}$$

again the second control cycle starts. Now braking pressure is immediately decreased without waiting for threshold $a_2$ to be reached (second phase 3). Running through such cycles, the wheel rotational speed is kept in an area where wheel slip $s_L$ is close to that of the maximum friction co-efficient. Thus *braking distance* can be minimized.

In case of a large moment of wheel inertia $J_W$, of small friction co-efficients $\mu_L(s_L)$ and of slow braking pressure increase (due to cautious braking, e.g. on icy roads), the wheel might lock without reaching deceleration threshold $-a_1$. Such a situation would endanger the *steerability* of the vehicle. Independent of the above control cycles, the braking pressure is therefore decreased if the rotational equivalent wheel speed goes below:

$$v_R < (1 - s_{L,max})v_W \qquad (10.9)$$

Under any conditions, a maximum slip $s_{L,max}$ will not be surpassed, even if the maximum friction co-efficient $\mu_L(s_L)$ cannot be reached.

The front wheels of a vehicle are independently controlled, whereas the rear wheels jointly get the lower braking pressure. This ensures driving stability.

### 10.1.3   ABS Cycle Detection

The estimation of the friction coefficient (see Section 9.4.1) is only reasonable, if the maximum friction is used, for example when braking. In order to assess the type of the road surface (dry, icy, snow-covered and so on) the state "maximum braking" must be recognized. For vehicles equipped with ABS, a panic braking situation causes ABS control cycles. That is, if an ABS control cycle occurs, then the maximum friction is used. On the other hand, the operativeness of ABS shall be assessed in general to see if malfunction of ABS systems have influenced the investigated accident.

To solve these two questions, a three step algorithm will be presented which is capable to detect the characteristic ABS control cycle pattern.

**Basic Approach of ABS-cycle detection**

An ABS-cycle pattern can be seen in Figure 10.4. Figure 10.4(a) shows a wheel speed signal of a real test run with a test car (Ford Scorpio). The ABS-cycles are marked respectively. In Figure 10.4(b) a cut-out of the wheel *acceleration* signal is displayed. It is the time-derivation of the first ABS-cycle of Figure 10.4(a). The method must be capable to detect these patterns. The shape of the pattern is hardly changing. However, the amplitude and time duration of the pattern varies.

**Prediction**

The prediction method employs the wheel *speed* measurements, in order to detect an ABS-cycle. Detecting the ABS-activity, the past two measurements of the wheel speed are utilized to extrapolate linearly (Equation 10.10) a value of the current wheel speed (see Figure 10.5). $v_R$ denotes the measured wheel speed signal $v_R$, $v_{R,est}$ is calculated by means of Equation 10.10.

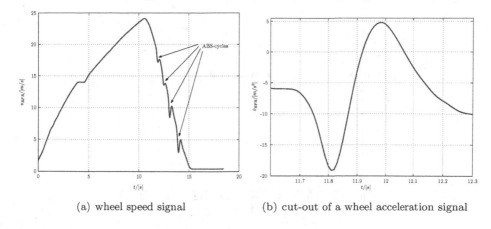

(a) wheel speed signal

(b) cut-out of a wheel acceleration signal

**Figure 10.4** Pattern of an ABS-cycle

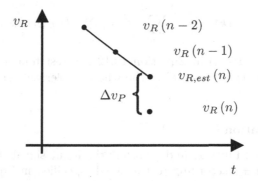

**Figure 10.5** Sketch of extrapolated wheel speed signal

$$v_{R,est}(n) = 2 \cdot v_R(n-1) - v_R(n-2) \qquad (10.10)$$

The estimated value $v_{R,est}(n)$ is compared with the currently measured value $v_R(n)$. The difference $\Delta v_P$ between these two values is almost zero for normal signal behavior. An ABS-cycle, however, causes a prediction difference $\Delta v_P$, which exceeds specific limits (for the Opel Vita, see Equation 10.11) for the beginning and end of the ABS-cycle. This approach detects ABS-cycles with a high probability.

$$\Delta v_P = v_R - v_{R,est} \geq 0.08 \qquad (10.11)$$

The input signals into the ABS-cycle detection system are not zero-mean. Thus, the conventional correlation is replaced by a "tri-state correlation". In Equation 10.12 the calculation of the tri-state correlation is presented. The difference to a polarity correlation, [72], is an additional state "0". The states of a tri-state correlation are +1, 0, −1.

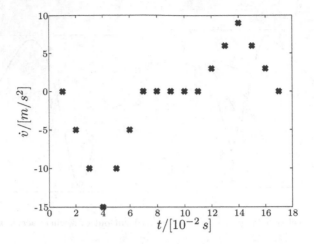

**Figure 10.6** Test signal for tri-state correlation

$$\hat{r}_{xy}(k) = \frac{1}{N} \sum_{n=0}^{N-1-k} T(x(n)) \cdot T(y(n+k)) \qquad (10.12)$$

The function $y(n+k)$ in Equation 10.12 is a test function shown in Figure 10.6, whereas $x(n)$ is a "cut-out" of the wheel acceleration signal $\dot{v}$ displayed in Figure 10.4(b).

### Tri-state correlation

$T(x)$ is a threshold function and transfers the input signals to signals with the only values 0 and $\pm 1$ according to thresholds specified in Equation 10.13. The thresholds are, however, depending on the ABS system installed in the car.

$$T(x) = \begin{cases} 1 & , \quad \text{if} \quad x \geq 2 \\ 0 & , \quad \text{if} \quad -10 < x < 2 \\ -1 & , \quad \text{if} \quad -10 \geq x \end{cases} \qquad (10.13)$$

The tri-state correlation can be carried out using the signal created by the threshold function. In Figure 10.7 the result of a tri-state correlation of an ABS-cycle pattern is sketched. This pattern results from the structure of the signals. The test signal possesses the following structure due to the threshold function:
$|-1|-1|\cdots|-1|0|0|\cdots|0|1|1|\cdots|1|$
Applying the threshold function to the wheel acceleration signal yields a similar structure. Thus, if the test signal shown in Figure 10.6 is "moving" over the measured wheel acceleration signal and matches an ABS-cycle, the structure of the signal shown in Figure 10.7 is gained. Tests show, that this method is robust and reliable.

Since the tri-state correlation utilizes a 2-bit input signal, it is more suitable for microcontroller applications than a "conventional" correlation. Due to this advantage and the reliability of this approach, it is used for cross-checking the result of the prediction.

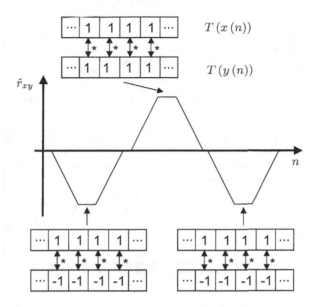

**Figure 10.7** Sequence of tri-state correlation coefficients of a single ABS-cycle pattern

The algorithm of ABS-cycle detection contains three steps: after preprocessing (step 1) a probable ABS-cycle is detected and marked with the prediction method (step 2). To increase reliability and robustness of the method, the result of the prediction is cross-checked with the tri-state correlation afterwards (step 3).

Figure 10.8 shows the results of an ABS braking situation on low $\mu$ road surface. The presented method detects all ABS-cycles above a velocity of $2m/s$. The method works well for different road surfaces. However, the threshold of the prediction method must be adapted to the respective car. For the Ford Scorpio, for example, the threshold specified in Equation 10.11 is 0.013.

## 10.2 Control of the Yaw Dynamics

The control of the yaw dynamics detailed here is based on the reduced nonlinear two-track model of Section 8.4.4. In linearized form the model is given by the following equation (Equation 8.90):

$$\underline{f}(\underline{x},\underline{u}) \approx \underline{f}(\underline{x}_0,\underline{u}_0) + \underbrace{\left.\frac{\partial \underline{f}(\underline{x},\underline{u})}{\partial \underline{x}}\right|_{\substack{\underline{x}=\underline{x}_0\\\underline{u}=\underline{u}_0}} \cdot (\underline{x}-\underline{x}_0)}_{\text{Jacobian}} + \underbrace{\left.\frac{\partial \underline{f}(\underline{x},\underline{u})}{\partial \underline{u}}\right|_{\substack{\underline{x}=\underline{x}_0\\\underline{u}=\underline{u}_0}} \cdot (\underline{u}-\underline{u}_0)}_{\text{Jacobian}}$$

(10.14)

The three state variables are the yaw rate $\dot{\psi}$ (Section 9.2), wheel speed $v_{CoG}$ (Section 9.1) and vehicle side slip angle $\beta$ (Section 9.6). The control inputs are:

$$\underline{u} = [F_{LFL}, \quad F_{LFR}, \quad F_{LRL}, \quad F_{LRR}, \quad \delta_W]^T \tag{10.15}$$

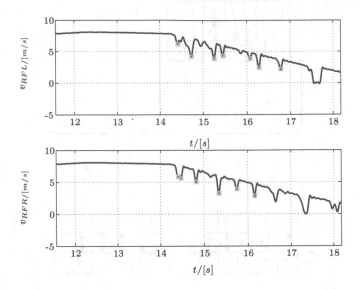

**Figure 10.8** ABS-cycle detection for a low $\mu$ braking maneuver with an Opel Vita

Note that, when only the brakes are available as actuators, then only braking forces $F_{LFL}$, $F_{LFR}$, $F_{LRL}$, and $F_{LRR}$ can be generated. In the case when each wheel were driven individually, e.g. by electrical motors, accelerating forces $F_{LFL}$, $F_{LFR}$, $F_{LRL}$, and $F_{LRR}$ could also be generated.

The wheel turn angle $\delta_W$ is influenced by the driver using the steering wheel. In the event of an inappropriate vehicle body side slip angle $\beta$, the driver attempts to correct the error using steering. When the driver is overloaded, i.e. in critical driving situations, the force inputs $F_{Lij}$ can be used to correct the vehicle body side slip angle, via short term braking of the individual wheels.

Figure 10.9 shows the cases where the braking forces could be used to correct the angular position of the vehicle relative to the desired course.

In the case of understeering, braking can be applied to the rear right wheel, causing a yaw torque about the CoG which corrects for the understeering. This is applied until $\beta$ falls below a given threshold, at which point the driver can once more control the yaw motion himself using the steering. A respective correction is applied at oversteering.

## 10.2.1   Derivation of Simplified Control Law

Note that the driver influences the steering (i.e. $\delta_W$), and the controller affects the brake forces $F_{Lij}$, thus the effect of the two can be separated. The Jacobian is:

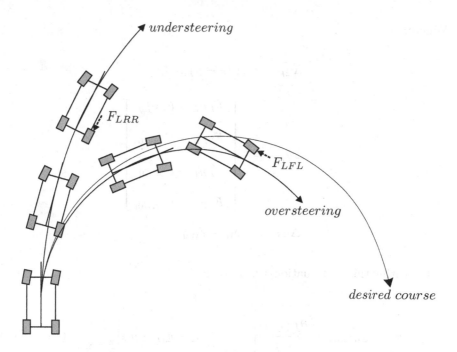

**Figure 10.9** Under- and oversteering

$$\left. \frac{\partial \underline{f}(\underline{x}, \underline{u})}{\partial \underline{u}} \right|_{\substack{\underline{x} = \underline{x}_0 \\ \underline{u} = \underline{u}_0}} \cdot \Delta \underline{u} =$$

$$= \underbrace{\left. \begin{bmatrix} \frac{\partial f_1}{\partial F_{LFL}} & \frac{\partial f_1}{\partial F_{LFR}} & \frac{\partial f_1}{\partial F_{LRL}} & \frac{\partial f_1}{\partial F_{LRR}} & \frac{\partial f_1}{\partial \delta_W} \\ \frac{\partial f_2}{\partial F_{LFL}} & \frac{\partial f_2}{\partial F_{LFR}} & \frac{\partial f_2}{\partial F_{LRL}} & \frac{\partial f_2}{\partial F_{LRR}} & \frac{\partial f_2}{\partial \delta_W} \\ \frac{\partial f_3}{\partial F_{LFL}} & \frac{\partial f_3}{\partial F_{LFR}} & \frac{\partial f_3}{\partial F_{LRL}} & \frac{\partial f_3}{\partial F_{LRR}} & \frac{\partial f_3}{\partial \delta_W} \end{bmatrix} \right|_{\substack{\underline{x} = \underline{x}_0 \\ \underline{u} = \underline{u}_0}}}_{\underline{M}_F |_{\substack{\underline{x} = \underline{x}_0 \\ \underline{u} = \underline{u}_0}} \qquad \underline{m}_\delta |_{\substack{\underline{x} = \underline{x}_0 \\ \underline{u} = \underline{u}_0}}} \cdot \begin{bmatrix} \Delta \underline{u}_F \\ \hline \Delta u_\delta \end{bmatrix} \qquad (10.16)$$

$$= \left. \underline{M}_F \right|_{\substack{\underline{x} = \underline{x}_0 \\ \underline{u} = \underline{u}_0}} \cdot (\underline{u}_F - \underline{u}_{F0}) + \left. m_\delta \right|_{\substack{\underline{x} = \underline{x}_0 \\ \underline{u} = \underline{u}_0}} \cdot (\delta_W - \delta_{W0}) \qquad (10.17)$$

Where,

$$\Delta \underline{u}_F = (\underline{u}_F - \underline{u}_{F0})$$

$$\Delta \underline{u}_F = \begin{bmatrix} F_{LFL} - F_{LFL0} \\ F_{LFR} - F_{LFR0} \\ F_{LRL} - F_{LRL0} \\ F_{LRR} - F_{LRR0} \end{bmatrix}$$

$$\Delta u_\delta = \delta_W - \delta_{W0}$$

Substituting this in Equation 10.14 gives:

$$\dot{\underline{x}} = \underbrace{\underline{f}(\underline{x}_0, \underline{u}_0) + \left.\frac{\partial \underline{f}(\underline{x}, \underline{u})}{\partial \underline{x}}\right|_{\substack{\underline{x} = \underline{x}_0 \\ \underline{u} = \underline{u}_0}} \cdot (\underline{x} - \underline{x}_0) + \underline{m}_\delta \Big|_{\substack{\underline{x} = \underline{x}_0 \\ \underline{u} = \underline{u}_0}} \cdot (\delta_W - \delta_{W0})}_{\text{Vehicle dynamics + driver input}}$$

$$+ \underbrace{\underline{M}_F \Big|_{\substack{\underline{x} = \underline{x}_0 \\ \underline{u} = \underline{u}_0}} \cdot (\underline{u}_F - \underline{u}_{F0})}_{\text{Yaw control input (forces)}} \tag{10.18}$$

To allow a good physical interpretation of the control parameters, the controller will be designed using a pole placement approach. Normally, the output vector $\underline{y}$ is fed back. This output vector however does not contain $\beta$, which is the main variable to be controlled. Because of this, the state vector $\underline{x}$ is obtained using data fusion in Sections 9.1 and 9.2 and the observer in Section 9.6). The controller design can then be based on the feedback of the state vector $\underline{x}$.

A control law is then defined for small deviations from the operating point $\underline{x} - \underline{x}_0$. This is justified because under normal, non-critical cornering, $\underline{x} - \underline{x}_0$ is equal to zero. Only when pre-determined thresholds are exceeded, in critical situations, is there a difference between the desired and actual state vectors which must be controlled.

The following control law is used:

$$\Delta \underline{u}_F = -\underline{K}_C \cdot \Delta \underline{x} \quad , \tag{10.19}$$

where $\underline{K}_C$ is the feedback matrix.

Substituting into the linearized state space description Equation 10.18 gives:

$$\Delta \underline{\dot{x}} = \left.\frac{\partial \underline{f}}{\partial \underline{x}}\right|_{\substack{\underline{x}=\underline{x}_0\\\underline{u}=\underline{u}_0}} \cdot \Delta \underline{x} - \left.\underline{M}_F\right|_{\substack{\underline{x}=\underline{x}_0\\\underline{u}=\underline{u}_0}} \cdot \underline{K}_C \cdot \Delta \underline{x} + \left.\underline{m}_\delta\right|_{\substack{\underline{x}=\underline{x}_0\\\underline{u}=\underline{u}_0}} \cdot \Delta \delta_W \quad,$$

$$\Delta \underline{\dot{x}} = \left( \left.\frac{\partial \underline{f}}{\partial \underline{x}}\right|_{\substack{\underline{x}=\underline{x}_0\\\underline{u}=\underline{u}_0}} - \left.\underline{M}_F\right|_{\substack{\underline{x}=\underline{x}_0\\\underline{u}=\underline{u}_0}} \cdot \underline{K}_C \right) \cdot \Delta \underline{x} + \left.\underline{m}_\delta\right|_{\substack{\underline{x}=\underline{x}_0\\\underline{u}=\underline{u}_0}} \cdot \Delta \delta_W \quad,$$

$$(10.20)$$

where the term inside the brackets represents the system matrix for the closed loop. The driver input shall be regarded as superimposed disturbance. The dynamic characteristics of this system can then be set using pole placement. A desired matrix $\underline{G}$ is defined, which has the desired system characteristics, i.e. the desired pole positions:

$$\left.\frac{\partial \underline{f}}{\partial \underline{x}}\right|_{\substack{\underline{x}=\underline{x}_0\\\underline{u}=\underline{u}_0}} - \underline{M}_F \cdot \underline{K}_C \overset{!}{=} \underline{G} \tag{10.21}$$

Solving for $\underline{K}_C$ gives the control law:

$$\underline{K}_C = [\underline{M}_F]^+ \cdot \left( \left.\frac{\partial \underline{f}}{\partial \underline{x}}\right|_{\substack{\underline{x}=\underline{x}_0\\\underline{u}=\underline{u}_0}} - \underline{G} \right) \quad, \tag{10.22}$$

where $[\underline{M}_F]^+$ is the Moore-Penrose pseudo inverse of $\underline{M}_F$.

Note that the above gives the feedback matrix $\underline{K}_C$ for the operating point $x_0$, $u_0$. Thus, $\underline{K}_C$ must be re-calculated for each new operating point. This involves:

- Calculation of the tire side slip constants (Section 9.5.2),

- Estimation of the vehicle velocity (Section 9.1), and of the yaw rate (Section 9.2),

- Observation of the vehicle body side slip angle (Section 9.6),

## 10.2.2   Derivation of Reference Values

The lateral acceleration $a_y$ is limited by the friction co-efficient $\mu_S$. Theoretically, a vehicle can turn with a lateral acceleration of 9.81 times the maximum lateral friction co-efficient, i.e. for $\mu_S = 1$, the lateral vehicle acceleration could be $9.81\,m/s^2$, if the vehicle body side slip angle were zero. For vehicle body side

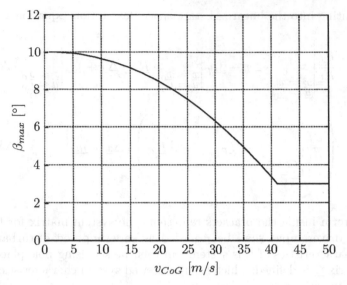

**Figure 10.10** Dependence of maximum vehicle body side slip angle on speed

slip angles greater than zero, a maximum acceleration of $8\,m/s^2$ is adopted. For friction coefficients below 1, the maximum lateral acceleration is given by:

$$a_{Ymax} = \mu_S \cdot 8\,m/s^2 \tag{10.23}$$

The vehicle body side slip angle $\beta$ is also limited. The predominant variable affecting this is the vehicle speed $v_{CoG}$ (see Figure 10.10):

$$\beta_{max} = 10° - 7° \cdot \frac{v_{CoG}^2}{(40\,m/s)^2} \tag{10.24}$$

The reference vehicle body side slip angle is thus given by:

$$\beta_{ref} = \begin{cases} \beta & , & |\beta| \le |\beta_{max}| \\ \pm\beta_{max} & , & \text{otherwise} \end{cases} \tag{10.25}$$

Note that $(\beta_{ref} - \beta) = \Delta x_2 = 0$ when the vehicle body side slip angle is below it's maximum limit.

In the case of **oversteering**, the yaw rate $\dot\psi$ must also be limited by means of the yaw dynamic control. If the derivative of the vehicle body side slip angle $\dot\beta$ is approximated to zero, Equation 8.76 yields:

$$\dot\psi \approx \frac{1}{v_{CoG} \cdot \cos\beta} \cdot \underbrace{\left( \frac{F_{YFL} + F_{YFR} + F_{YRL} + F_{YRR}}{m_{CoG}} - \dot v_{CoG} \cdot \sin\beta \right)}_{a_Y} \tag{10.26}$$

$$\dot\psi_{max} = \frac{1}{v_{CoG} \cdot \cos\beta} (a_{Ymax} - \dot v_{CoG} \cdot \sin\beta_{ref}) \tag{10.27}$$

The reference yaw rate in the case of **oversteering** is then,

$$\dot{\psi}_{ref} = \begin{cases} \dot{\psi} & , & |\dot{\psi}| \leq |\dot{\psi}_{max}| \\ \pm\dot{\psi}_{max} & , & \text{otherwise} \end{cases} \tag{10.28}$$

When the yaw rate is below it's maximum limit, the difference $\dot{\psi}_{ref} - \dot{\psi} = \Delta x_3 = 0$.

In the case of **understeering**, the vehicle body side slip angle $\beta$ and the yaw rate $\dot{\psi}$ are well below their maximum allowable values. The driver tries to maintain the vehicle on the desired course by increasing the steering angle. If the tire side slip angle $\alpha$ and therefore the lateral wheel slip $s_S$ become too large, the lateral friction coefficient exceeds the maximum. The vehicle would then leave the set course. In order to prevent such situations, the vehicle body must turn into the curve at a yaw rate greater than the actual one.

The rear tire side slip angles $\alpha_R$ shall be used as a reference to determine when the front tire side slip angles $\alpha_F$ reach a critical value. A critical ratio of $|\alpha_F/\alpha_R| = 1.5$ is adopted here.

The reference yaw rate in the case of **understeering** is:

$$\dot{\psi}_{ref} = \begin{cases} \pm\dot{\psi}_{max} & , & |\alpha_F/\alpha_R| \geq 1.5 \\ \dot{\psi} & , & \text{otherwise} \end{cases} \tag{10.29}$$

In **non-critical** driving situations, the differences between reference and actual values are zero:

$$\begin{aligned} \Delta x_1 &= v_{CoGref} - v_{CoG} = 0 \\ \Delta x_2 &= \beta_{ref} - \beta = 0 \\ \Delta x_3 &= \dot{\psi}_{ref} - \dot{\psi} = 0 \end{aligned} \tag{10.30}$$

In these cases the control input is equal to zero. Only for $|\beta| > |\beta_{max}|$, $|\dot{\psi}| > |\dot{\psi}_{max}|$ or $|\alpha_F/\alpha_R| > 1.5$, the control becomes active.

The vehicle speed $v_{CoG}$ has not been considered up to now. In an enhanced approach, a maximum vehicle velocity could be derived e.g. from an image processing of the future highway course. If the vehicle speed would exceed this maximum, a corresponding braking force would be generated by the above control law.

Electrical brake systems can directly generate a braking force $F_{Br}$. Hydraulic brakes have a nonlinear characteristic. The brake force $F_{Br}$ must then be converted into corresponding drive signals for the hydraulic brake valves. If $\dot{\omega}$ is neglected, Equation 10.2 becomes:

$$F_{WL} = \mu_L \cdot F_Z = k_{Br} \cdot p_{Br} \tag{10.31}$$

The hydraulic brake pressure is generated by an underlying control loop, as shown in Figure 10.11.

As a simulation example, the case is considered where the driver must carry out an evasive action manoeuvre at high speed. Figure 10.12 shows the uncontrolled behavior of the vehicle. After the first steering action, the vehicle starts to turn around and is no longer steerable.

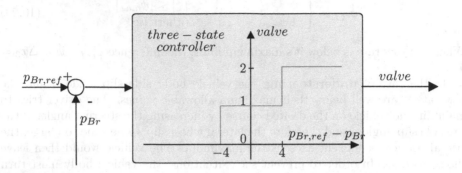

**Figure 10.11**  Underlying brake pressure control loop

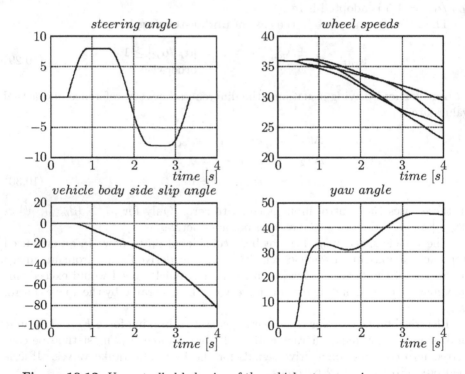

**Figure 10.12**  Uncontrolled behavior of the vehicle at an evasive manoeuvre

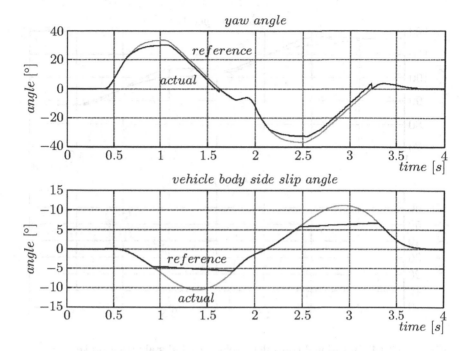

**Figure 10.13** Behavior of the controlled state variables

Figure 10.13 shows the behavior of the controlled state variables. The vehicle body side slip angle cannot be limited to the desired value, however the states remain stable and the vehicle steerable. The deviation of the yaw angle from the desired maximum value is smaller than the deviation of the vehicle body side slip angle.

Figure 10.14 shows the wheel rotational equivalent velocities and the brake pressures. Because the vehicle is concerned in a non-braking drive situation, hence no braking pressure can be reduced, only one wheel is controlled at a time. During the first steering action, the brake pressure on the front right wheel is increased, and during the reverse steering the front left brake pressure is increased. In the second control cycle, the pulsating of the brake pressure shows that the underlying ABS system is activated, to ensure that the maximum value of the friction co-efficient is not exceeded.

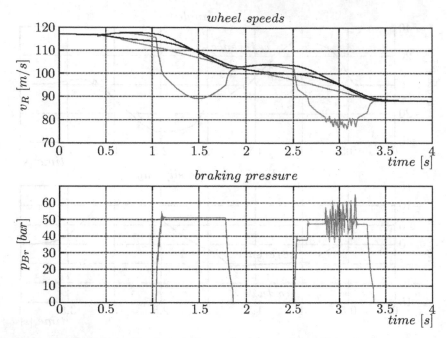

**Figure 10.14** Rotational equivalent velocities and stabilizing brake pressure

# 11 Road and Driver Models

## 11.1 Road Model

### 11.1.1 Requirements of the Road Model

The construction of a course is a fundamental prerequisite for all simulations with driver-vehicle models. In order for practically relevant applications to be carried out, the course must first be defined and then must be connected to the complete model in a meaningful manner.

Naturally, as detailed a reproduction of the road is required as possible, i.e. environmental influences such as wind and weather should be reproduced as realistically as is necessary and possible. Furthermore, the road data under different road gradients in the longitudinal and lateral directions, and different road surfaces, on top of the actual course co-ordinates must not be forgotten. The most important criteria to be considered however is that the course must be one which can, in reality, actually be driven. The planned desired course must therefore satisfy the following criteria [62]:

- **Continuity of the path**
  No car can move instantaneously from one point to another. Thus the continuity of the course is an elementary criterion. The path must be complete (without gaps) in order that the vehicle can travel from the start to the end point.

- **Continuity of the curvature**
  The physical relationship between the rotational equivalent wheel velocity $v_R$ and the rotational velocity of the vehicle in a curve $\omega_C$ is given by

$$\omega_C = \frac{v_R}{\rho} \ ,$$

(11.1)

where $\rho$ corresponds to the curve radius. From this it can be seen that a step-wise change in curve radius $\rho$ would have to result in a step-wise change in the curve rotational velocity $\omega_C$; this is not possible in a real moving vehicle.

- **Differentiability of the set course**
  If a course path is carried out without steps, it is essential that the course path is differentiable, because vehicles cannot in general move in a zigzag fashion. A sharp bend in the set course corresponds to a point with infinitely large curvature $1/\rho$, which again cannot be carried out by a real vehicle.

**Essential Data**

The important values which a road model must provide are:

- Co-ordinates in the x- and y directions

- Curve radius $\rho$

- Desired yaw angle $\psi_{road}$ and yaw rate $\dot{\psi}_{road}$

- Recommended speed, which must be given for guidance $v_{ref}$

- Road gradient in longitudinal and lateral directions $\chi_{road}$, $\varphi_{road}$

- Road surface information $\mu(s)$

- Wind velocity $v_{wind}$

These values are calculated during the simulation using the previously defined set course, and made available both to the driver model and, partially, to the vehicle model. The driver takes in the road information with help of his sensory organs, judges them, and uses them as guidance and system measurements for the steering and control. The vehicle model uses the available values to derive the dynamic changes of the drive states.

## 11.1.2   Definition of the Course Path

A relatively simple and easy to understand method of producing a course path is the division of the whole road path into separate sections. It is possible to reproduce all road courses which occur in real road systems using only three different path forms, namely straight line $s$, spirals, and circular arc segments. As many individual segments as required can be strung together to form a complete course. Each segment is defined using a parameter set. As well as the start and end-positions of each segment, additional parameters such as wind strength and recommended velocity can be defined. Care must be taken however, to ensure that the criteria given in the previous section for a realistic course are satisfied, i.e. the endpoint of one segment must correspond to the beginning point of the next. In order to simplify matters, it is assumed that the given parameters such as recommended velocity, road inclination and road surface remain constant

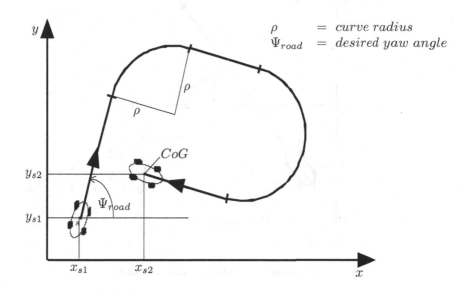

**Figure 11.1** Course definition using segmentation

within the individual segments. This simplification does not put any limits on course design, as a change in parameters is possible via the division of the course into smaller segments.

Figure 11.1 shows an example course constructed from 5 segments. The straight-line segments are at an angle $\psi_{road}$ to the fixed co-ordinate system. The circular arc segments have radius $\rho$, which corresponds to curvature $\kappa = 1/\rho$.

In the following the various forms will be considered in more detail, and the parameters necessary for the definition of the segments given.

### Straight line segments

Straight line segments form the simplest section of the road path. Formally they are described in the following way:

The curvature is equal to zero and the curve radius correspondingly infinite. The following parameters are necessary to define the segment:

| Straight line segment parameters | |
|---|---|
| s_0 | initial value of arc $(m)$ |
| s_end | final value of arc $(m)$ |
| $v_{ref}$ | recommended speed in current segment $(m/s)$ |
| $\psi$_0 | initial yaw angle $(degrees)$ |
| $\psi$_end | final yaw angle $(degrees)$ |

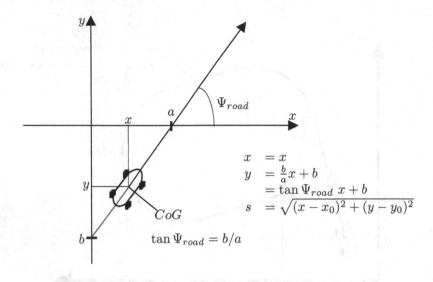

**Figure 11.2** Straight line segments

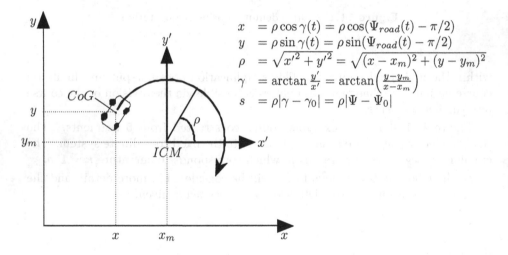

**Figure 11.3** Circular arc segment

## Circular arc segment

Here the radius $\rho$ is constant. The mathematical representation is given in polar co-ordinates in Figure 11.3.

The required parameters for a circular arc segment are:

| Circular arc segment parameters | |
|---|---|
| a/c | Direction (*anticlockwise/clockwise*) |
| s_0 | initial value of arc (*m*) |
| s_end | final value of arc (*m*) |
| $v_{ref}$ | recommended speed in current segment (*m/s*) |
| $\psi$_0 | initial yaw angle (*degrees*) |
| $\psi$_end | final yaw angle (*degrees*) |
| $\rho$ | curve radius (*m*) |

### Clothoide Spiral

The curves of roads built today are generally made up of entrance curves, a section of constant radius and an exit curve. The transition between straight line and circular arc segments is made possible by clothoide spirals, as in this way the demanded condition of continually variable curvature can be satisfied. Clothoide spirals are curves where the radius decreases inversely proportional to increasing curve length $s$:

$$\rho(s) = \frac{v_{ref}^2}{\pi s} T^2 \tag{11.2}$$

| | | |
|---|---|---|
| $a_N$ | : | Normal acceleration of the CoG trajectory |
| $T$ | : | Spiral co-efficient (unit: time) |
| $v_{ref}$ | : | Vehicle velocity |
| $\rho(s)$ | : | Curve radius |

The normal acceleration of the CoG increases with the curve length $s$:

$$a_N(s) = \frac{v_{ref}^2}{\rho(s)} = \frac{\pi s}{T^2} \tag{11.3}$$

The distance of the asymptotic points from the center point $(x_m, y_m)$ is determined by the spiral coefficient $T$.

$$x(s) = x_m + \int_0^s \cos\left(\frac{\pi s^2}{2T^2 v_{ref}^2}\right) ds \quad ; \quad y(s) = y_m + \int_0^s \sin\left(\frac{\pi s^2}{2T^2 v_{ref}^2}\right) ds \tag{11.4}$$

One problem when dealing with spirals is the calculation of the Cartesian coordinates of the individual path points from Eq. 11.4, as the integrals are not closed solutions. The derivation of the curve radius $\rho(s)$ is thus only numerically possible; for this reason observance of the criteria from Section 11.1.1 becomes difficult.

## 11.1.3 Road Surfaces and Wind Strength

It has already been stated that as exact a reproduction as possible of real roads with different road surface qualities such as weather behavior should be considered. Thus for a complete definition of the course some additional parameters are required.

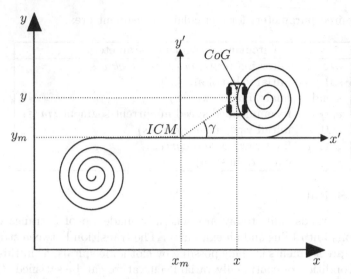

**Figure 11.4** Clothoide Spiral

The separate definition of road surface and weather is not meaningful here, because for the vehicle model, and in particular the wheel model, only a friction co-efficient $\mu_{Res}$ resulting from a combination of the two factors is relevant. Another important factor for the reconstruction of real environmental conditions is the wind force. The wind force is derived during the simulation from the previously defined wind velocities. Thus the wind velocities for the individual $x$, $y$ and $z$ directions should be defined separately for each segment.

## 11.2   PID Driver Model

In order to close the loop in a simulation, knowledge about human control behavior is required. To date, descriptions of driver behavior have not been sufficiently realistic. The main problem lies in the fact that the behavior is dependant upon physical and psychological factors, as well as the demands of the driving situation. Despite these difficulties, many attempts have been made. The driving behavior of humans can, for example, be modeled using classical control theory or fuzzy theory. In the classical approach taken in this section, the driver is modeled as a continuous PID controller. The PID controller reduces the lateral errors very well, but does not behave like a human driver. For a closed loop vehicle-driver simulation this is often sufficient.

The PID controller consists of two almost independent sub-controllers for the longitudinal and lateral dynamics. Coupling occurs only during critical driving situations, if the longitudinal controller reduces the throttle angle. Fig 11.5 shows the structure of the longitudinal controller. As the driver knows the engine characteristics he can set either the desired acceleration $a_{x,des}$ or the corresponding throttle angle $\alpha_{t,des}$.

A proportional controller compensates for errors between the desired and actual acceleration, which occur for example during hill driving. If the engine

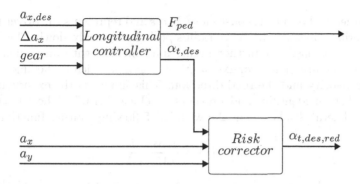

**Figure 11.5** Structure of the longitudinal controller

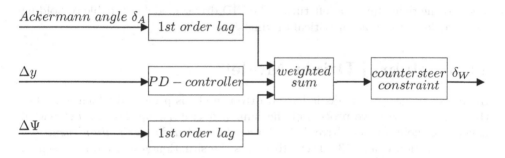

**Figure 11.6** Structure of the lateral controller

torque is still too small, despite an open throttle, the transmission control moves down a gear to increase the torque. If the desired acceleration is zero, the longitudinal controller does not accelerate even in the event of a negative actual acceleration. Thus, during cornering a throttle angle of zero is stipulated. This is mainly important for simulating critical driving situations, where the curve resistance and hence the brake deceleration can become relatively large. The risk corrector (Figure 11.5) mirrors the behavior of the driver, in that no acceleration is carried out when driving quickly and cornering at the same time. The risk corrector comes into play between 60 and 90 % of the exploited traction potential, and reduces the drive torque proportionally. In this approach the stipulation of suitable speed is not the task of the driver (dependant on road geometry), but of the course designer. The longitudinal dynamics driver within the model has just the task to maintain the previously defined acceleration and speed.

The lateral controller is more complex. It consists of three parts, as shown in Figure11.6.

The inputs are three pre-processed variables. Under the assumption that no vehicle body side slip angle $\beta$ or tire side slip angle $\alpha$ is present, the wheel turn angle can be approximated using the Ackermann angle (see [84]):

$$\delta_A \approx \frac{l}{\rho}\left(1 + \frac{v_{CoG}^2}{v_{char}^2}\right) \approx \frac{l_R + l_F}{\rho} \tag{11.5}$$

The first order lag has a time constant of $30\,ms$ and represents the driver behavior during steering. Because the curve radius $\rho(s)$ changes only slowly, this fast first order lag can be used. A further input, implemented with an equivalent first order lag, is the direction error $\Delta\psi = \psi + \beta - \psi_{road}$, which is the angle between the vehicle velocity and the road direction. This simulates the experience of the driver, in that he identifies and prevents a vehicle "drift". The PD controller reduces the lateral displacement $\Delta y$ with the following transfer function:

$$G_{\Delta y}(s) = \frac{0.3s + 1}{0.15s + 1} \qquad (11.6)$$

All three controller outputs are weighted and summed together. The resulting steering angle is constrained against counter-steering, which is important for the vehicle stability during changing curve conditions. Here, the driver should not steer to the right during a left turn. The PID driver model is capable of holding the vehicle on the road in critical driving situations.

# 11.3  Hybrid Driver Model

In this section a more realistic hybrid driver model is presented which describes the complete cognitive process of the human operator [79]. From a behavioral psychology point of view driver behavior can be modelled as a *situation-cognition-action* reaction chain [53]. Basic to this is the situation-action model designed by [15], which is based on the following assumptions:

- The course of travel can be stated as a sequence of situations,

- Every driver has several alternative actions at his disposal for any given situation, each of which will result in alternative consequential situations,

- Each driver chooses one of the possible actions, giving consideration to individual intentions.

## 11.3.1  Vehicle Control Tasks

In human vehicle control tasks can be divided into approximately 65 main tasks and 1799 elementary tasks which a vehicle guidance system must carry out. In [108] and [109] the functions are split into six different task groups.

- Strategic tasks (choice of route, time of departure)

- Navigational tasks (adherence to the chosen route during travel)

- Tasks relating to the road (chosen position within traffic, course)

- Speed control (choice of speed according to situation)

- Traffic related tasks (interacting with other road users in such a way that the traffic is not obstructed and collisions with other road users are avoided)

- Adherence to rules (traffic signs, signals etc.)

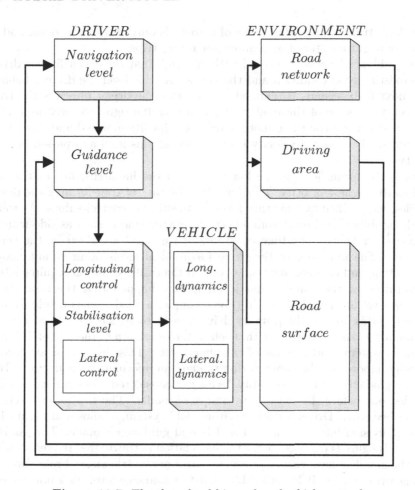

**Figure 11.7** The three level hierarchy of vehicle control

In general drivers have to cope with all of these tasks. The importance of the different tasks changes according to the driver type. An experienced driver for instance will weight the tasks differently than a learner. The six driving task groups above can be further compressed into three main activities, as shown in Fig. 11.7. These are:

1. *Navigational level* - Macro-activities (way finding)

2. *Guidance level* - Situation dependant activities (collision avoidance)

3. *Stabilization level* - Micro-activities (manoeuvring the vehicle)

The navigation level involves the choice of an optimal route from A to B from a range of alternative routes. This is normally carried out before the start of the journey, and depending on the distance between the two points, can involve forward planning for a long time period. For frequently travelled routes (such as the daily trip to work) the choice of a route is only carried out once, and the chosen route is then stored in memory for future trips.

The guidance level is concerned with ensuring that the driver fits his driving to the existing road-condition and the traffic. At this level the driving behavior of the next 5-10 seconds is planned. To do this the driver observes the traffic situation, the course of the road, and possible traffic signs. As well as this the driver must control the three motion variables (position, speed and direction of the vehicle). The desired velocity is calculated at this level and passed onto the stability level.

The stabilization level is the lowest level in the hierarchy, but at the same time has the highest priority. This level has the task of transforming the desired variables, as specified by the other levels, into suitable control values. As well as this, the stability level must compensate for disturbances, such as side-winds or changes in gradients. The time constants at this level are given by the vehicle dynamics. The processes of the three hierarchical levels occur simultaneously during a trip, and thus conflicts can occur. If the traffic situation requires intense concentration at the stabilization level, it may happen that the tasks of the guidance level are not dealt with. For example, a road sign is overlooked. In addition to this, and to the primary driving tasks today's drivers have a myriad of secondary tasks to contend with, such as the operation of the air conditioning, the audio system, and the use of telephones etc. This division of attention is especially important when considering the human information acquisition. Here, only the guidance and the stability levels are considered. The navigation level is carried out before the journey by the road model. The two lower levels can be described with "Driver-vehicle-environment" system, as shown in Figure 11.8. Again, a division into longitudinal and lateral guidance is made. The guidance level in this figure is responsible for the calculation of the desired reference values, while the stability level carries out the control of throttle angle, brake force and steering wheel angle. It is noticeable that the guidance level does not calculate a reference value for the lateral guidance. This is because the task of the lateral controller is well defined - i.e. to keep the lateral error to zero. The instantaneous lateral error can be calculated using the current drive states. For the longitudinal control the situation is somewhat different. The task is to maintain a velocity as exactly as possible. The reference velocity is however strongly dependent upon the street course and the characteristics of the driver. The guidance values for the longitudinal controller change relatively frequently and must be calculated at the guidance level.

## 11.3.2  Characteristics of the Human as a Controller

The human controller is not a technical controller with mathematically determined characteristics. [126] describes a good, safe driver as follows: "The good driver ....

- has complete command of the vehicle equipment

- keeps fairly well to the rules

- takes no unnecessary risks

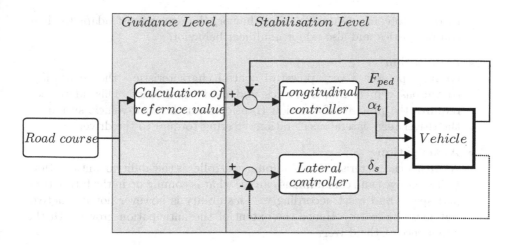

**Figure 11.8** The driver-vehicle-environment system

- has good anticipation ability of the traffic situation in the immediate future

- shows consideration for the mistakes of others

- keeps his temper under control."

This is of course no scientifically correct definition of the characteristics of human control. It shows however that the control behavior is also dependant upon psychological factors. Drivers possess characteristics which are common to all humans. These physiological and psychological factors are considered together as personal factors. These factors affect the characteristics of the human as a controller. The general characteristics of human controllers are:

- **Operating states**
  [38] studied the steering behavior of humans and determined that there are two different states which humans switch between. The first state is the so-called *error correction mode*, in which the human acts as a classical controller and seeks to control the lateral error without interruption. If the lateral error falls below a certain threshold the driver switches to an *error ignoring mode*, whereby the driver tolerates a certain lateral offset and keeps the steering angle constant. Only when the lateral error exceeds a certain threshold will the driver return to error-correcting mode. The thresholds are driver dependent and vary greatly. The transfer from error ignoring to error correction mode can be modeled as an event. The different event rates shall later be joined at prioritized queuing systems, which model the limited human resources to cope with such events.

- **Nonlinearity**
  The nonlinearity of the human controller is mainly due to the many nonlinear processes within the vehicle. The acceleration of the vehicle depends for example in a highly nonlinear manner upon the position of the gas pedal.

In order to compensate for this behavior, the driver must adapt to these characteristics and also adopt nonlinear behavior.

- **Adaptation**
  Humans possess a very powerful control characteristic - they can adapt to various vehicle characteristics and road conditions.  This adaptation requires the precise information transfer from the road-vehicle system to the driver (e.g. lateral acceleration, steering torque) to the driver.

- **Anticipation**
  A further characteristic of the human controller is the ability of anticipation. Using the eyes and ears humans can see what is coming up in the future time and space, and react accordingly.  This ability is however not instinctive, but must be learnt.  Hence the extent of the anticipation grows with the experience of the driver.

- **Time variance**
  Time variance of a controller means that identical input variables at two different time instants may result in two different outputs.  The change of the calculated values for the control inputs is explained by the fact that the parameters or characteristics of the controller change with time.  In human controllers, the personal factors are responsible for the changes in characteristics.

## Mapping of Driving Tasks to the Human Cognitive Levels

In order to model the human brain the three driving task levels must be mapped to the three human cognitive ability levels.  For this, Rasmussen [99] differentiates between:

- Skill-based behavior

- Rule-based behavior

- Knowledge-based behavior

These three levels are shown schematically in Figure 11.9.

In the skill level, the execution of mainly automatic and routine tasks are carried out.  Sampled signals from the environment are compared with previously acquired patterns.  Skill based tasks can be subject to information reduction, followed by characterization and "discovery" of events.  This stimulus-reaction mechanism is first trained in a learning process and from then on runs unconsciously.  In Section 11.3.5 this is modeled using queuing theory.  Examples of the action at this level are the maintenance of a course on a straight road, adherence to a particular speed, or execution of standard actions when overtaking.

Rule-based behavior occurs in known situations.  When the circumstances of a given situation have occurred previously, and it can be dealt with using previously learned rules.  The incoming information is interpreted only after a process of selective information reduction.  It characterizes the different states of the environment.  Fuzzy logic is an obvious choice for the modeling of this level, hence these states are linked to the actions to be performed by IF-THEN rules.

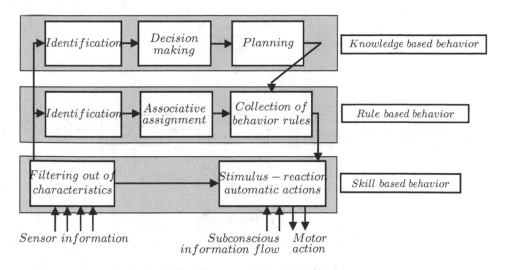

**Figure 11.9** The three human behavior categories (of Rasmussen)

Knowledge based behavior is used in previously unknown situations. Appropriate rules are derived from a modification and combination of existing rules.

Because the execution of unconscious activities takes place in the skill level, allowing the higher levels to concentrate on other matters, there is only a limited amount of processing capacity available to the higher levels. The activities at the higher level run more slowly, but have the advantage of flexibility. In contrast, the lower level has a high operating speed, but is inflexible. With increasing driver experience more and more conscious activities become unconscious ones.

### 11.3.3 Information Handling

The previous section discussed which tasks humans have to carry out within the driver-vehicle-environment system. Now the processing steps within the human will be discussed in more detail. These can be divided into three subsequent steps: information acquisition - information processing - control action. This is shown schematically in Figure 11.10.

#### Information Acquisition of Humans

The driver requires first of all information about the current drive state and the future course. The driver obtains this information via the eyes. The most important perception types are the optical (i.e. information obtained via the eyes) and the vestibular (linear and rotary motion perceived by the sense of balance). The eyes supply the main part of the information. In the literature, values above 90 % are given as the amount of information the eyes supply. The visual sense is distinguished not only because of the wealth of information that it provides, but also in that it provides information about future events (in terms of time and space), which the driver must deal with. The other sensory organs only provide information about the current drive states.

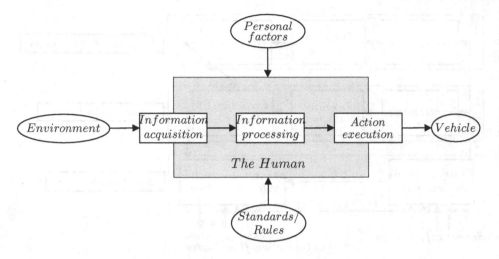

**Figure 11.10**  Information handling of the driver

## Information Processing and Control Action

After the information acquisition, data reduction takes place, as not all the obtained data can be processed. This takes place partly in the sensory organs themselves and partly in the brain. Every sensory organ is equipped with a mechanism whereby physical stimuli can be stored for a short period of time. Thus a physical stimulus can be perceived even though it is no longer present. When, for example, the attention is directed on one object, the sensory short term memory allows information from other physical stimulation to be obtained, for later processing. The information from the sensory organs are selected, assessed and formed into a meaningful complete picture, so that the driver can understand the whole situation and react accordingly.

The various stages of information processing can be seen in Figure 11.11 ([127]). Each step of the processing reshapes the incoming signals in a suitable way. The individual transformations require a certain amount of time and a certain portion of the limited processing resources. Thus the speed of the processing and the amount of data which can be processed is limited. The number of alternatives is dependent upon the experience of the driver. The driver selects one course of action at a time which he considers suitable. He then implements the required changes to throttle, brake and steering wheel to carry out the chosen plan. The reactions of the vehicle and the environment can then be seen again by the driver and so the loop is complete.

## Perception

The perception process is a mapping of many different physical stimuli to perception categories. For example the various ways of writing the letter A,a,*a*, **a** all

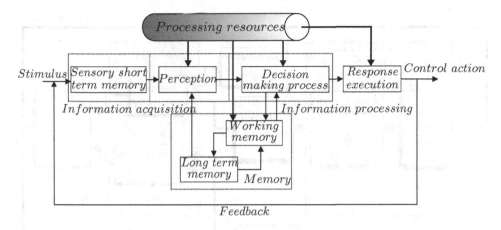

**Figure 11.11** Model of the human information processing

belong to one perception category. Through this division of the many physical stimuli into fewer categories the huge quantity of data can be reduced. However, important information can be lost. For this reason the original physical stimuli are additionally saved at a higher processing level.

### Decision Making Process

The process of decision making is carried out in accordance with real time requirements. On the one hand decisions are made in a very fast reflexive manner. For example, a driver approaching a traffic light which is changing from green to amber has two possibilities. He can either brake and stop before the lights or accelerate and attempt to pass the lights before they change to red. The decision however must be made in a fraction of a second. This critical situation thus requires reflexive decision making. On the other hand the planning of an overtaking manoeuvre is a comparatively longer process. Here a decision is first made as to whether the perceived information will be saved in memory, before a response is selected. If the information is saved, it enters working memory. A further decision is then made as to whether the information is only required for the short term, or if the current situation contains information which is so important that it should be saved in long term memory and thus "learnt". After a long decision period a suitable response to the current situation is chosen, using the just saved information and the various alternative actions called forth from long term memory.

### The Response Execution

Once a response is produced by the decision making procedure it is transformed into corresponding control outputs to the steering wheel, throttle and brake pedal.

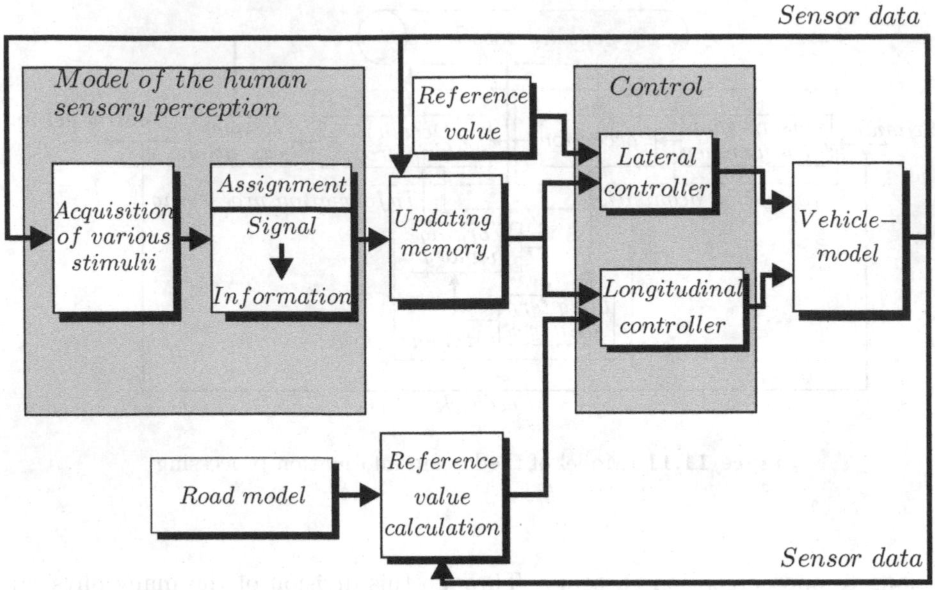

**Figure 11.12** Adaptation of the cognitive process to the driver model

**Processing Resources**

All parts of the information processing procedure, with the exception of the sensory short term memory, require processing resources in order to function effectively. The supply of resources is shown at the top in Figure 11.11. If a processing level requires more capacity, this has to be taken from the remaining processing blocks, so that their performance falls. This may result in the tasks of these blocks being incomplete or faulty.

Through constant learning and practice of recurring tasks the resources required for the execution of tasks can be reduced. Thus an experienced driver requires much less resources than a learner driver, placed in the same situation, would require.

## 11.3.4  Complete Driver Model

The characteristics of humans as controllers must now be transferred with suitable modeling approaches into the a driver model. The individual blocks for the representation of the human cognitive processes are replaced by suitable function blocks. The resulting driver model is shown in Figure 11.12.

Basic to this driver model is that humans can only reach a high control quality when all drive-relevant information is available to them. The control quality depends upon the experience and personality of the driver. Although both the lateral and longitudinal controllers always have access to the current guidance variables, the state variables (i.e. the actual values) are taken from memory (as represented by the block "Updating memory" in Figure 11.12). The updating of

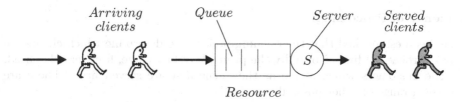

**Figure 11.13** Queuing system

the memory values is dependant upon the results of the information acquisition. Thus it is possible that the controller uses outdated rather than current values. In a situation where the precessing resources of the driver are heavily overloaded, e.g. by servicing the telephone simultaneously to driving, the control performance will significantly degrade.

## 11.3.5   Model of Human Information Acquisition

The information intake of humans is a very complex task and thus some simplifications are necessary in order to model the process. Two major simplifications are carried out:

- Only two perception channels are considered: the visual and the vestibular.

- Other traffic users are not considered.

**Queuing Theory**

Queuing systems are used for the reproduction of the acquisition processes with limited resources. In the human information acquisition, the intake and processing abilities are limited.

Queuing systems basically consist of two parts: a queue and a server. This construction is shown in Figure 11.13.

The incoming clients go first into the queue, which acts as a temporary storage for the clients and determines which client is passed into the server next. The sequence of processing is determined by priority or by the arrival order of the clients. The chosen client passes into the server and is processed. The processing lasts for a certain amount of time, the service time $\tau_S$, which in most cases has a stochastic distribution. The time between the arrival of two clients is termed the inter-arrival time $\tau_A$, which can also be regarded as a stochastic variable. The aim of the queuing theory is to describe analytically the waiting time $\tau_W$ which a client spends in the queue, based on the distribution of the inter-arrival time and the service time. Furthermore, from the inter-arrival times and the service time the number of clients in the queue can be determined. If this number increases with time then the system becomes unstable, i.e. the resource is overloaded [65].

## Sequence of Processing

It has been established that the queuing policy is to determine which clients are served (processed) first. Normally the principle of first come, first served is used, i.e. the customers which arrive at the queue first are served first. There are however a range of other possibilities:

### Priority Queuing

Each client has a predefined priority assigned to it. In this case the order of processing is dependant not upon the order in which the clients arrive in the queue, but upon the priority: higher priority clients are always served first.

### Limited capacity Queuing

A limited capacity queue is one in which only a limited number of clients can be stored prior to processing. If the maximum number of clients is reached then no new clients are taken into the queue. As well as this number-based limit there are also time based limits, whereby a client which has been in the queue longer than a specified time is removed from the queue, whether it has been processed or not.

### Pre-emptive / Non-pre-emptive

If a client appears in the queue with a higher priority than the client which is currently being processed, there are two possibilities. Firstly the lower priority client can be removed from the server and the higher priority client processed immediately. Only when the higher priority client has been served can the lower priority client continue being processed. Such a system is termed *pre-emptive*. If on the other hand the higher priority client has to wait until the lower priority client is finished, the system is termed *non-pre-emptive*.

The human information intake is modeled using two separate queues (Figure 11.14).

One queue is used to model the visual information intake. Every possible viewpoint on the road and within the vehicle is reproduced as a source node. A source node becomes a client or event when the driver should consider the corresponding viewpoint. A client thus represents a viewing event. The viewpoints can be divided into two groups: those on the road and those within the vehicle. This latter group are considered as disturbances as they provide no relevant driving information. Each event, as it appears, is given a number, which denotes which source node the event came from, and a priority, which depends on the source node. The level of priority represents the importance of the corresponding source node. The time of appearance of each event is also recorded. As well as deciding when a view point will be seen, it must also be decided the length of viewing time. Thus each event is assigned a service time which represents the normal time a driver must consider the corresponding viewpoint in order to obtain all the necessary information. Once these values have been assigned the clients enter the priority queuing system of the visual information intake. When a request is processed by the server this means that the driver looks at the corresponding viewpoint for the duration of the assigned service time. Once the client has been

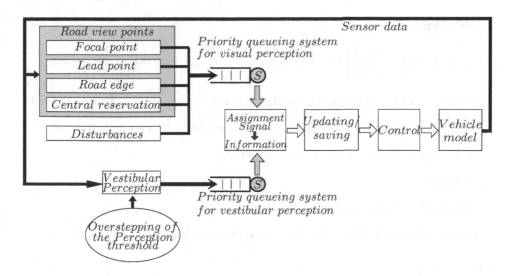

**Figure 11.14** Queuing model of human information intake

processed in this way it passes to a so-called destination node, where it is erased. The system operates in pre-emptive mode. Previously it was stated that all the sensory organs have a short term memory capability. This is reproduced in the queuing system by limiting the amount of time a client may wait in the queue. If a client has not been serviced within this time period then it is removed from the queue and erased. A further source node is necessary to prevent the case that there is no client either in the queue or in the server - this would in effect correspond to the driver having closed eyes. To prevent this there is always an event in the category "lead point", if there are no other clients in the queue.

*Assignment* of signals to an information category takes place in the block "Assignment Signal - Information". The assignment is necessary because a viewpoint can contain several pieces of information. The information from this is then updated in memory, and the controller has access to the current values with the event arrival rate of the "clients" (i.e. perception events).

The vestibular information acquisition is modeled with a separate queue in parallel. In the source nodes clients are created when the state values exceed the corresponding perception thresholds. The clients then land in the queue and are served, and the values assigned and updated in the same way as with the visual intake.

## 11.3.6  Inter-event Arrival and Service Times

In this section the time interval between the creation of customers, *the inter-event arrival time*, will be defined. The viewpoints are divided into two main groups, those outside the vehicle and those within the vehicle (disturbances). The overall viewing frequency $f_{view}$ of a driver is in the range of 2-6 viewpoints per second, and is given by:

$$f_{view}(v_{CoG}, a_X) = \frac{1}{3}\left[\frac{s}{m}\right]|a_x| - 0.001401\left[\frac{s}{m^2}\right]v_{CoG}^2 - 0.09384\left[\frac{1}{m}\right]v_{CoG} + 7\left[\frac{1}{s}\right]$$

(11.7)

This overall viewing frequency is divided between various viewpoints on the road:

- Visual focus (point corresponding to approximately 3 seconds drive away)

- Leadpoint (furthest point of driver view)

- Road edge

- Center of road (i.e. central reservation)

The time intervals for the individual viewpoints are given by:

**Visual focus**

$$\tau_f = \frac{35\left[\frac{m}{s}\right]f_1 + 6\left[\frac{m}{s^2}\right]f_2 + 17°\left[\frac{1}{s}\right]f_3 + 4.5°\left[\frac{1}{s^2}\right]f_4 + 90°f_5}{C^* \cdot \left(f_1 v_{CoG} + f_2|a_Y| + f_3|\dot\psi| + f_4|\ddot\psi| + f_5|\Delta\psi|\right) \cdot p_f}$$

(11.8)

**Lead point**

$$\tau_l = \frac{35\left[\frac{m}{s}\right]l_1 + 6\left[\frac{m}{s^2}\right]l_2 + 17°\left[\frac{1}{s}\right]l_3 + 4.5°\left[\frac{1}{s^2}\right]l_4 + 90°l_5}{C^* \cdot \left(l_1 v_{CoG} + l_2|a_Y| + l_3|\dot\psi| + l_4|\ddot\psi| + l_5|\Delta\psi|\right) \cdot p_l}$$

(11.9)

**Road edge**

$$\tau_r = \frac{0.4\,[m]\,r_1 + \left[\frac{m}{s}\right]r_2 + 17°\left[\frac{1}{s}\right]r_3 + 4.5°\left[\frac{1}{s^2}\right]r_4 + 90°r_5 + 35\left[\frac{m}{s}\right]r_6}{C^* \cdot \left(r_1|y| + r_2|\dot y| + r_3|\dot\psi| + r_4|\ddot\psi| + r_5|\Delta\psi| + r_6 v_{CoG}\right) \cdot p_r}$$

(11.10)

**Road center**

$$\tau_c = \frac{0.4\,[m]\,m_1 + \left[\frac{m}{s}\right]m_2 + 17°\left[\frac{1}{s}\right]m_3 + 4.5°\left[\frac{1}{s^2}\right]m_4 + 90°m_5 + 35\left[\frac{m}{s}\right]m_6}{C^* \cdot \left(m_1|y| + m_2|\dot y| + m_3|\dot\psi| + m_4|\ddot\psi| + m_5|\Delta\psi| + m_6 v_{CoG}\right) \cdot p_m}$$

(11.11)

with the abbreviation

$$C^* = \left(c_1|a_X| + c_2 v_{CoG}^2 + c_3 v_{CoG} + 7\right) \quad .$$

The variables $f_1 \ldots f_5$, $l_1 \ldots l_5$, $r_1 \ldots r_6$, $m_1 \ldots m_6$, $c_1 \ldots c_3$, $p_f$, $p_l$, $p_r$ and $p_m$ are weighting factors, values for which are given in Appendix A.4. Values for the constants $c_1 \ldots c_3$ are also given in Appendix A.4.

Using Eqs 11.8 to 11.11, the event inter-arrival time for viewpoints outside the vehicle can now be calculated.

The inter-arrival time of disturbances has been determined by [129], who lists mean values and standard deviations for the viewing frequencies of viewpoints within a vehicle. The viewing frequencies are assumed to have a normal distribution over time. An exception is the arrival time of the revolution counter, which is mainly used during acceleration manoeuvres:

$$\tau_{tacho} = \frac{3\left[\frac{m}{s}\right]}{|a_X|} \tag{11.12}$$

Values for the event inter-arrival times of some common disturbances are given in Appendix A.4.

**Determination of the Service Times**

The service time is the amount of time which the driver spends looking at a particular point.

In the literature, the service times for the road viewpoints are given as between 0.25 and 1.8 seconds [51]. Here a middle value of 0.5 to 1 second is used. The viewing length is however dependent upon the vehicle speed; it increases with increasing speed. A linear relationship between the service time and the vehicle velocity is defined for the lead point, road edge and road center as:

$$\tau_S = \frac{1}{70}\left[\frac{s^2}{m}\right] v_{CoG} + \frac{1}{2}[s] \tag{11.13}$$

The viewing time for the visual focus lies between 1.2 and 1.9 seconds [128].

The service times of the disturbances are taken from [129], and some are listed in Appendix A.4.

**Viewpoint Priorities**

The order of processing is dependant upon the priorities of the customers. Each transaction is assigned a priority depending upon the importance of the corresponding view point. The higher the priority is, the more important the viewpoint is. The priorities of some common viewing points are given in Appendix A.4.

**Simulation Results**

Simulations were carried out using typical road courses. The event inter-arrival times and service times for view points within the car were compared with the measured values found by [129]. Figure 11.15 shows the values obtained for the event inter-arrival times for the disturbances. The average values calculated using the queuing theory model correspond very well to the measured values. Figure 11.16 shows a comparison of calculated and measured service times for the disturbances. Again, the correspondence is good.

## 11.3.7 Reference Value Calculation

The longitudinal controller requires a reference value (desired velocity) which it can compare with the actual value (actual velocity). The calculation of this desired value, which is dependant upon the road layout and the individual driver, is

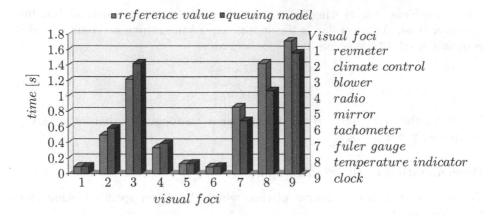

**Figure 11.15** Average event inter-arrival times of disturbances

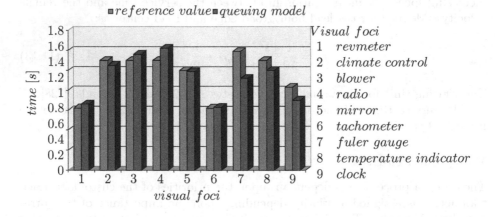

**Figure 11.16** Average service times of disturbances

the task of the "Reference value calculation" block of the driver model. In the following a pure acceleration controller is considered. Thus a reference acceleration must be considered, rather than a reference velocity. The task first requires the construction of a velocity profile for the upcoming road. This changes throughout the journey. Thus only the current, and short term future reference velocity can be calculated. This velocity profile must then be tuned according to the driver type.

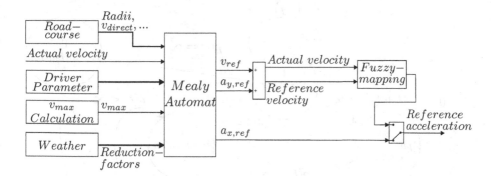

**Figure 11.17** Structure of reference value calculation

A reference velocity is derived for all subsequent course segments (Section 11.1.2). For the modeling of such a system of different course segments, automata are particularly suitable.

### Structure of the Desired Value Calculation

The structure of the desired value calculation is shown in Figure 11.17.

A Mealy automat forms the core of this function block. From the "Road course" block come the current and future values for the course radius, velocity etc. The automat also requires the parameters of the driver in order to fit the velocity to the driver type. It further requires the maximum possible velocity, which is calculated from the friction characteristic between the tires and road. The weather conditions along the course path also have a strong influence on the maximum permitted velocity - this information being provided by the "weather" block. From these inputs the automat calculates its own states and the outputs $v_{ref}$, $a_{y,ref}$ and $a_{x,ref}$. The actual and reference velocities must then be converted into reference accelerations, which are then passed to the longitudinal controller.

### The Mealy Automat

All physically realizable automata react in a particular way to inputs from the environment. For each input the automat moves from its current state to a new state (state transfer on external event). The output is then calculated from the current state of the automat and the current input variables. There are various types of automata - here we are concerned with the Mealy Automat. Figure 11.18 shows the states and transitions of a finite Mealy automat for the reference value calculation.

In describing the individual states two times are used: $t$ is the current simulation time and $t_f$ is the so-called foresight time of the driver. This foresight time is a drive parameter and is dependent upon the driver type.

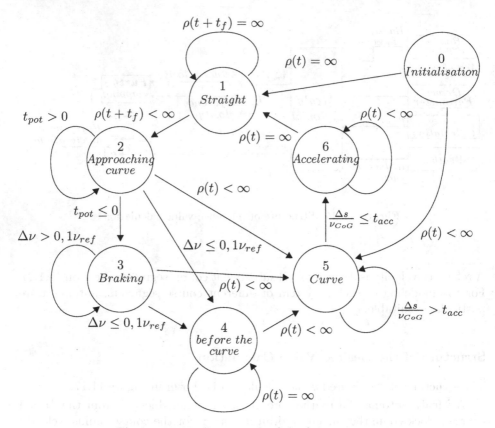

**Figure 11.18**  State graph of the finite automat

## State 0: Initialization

The first time the automat is called, it is in state 0. The automat does not return to state 0 during the course of travel.

## State 1: Straight Line Segment

This state represents straight ahead driving with no curve in the foreseeable future (i.e. $\rho(t + t_f) = \infty$ ). In this stage the reference velocity for an average driver $v$ is multiplied by a factor $fact$ which represents the driver type:

$$v_{ref} = v \cdot fact \tag{11.14}$$

If the vehicle approaches a curve, (i.e. $\rho(t+t_f) < \infty$ ) then the automat switches to state 2.

## State 2: Approaching a Curve

The automat is in this stage when the vehicle is driving on a straight patch of road and the driver is in view of a curve. Normally a curve cannot be driven at the same speed as a straight section of road and so the driver must reduce speed before entering the curve, i.e. he must brake before he reaches the curve. The extent and time of the braking must be determined. The driver maintains the current speed until the start of the braking manoeuvre. The calculation of the curve velocity is carried out using the equations from [75]:

$$
v_{ref} = \left( 202.33 - 104.70 \cosh \left( \frac{\chi^*_{road} + 1}{10} \right) \right)
$$

$$
\cdot \tanh \left( \frac{\rho^*}{64.0 + (0.6\rho)^{0.99}} \right) \left[ \frac{m}{s} \right] + 6.375 \tanh \left[ 1.10(B_F - 7.70m) \right] \left[ \frac{1}{s} \right]
$$

$$(11.15)$$

$$
\rho^* = \rho + (1.75 - 0.4 B_F) \frac{\sin \frac{\psi_{road}}{2}}{1 + \sin \frac{\psi_{road}}{2}} \tag{11.16}
$$

$$
\chi^*_{road} = \left[ 1 - e^{-(0.014\rho^*)^3} \right] \cdot \chi_{road} \tag{11.17}
$$

The variables have the following meaning:

| | |
|---|---|
| $\chi_{road}$ | longitudinal gradient |
| $\chi^*_{road}$ | equivalent gradient |
| $\rho$ | curve radius |
| $B_F$ | road width |
| $\psi_{road}$ | road yaw angle |
| $\rho^*$ | equivalent radius |

The calculated curve velocity $v_{ref}$ represents the velocity for optimal weather and road surface conditions for an average driver. In the calculation the actual curve radius $\rho$ is replaced by the equivalent radius $\rho^*$, which considers "cutting" of the curve. In the simulations done in this book the longitudinal gradient $\chi_{road}$ was always considered to be zero. The velocity is then fitted to the driver type using a factor $fact$. To calculate the time for the start of braking, the individual driver is again important. Each driver has a "potential braking acceleration", $a_{xpot}$, with which they will comfortably reach the reference speed at the start of the curve. In addition to this non-critical situation (i.e. the driver has time to brake comfortably) there is a "maximum braking" situation - i.e. full brakes $a_{xmax}$. Thus two braking times can be calculated:

$$
t_{pot} = \frac{\Delta s}{v_{CoG}} + \frac{\Delta v}{a_{xpot}} - \frac{1}{2} \left( \frac{\Delta v^2}{v_{CoG} \cdot a_{xpot}} \right) \tag{11.18}
$$

$$
t_{max} = \frac{\Delta s}{v_{CoG}} + \frac{\Delta v}{a_{xmax}} - \frac{1}{2} \left( \frac{\Delta v^2}{v_{CoG} \cdot a_{xmax}} \right) \tag{11.19}
$$

There, $\Delta v$ is the difference between the current and future reference speeds, $\Delta s$ the distance to the curve. Normally the potential time is greater than the maximum time. If either of the above times reaches the value zero then the braking manoeuvre begins. If the braking manoeuvre begins because $t_{pot}$ is equal to zero then the driver can brake with the desired "comfortable" brake force. Otherwise, a higher braking force must be used. Some drivers wish to obtain the desired curve speed before entering the curve, whilst others only reach the cornering speed within the curve itself. To allow for these driver differences a further time $t_{antic}$ may be considered, which represents the time span where the driver already wishes to have reached the cornering velocity before entering the curve. This value must be subtracted from $t_{pot}$ and $t_{max}$. Note that $t_{antic}$ can be negative. Once braking commences the automat switches to state 3.

## State 3: Braking

In this state the braking manoeuvre must be carried out. The output of this state is a reference deceleration. Firstly it must be decided if full braking is necessary or if lighter braking will suffice. If lighter braking can be used then the following equation is used:

$$ax\_ref = -\frac{v_{CoG} \cdot \Delta v - \frac{1}{2}\Delta v^2}{\Delta s_1} \tag{11.20}$$

There, $\Delta s_1 = \Delta s - v_{ref} \cdot t_{antic}$ is the distance to the point at which the driver wishes to reach the curve speed. Braking is only applied until:

$$\Delta v \leq \frac{1}{10}v_{ref} \tag{11.21}$$

If $\Delta v$ falls below this threshold before the curve is reached, the automat switches to state 4, otherwise the automat enters state 5.

## State 4: Before the Curve

In state 4 the vehicle is immediately before the curve, having almost reached the reference curve velocity. The necessary light braking required to obtain the reference curve speed can be calculated. Once the vehicle enters the curve, the automat enters state 5.

## State 5: Curve (Circular Arc Segment or Spiral)

In this state the vehicle is traveling at the reference speed determined by the driver using his perception of the curve radius and the weather and road surface conditions. The driver can now determine from the lateral acceleration whether this speed is indeed correct or if the speed must be adjusted. The corrected reference speed is calculated based on the lateral acceleration - different drivers will tolerate different levels of lateral acceleration $a_{ypot}$ and so again a driver factor must be included.

$$v_{ref} = \sqrt{|a_{ypot} \cdot \rho|} \tag{11.22}$$

Here, $\rho$ is the curve radius and $a_{ypot}$ represents the driver-dependant tolerable lateral acceleration.

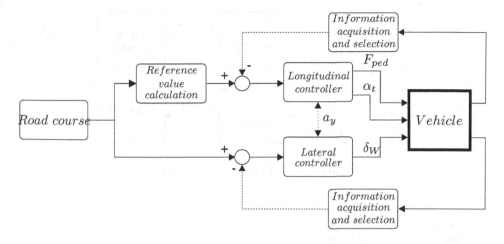

**Figure 11.19** Controller structure

### State 6: Accelerating

Some drivers begin to accelerate before the end of the curve. A driver dependent time $t_{acc}$ is introduced which represents the time before the end of the curve at which the driver begins to accelerate. If the residual curve length is $\Delta s \leq v_{CoG} \cdot t_{acc}$, then state 6 is entered. The reference acceleration is calculated as:

$$a_{ref} = min \left\{ \begin{array}{l} (v_{ref} - v_{CoG})/t_{acc} \quad , \\ a_{max} \end{array} \right. \tag{11.23}$$

## 11.3.8  Longitudinal and Lateral Control

The controller represents the last unit of the driver model. This block has the task of calculating the suitable control variables from the information about the actual driving state, provided by the information acquisition and information selection, the information about the future course path and the reference values. The control variables are the gas pedal position, the brake pedal force and the steering angle.

The controller consists of two connected blocks as shown in Figure 11.19. The first block serves to control the longitudinal dynamics and calculates the control variables for the gas- or brake-pedal. The second calculates the suitable steering angle and thus influences the lateral vehicle dynamics. The two controllers are coupled, since a rising vehicle speed $v_{CoG}$ increases the lateral acceleration $a_Y$ .

$$a_y = v_{CoG}^2 \left[ \frac{\delta_W}{l(1 + K v_{CoG}^2)} \left[ \frac{1}{deg} \right] - \frac{1}{\rho} \right] \tag{11.24}$$

where:

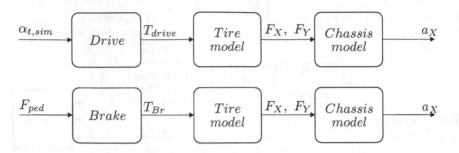

**Figure 11.20**  Drive and brake plants

$v_{CoG}$ : Vehicle velocity
$\delta_W$ : Wheel turn angle
$l$ : Wheel base $l_F + l_R$
$K$ : Stability factor (units : $(h/km)^2$)
$\rho$ : Curve radius

In order to be able to design a longitudinal and lateral controller for the vehicle model described in Chapter 8.2, simplified linear longitudinal and lateral closed loop systems must be identified. The identification of the longitudinal control plant is divided into drive and brake control.

The input signals of the two systems are either the throttle angle $\alpha_t$ or brake pedal force $F_{ped}$, and the output is the longitudinal acceleration $a_X$ (see Figure 11.20).

## Identification and Validation of the Drive System

The identification is limited to the drive system of Figure11.20, with the throttle angle as the input and the drive torque $T_{Drive}$ as the output. For this a linear second order ARMA transfer function is assumed. For the identification a test drive on a straight dry road was carried out, whereby the initial velocity corresponds to $24\,m/s$. The transmission ratio was $i_{gear} = 1.41$. The test signals were a minimum throttle angle $\alpha_{t,sim,1} = 1\,°$ and a maximum $\alpha_{t,sim,2} = 10\,°$ (Figure 11.21). An average driver produces frequencies of between 0 and $2\,Hz$ by acceleration. The considered bandwidth of the system is therefore $5\,Hz$ and the settling time lies at approximately $0.3\,s$. Thus a sampling frequency of $100\,Hz$ will suffice. Figure 11.21 shows the behavior of the throttle angle with time and the resulting drive torque, which are required as inputs and outputs for the identification.

The transfer function for the drive system was estimated as:

$$G_D(z^{-1}) = \frac{0.112295 + 0.112314\,z^{-1}}{1 - 1.7713\,z^{-1} + 0.810704\,z^{-2}} \tag{11.25}$$

When the outputs of the real system and of the identified linear second order ARMA model are plotted the signals are almost identical, thus the system behavior was very well reproduced (Figure 11.22).

Figure 11.24 shows the validation of the estimated 2nd order drive model in the frequency range. There is a good correspondence in terms of both amplitude

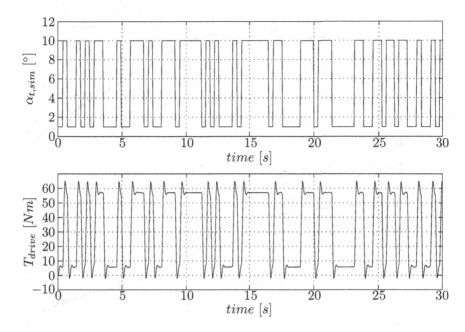

**Figure 11.21** Input and output signals for the identification of the drive system

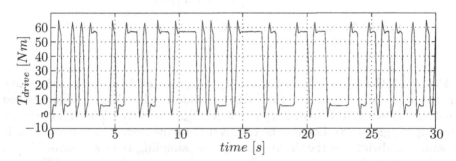

**Figure 11.22** Results of the validation of the 2nd order drive model in the frequency range

and frequency between the estimated model (dashed line) and the real drive system (solid line).

### Identification and Validation of the Brake System

Again in the case of the brake system a model is identified only for a part of the system. This reduced system takes as its input the brake pedal force $F_{ped}$ and outputs the brake torque $T_{Br}$. Again a test drive on a straight dry road was carried out, this time with initial velocity of $35\,m/s$. The reason for the higher initial speed is to avoid the speed falling below $5\,m/s$. A linear third order ARMA model is assumed for the brake control system.

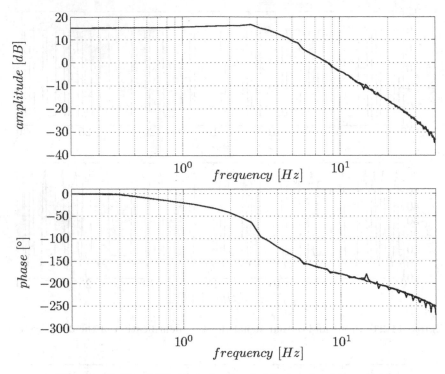

**Figure 11.23**  Validation of estimated 2nd order drive model

Figure 11.23 shows the input and output signals used for the identification. For upper and lower values of the PRBS (pseudo random binary sequence)test signal the following were used: $F_{ped,1} = 30\,N$ , $F_{ped,2} = 80\,N$ . The impulse width of the input signals was chosen as before, as the braking system has approximately the same bandwidth as the drive system. The sampling time was again chosen as $10\,ms$. Also, the sampling time for the brake system should not differ from that of the drive system.

The 3rd order transfer function for the brake system was identified as:

$$G_B(z^{-1}) = \frac{0.0909361 - 0.0861402\,z^{-1} + 0.0502816\,z^{-2}}{1 - 2.06324\,z^{-1} + 1.44986\,z^{-2} - 0.34446\,z^{-3}} \qquad (11.26)$$

A plot of the output of the real break system and the output of the 3rd order identified model in response to a PRBS input signal shows the two to be almost identical.

The frequency responses of the real system and the identified 3rd order model are shown in Figure 11.25. The plots cannot be differentiated for the main part. Hence the real brake system is well described in the frequency range of interest.

As already stated, the entire longitudinal control system cannot be identified because of the nonlinearities in the tire model. In order to design a controller for the entire system, these nonlinearities must be approximated. The two already obtained drive and brake models allow controllers to be designed for the brake

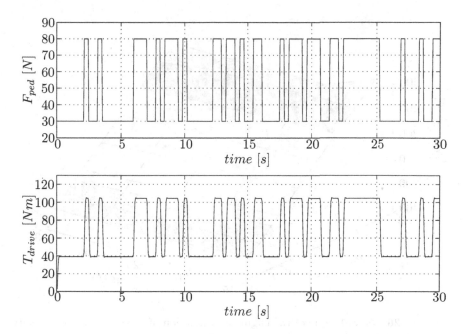

**Figure 11.24** Input and output signals used for the identification of the brake system

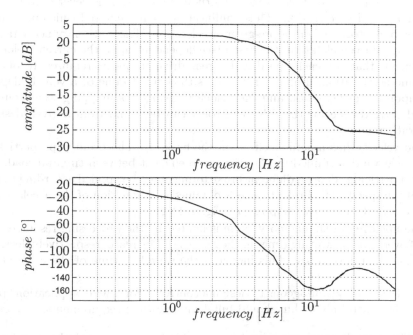

**Figure 11.25** Frequency response of the real brake system (solid line) and the identified 3rd order model (dashed line)

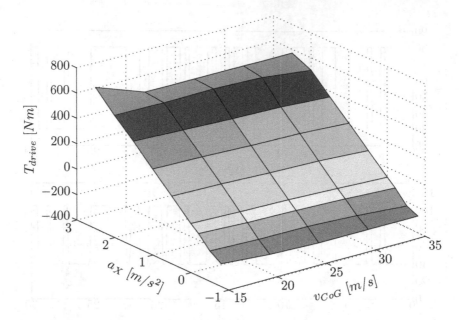

**Figure 11.26** Drive torque of the engine as a function of acceleration and velocity

and drive torques. The task of the longitudinal controller however is to control the longitudinal acceleration. Thus nonlinear maps are required which produce a drive or brake torque from a pre-given acceleration. The stationary acceleration properties of a vehicle are in the first place dependent upon the current velocity. This means that for each speed a different drive torque is required to obtain the same acceleration. Figure 11.26 shows the drive torque of a vehicle dependant upon the acceleration and the velocity. Using this mapping, the required desired drive torque for the drive controller can be found from a given desired acceleration.

The velocity has a similar effect on the braking. The braking properties of the vehicle are determined by the friction co-efficient between tire and road. To obtain a particular delay a different brake torque must be applied according to the level of the speed. The brake torque is obtained as a function of the acceleration and the velocity from the mapping of Figure 11.27.

The updating of the memory value is limited by the processing resources. In case of overload the waiting times in the queues will increase, respectively decreasing memory update rates. This increases the sampling time of the controllers.

It is therefore advantageous to design a controller for a wide application spectrum. The control algorithm should fulfill the following requirements:

- The controller concept must be applicable to non-minimal phase systems

- It must be suitable for the control of unstable systems and systems with poorly damped poles

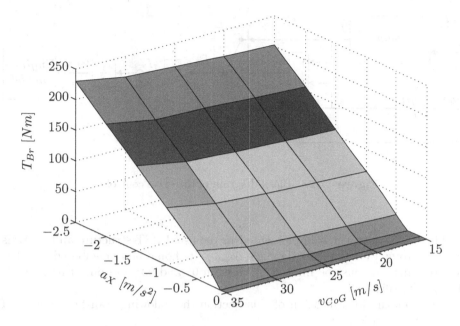

**Figure 11.27** Brake torque as a function of acceleration and velocity

- It must be capable of controlling systems with unknown and variable dead times

- It must be applicable to systems with unknown order.

These requirements are fulfilled by the General predictive controller (GPC) [19]. An introduction can be found in appendix A.3.

### Drive Controller

The control of the longitudinal dynamics is divided into two parts, the drive and the brake control. Both controllers are acceleration controllers, i.e. they take as inputs the desired and actual acceleration and as output the throttle angle or brake pedal force. It is important to note that the two controllers are never active both at the same time. The most important parts of the drive controller, as shown in Figure 11.28 are the velocity dependant map, a switch which chooses the controller input, and the GPC controller itself (R-, S-, T-polynomials).

The velocity dependant map derives the drive torque $T_{ref}$ from the desired acceleration $a_{ref}$ and the actual velocity $v_{CoG}$. This is required by the GPC controller as a set point. In the switch a choice is made between this desired drive torque and the maximum engine brake torque which can be delivered by the vehicle given the current transmission ratio. If the desired drive torque is larger than the maximum brake torque then this desired value is passed to the controller (this is the situation depicted in Figure 11.28). If this is not the case then the maximum brake torque is passed. This has the advantage that small decelerations can appear exclusively via the braking action of the engine.

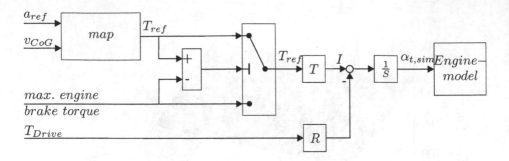

**Figure 11.28**  Block diagram of the drive controller

The GPC controller is implemented in the usual RST-structure and controls the drive torque via the throttle angle $\alpha_{t,sim}$. It should be remembered that the throttle angle $\alpha_t$ must be limited to between $0°$ and $90°$ and only this range is physically realizable.

Based on the identification of this Section the following transfer function of the drive system is given:

$$G_D(z^{-1}) = \frac{B(z^{-1})}{A(z^{-1})} = \frac{0.112295 + 0.112314\,z^{-1}}{1 - 1.7713\,z^{-1} + 0.810704\,z^{-2}} \qquad (11.27)$$

Based on this transfer function and consideration of the tuning rules, the control parameters are chosen as:

- $N_1 = 1$

- $N_2 = 40$

- $N_u = 3$

- $\lambda_u = 10$

The poles of the T-polynomial filter lie between 0.5 and 0.6 so that a better damping of the poles of the estimated model is obtained. This gives for the polynomials $R(z^{-1})$, $S(z^{-1})$ and $T(z^{-1})$ of the drive controller:

$$
\begin{aligned}
R_D(z^{-1}) &= 1.14153 - 1.90449\,z^{-1} + 0.811293\,z^{-2} & (11.28)\\
S_D(z^{-1}) &= 1 - 0.50317\,z^{-1} + 0.140577\,z^{-2} & (11.29)\\
T_D(z^{-1}) &= 0.0483358 & (11.30)
\end{aligned}
$$

**Brake Controller**

The structure of the brake controller is similar to that of the drive controller, and is shown in Figure 11.29.

The uppermost of the two maps in Figure 11.29 produces a desired brake torque $T_{ref}$ as a function of the velocity $v_{CoG}$ and a predefined desired acceleration $a_{ref}$, which in the case of the brake controller is negative. Map 2 has

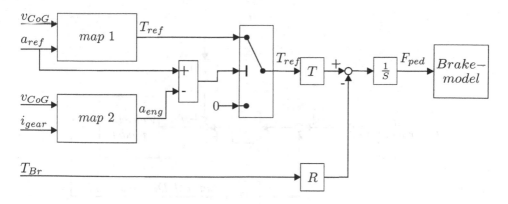

**Figure 11.29** Simplified block diagram of the brake controller

the task of giving the negative acceleration $a_{eng}$ due to the braking action of the engine. This negative acceleration is dependant upon the velocity $v_{CoG}$ and the transmission ratio $i_{gear}$. In the switch a comparison takes place between the desired acceleration $a_{ref}$ and the acceleration $a_{eng}$. If the brake action of the engine is sufficient to achieve the desired deceleration then $T_{ref} = 0$ is taken as the control value, and the brake control remains inactive. If the required deceleration exceeds $a_{eng}$ then the brakes must be activated, i.e. the desired brake torque from the uppermost map of Figure 11.29 is passed on.

The transfer function of the brake control system is:

$$G_B(z^{-1}) = \frac{A(z^{-1})}{B(z^{-1})} = \frac{0.0909361 - 0.0861402\,z^{-1} + 0.0502816\,z^{-2}}{1 - 2.06324\,z^{-1} + 1.44986\,z^{-2} - 0.34446\,z^{-3}} \quad (11.31)$$

The tuning parameters are chosen as the following:

- $N_1 = 1$

- $N_2 = 40$

- $N_u = 3$

- $\lambda_u = 10$

This gives the following polynomials for the brake controller in RST structure:

$$
\begin{aligned}
R_B(z^{-1}) &= 6.51905 - 13.1809\,z^{-1} + 9.1206\,z^{-2} - 2.13962\,z^{-3} &(11.32)\\
S_B(z^{-1}) &= 1 - 0.239487\,z^{-1} + 0.312326\,z^{-2} &(11.33)\\
T_B(z^{-1}) &= 0.319119 &(11.34)
\end{aligned}
$$

**Lateral Controller**

The task of the lateral controller is to minimize the lateral offset of the vehicle to the ideal line. The control output is the steering angle $\delta_S$ (wheel turn angle $\delta_W$). The GPC controller uses the offset of the center point of the front axle. The

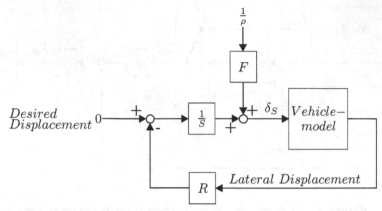

**Figure 11.30** Simplified block diagram of the lateral control with feed forward

controller has no information about the vehicle body side slip angle error, and does not control this. The controller structure is shown in Figure 11.30. Note that only R- and S-polynomials are employed.

To improve the performance of lateral the controller a static feed-forward component $F$ is added, so that in curves the steering angle $\delta_S$ corresponds to the inverse of the curve radius $\rho$, using a constant factor $F = 3.6$.

An identification of the lateral control system gives the following fifth order ARMA transfer function [79]:

$$G_L(z^{-1}) = \frac{-0.0033917 + 0.0075914\,z^{-1} - 0.0037304\,z^{-2}}{1 - 4.7847z^{-1} + 9.1727z^{-2} - 8.8077z^{-3} + 4.2360z^{-4} - 0.81629z^{-5}}$$

$$(11.35)$$

The tuning parameters are chosen as:

- $N_1 = 1$

- $N_2 = 60$

- $N_u = 3$

- $\lambda_u = 50$

The poles of the S-polynomial filter are 0.5, 0.6, 0.7, 0.8, 0.9 and 0.95.

This gives the following polynomials for the lateral controller:

$$
\begin{aligned}
R_L(z^{-1}) = \; & -0.737212 + 2.42144\,z^{-1} - 1 - 64676\,z^{-2} - 2.65418\,z^{-3} \\
& + 4.17551\,z^{-4} - 0.78585\,z^{-5} - 1.66839\,z^{-6} + 1.09109\,z^{-7} \\
& - 0.195652\,z^{-8}
\end{aligned}
$$

$$(11.36)$$

$$
\begin{aligned}
S_L(z^{-1}) = \; & 1 - 4.42949\,z^{-1} + 8.36477\,z^{-2} - 8.73042\,z^{-3} + 5.43316\,z^{-4} \\
& - 2.01127\,z^{-5} + 0.409556\,z^{-6} - 0.0353928\,z^{-7}
\end{aligned}
$$

$$(11.37)$$

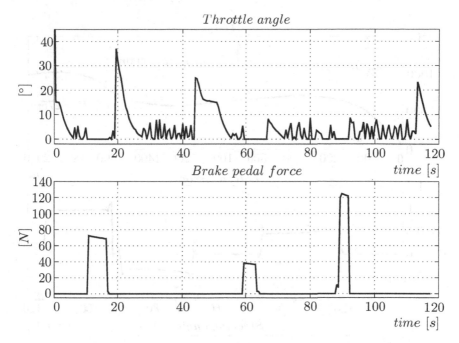

**Figure 11.31** Simulation results for the longitudinal controller

## Simulation Results

In order to prove the robustness and performance of the longitudinal and lateral controllers, various drive situations are simulated. Various curve radii from $\rho = 275\,m$ to $\rho = 400\,m$ are included, and the road conditions change from dry to wet. This course is then used as the reference for the following simulations.

The reference velocity of Figure 11.33 is calculated at each time instant dependent upon the weather conditions and the curve of the road. On top of this, according to the drive state (entering or leaving the curve) a suitable velocity profile is determined, which approximates the human acceleration behavior.

The first three plots in Figure 11.33 show reference and actual values for drive and brake torques and the acceleration. In all three plots it can be seen that the actual value tracks the reference value very well, even in the case of high acceleration or deceleration. The large spread of velocities (over $30\,km/h$) also had no negative effect on the controller performances. The desired velocity is also very closely tracked after transients.

Figure 11.31 shows the throttle angle and the brake pedal force of the longitudinal controller. It can be seen that small decelerations are obtained just by engine braking, without activating the brakes (i.e. when the throttle angle and the brake pedal force are both equal to zero).

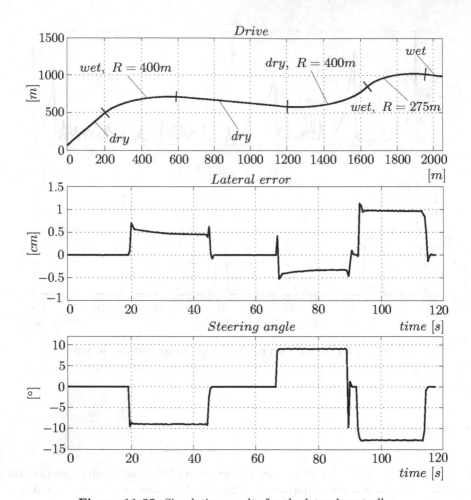

**Figure 11.32** Simulation results for the lateral controller

Figure 11.32 shows the results of the lateral controller, as well as the course traveled by the vehicle, which due to the low lateral error is identical to the desired course. The second plot of Figure 11.34 shows the lateral offset of the vehicle. Despite large variations in speed, different curve radii and road surfaces, the maximum offset throughout the whole simulation is only approximately 1.2 cm. It shows no oscillating behavior and does not need to be corrected.

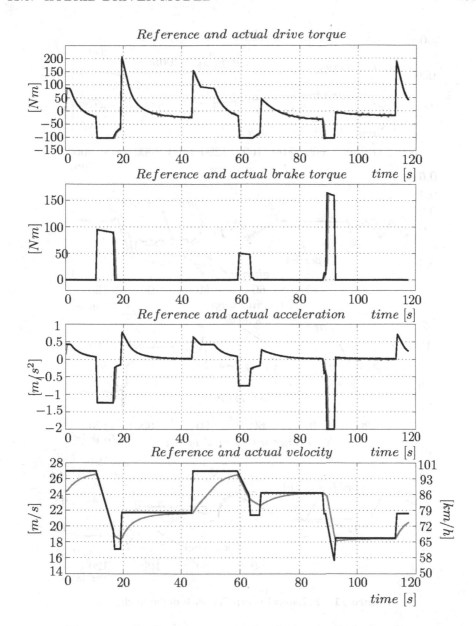

**Figure 11.33** Simulation results for the longitudinal controller

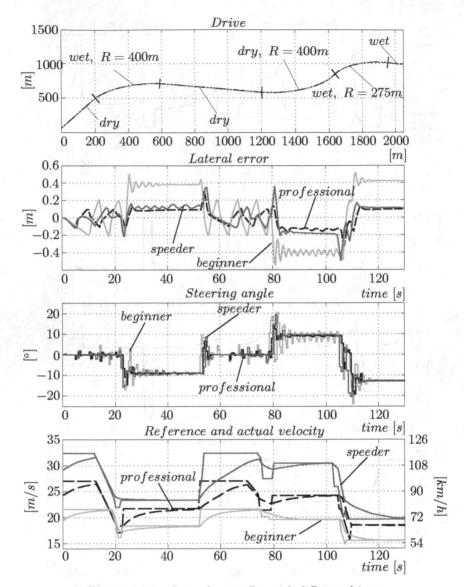

**Figure 11.34** Lateral controller with different drivers

# A Appendix

## A.1 Jacobian Matrices / Nonlinear Two-track Model

The elements of the Jacobian matrices for the nonlinear reduced 3rd order model in Section 8.4.4 are:

$$\frac{\partial \dot{v}_{CoG}}{\partial v_{CoG}} = \frac{1}{m_{CoG}} \cdot \left\{ -c_{aer} A_L \rho\, v_{CoG} \cos\beta - \frac{l_F\, \dot\psi}{v_{CoG}^2}(c_{FL} + c_{FR}) \sin(\delta_W - \beta) \right.$$
$$\left. -(c_{RL} + c_{RR})\frac{l_R\, \dot\psi}{v_{CoG}^2}\, \sin\beta \right\} \tag{A.1}$$

$$\frac{\partial \dot{v}_{CoG}}{\partial \beta} = \frac{1}{m_{CoG}} \cdot \left\{ (F_{LFL} + F_{LFR} + c_{FL} + c_{FR}) \cdot \sin(\delta_W - \beta) \right.$$
$$-\left(c_{RL} + c_{RR} + F_{LRL} + F_{LRR} - c_{aer} A_L \frac{\rho}{2} \cdot v_{CoG}^2\right) \sin\beta$$
$$+(c_{FL} + c_{FR})(\delta_W - \beta - \frac{l_F\, \dot\psi}{v_{CoG}}) \cos(\delta_W - \beta)$$
$$\left. +(c_{RL} + c_{RR})(-\beta + \frac{l_R\, \dot\psi}{v_{CoG}}) \cos\beta \right\} \tag{A.2}$$

$$\frac{\partial \dot{v}_{CoG}}{\partial \dot\psi} = \frac{1}{m_{CoG} v_{CoG}} \cdot \left\{ l_F \cdot \sin(\delta_W - \beta) \cdot (c_{FL} + c_{FR}) \right.$$
$$\left. +l_R \cdot \sin\beta \cdot (c_{RL} + c_{RR}) \right\} \tag{A.3}$$

$$\frac{\partial \dot{\beta}}{\partial v_{CoG}} = -\frac{1}{m_{CoG}v_{CoG}^2} \cdot \Big\{ (c_{FL} + c_{FR})(\delta_W - \beta - 2\frac{l_F\,\dot{\psi}}{v_{CoG}})\cos(\delta_W - \beta)$$

$$+(F_{LFL} + F_{LFR}) \cdot \sin(\delta_W - \beta)$$

$$-(F_{LRL} + F_{LRR} + c_{aer}A_L\frac{\rho}{2} \cdot v_{CoG}^2) \cdot \sin\beta$$

$$+(c_{RL} + c_{RR})(-\beta + 2\frac{l_R\,\dot{\psi}}{v_{CoG}})\cos\beta \Big\} \qquad \text{(A.4)}$$

$$\frac{\partial \dot{\beta}}{\partial \beta} = \frac{1}{m_{CoG}v_{CoG}} \cdot \Big\{ (c_{FL} + c_{FR})(\delta_W - \beta - \frac{l_F\,\dot{\psi}}{v_{CoG}})\sin(\delta_W - \beta)$$

$$-(c_{FL} + c_{FR} + F_{LFL} + F_{LFR})\cos(\delta_W - \beta)$$

$$-(c_{RL} + c_{RR})(-\beta + \frac{l_R\,\dot{\psi}}{v_{CoG}})\sin\beta$$

$$-(c_{RL} + c_{RR} + F_{LRL} + F_{LRR}$$

$$-c_{aer}A_L\frac{\rho}{2} \cdot v_{CoG}^2)\cos\beta \Big\} \qquad \text{(A.5)}$$

$$\frac{\partial \dot{\beta}}{\partial \dot{\psi}} = \frac{1}{m_{CoG}v_{CoG}^2} \cdot \Big\{ l_R\,(c_{RL} + c_{RR})\cos\beta$$

$$-l_F(c_{FL} + c_{FR})\cos(\delta_W - \beta) \Big\} - 1 \qquad \text{(A.6)}$$

$$\frac{\partial \ddot{\psi}}{\partial v_{CoG}} = \frac{1}{J_Z\,v_{CoG}^2} \Big\{ l_F\,\dot{\psi}\,(c_{FL} + c_{FR})(l_F - n_{LF}\cos\delta_W)\cos\delta_W$$

$$+\frac{1}{2}\,l_F\,\dot{\psi}\,b_F(c_{FL} - c_{FR})\sin\delta_W$$

$$+l_R\,\dot{\psi}\,(c_{RL} + c_{RR})(l_R + n_{LR}) \Big\} \qquad \text{(A.7)}$$

$$\frac{\partial \ddot{\psi}}{\partial \beta} = \frac{1}{J_Z} \cdot \Big\{ -(c_{FL} + c_{FR})(l_F - n_{LF}\cos\delta_W)\cos\delta_W$$

$$-\frac{b_F}{2}(c_{FL} - c_{FR})\sin\delta_W$$

$$+(c_{RL} + c_{RR})(l_R + n_{LR}) \Big\} \qquad \text{(A.8)}$$

$$\frac{\partial \ddot{\psi}}{\partial \dot{\psi}} = \frac{1}{J_Z\,v_{CoG}} \cdot \Big\{ -l_F\,(c_{FL} + c_{FR})(l_F - n_{LF}\cos\delta_W)\cos\delta_W$$

$$-\frac{l_F b_F}{2}(c_{FL} - c_{FR})\sin\delta_W$$

$$-l_R\,(c_{RL} + c_{RR})(l_R + n_{LR}) \Big\} \qquad \text{(A.9)}$$

$$\frac{\partial \dot{v}_{CoG}}{\partial F_{LFL}} = \frac{\cos(\delta_W - \beta)}{m_{CoG}} \tag{A.10}$$

$$\frac{\partial \dot{v}_{CoG}}{\partial F_{LFR}} = \frac{\cos(\delta_W - \beta)}{m_{CoG}} \tag{A.11}$$

$$\frac{\partial \dot{v}_{CoG}}{\partial F_{LRL}} = \frac{\cos \beta}{m_{CoG}} \tag{A.12}$$

$$\frac{\partial \dot{v}_{CoG}}{\partial F_{LRR}} = \frac{\cos \beta}{m_{CoG}} \tag{A.13}$$

$$\frac{\partial \dot{v}_{CoG}}{\partial \delta_W} = -\frac{1}{m_{CoG}} \cdot \Big[ (F_{LFL} + F_{LFR} + c_{FL} + c_{FR}) \cdot \sin(\delta_W - \beta) $$
$$ + (c_{FL} + c_{FR}) \cdot \left( \delta_W - \beta - \frac{l_F \cdot \dot{\psi}}{v_{CoG}} \right) \cdot \cos(\delta_w - \beta) \Big] \tag{A.14}$$

$$\frac{\partial \dot{\beta}}{\partial F_{LFL}} = \frac{\sin(\delta_w - \beta)}{m_{CoG} \cdot v_{CoG}} \tag{A.15}$$

$$\frac{\partial \dot{\beta}}{\partial F_{LFR}} = \frac{\sin(\delta_w - \beta)}{m_{CoG} \cdot v_{CoG}} \tag{A.16}$$

$$\frac{\partial \dot{\beta}}{\partial F_{LRL}} = -\frac{\sin \beta}{m_{CoG} \cdot v_{CoG}} \tag{A.17}$$

$$\frac{\partial \dot{\beta}}{\partial F_{LRR}} = -\frac{\sin \beta}{m_{CoG} \cdot v_{CoG}} \tag{A.18}$$

$$\frac{\partial \dot{\beta}}{\partial \delta_W} = \frac{1}{m_{CoG} \cdot v_{CoG}} \cdot \Big[ (c_{FL} + c_{FR} + F_{LFL} + F_{LFR}) \cdot \cos(\delta_w - \beta) $$
$$ - (c_{FL} + c_{FR}) \cdot \left( \delta_W - \beta - \frac{l_F \cdot \dot{\psi}}{v_{CoG}} \right) \cdot \sin(\delta_W - \beta) \Big] \tag{A.19}$$

$$\frac{\partial \ddot{\psi}}{\partial F_{LFL}} = \frac{1}{J_Z} \cdot \left\{ (l_F - n_{LF} \cos \delta_W) \cdot \sin \delta_W - \frac{b_F}{2} \cdot \cos \delta_W \right\} \tag{A.20}$$

$$\frac{\partial \ddot{\psi}}{\partial F_{LFR}} \quad = \quad \frac{1}{J_Z} \cdot \left\{ (l_F - n_{LF} \cos \delta_W) \cdot \sin \delta_W + \frac{b_F}{2} \cdot \cos \delta_W \right\} \quad \text{(A.21)}$$

$$\frac{\partial \ddot{\psi}}{\partial F_{LRL}} \quad = \quad -\frac{b_R}{2 \cdot J_Z} \quad \text{(A.22)}$$

$$\frac{\partial \ddot{\psi}}{\partial F_{LRR}} \quad = \quad \frac{b_R}{2 \cdot J_Z} \quad \text{(A.23)}$$

$$
\begin{aligned}
\frac{\partial \ddot{\psi}}{\partial \delta_W} \quad = \quad & \frac{1}{J_Z} \cdot \Bigg\{ (c_{FL} + c_{FR}) \cdot \big[ (l_F - n_{LF} \cdot \cos \delta_W) \\
& + \left( \delta_W - \beta - \frac{l_F \cdot \dot{\psi}}{v_{CoG}} \right) \cdot n_{LF} \cdot \sin \delta_W \big] \cdot \cos \delta_W \\
& - \Big[ (c_{FL} + c_{FR}) \cdot \left( \delta_W - \beta - \frac{l_F \cdot \dot{\psi}}{v_{CoG}} \right) \cdot (l_F - n_{LF} \cdot \cos \delta_W) \\
& + (F_{LFR} - F_{LFL}) \cdot \frac{b_F}{2} \Big] \cdot \sin \delta_W \\
& + \Big[ (c_{FL} - c_{FR}) \cdot \frac{b_F}{2} + n_{LF} \cdot \sin \delta_W \cdot (F_{LFR} + F_{LFL}) \Big] \cdot \sin \delta_W \\
& + \Big[ (c_{FL} - c_{FR}) \cdot \left( \delta_W - \beta - \frac{l_F \cdot \dot{\psi}}{v_{CoG}} \right) \cdot \frac{b_F}{2} \\
& + (F_{LFR} + F_{LFL}) \cdot (l_F - n_{LF} \cdot \cos \delta_W) \Big] \cdot \cos \delta_W ) \\
& - (c_{RL} + c_{RR}) \cdot (l_R + n_{LR}) \Bigg\}
\end{aligned}
\quad \text{(A.24)}
$$

# A.2 Observability of the reduced nonlinear two-track model

In this section, the observability of the nonlinear two-track model presented in Section 8.4.4 shall be shown.

Knowing the input $u(t)$ of a system and measuring its output $y(t)$, a *linear* system is called observable, if the initial system state $x(t_0)$ can be reconstructed uniquely, [25].

Observability for *nonlinear* systems, though, is a more complex issue. For most nonlinear systems, observability can only be proved in the neighborhood of the actual point of operation ("*local* observability").

The proof of local observability carried out in this section is split up into two steps. Firstly, the system output $y(t)$ is developed into a Taylor expansion with respect to time. Afterwards, the output is linearized around the actual point of operation:

## A.2.1 Step 1: Taylor expansion with respect to time

The function

$$\underline{y} = \underline{c}(\underline{x}, \underline{u}) \tag{A.25}$$

provides the relationship between inputs, states and outputs. The system equations are developed into a Taylor-expansion with respect to time

$$\underline{y}(t) \approx y(0) + t\underline{\dot{y}}(0) + \frac{t^2}{2!}\underline{\ddot{y}}(0) + \cdots + \frac{t^{n-1}}{(n-1)!}\overset{(n-1)}{\underline{y}}(0) \qquad . \tag{A.26}$$

In order to describe the output of a dynamical system by means of equation A.26, the time derivatives $\underline{\dot{y}}, ..., \overset{(n-1)}{\underline{y}}$ must be determined. Using equation A.25 this is a vector

$$\begin{pmatrix} \underline{y} \\ \underline{\dot{y}} \\ \vdots \\ \overset{(n-1)}{\underline{y}} \end{pmatrix} = \begin{pmatrix} \underline{c}(\underline{x}, \underline{u}) \\ \frac{d}{dt}\underline{c}(\underline{x}, \underline{u}) \\ \vdots \\ \frac{d^{n-1}}{dt^{n-1}}\underline{c}(\underline{x}, \underline{u}) \end{pmatrix} \qquad . \tag{A.27}$$

The first time derivative is

$$\underline{\dot{y}} = \frac{\partial \underline{c}}{\partial t}$$

$$= \frac{\partial \underline{c}}{\partial \underline{x}} \cdot \frac{\partial \underline{x}}{\partial t} + \frac{\partial \underline{c}}{\partial \underline{u}} \cdot \frac{\partial \underline{u}}{\partial t}$$

$$\overset{\underline{\dot{x}}=\underline{f}(\underline{x},\underline{u})}{=} \frac{\partial \underline{c}}{\partial \underline{x}}\underline{f}(\underline{x}, \underline{u}) + \frac{\partial \underline{c}}{\partial \underline{u}}\underline{\dot{u}} \qquad . \tag{A.28}$$

The output of the nonlinear two-track system (section 8.4.4) does not directly depend on the input. That is why only the first term in equation A.28 remains.

$$\underline{\dot{y}} = \frac{\partial \underline{c}}{\partial \underline{x}}\underline{f}(\underline{x}, \underline{u}) \tag{A.29}$$

Higher time derivatives of equation A.29 would become very complex. In order
to reduce the complexity, a linear differential operator $L_f$ is defined. For systems
without direct feedthrough, the operator $L_f$ for a certain element $c_i$ becomes

$$L_f \cdot c_i = \frac{\partial c_i}{\partial \underline{x}} \cdot \underline{f}(\underline{x}, \underline{u}) \tag{A.30}$$

Application of operator $L_f$ to the vector $\underline{c}$ yields

$$L_f \cdot \underline{c}(\underline{x}, \underline{u}) = [L_f \cdot c_1(\underline{x}, \underline{u}) \dots L_f \cdot c_r(\underline{x}, \underline{u})]^T \tag{A.31}$$

where $r$ denotes the number of output variables. Multiple application of operator
$L$ provides a recursive calculation procedure

$$L_f^k \cdot \underline{c}(\underline{x}, \underline{u}) = L_f \left( L_f^{k-1} \cdot c(\underline{x}, \underline{u}) \right) \tag{A.32}$$

The "initial condition" thereby is

$$L_f^0 \cdot \underline{c}(\underline{x}, \underline{u}) = \underline{c}(\underline{x}, \underline{u}) \tag{A.33}$$

Using the differential operator $L_f$ equation A.27 can be rewritten as

$$\begin{pmatrix} \underline{y} \\ \dot{\underline{y}} \\ \vdots \\ \overset{(n-1)}{\underline{y}} \end{pmatrix} = \begin{pmatrix} L_f^0 \\ L_f \\ \vdots \\ L_f^{n-1} \end{pmatrix} \underline{c}(\underline{x}, \underline{u}) \quad . \tag{A.34}$$

## A.2.2   Step 2: Linearization around actual point of operation

As mentioned above, the goal is to prove *local* observability of the nonlinear
two-track model in section 8.4.4. Therefore, only the neighborhood around the
actual point of operation must be regarded. The (n-1)-th derivative with respect
to time in equation A.27 is linearized around the actual point of operation

$$\overset{(k)}{\underline{y}} \approx \underline{c}(\underline{x}_p, \underline{u}) + \left.\frac{\partial \underline{c}(\underline{x}, \underline{u})}{\partial \underline{x}}\right|_{x=x_p} \cdot (x - x_p) \qquad k = 1 \dots n - 1 \tag{A.35}$$

The system is observable, if measurement of $\overset{(k)}{\underline{y}}, k = 1 \dots n - 1$ yields the system
state. That means that equation A.35 must be invertible. That's the case if the
nonlinear observability matrix

$$\underline{Q}_{NL}(\underline{x}, \underline{u}) = \frac{\partial \underline{y}}{\partial \underline{x}} = \frac{\partial \underline{c}(\underline{x}, \underline{u})}{\partial \underline{x}} \overset{eq. A.27}{=} \begin{pmatrix} \underline{y} \\ \dot{\underline{y}} \\ \vdots \\ \overset{(n-1)}{\underline{y}} \end{pmatrix} = \begin{pmatrix} L_f^0 \\ L_f \\ \vdots \\ L_f^{n-1} \end{pmatrix} \frac{\partial \underline{c}(\underline{x}, \underline{u})}{\partial \underline{x}} \tag{A.36}$$

has maximum rank ("observability rank condition")

$$\text{rank}\left\{\underline{Q}_{NL}(\underline{x},\underline{u})\right\} = n \qquad . \tag{A.37}$$

For the three-dimensional system in equations 8.88 and 8.89 and using the differential operator $L_f$ the observability matrix becomes

$$\underline{Q}_{NL}(\underline{x},\underline{u}) = \begin{bmatrix} L_f^0 \\ L_f^1 \\ L_f^2 \end{bmatrix} \cdot \frac{\partial \underline{c}(\underline{x},\underline{u})}{\partial \underline{x}} \qquad . \tag{A.38}$$

Note, that only the neighborhood of the operational point $\underline{x}_p$ must be regarded and that therefore operator $L_f$ is now applied to $\frac{\partial \underline{c}}{\partial \underline{x}}$.

## A.2.3   Proof of observability

In order to analyze observability, first the functions $L_f^0 \frac{\partial \underline{c}}{\partial \underline{x}}, L_f^1 \frac{\partial \underline{c}}{\partial \underline{x}}$ and $L_f^2 \frac{\partial \underline{c}}{\partial \underline{x}}$ have to be determined. In the following the expression $\underline{c}(\underline{x},\underline{u})$ is replaced by $\underline{c}$ for simplification.

**Step 1:** $L_f^0 \frac{\partial \underline{c}}{\partial \underline{x}}$

$$\underline{c} = \begin{bmatrix} x_1 \\ x_3 \end{bmatrix} \qquad \Rightarrow \qquad \frac{\partial \underline{c}}{\partial \underline{x}} = \begin{bmatrix} 1 & 0 & 0 \\ 0 & 0 & 1 \end{bmatrix} \tag{A.39}$$

Therefore, the first two rows of matrix $\underline{Q}_{NL}$ are:

$$L_f^0 \frac{\partial \underline{c}}{\partial \underline{x}} := \frac{\partial \underline{c}}{\partial \underline{x}} = \begin{bmatrix} 1 & 0 & 0 \\ 0 & 0 & 1 \end{bmatrix} \tag{A.40}$$

**Step 2:** $L_f^1 \frac{\partial \underline{c}}{\partial \underline{x}}$

$$L_f^1 \frac{\partial \underline{c}}{\partial \underline{x}} = L_f \cdot \left( L_f^0 \frac{\partial \underline{c}}{\partial \underline{x}} \right) = L_f \frac{\partial \underline{c}}{\partial \underline{x}} \tag{A.41}$$

Element-wise application of eqn. A.41 to $c_1(\underline{x})$ and $c_2(\underline{x})$:

$$
\begin{aligned}
L_f^1 \frac{\partial c_1}{\partial \underline{x}} &= L_f \frac{\partial c_1}{\partial \underline{x}} = \frac{\partial}{\partial \underline{x}}(L_f \cdot c_1) = \frac{\partial}{\partial \underline{x}}\left( \frac{\partial c_1}{\partial \underline{x}} \cdot \begin{pmatrix} f_1 \\ f_2 \\ f_3 \end{pmatrix} \right) \\
&= \frac{\partial}{\partial \underline{x}}\left( \begin{bmatrix} 1 & 0 & 0 \end{bmatrix} \cdot \begin{pmatrix} f_1 \\ f_2 \\ f_3 \end{pmatrix} \right) = \frac{\partial}{\partial \underline{x}} \cdot f_1(\underline{x},\underline{u}) \\
&= \begin{bmatrix} \frac{\partial f_1}{\partial x_1} & \frac{\partial f_1}{\partial x_2} & \frac{\partial f_1}{\partial x_3} \end{bmatrix}
\end{aligned}
\tag{A.42}
$$

$$L_f^1 \frac{\partial c_2}{\partial \underline{x}} = L_f \frac{\partial c_2}{\partial \underline{x}} = \frac{\partial}{\partial \underline{x}} (L_f \cdot c_2) = \frac{\partial}{\partial \underline{x}} \left( \frac{\partial c_2}{\partial \underline{x}} \cdot \begin{pmatrix} f_1 \\ f_2 \\ f_3 \end{pmatrix} \right)$$

$$= \frac{\partial}{\partial \underline{x}} \left( \begin{bmatrix} 0 & 0 & 1 \end{bmatrix} \cdot \begin{pmatrix} f_1 \\ f_2 \\ f_3 \end{pmatrix} \right) = \frac{\partial}{\partial \underline{x}} \cdot f_3(\underline{x}, \underline{u})$$

$$= \begin{bmatrix} \frac{\partial f_3}{\partial x_1} & \frac{\partial f_3}{\partial x_2} & \frac{\partial f_3}{\partial x_3} \end{bmatrix} \tag{A.43}$$

**Step 3:** $L_f^2 \frac{\partial c}{\partial \underline{x}}$

$$L_f^2 \frac{\partial \underline{c}}{\partial \underline{x}} = L_f \cdot \left( L_f \frac{\partial \underline{c}}{\partial \underline{x}} \right) \tag{A.44}$$

Element-wise application of eqn. A.44 to $c_1(\underline{x})$ and $c_2(\underline{x})$:

$$L_f^2 \frac{\partial c_1}{\partial \underline{x}} = L_f \cdot \left( L_f \frac{\partial c_1}{\partial \underline{x}} \right) = L_f \cdot \left( \frac{\partial f_1}{\partial \underline{x}} \right)$$

$$= \frac{\partial}{\partial \underline{x}} (L_f \cdot f_1) = \frac{\partial}{\partial \underline{x}} \left( \frac{\partial f_1}{\partial \underline{x}} \cdot \begin{pmatrix} f_1 \\ f_2 \\ f_3 \end{pmatrix} \right)$$

$$= \frac{\partial}{\partial \underline{x}} \underbrace{\left( \frac{\partial f_1}{\partial x_1} \cdot f_1 + \frac{\partial f_1}{\partial x_2} \cdot f_2 + \frac{\partial f_1}{\partial x_3} \cdot f_3 \right)}_{A(\underline{x}, \underline{u})}$$

$$= \begin{bmatrix} \frac{\partial A(\underline{x}, \underline{u})}{\partial x_1} & \frac{\partial A(\underline{x}, \underline{u})}{\partial x_2} & \frac{\partial A(\underline{x}, \underline{u})}{\partial x_3} \end{bmatrix} \tag{A.45}$$

$$L_f^2 \frac{\partial c_2}{\partial \underline{x}} = L_f \cdot \left( L_f \frac{\partial c_2}{\partial \underline{x}} \right) = L_f \cdot \left( \frac{\partial f_3}{\partial \underline{x}} \right)$$

$$= \frac{\partial}{\partial \underline{x}} (L_f \cdot f_3) = \frac{\partial}{\partial \underline{x}} \left( \frac{\partial f_3}{\partial \underline{x}} \cdot \begin{pmatrix} f_1 \\ f_2 \\ f_3 \end{pmatrix} \right)$$

$$= \frac{\partial}{\partial \underline{x}} \underbrace{\left( \frac{\partial f_3}{\partial x_1} \cdot f_1 + \frac{\partial f_3}{\partial x_2} \cdot f_2 + \frac{\partial f_3}{\partial x_3} \cdot f_3 \right)}_{B(\underline{x}, \underline{u})}$$

$$\overline{\overline{\partial x_1}} \begin{bmatrix} \frac{\partial B(\underline{x}, \underline{u})}{\partial x_2} & \frac{\partial B(\underline{x}, \underline{u})}{\partial x_3} \end{bmatrix} \tag{A.46}$$

Using the calculated results for $L_f^j \frac{\partial c}{\partial x}$, $j = 0...2$, the observability matrix $\underline{Q}_{NL}$ becomes

$$
\underline{Q}_{NL} = \begin{pmatrix}
1 & 0 & 0 \\
0 & 0 & 1 \\
\frac{\partial f_1}{\partial x_1} & \frac{\partial f_1}{\partial x_2} & \frac{\partial f_1}{\partial x_3} \\
\frac{\partial f_3}{\partial x_1} & \frac{\partial f_3}{\partial x_2} & \frac{\partial f_3}{\partial x_3} \\
\frac{\partial A(x,u)}{\partial x_1} & \frac{\partial A(x,u)}{\partial x_2} & \frac{\partial A(x,u)}{\partial x_3} \\
\frac{\partial B(x,u)}{\partial x_1} & \frac{\partial B(x,u)}{\partial x_2} & \frac{\partial B(x,u)}{\partial x_3}
\end{pmatrix} .
\tag{A.47}
$$

In the next step, the observability rank condition is checked, i.e. whether $\underline{Q}_{NL}$ is of maximum rank. This is true if all the columns or if at least three of the rows are linearly independent. We prove here, that three rows are linearly independent. Obviously the first two rows are linearly independent. Therefore, we must show that one of the remaining rows is not a linear combination of the two first rows. According to [25], the rank of a matrix will not change, if linear combinations of rows or columns are regarded instead of the original matrix. Matrix $\underline{Q}_{NL}$ can be transformed as follows without changing its rank:

$$
\underline{Q}_{NL} = \begin{pmatrix}
1 & 0 & 0 \\
0 & 0 & 1 \\
0 & \frac{\partial f_1}{\partial x_2} & 0 \\
0 & \frac{\partial f_3}{\partial x_2} & 0 \\
0 & \frac{\partial A(x,u)}{\partial x_2} & 0 \\
0 & \frac{\partial B(x,u)}{\partial x_2} & 0
\end{pmatrix}
\tag{A.48}
$$

It is now checked, whether the middle elements of rows 3 to 6 become zero for the same inputs. This is reduced to the middle elements of rows 3 and 4. Using equations A.2 and A.8, we get the conditions

$$
\frac{\partial f_1}{\partial x_2} = \frac{\partial \dot{v}_{CoG}}{\partial \beta} = \frac{1}{m_{CoG}} \cdot \Big\{ (F_{LFL} + F_{LFR} + c_{FL} + c_{FR}) \cdot \sin(\delta_W - \beta)
$$
$$
- \Big( c_{RL} + c_{RR} + F_{LRL} + F_{LRR} - c_{aer} A_L \frac{\rho}{2} \cdot v_{CoG}^2 \Big) \cdot \sin \beta
$$
$$
+ (c_{FL} + c_{FR}) \cdot \Big( \delta_W - \beta - \frac{l_F \cdot \dot{\psi}}{v_{CoG}} \Big) \cdot \cos(\delta_W - \beta)
$$
$$
+ (c_{RL} + c_{RR}) \cdot \Big( -\beta + \frac{l_R \cdot \dot{\psi}}{v_{CoG}} \Big) \cdot \cos \beta \Big\} \neq 0
\tag{A.49}
$$

$$\frac{\partial f_3}{\partial x_2} = \frac{\partial \ddot{\psi}}{\partial \beta} = \frac{1}{J_Z} \cdot \Big\{ -(c_{FL} + c_{FR}) \cdot (l_F - n_{LF} \cdot \cos \delta_W) \cdot \cos \delta_W$$

$$-(c_{FL} - c_{FR}) \cdot \frac{b_F}{2} \cdot \sin \delta_W + (c_{RL} + c_{RR}) \cdot (l_R + n_{LR}) \Big\} \neq 0 \quad (A.50)$$

The middle elements of row 3 and 4 in matrix A.48 are never simultaneously zero.

Therefore the observability rank condition is fulfilled and the system 8.88 and 8.89 is locally observable.

# A.3 Design of GPC Controllers

The GPC controller specifications are formally described using the following cost function:

$$J_{GPC} = E \left\{ \sum_{j=N_1}^{N_2} [y(t+j) - r(t+j)]^2 + \lambda_u \sum_{j=0}^{N_u} [\Delta u(t+j-1)]^2 \right\} \quad \text{(A.51)}$$

Where,

| | | |
|---|---|---|
| $y(t+j)$ | : | future control value |
| $r(t+j)$ | : | reference trajectory |
| $\Delta u(t+j-1)$ | : | future change in control input ($\Delta = 1 - z^{-1}$) |
| $N_1$ | : | lower prediction horizon |
| $N_2$ | : | upper prediction horizon |
| $N_u$ | : | control horizon |
| $\lambda_u$ | : | weighting factor for the control input changes. |

The first term in the cost function represents the predicted squared control error over a limited period of time between $N_1$ and $N_2$. The second term represents the weighted sum of the first $N_u$ future control values. The larger the weighting factor $\lambda_u$, the more gently the control value changes. To reduce the computational load, the control vector is considered to be constant after time $N_u$, i.e. the changes are

$$\Delta u(t+i) = 0 \quad \text{for} \quad i \geq N_u \quad . \quad \text{(A.52)}$$

At each sampling instant, the following calculations are carried out:

- The reference trajectory $r(t+j)$ is determined.

- The process outputs $y(t+j)$ are predicted, along with the future predicted control errors and controller input change for the time $N_2 - N_1$.

- Using the cost function, the future controller error and control input value changes are minimized over the considered time period.

- The first value in the optimized control vector $u(t+j)$, $0 \leq j \leq N_u - 1$ is now applied to the real system, and the process repeated.

## A.3.1 The Process Model

Consider a model of the following form:

$$y(t) = \frac{B(z^{-1})}{A(z^{-1})} u(t-1) + \frac{C(z^{-1})}{A(z^{-1})\Delta} \xi(t) \quad \text{(A.53)}$$

or

$$A(z^{-1})\Delta y(t) = B(z^{-1})\Delta u(t-1) + C(z^{-1})\xi(t) \quad \text{(A.54)}$$

Where,

$$
\begin{aligned}
A(z^{-1}) &= 1 + a_1 z^{-1} + \ldots + a_{na} z^{-na} \\
B(z^{-1}) &= b_0 + b_1 z^{-1} + \ldots + b_{nb} z^{-nb} \\
C(z^{-1}) &= 1 + c_1 z^{-1} + \ldots + c_{nc} z^{-nc} \\
\Delta &= 1 - z^{-1}
\end{aligned}
$$

Variable $\xi(t)$ represents the system disturbance. For simplicity, $C(z^{-1})$ is chosen to be 1.

In order to minimize the cost function, the outputs of the system $y(t+j)$ must be predicted for $j = N_1, \ldots, N_2$. This requires the following *Diophantine Equation*:

$$
1 = E_j(z^{-1}) A(z^{-1}) \Delta + z^{-j} F_j(z^{-1}) \tag{A.55}
$$

where,

$$
\begin{aligned}
E_j(z^{-1}) &= e_{j,0} + e_{j,1} z^{-1} + \ldots + e_{j,j} z^{-j+1} \\
F_j(z^{-1}) &= f_{j,0} + f_{j,1} z^{-1} + \ldots + f_{j,j} z^{-j+1}
\end{aligned}
$$

$E_j(z^{-1})$ and $F_j(z^{-1})$ can be uniquely determined given $A_j(z^{-1})$ and $j$.

Multiplying Equation A.54 by $E_j(z^{-1})z^j$ gives;

$$
E_j(z^{-1}) A(z^{-1}) \Delta y(t+j) = E_j(z^{-1}) B(z^{-1}) \Delta u(t+j-1) + E_j(z^{-1}) \xi(t+j) \tag{A.56}
$$

Substituting $E_j(z^{-1}) A(z^{-1}) \Delta$ from Equation A.55 gives,

$$
y(t+j) = F_j(z^{-1}) \cdot y(t) + E_j(z^{-1}) B(z^{-1}) \cdot \Delta u(t+j-1) + E_j(z^{-1}) \xi(t+j) \tag{A.57}
$$

The last term in Equation A.57 represents the system disturbance, and is dependant on future values which are not available at time $t$. Thus this term is set equal to zero, giving the following prediction:

$$
\hat{y}(t+j) = F_j(z^{-1}) \cdot y(t) + E_j(z^{-1}) B(z^{-1}) \cdot \Delta u(t+j-1) \tag{A.58}
$$

This gives the predicted outputs $\hat{y}(t+j)$ dependant upon previous values of the output and previous and future values of the control change.

## A.3.2   Recursion of the Diophantine Equation

The determination of the coefficients of the predictor polynomials is carried out recursively using the Diophantine equation. The solution of Equation A.55 is obtained via the recursive calculation of the two polynomials $E_j(z^{-1})$ and $F_j(z^{-1})$ from the two previous values $E_{j-1}(z^{-1})$ and $F_{j-1}(z^{-1})$:

$$
\begin{aligned}
1 &= E_j(z^{-1}) A(z^{-1}) \Delta + z^{-j} F_j(z^{-1}) \tag{A.59} \\
1 &= E_{j+1}(z^{-1}) A(z^{-1}) \Delta + z^{-j-1} F_{j+1}(z^{-1}) \tag{A.60}
\end{aligned}
$$

Subtracting Equation A.59 from Equation A.60 gives:

$$
0 = A(z^{-1}) \Delta \left[ E_{j+1}(z^{-1}) - E_j(z^{-1}) \right] + z^{-j} \left[ z^{-1} F_{j+1}(z^{-1}) - F_j(z^{-1}) \right] \tag{A.61}
$$

Substituting,

$$E_{j+1}(z^{-1}) - E_j(z^{-1}) = e_{j+1,j} \cdot z^{-1} + \tilde{E}(z^{-1}) \qquad (A.62)$$

where $\tilde{E}(z^{-1})$ is a polynomial of order $(j-1)$ and $e_{j+1,j}$ is the $j$-th coefficient of $E_{j+1}(z^{-1})$ into Equation A.61 gives,

$$0 = A(z^{-1})\Delta\tilde{E}_{j+1}(z^{-1}) + z^{-j}\left[z^{-1}F_{j+1}(z^{-1}) - F_j(z^{-1}) + A(z^{-1})\Delta e_{j+1,j}\right] \qquad (A.63)$$

From Equation A.63 it then follows that,

$$\tilde{E}_{j+1}(z^{-1}) = 0$$
$$z^{-1}F_{j+1}(z^{-1}) = F_j(z^{-1}) - A(z^{-1})\Delta e_{j+1,j} \qquad (A.64)$$

The coefficients of $E_{j+1}(z^{-1})$ and $F_{j+1}(z^{-1})$ are then calculated in the following way:

$$e_{j+1,j} = f_{j,0} \qquad (A.65)$$
$$f_{j+1,j} = f_{j,j+1} - \tilde{a}_{j+1}e_{j+1,j} \quad , \quad j = 0, \ldots, \text{degree of } F_j(z^{-1}) \qquad (A.66)$$

where $\tilde{a}_{j+1}$ is the $(j+1)$-th coefficient of the polynomial $\tilde{A}(z^{-1}) = A(z^{-1})\Delta$. The polynomial $E_{j+1}(z^{-1})$ is given by,

$$E_{j+1}(z^{-1}) = E_j(z^{-1}) + z^{-1}e_{j+1,j} \qquad (A.67)$$

and,

$$G_{j+1}(z^{-1}) = B(z^{-1}) \cdot E_{j+1}(z^{-1}) \qquad (A.68)$$

The initial values for the recursion are found from Equation A.59 for $j = 1$:

$$1 = E_1(z^{-1})\tilde{A}(z^{-1}) + z^{-1}F_1(z^{-1}) \qquad (A.69)$$

As the leading element of $\tilde{A}(z^{-1})$ is 1, then,

$$E_1 = 1$$
$$F_1 = q\left(1 - \tilde{A}(z^{-1})\right) \qquad (A.70)$$

Starting from the solution for $j = 1$, the later solutions can be recursively calculated using Equations A.65 and A.66.

## A.3.3 The Control Law

The minimization of the cost function results in a predicted optimal control change vector $\underline{\tilde{u}}$,

$$\underline{\tilde{u}} = \left(\underline{G}^T\underline{G} + \lambda_u\underline{I}\right)^{-1}\underline{G}^T(\underline{r} - \underline{f}) \qquad (A.71)$$

where,

$$\underline{\tilde{u}} = [\Delta u(t), \Delta u(t+1), \ldots, \Delta u(t+N_u-1)]^T$$
$$\underline{r} = [r(t+N_1), r(t+N_1+1), \ldots, r(t+N_2)]^T$$
$$\underline{f} = [f(t+N_1), f(t+N_1+1), \ldots, f(t+N_2)]^T \qquad (A.72)$$

$\underline{r}$ is the reference value sequence.

The following controller structure results (see Figure 11.28),

$$S(z^{-1})\Delta u(t) = T(z^{-1})r(t+N_2) - R(z^{-1})y(t) \qquad (A.73)$$

## A.3.4   Choice of the Controller Parameters

- The lower prediction horizon $N_1$
  This is normally chosen as 1. If the dead time of the process is known to be $d$ sampling instants, then $N_1$ should be chosen to be at least as large as $d$.

- The upper prediction horizon $N_2$
  In general, $N_2$ should be chosen so that the result can track a controller action. Thus, $N_2$ should at least be larger than the order of $B(z^{-1})$. In practice, $N_2$ is chosen much larger than this, and corresponds to the rise time of the system.

- The control horizon $N_u$
  For simple systems, $N_u = 1$ can lead to good results, though higher values may be necessary for more complex systems (i.e. unstable systems). An increase in $N_u$ (up to a certain point) leads to faster control and reduced effect of disturbances.

- The weighting factor $\lambda_u$
  This is normally chosen to be small, as larger values of $\lambda_u$ leads to a slower overall system response and worse disturbance handling. In practice, $\lambda_u$ is taken as zero, or a small value.

- The T-filter polynomial is the so-called disturbance filter. It is used to improve the robustness of the controller to not-modelled disturbances. For well damped stable processes, the following is suitable:

$$T(z^{-1}) = A(z^{-1})(1 - \mu z^{-1}) \quad , \quad 0 \leq \mu < 1 \qquad (A.74)$$

For smaller values of $\mu$, the disturbance behavior is improved, however the influence of the disturbance at higher frequencies increases. For higher values of $\mu$, the stability of the control loop is affected. $\mu = 0$ is a compromise between the two extremes.

For badly damped processes however, $\mu = 0$ cannot be used, and the following is applied,

$$T(z^{-1}) = (1 - \mu z^{-1})^{na} \quad , \quad 0 \leq \mu < 1 \qquad (A.75)$$

where $na$ is the order of the polynomial $A(z^{-1})$. $\mu$ of between 0.6 and 0.9 is suitable.

# A.4  Driver Model: Constants and Weighting Factors

Table A.1 gives the constants required for the inter-arrival time calculations of Section 11.3.6.

| Constant | Value | Constant | Value |
|----------|-------|----------|-------|
| $p_f$ | 0.3 | $p_l$ | 0.3 |
| $f_1$ | $9\,s/m$ | $l_1$ | $9\,s/m$ |
| $f_2$ | $20\,s^2/m$ | $l_2$ | $20\,s^2/m$ |
| $f_3$ | $7\,s/deg$ | $l_3$ | $7\,s/deg$ |
| $f_4$ | $30\,s^2/deg$ | $l_4$ | $30\,s^2/deg$ |
| $f_5$ | $5\,1/deg$ | $l_5$ | $5\,1/deg$ |
| $p_r$ | 0.2 | $p_m$ | 0.2 |
| $r_1$ | $40\,1/m$ | $m_1$ | $40\,1/m$ |
| $r_2$ | $0.1\,s/m$ | $m_2$ | $0.1\,s/m$ |
| $r_3$ | $5\,s/deg$ | $m_3$ | $5\,s/deg$ |
| $r_4$ | $20\,s^2/deg$ | $m_4$ | $20\,s^2/deg$ |
| $r_5$ | $0.1\,1/deg$ | $m_5$ | $0.1\,1/deg$ |
| $r_6$ | $2.5\,s/m$ | $m_6$ | $2.5\,s/m$ |
| $c_1$ | $1/3\,s^3/m$ | $c_3$ | $-0.09384\,1/m$ |
| $c_2$ | $0.001401\,s/m^2$ | | |

**Table A.1**  Constants for the arrival time calculations

Table A.2 shows the inter-arrival times, the service times (in seconds) and the priorities of the different perception categories.

| Category | inter-arrival time | Service time | Priority |
|----------|--------------------|--------------|----------|
| Visual focus | (see Equation 11.8) | N(1.55 / 0.35) | 9 |
| Lead point | (see Equation 11.9) | (see Eq. 11.13) | 8 |
| Road edge | (see Equation 11.10) | (see Eq. 11.13) | 7 |
| Road center | (see Equation 11.11) | (see Eq. 11.13) | 7 |
| Warning lights | 100 | N(0.83 / 0.3) | 10 |
| Speed indicator | N(86 / 27.3) | N(0.62 / 0.48) | 5 |
| Rev counter | (see Equation 11.12) | N(0.62 / 0.48) | 4 |
| Mirror | N(130 / 3) | N(1 / 0.37) | 6 |
| Radio | N(331.7 / 155.4) | N(1.1 / 0.47) | 2 |
| Air conditioning control | N(493.8 / 277.3) | N(0.92 / 0.41) | 2 |
| Petrol gauge | N(855.5s / 399.6) | N(1.04 / 0.5) | 3 |
| Clock | N(1420 / 627.94) | N(0.83 / 0.38) | 2 |
| Temperature indicator | N(1025.6 / 666) | N(1.1 / 0.52) | 3 |
| Ventilation system | N(86 / 27.3) | N(1.1 / 0.48) | 2 |

**Table A.2**  Priorities, inter-arrival and service times for different perception categories

# A.5   Least Squares Parameter Estimation

A process with outputs $y(t)$ and unknown process parameters $\Theta_1$ to $\Theta_N$ is assumed. The output variables are however not directly measurable - only the measurement vector $\underline{y}_p(t)$ is available which is corrupted with the disturbance signal $n(t)$.

For parameter estimation, the model structure,

$$\underline{y}_m = \dot{F}\left[\underline{\Theta}, \underline{u}(t)\right] \tag{A.76}$$

is assumed to be known, and the inputs are known or can be measured without errors.

The task of the parameter estimation is to determine the process parameters,

$$\underline{\Theta}^T = [\Theta_1, \Theta_2, \ldots, \Theta_N] \tag{A.77}$$

such that the modeled outputs $\underline{y}_m(t)$ correspond as exactly as possible with the measured outputs $\underline{y}_p(t)$ of the system. The quality of the correspondence is defined [55] in that the sum of the squares of the observed errors at a particular time instant $k$ is minimized:

$$\sum_{k=1}^{M} e^2(k) = \sum_{k=1}^{M} \left[\underline{y}_p(k) - \underline{y}_m(k)\right]^2 \to MIN \tag{A.78}$$

## A.5.1   Parameter Estimation by means of Least Squares Method

If it is assumed that the model is linear with respect to the parameters $\Theta_1 \ldots \Theta_N$, i.e.:

$$y_m(t) = \Theta_1 \cdot F_1\left[u(t)\right] + \Theta_2 \cdot F_2\left[u(t)\right] + \ldots + \Theta_N \cdot F_N\left[u(t)\right] \tag{A.79}$$

then a unique algebraic solution exists for the optimization problem [55]. For $M$ process outputs, the following solution is obtained:

$$\begin{bmatrix} \hat{\Theta}_1 \\ \hat{\Theta}_2 \\ \vdots \\ \hat{\Theta}_N \end{bmatrix} = \left[\ \underline{\Psi}^T\ \underline{\Psi}\ \right]^{-1} \cdot \underline{\Psi}^T \begin{bmatrix} y_p(t_1) \\ y_p(t_2) \\ \vdots \\ y_p(t_M) \end{bmatrix} \tag{A.80}$$

with the observation matrix:

$$\underline{\Psi} = \begin{bmatrix} F_1\left[u(t_1)\right] & F_2\left[u(t_1)\right] & \cdots & F_N\left[u(t_1)\right] \\ F_1\left[u(t_2)\right] & F_2\left[u(t_2)\right] & \cdots & F_N\left[u(t_2)\right] \\ \vdots & \vdots & \ddots & \vdots \\ F_1\left[u(t_M)\right] & F_2\left[u(t_M)\right] & \cdots & F_N\left[u(t_M)\right] \end{bmatrix} \tag{A.81}$$

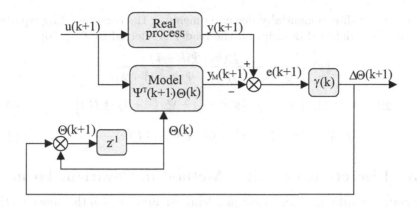

**Figure A.1** The recursive least squares method

## A.5.2 Parameter Estimation by means of Recursive Least Squares

In order to adapt the vehicle model to the current conditions and, should the occasion arise, to design adaptive controllers, changing parameters must be identified online. Whilst with non-recursive methods the estimated parameter is only available at the end of the measurement time, with dynamic parameters it is the changing parameter values after each sampling instant which are of interest. In order to prevent the saving of all past measurement values, and thus save computation time, the recursive method is used for online identification.

For the recursive least squares (RLS) the following equations apply [55]:

$$\underline{P}(k) = \left[\underline{\Psi}^T(k) \cdot \underline{\Psi}(k)\right]^{-1} \tag{A.82}$$

$$\underline{\gamma}(k) = \underline{P}(k+1) \cdot \underline{\Psi}(k+1) = \frac{\underline{P}(k) \cdot \underline{\Psi}(k+1)}{\underline{\Psi}^T(k+1)\underline{P}(k) \cdot \underline{\Psi}(k+1) + 1} \tag{A.83}$$

$$\underline{\hat{\Theta}}(k+1) = \underline{\hat{\Theta}}(k) + \underline{\gamma}(k) \cdot \left[y(k+1) - \underline{\Psi}^T(k+1) \cdot \underline{\hat{\Theta}}(k)\right] \tag{A.84}$$

$$\underline{P}(k+1) = \underline{P}(k) - \underline{\gamma}(k) \cdot \underline{\Psi}^T(k+1) \cdot \underline{P}(k) \tag{A.85}$$

Figure A.1 shows the principle of the recursive least squares method.

The basic equations of the RLS estimator (Equations A.82 - A.85) represent a versatile tool for online parameter identification. Further related RLS-estimators can be developed from these equations, such as RLS with weighted memory, generalized RLS, extended RLS etc. [55].

Via the introduction of a forgetting factor $\lambda_{RLS}$, which increases the elements in the covariance matrix $\underline{P}(k)$ for each iteration and thus weights the new data higher than the older data, one is able to slowly *forget* past values. This overcomes the consistency property of least squares estimators for time variant parameters. $\lambda_{RLS}$ should not be chosen too small, or the influence of disturbances may not be satisfactorily removed. Good results have been obtained with values of $0.95 < \lambda_{RLS} < 0.995$. Because a large part of the identification with this RLS method is carried out with exponentially decaying memory, the corresponding equations are given. For a detailed description the reader is referred to [61], [76].

$$\underline{\gamma}(k) = \frac{\underline{P}(k) \cdot \underline{\Psi}(k+1)}{\underline{\Psi}^T(k+1)\underline{P}(k) \cdot \underline{\Psi}(k+1) + \lambda_{RLS}} \tag{A.86}$$

$$\hat{\underline{\Theta}}(k+1) = \hat{\underline{\Theta}}(k) + \underline{\gamma}(k) \cdot \left[y(k+1) - \underline{\Psi}^T(k+1) \cdot \hat{\underline{\Theta}}(k)\right] \tag{A.87}$$

$$\underline{P}(k+1) = \left[I - \underline{\gamma}(k) \cdot \underline{\Psi}^T(k+1)\right] \cdot \underline{P}(k)\frac{1}{\lambda_{RLS}} \tag{A.88}$$

## A.5.3   Discrete Root Filter Method in Covariant Form

Under certain conditions, numerical problems can appear with the above method due to an ill-conditioned covariance matrix $\underline{P}(k)$. The reasons for this can be a too high sampling rate [55] (i.e. input signals which change too slowly) or too much similarity in the signal values.

Via suitable numerically improved methods round off failures in computers with short word length ($\leq 16\,bit$) can be reduced, and the influence of initial values can be minimized.

One possible method, with which the ill conditioning of the equation system can be avoided, is the method of the discrete root filtering (DSFC), in which two triangular matrices $\underline{S}$ are used instead of the symmetric matrix $\underline{P}$:

$$\underline{P} = \underline{S} \cdot \underline{S}^T \tag{A.89}$$

As opposed to the covariance matrix, where products of the original data appear, the triangular matrices contain the roots of the products of the original data. Although the method is called the root filter method, a real construction of the roots does not occur.

Following the least squares method, substitution of the covariance matrix according to Equation A.89 yields:

$$\underline{f}(k) = \underline{S}^T(k) \cdot \underline{\Psi}(k+1) \tag{A.90}$$

$$a(k) = \frac{1}{\underline{f}^T(k) \cdot \underline{f}(k) + \lambda_{RLS}} \tag{A.91}$$

Equations A.86 - A.88 are now rewritten as:

$$\underline{\gamma}(k) = a(k) \cdot \underline{S}(k) \cdot \underline{f}(k) \tag{A.92}$$

$$\hat{\underline{\Theta}}(k+1) = \hat{\underline{\Theta}}(k) + \underline{\gamma}(k) \cdot \underbrace{\left[y(k+1) - \underline{\Psi}^T(k+1) \cdot \hat{\underline{\Theta}}(k)\right]}_{e(k+1)} \tag{A.93}$$

$$\underline{S}(k+1) = \left[\underline{S}(k) - \frac{1}{1 + \sqrt{\lambda_{RLS} \cdot a(k)}} \cdot \underline{\gamma}(k) \cdot \underline{f}^T(k)\right] \cdot \frac{1}{\sqrt{\lambda_{RLS}}} \tag{A.94}$$

The initial values are taken as $\underline{S}(0) = \sqrt{\lambda_{RLS}} \cdot \underline{I}$ and $\hat{\underline{\Theta}}(0) = 0$, i.e. no more thought must be given to the initial values of the parameter vectors. $\lambda_{RLS}$ is the pre-defined forgetting factor.

The implementation of this method brings improvement over the RLS- estimator if microcomputers are used, as is the case in the vehicle itself.

# B Nomenclature

## B.1 Mathematical Definitions

**Indices:**

$(.)_{FL}$ , $(.)_{FR}$ , $(.)_{RL}$ , $(.)_{RR}$      Wheel indices:
front left, front right, rear left, rear right

$(.)_F$ , $(.)_R$      front and rear axle

**Dimensions of vectors and matrices:**

| | |
|---|---|
| $n$ | Number of states |
| $m$ | Number of control inputs |
| $k$ | Number of measurements |

**Vectors:**

| | | |
|---|---|---|
| $\underline{x}$ | $\in \Re^n$ | State vector |
| $\underline{u}$ | $\in \Re^m$ | Input vector |
| $\underline{y}$ | $\in \Re^k$ | Measurement vector |

**Matrices:**

| | | |
|---|---|---|
| $\underline{A}$ | $\in \Re^{n,n}$ | System matrix |
| $\underline{B}$ | $\in \Re^{n,m}$ | Input matrix |
| $\underline{C}$ | $\in \Re^{k,n}$ | Output matrix |
| $\underline{D}$ | $\in \Re^{k,m}$ | Throughput matrix |
| $\underline{K}_C$ | | feedback matrix (control law) |

# B.2 Physical variables

| Symbol | Units | Physical variable |
|---|---|---|
| **A** | | |
| $a_{eng}$ | $[m/s^2]$ | negative acceleration at engine coasting |
| $a_N$ | $[m/s^2]$ | normal acceleration |
| $a_X$ | $[m/s^2]$ | longitudinal acceleration of CoG |
| $a_{xpot}$ | $[m/s^2]$ | potential braking acceleration of drivers |
| $a_Y$ | $[m/s^2]$ | lateral acceleration of CoG |
| $a_{ypot}$ | $[m/s^2]$ | driver-dependant tolerable lateral acceleration |
| $a_Z$ | $[m/s^2]$ | vertical acceleration of CoG |
| $A$ | | relative amount of effective collisions |
| $A_B$ | | constant of soot accruement |
| $A_{Br}$ | $[m^2]$ | area of wheel brake cylinder |
| $A_{ch}$ | $[m^2]$ | combustion chamber surface |
| $A_{CYL}$ | $[m^2]$ | cylinder head surface |
| $A_{eff}$ | $[m^2]$ | opening area of valve, cross sectional area |
| $A_L$ | $[m^2]$ | max. vehicle cross sectional area |
| $A_n$ | $[m^2]$ | effective cross sectional area of the nozzles |
| $A_0$ | | constant of soot oxidation |
| $A_p$ | $[m^2]$ | piston surface area |
| $A_{piston}$ | $[m^2]$ | piston head surface |
| $A_{L,S}$ | $[m^2]$ | front and side vehicle areas |
| **B** | | |
| $b_F, b_R$ | $[m]$ | distance between wheels on front / rear axles |
| $B_F$ | $[m]$ | road width |
| **C** | | |
| $c$ | $[J/(kg\,K)]$ | specific heat constant |
| $c_{air}$ | | coefficient of aerodynamic drag |
| $c_{BrF,R}$ | | front and rear brake transmission factors |
| $c_{F,R}$ | | front and rear tire side slip constants |
| $c_{hh}$ | | covariance of random road profile |
| $c_{ij}$ | | tire side slip constant |
| $c_p$ | $[m^2/(s^2 K)]$ | specific heat capacity by constant pressure |
| $c_r$ | | conversion ratio |
| $c_v$ | $[m^2/(s^2 K)]$ | specific heat capacity by constant volume |
| $c_{W,U}$ | | spring constants |
| $c_0$ | $[m/s]$ | sound propagation velocity at $273\,°K$ |
| $c_{1...5}$ | | co-efficients of Burkhard's tire equation |
| $C$ | $[F]$ | capacity |

| | | |
|---|---|---|
| $C_{diff}$ | | diffusion constant |
| $CO$ | $[1/s]$ | carbon monoxide emission level per time |
| $CYL$ | | number of cylinders |

**D**

| | | |
|---|---|---|
| $d$ | $[m]$ | cylinder diameter |
| $d_d$ | | internal damping of driveshaft |
| $d_{dr}$ | $[m]$ | fuel drop diameter |
| $d_f$ | | damping co-efficient at final drive for calculating the friction torque |
| $d_{Fl,Fnl}$ | | damping constants |
| $d_n$ | $[m]$ | nozzle diameter |
| $d_p$ | | propeller shaft damping |
| $d_v$ | $[m]$ | valve seat diameter |
| $d_w$ | | damping co-efficient at wheel for calculating the friction torque |
| $d_{W,U}$ | | damping constants |
| $d_{32}$ | $[m]$ | Sauter fuel drop diameter |
| $DB$ | | brake force distribution factor |

**E**

| | | |
|---|---|---|
| $e_r$ | $[J/kg]$ | specific external energy |
| $E$ | $[J]$ | activation energy |
| $E_{ext}$ | $[J]$ | external energy |
| $E_{mass}$ | $[J]$ | kinetic energy of the engine masses in motion |
| $E_t$ | $[J]$ | thermal energy |
| $E_y$ | $[J]$ | energy of signal y |
| $E\{\ \}$ | | expectation operator |

**F**

| | | |
|---|---|---|
| $f_a$ | | additive actuator fault |
| $f_{act}$ | | driver dependent factor |
| $f_{mass}$ | | mass factor considering driveline rotation |
| $f_{m,n}$ | | radial resonances with frequency modes $m, n$ |
| $f_p$ | $[Hz]$ | frequency of air pulsation |
| $f_p$ | | general process fault |
| $f_s$ | | additive sensor fault |
| $F$ | $[N]$ | force |
| $F_C$ | $[N]$ | spring force |
| $F_{fric}$ | $[N]$ | horizontal friction force |
| $F_D$ | $[N]$ | damping force |
| $F_G$ | $[N]$ | gravitational force |
| $F_{Hi}$ | | lambda control output factor |

|                |              | at high engine power |
| $F_L$          | $[N]$        | horizontal wheel force in direction $x_W$ |
| $F_{Lo}$       |              | lambda control output factor at |
|                |              | low engine power |
| $F_{ped}$      | $[N]$        | brake pedal force |
| $F_R$          | $[N]$        | rolling resistance force |
| $F_S$          | $[N]$        | horizontal wheel force in direction $y_W$ |
| $F_{wind}$     | $[N]$        | wind force |
| $F_{WL}$       | $[N]$        | horizontal wheel force in direction $v_W$ |
| $F_{WS}$       | $[N]$        | horizontal wheel force in direction |
|                |              | perpendicular to $v_W$ |
| $F_X$          | $[N]$        | longitudinal wheel force in direction $x_{Un}$ |
| $F_Y$          | $[N]$        | lateral wheel force in direction $y_{Un}$ |
| $F_Z$          | $[N]$        | vertical wheel force |
| $F_{ZC}$       | $[N]$        | vertical chassis force |
| $F_{Z0}$       | $[N]$        | nominal vertical wheel force |
| $F_\lambda$    |              | control output factor of the lambda |
|                |              | controller |

**G**

| $g$      | $[m/s^2]$ | gravitational constant |
| $G(s)$   |           | transfer function |

**H**

| $h$         |           | enthalpy of gas |
| $h_{ac}$    |           | mass related enthalpy of charge air/gas |
| $h_{CoG}$   | $[m]$     | height of CoG |
| $h_{in}$    |           | inlet enthalpy |
| $h_{out}$   |           | outlet enthalpy |
| $h_{road}$  | $[m]$     | road height |
| $h'$        | $[m]$     | distance from roll axis to CoG |
| $h_\nu$     |           | valve stroke |
| $\dot{H}C$  | $[1/s]$   | hydrocarbons emission level per time |
| $H_f$       | $[J/kg]$  | specific energy of the fuel released in |
|             |           | the combustion |
| $H_{low}$   | $[J]$     | lower calorific value |
| $H_r$       |           | number of drops with diameter $d_{dr}$ |

**I**

| $i$         | $[A]$  | current |
| $i_{diff}$  |        | transmission ratio of differential |
| $i_f$       |        | gear ratio at final drive |
| $i_{gear}$  |        | transmission ratio of gear |
| $i_s$       |        | transmission ratio of steering angle |
| $i_t$       |        | gear ratio at transmission |
| $i_{X,Y,Z}$ | $[m]$  | radii of gyration |

**J**

| | | |
|---|---|---|
| $J$ | $[kg\,m^2]$ | moment of inertia |
| $J_{crank}$ | $[kg\,m^2]$ | crankshaft moment of inertia |
| $J_{DT}$ | $[kg\,m^2]$ | drive-train moment of inertia |
| $J_e$ | $[kg\,m^2]$ | engine mass moment of inertia |
| $J_f$ | $[kg\,m^2]$ | moment of inertia of final drive |
| $J_t$ | $[kg\,m^2]$ | transmission moment of inertia |
| $J_Z$ | $[kg\,m^2]$ | moment of inertia about vertical axis |
| $J_X$ | $[kg\,m^2]$ | moment of inertia about longitudinal axis |
| $J_Y$ | $[kg\,m^2]$ | moment of inertia about lateral axis |
| $J_W$ | $[kg\,m^2]$ | wheel moment of inertia |

**K**

| | | |
|---|---|---|
| $k$ | | concentration ratio |
| $k_{camb}$ | | shifting factor of tire side slip angle by camber angle |
| $k_d$ | | driveshaft stiffness |
| $k_i$ | | weighting factor at fuzzy vehicle velocity estimation |
| $k_k$ | | heat release constant |
| $k_l$ | | adaptation factor |
| $k_p$ | | propeller shaft stiffness |
| $k_s$ | | attenuation factor of tire tread profile |
| $k_S$ | $[N/m]$ | rotational spring constant of steering |
| $k_{Br}$ | | brake transmission factor |
| $k_T$ | | tire spring stiffness |
| $k_u$ | $[N/m]$ | spring damper stiffness |
| $K_c$ | | amplifier gain control parameter |
| $K_{l,e}$ | | amplifier gain |
| $K_P,\ K_N$ | | idle speed control parameters |

**L**

| | | |
|---|---|---|
| $l$ | $[m]$ | wheel base (distance between front and rear axles) |
| $l$ | $[m]$ | connecting rod length |
| $l_F$ | $[m]$ | distance from CoG to front axle |
| $l_R$ | $[m]$ | distance from CoG to rear axle |
| $L_1,\ L_2$ | $[H]$ | inductances |
| $L_{HC},\ L_{CO},\ L_{NO_x}$ | | Lagrange factors |
| $L_{st}$ | | stoichiometric ratio |

**M**

| | | |
|---|---|---|
| $m$ | $[kg]$ | mass |
| $m^*$ | $[kg]$ | virtual mass |
| $m_a$ | $[kg]$ | air mass |
| $\dot{m}_a$ | $[kg/s]$ | mass air flow |
| $\Delta\dot{m}_a$ | $[kg/s]$ | offset error air flow per time |

| | | |
|---|---|---|
| $\dot{m}_{a,in}$ | $[kg]$ | mass air flow into the manifold |
| $m_{a,th}$ | $[kg]$ | theoretical air mass |
| $\dot{m}_{ac,air}$ | $[kg]$ | charge air |
| $\dot{m}_{ac,air}$ | $[kg/s]$ | charge air flow |
| $\dot{m}^*_{ac,air}$ | $[kg/s]$ | charge air flow, corrected for temperature ratio |
| $m_{CoG}$ | $[kg]$ | vehicle mass |
| $m_{crank}$ | $[kg]$ | crankshaft mass |
| $\dot{m}_D$ | $[kg/s]$ | depositing fuel flow |
| $\dot{m}_E$ | $[kg/s]$ | fuel flow from evaporation |
| $m_{EGR}$ | $[kg]$ | recirculated exhaust gas |
| $m_{EGR,air}$ | $[kg]$ | recirculated air mass |
| $m_{EGR,burnt}$ | $[kg]$ | recirculated burnt gas |
| $m_f$ | $[kg]$ | measured fuel mass per cylinder |
| $\dot{m}_f$ | $[kg/s]$ | fuel flow |
| $m_{f,burn}$ | $[kg]$ | burnt fuel mass |
| $m_{f,th}$ | $[kg]$ | theoretical fuel mass |
| $\dot{m}_{f,in}$ | $[kg/s]$ | fuel flow injected into the manifold |
| $\dot{m}_{f,out}$ | $[kg/s]$ | fuel flow into the cylinders |
| $m_m$ | $[kg]$ | air mass in the intake manifold |
| $\dot{m}_{min}$ | $[kg/s]$ | minimum fuel flow |
| $\dot{m}_{max}$ | $[kg/s]$ | maximum fuel flow |
| $m_{out}$ | $[kg]$ | exhaust mass |
| $m_{osc}$ | $[kg]$ | oscillating mass |
| $m_{piston}$ | $[kg]$ | piston mass |
| $m_{rem}$ | $[kg]$ | remaining gas mass |
| $m_{rem,air}$ | $[kg]$ | remaining air mass |
| $m_{rem,fuel}$ | $[kg]$ | remaining fuel mass |
| $m_{rod}$ | $[kg]$ | rod mass |
| $m_{rod,osc}$ | $[kg]$ | oscillating rod mass |
| $m_{rod,rot}$ | $[kg]$ | rotational rod mass |
| $m_{soot}$ | $[kg]$ | soot mass |
| $m_{St}$ | $[kg]$ | static wheel load |
| $m_W$ | $[kg]$ | wheel mass |
| $m_W$ | $[kg]$ | wall fuel mass |

**N**

| | | |
|---|---|---|
| $n$ | $[1/min]$ | number of crankshaft revolutions |
| $n_{Comb}$ | | combustion cycles |
| $n_L$ | $[m]$ | longitudinal (dynamic) caster |
| $n_0$ | | number of input/output valves |
| $n_S$ | $[m]$ | lateral (side) caster |
| $N_{d,i}$ | | number of drops in fuel package i |
| $N_1$ | | lower prediction horizon of GPC |
| $N_2$ | | upper prediction horizon of GPC |
| $N_u$ | | control horizon of GPC |
| $NO_x$ | $[1/s]$ | nitrogen oxides emission level per time |

**P**

| | | |
|---|---|---|
| $p$ | $[N/m^2]$ | pressure, combustion pressure |
| $p_{Br}$ | $[N/m^2]$ | braking pressure |
| $p_m$ | $[N/m^2]$ | intake manifold pressure |
| $p_o$ | $[N/m^2]$ | barometric pressure |
| $p_{rail}$ | $[N/m^2]$ | injection rail pressure |
| $P_e$ | $[W]$ | effective power |
| $P_i$ | $[W]$ | indicated power |
| $P_{max}$ | $[W]$ | maximum power |
| $P_{min}$ | $[W]$ | minimum power |
| $\underline{P}(k)$ | | covariance matrix in RLS estimation |

**Q**

| | | |
|---|---|---|
| $q$ | $[J]$ | thermal energy |
| $q_{hl,r}$ | $[J]$ | heat loss caused by incomplete combustion |
| $q_{hl,th}$ | $[J]$ | theoretical heat loss |
| $Q_{comb}$ | $[J]$ | combustion energy |
| $Q_w$ | $[J]$ | heat transfer through combustion chamber wall |

**R**

| | | |
|---|---|---|
| $r$ | $[m]$ | crankshaft radius |
| $r_{Br}$ | $[m]$ | effective frictional radius of brake disc |
| $r_{eff}$ | $[m]$ | effective dynamic rolling radius |
| $r_{ij}$ | $[m]$ | distance from CoG to wheel ground contact point |
| $r_{Stat}$ | $[m]$ | static rolling radius |
| $r_T(t)$ | | rectangular time window (width T) |
| $r_{\Delta f}(f)$ | | rectangular frequency window (width $\Delta f$) |
| $r_0$ | $[m]$ | original wheel radius |
| $R$ | $[m^2/(s^2 K)]$ | ideal gas constant |
| $R$ | $[m]$ | radius from wheel to instantaneous center of motion |
| $R_e$ | | Reynolds number |
| $R_i$ | $[\Omega]$ | internal resistance |
| $R_p$ | $[\Omega]$ | parallel resistance |

**S**

| | | |
|---|---|---|
| $s$ | $[m]$ | path length of road course segment |
| $s$ | $[m]$ | piston stroke |
| $\dot{s}$ | $[m/s]$ | piston velocity |
| $s_{ch}$ | $[m]$ | combustion chamber thickness |
| $s_j$ | $[m]$ | piston stroke of cylinder $j$ |
| $s_L$ | | longitudinal tire slip (in direction $v_W$) |
| $s_{max}$ | $[m]$ | maximum piston stroke |

| | | |
|---|---|---|
| $s_{Res}$ | | resultant tire slip |
| $s_S$ | | lateral tire slip (in direction perpendicular to $v_W$) |
| $S$ | | entropy |
| **T** | | |
| $t$ | $[s]$ | time |
| $t_{acc}$ | $[s]$ | lead time for acceleration before end of curve |
| $t_{antic}$ | $[s]$ | time before a curve where the cornering velocity is already reduced |
| $t_f$ | $[s]$ | foresight time |
| $t_{inj}$ | $[s]$ | injection time |
| $t_i$ | $[s]$ | ignition time |
| $t_{pot}$ | $[s]$ | potential lead time before braking to reach a lower speed |
| $t_s$ | $[s]$ | starting time |
| $t_{tacho}$ | $[s]$ | average time interval between tachometer readings |
| $T_{Br}$ | $[Nm]$ | braking torque |
| $T_{burn}$ | $[s]$ | time between opening of inlet and exhaust valve |
| $T_c, T_i$ | $[s]$ | time controller parameters |
| $T_c$ | $[Nm]$ | load torque between clutch and engine |
| $T_{comb}$ | $[Nm]$ | combustion torque |
| $\overline{T}_{comb}$ | $[Nm]$ | average combustion torque |
| $T_d$ | $[Nm]$ | torque between final drive and driveshaft |
| $T_{d,e}$ | $[s]$ | delay time (engine model) |
| $T_{Drive}$ | $[Nm]$ | drive torque |
| $T_e$ | $[Nm]$ | engine torque |
| $T_{exh}$ | $[s]$ | delay time between exhaust valve and lambda sensor |
| $T_f$ | $[Nm]$ | load torque between propeller shaft and final drive |
| $T_{fric}$ | $[Nm]$ | friction torque |
| $T_{fric,e}$ | $[Nm]$ | internal friction torque from the engine |
| $T_{fric,f}$ | $[Nm]$ | internal friction torque of final drive |
| $T_{fric,t}$ | $[Nm]$ | internal friction torque of the transmission |
| $T_J$ | $[s]$ | time constant due to engine moment of inertia |
| $T_{kc}$ | $[Nm]$ | torque from clutch nonlinearity |
| $T_L$ | $[Nm]$ | longitudinal friction torque at wheel in $x_W$ direction |
| $T_l$ | $[s]$ | adaptation time constant |
| $T_{l,e}$ | $[s]$ | lag time (engine model) |
| $T_{load}$ | $[Nm]$ | load torque |
| $T_{load}^*$ | $[Nm]$ | extended load torque (with friction torque) |

| | | |
|---|---|---|
| $T_{mass}$ | $[Nm]$ | mass torque |
| $T_s$ | $[s]$ | sample time |
| $T_p$ | $[Nm]$ | torque at transmission output into propeller shaft |
| $T_{self}$ | $[Nm]$ | self righting torque at steering |
| $T_t$ | $[Nm]$ | load torque between clutch and transmission |
| $T_{th}$ | $[Nm]$ | normalized torque on the throttle plate |
| $T_{U,In}$ | | transformation matrix (from undercarriage to inertial co-ordinate systems) |
| $T_w$ | $[Nm]$ | drive torque at wheel |

**U**

| | | |
|---|---|---|
| $u$ | $[J]$ | internal energy of a gas |
| $u$ | $[J/kg]$ | specific internal energy |
| $u$ | | control input variable |
| $U$ | $[V]$ | voltage |
| $U$ | $[J]$ | total internal energy |
| $U_b$ | $[V]$ | battery voltage |
| $U_\lambda$ | $[V]$ | output voltage lambda sensor |

**V**

| | | |
|---|---|---|
| $v_{CoG}$ | $[m/s]$ | vehicle CoG velocity |
| $v_{fl}$ | $[m/s]$ | flame propagation velocity |
| $v_n$ | $[m/s]$ | fuel flow velocity at injector nozzle |
| $v_{osc,j}$ | $[m/s]$ | speed of oscillating mass in cylinder $j$ |
| $v_R$ | $[m/s]$ | rotational equivalent wheel velocity |
| $v_{rot,j}$ | $[m/s]$ | rotational speed in cylinder $j$ |
| $v_W$ | $[m/s]$ | wheel ground contact point velocity |
| $v_{wind}$ | $[m/s]$ | wind velocity |
| $V$ | $[m^3]$ | volume |
| $V$ | $[kg/W]$ | Fuel Consumption |
| $\dot{V}$ | $[kg/(Ws)]$ | Fuel Consumption over time |
| $V_d$ | $[m^3]$ | displacement volume |
| $V_m$ | $[m^3]$ | manifold volume |

**W**

| | | |
|---|---|---|
| $w_e$ | $[J/m^3]$ | effective specific work per cycle |
| $w_{fr}$ | $[J/m^3]$ | frictional work |
| $w_i$ | $[J/m^3]$ | indicated specific work |
| $w_{i,hp}$ | $[J/m^3]$ | high pressure work |
| $w_{i,lp}$ | $[J/m^3]$ | low pressure work |
| $w_t$ | $[J]$ | technical work |
| $w_{th}$ | $[J/m^3]$ | theoretical work |
| $W_e$ | | Weber number |

**X**

| | | |
|---|---|---|
| $x_{CoG}$ | $[m]$ | CoG co-ordinate axis |
| $x_{In}$ | $[m]$ | inertial co-ordinate axis |
| $x_{Un}$ | $[m]$ | undercarriage co-ordinate axis |
| $x_W$ | $[m]$ | wheel co-ordinate axis |
| $X_{EGR}$ | | EGR-rate |

**Y**

| | | |
|---|---|---|
| $y_{CoG}$ | $[m]$ | CoG co-ordinate axis |
| $y_{In}$ | $[m]$ | inertial co-ordinate axis |
| $y_{Un}$ | $[m]$ | undercarriage co-ordinate axis |
| $y_W$ | $[m]$ | wheel co-ordinate axis |

**Z**

| | | |
|---|---|---|
| $z$ | | complex variable in z-transformation |
| $z_{CoG}$ | $[m]$ | CoG co-ordinate axis |
| $z_{In}$ | $[m]$ | inertial co-ordinate axis |
| $z_C$ | $[m]$ | chassis height |
| $z_U$ | $[m]$ | vertical chassis position |
| $z_{Un}$ | $[m]$ | undercarriage co-ordinate axis |
| $z_W$ | $[m]$ | vertical wheel-chassis contact point height, wheel co-ordinate axis |

**Greek**

**$\alpha$**

| | | |
|---|---|---|
| $\alpha$ | $[rad]$ | tire side slip angle |
| $\alpha_a$ | $[rad]$ | fixed advance angle |
| $\alpha_c$ | $[rad]$ | driveshaft angle between clutch and transmission |
| $\alpha_{ch}$ | $[1/s]$ | heat transfer rate of combustion chamber wall |
| $\alpha_{co}$ | $[1/s]$ | heat transfer rate of coolant |
| $\alpha_{CS}$ | $[rad]$ | crankshaft angle |
| $\alpha_e$ | $[rad]$ | effective ignition angle |
| $\alpha_f$ | $[rad]$ | driveshaft angle between final drive and drive shaft |
| $\alpha_i$ | $[rad]$ | ignition angle (from engine map) |
| $\alpha_k$ | $[rad]$ | knock control ignition angle |
| $\alpha_l$ | $[rad]$ | learned ignition angle from adaptive map |
| $\alpha_p$ | $[rad]$ | driveshaft angle between propeller shaft and final drive |
| $\alpha_Q$ | | percentage of combusted fuel |
| $\alpha_{RLS}$ | | effective term in RLS estimation |
| $\alpha_t$ | $[rad]$ | throttle angle |
| $\alpha_t$ | $[rad]$ | driveshaft angle between transmission and propeller shaft |

| | | |
|---|---|---|
| $\alpha_w$ | $[rad]$ | rotation angle of wheel |
| $\dot{\alpha}_w$ | $[rad/s]$ | wheel angular velocity |

**$\beta$**
| | | |
|---|---|---|
| $\beta$ | $[rad]$ | vehicle body side slip angle |
| $\beta_{m,n}$ | | Bessel functions |

**$\gamma$**
| | | |
|---|---|---|
| $\gamma$ | $[rad]$ | camber angle |

**$\delta$**
| | | |
|---|---|---|
| $\delta_A$ | $[rad]$ | Ackermann angle |
| $\delta_S$ | $[rad]$ | steering wheel angle |
| $\delta_W$ | $[rad]$ | wheel turn angle |

**$\varepsilon$**
| | | |
|---|---|---|
| $\varepsilon$ | | compression ratio |

**$\eta$**
| | | |
|---|---|---|
| $\eta$ | | thermal efficiency |
| $\eta_e$ | | effective thermodynamic efficiency |
| $\eta_{eff}$ | | effective thermodynamic efficiency |
| $\eta_{th}$ | | theoretical thermodynamic efficiency |

**$\vartheta$**
| | | |
|---|---|---|
| $\vartheta$ | $[K]$ | in-cylinder temperature |
| $\vartheta$ | $[K]$ | temperature |
| $\vartheta_a$ | $[K]$ | ambient air temperature |
| $\vartheta_f$ | $[kg/(m\,s)]$ | kinematic fuel viscosity |
| $\vartheta_e$ | $[K]$ | engine temperature |
| $\vartheta_{ij}$ | $[m]$ | angle between CoG co-ord. system and line from CoG to wheel ground contact point |
| $\vartheta_m$ | $[K]$ | manifold air temperature |
| $\vartheta_{oxidation}$ | $[K]$ | activation temperature of soot oxidation |
| $\vartheta_{Sensor}$ | $[K]$ | sensor temperature |
| $\vartheta_{soot,accrue}$ | $[K]$ | activation temperature |

**$\kappa$**
| | | |
|---|---|---|
| $\kappa$ | | adiabatic exponent |
| $\kappa$ | $[m]$ | curvature of road |

**$\lambda$**
| | | |
|---|---|---|
| $\lambda$ | | air-fuel ratio lambda |
| $\lambda_a$ | | relative air supply |
| $\lambda_c$ | | air fuel ratio within combustion chamber |

| | | |
|---|---|---|
| $\lambda_{ch}$ | $[m/s]$ | heat conductivity of combustion chamber wall |
| $\lambda_f$ | | relative fuel supply |
| $\Delta\lambda_g$ | | limit cycle of lambda |
| $\lambda_{RLS}$ | | forgetting factor in RLS estimation |
| $\lambda_u$ | | weighting factor for control input changes |

**$\mu$**

| | | |
|---|---|---|
| $\mu_{Br}$ | | friction coefficient between brake pedal and brake disc |
| $\mu_{flow}$ | | flow ratio co-efficient |
| $\mu_{inj}$ | | fitting parameter for fuel injector characteristic |
| $\mu_L$ | | friction co-efficient in direction of $v_W$ |
| $\mu_{Res}$ | | resultant friction co-efficient |
| $\mu_S$ | | friction co-efficient in direction perpendicular to $v_W$ |

**$\rho$**

| | | |
|---|---|---|
| $\rho_a$ | $[kg/m^3]$ | air density |
| $\rho$ | | injection ratio or load |
| $\rho$ | $[rad]$ | road curve radius |
| $\rho_{charge}$ | $[kg/m^3]$ | density of combustion chamber charge |
| $\rho_f$ | $[kg/m^3]$ | fuel density |

**$\sigma$**

| | | |
|---|---|---|
| $\sigma$ | $[rad]$ | valve seat angle |
| $\sigma_f$ | | fuel surface tension |

**$\tau$**

| | | |
|---|---|---|
| $\tau_a$ | $[s]$ | inter arrival time |
| $\tau_{char}$ | $[s]$ | characteristic mixture time |
| $\tau_{id}$ | $[s]$ | inflammation delay time |
| $\tau_s$ | $[s]$ | self inflammation time |
| $\tau_s$ | $[s]$ | service time |
| $\tau_w$ | $[s]$ | waiting time |

**$\varphi$**

| | | |
|---|---|---|
| $\varphi$ | $[rad]$ | roll angle |
| $\varphi_{road}$ | $[rad]$ | road camber |

**$\chi$**

| | | |
|---|---|---|
| $\chi$ | | pressure ratio |
| $\chi$ | $[rad]$ | pitch angle |
| $\chi_{road}$ | $[rad]$ | road gradient |

$\boldsymbol{\psi}$

| | | |
|---|---|---|
| $\psi$ | $[rad]$ | yaw angle |
| $\dot{\psi}_F$ | $[rad/s]$ | yaw rate calculated from front wheels |
| $\dot{\psi}_R$ | $[rad/s]$ | yaw rate calculated from rear wheels |
| $\dot{\psi}_S$ | $[rad/s]$ | measured yaw rate from sensor |
| $\dot{\psi}_{Fuz}$ | $[rad/s]$ | yaw rate from fuzzy estimator |

$\boldsymbol{\omega}$

| | | |
|---|---|---|
| $\omega$ | $[rad/s]$ | wheel angular velocity |
| $\omega_c$ | $[rad/s]$ | rotational velocity in curves |

# B.3   Abbreviations

| | |
|---|---|
| *ABS* | Anti-lock Braking System |
| *BDC* | Bottom Dead Center of piston movement |
| *C* | carbon |
| *CO* | carbon monoxide |
| *CoG* | Center of Gravity |
| *ECE* | Economic Commission for Europe, |
| | driving cycle in Europe |
| *EGR* | Exhaust Gas Reciruclation |
| *FIR* | Finite Impulse Response, |
| | filter algorithm which sums weighted values |
| | of the input variable over a time interval |
| *FTP* | Federal Test Procedure, |
| | driving cycle in the USA |
| *GPC* | General Predictive Control |
| *H* | Hydrogen |
| *HC* | Hydrocarbons |
| *ICM* | Instantaneous center of motion |
| *LQG* | Linear Quadratic Gaussian |
| *LTR* | Loop Transfer Recovery |
| $NO_x$ | Nitrogen oxides |
| $O_2$ | Oxygen |
| *OBDII* | On-Board Diagnostics-II |
| *PI* | Proportional-Integral Controller |
| *PID* | Proportional-Integral-Differential Controller |
| *RQ* | Minimum-maximum-speed governor (mechanical) |
| *RQV* | Variable speed governor (mechanical) |
| *SI* | Spark-Ignited |
| *TDC* | Top Dead Center of piston movement |

# B.4   Units

$$1\,bar = 10^5\,Pa$$
$$1\,bar = 10^5\,\frac{N}{m^2}$$
$$1\,\frac{km}{h} = 0,28\,\frac{m}{s}$$
$$1\,\frac{m}{s} = 3,6\,\frac{km}{h}$$

# Bibliography

[1] D. Ammon. *Radlastschwankungen, Seitenführungsvermögen und Fahrsicherheit*, volume 1088 of *VDI Fortschrittberichte*. VDI Verlag, Düsseldorf, 1993.

[2] C.F. Aquino. Transient a/f control characteristics of the 5 liter control fuel injection engine. *SAE Technical Paper*, (810494), 1981.

[3] J. Auzins, H. Johansson, and J. Nytomt. Ion-gap sense in misfire detection, knock and engine control. *SAE–Technical Paper Series*, (950004), 1995.

[4] M. Basseville. On fault detectability and isolability. *European Control Conference*, pages 567–573, 1999.

[5] M. Basseville and I.V. Nikiforov. *Detection of Abrupt Changes*. PTR Prentice-Hall, Inc, 1993.

[6] H. Bauer et al. *Automotive Handbook*. Robert Bosch GmbH, Stuttgart, Germany, 1993.

[7] M.C. Best. Nonlinear optimal control of vehicle driveline vibrations. *UKACC International Conference on CONTROL '98*, 455:658–663, 1998.

[8] A. Björnberg, M. Pettersson, and L. Nielsen. Nonlinear driveline oscillations at low clutch torques in heavy trucks. *Presented at Reglermötet in Luleå, Sweden*, 1996.

[9] Böning. Improvement of fuel economy by systematic computer-aided-control optimization. *Procedings of ISATA, Torino, Italy*, 1980.

[10] *Diesel fuel injection*. Robert Bosch GmbH, Stuttgart, Germany, 1994.

[11] K. Brammer and G. Siffling. *Kalman-Bucy-Filter.* R. Oldenbourg Verlag, München and Wien, 1994.

[12] W. Breuer. State- and parameter-estimation for four-wheel-drive passenger-cars. *European Control Conference*, 2, 1993.

[13] N. Bronstein and K. Semendjaev. *Taschenbuch der Mathematik.* Verlag Harri Deutsch, Frankfurt, 1987.

[14] M. Burckhardt. *Fahrwerktechnik: Radschlupf-Regelsysteme.* Vogel Fachbuch, Würzburg, 1993.

[15] G. Büschges. *Systemanalyse Straßenverkehrssicherheit.* Bundesanstalt für Straßenwesen, Bereich Unfallforschung, Köln, 1972.

[16] J. Bußhardt, J. Führer, and R. Isermann. *Ein elektronisches System zur parameteradaptiven Regelung und Diagnose von Kraftfahrzeugstoßdämpfern*, volume 1009 of *VDI FortschrittBericht 12.* VDI Verlag, Düsseldorf, 1992.

[17] CARB. California's OBD-II regulation (section 1968.1, title 13, california code of regulations), resolution 93-40, july 9. pages 220.7 – 220.12(h), 1993.

[18] D. Cho and P. Paolella. Model-based failure detection and isolation of automotive powertrain systems. *Proc. of ACC*, pages 2898–2905, 1990.

[19] D.W. Clarke and C. Mohtadi. Generalized predictive control. *Automatica*, 23(2), 1987.

[20] M. Constien. *Bestimmung von Einspritz- und Brennverlauf eines direkt einspritzenden Dieselmotors.* PhD thesis, Technische Universität München, 1991.

[21] A. Daiß. *Beobachtung fahrdynamischer Zustände und Verbesserung einer ABS- und Fahrdynamikregelung.* PhD thesis, Institute of Industrial Information Technology, Universität Karlsruhe, 1996.

[22] D.J. Dobner. A mathematical engine model for development of dynamic engine control. *SAE Technical Paper*, (800054), 1980.

[23] H. Fehrenbach. Model-based combustion pressure computation through crankshaft angular acceleration analysis. *Proceedings of 22nd International Symposium on Automotive Technology and Automation*, 1, 1990.

[24] O. Föllinger. *Nichtlineare Regelungen*, volume 1. R. Oldenbourg Verlag, München and Wien, 1987.

[25] O. Föllinger. *Regelungstechnik.* Hüthig Verlag, Heidelberg, 8. edition, 1994.

[26] P.M. Frank. Advances in observer-based fault diagnosis. Proc. TOOLDIAG'93, pages 817–836, Toulouse, France, 1993. CERT.

[27] P.M. Frank. Application of fuzzy logic to process supervision and fault diagnosis. Fault Detection, Supervision and Safety for Technical Processes, pages 507–514, Espoo, Finland, 1994. IFAC.

[28] A. Fürhapter, W.F. Piock, and G.K.Fraidl. Homogene Selbszündung: Die praktische Umsetzung am transienten Vollmotor. *Motortechnische Zeitschrift*, 65:94–101, 2 2004.

[29] S. Germann. *Modellbildung und modellgestützte Regelung der Fahrzeuglängsdynamik*, volume 309 of *VDI Fortschrittberichte, Reihe 12: Verkehrstechnik/Fahrzeugtechnik*. VDI Verlag, Düsseldorf, 1997.

[30] S. Germann, M. Würtenburger, and A. Daiß. Monitoring of friction between tyre and road surface. *3rd IEEE Conference on control Application*, 1994.

[31] J. Gertler. Analytical redundancy methods in fault detection and isolation; survey and synthesis. IFAC Fault Detection, Supervision and Safety for Technical Processes, pages 9–21, Baden-Baden, Germany, 1991.

[32] J. Gertler and M. Costin. Model-based diagnosis of automotive engines. IFAC Fault Detection, Supervision and Safety for Technical Processes, pages 393–402, Espoo, Finland, 1994.

[33] J. Gertler, M. Costin, X. Fang, R. Hira, Z. Kowalalczuk, M. Kunwer, and R. Monajemy. Model based diagnosis for automotive engines - algoritm development and testing on a production vehicle. *IEEE Trans. on Control Systems Technology*, 3(1):61–69, 1995.

[34] J. Gertler, M. Costin, X. Fang, R. Hira, Z. Kowalalczuk, and Q. Luo. Model-based on-board fault detection and diagnosis for automotive engines. *Control Engineering Practice*, 1(1):3–17, 1993.

[35] J. Gertler, M. Costin, X. Fang, R Hira, Z. Kowalczuk, and Q. Luo. Model-based on-board fault detection and diagnosis for automotive engines. IFAC Fault Detection, Supervision and Safety for Technical Processes, pages 503–508, Baden-Baden, Germany, 1991.

[36] P. R. Gill, W. Murray, and M. H. Wright. *Practical Optimization*. Academic Press, London, 1981.

[37] T.D. Gillespie. *Fundamentals of Vehicle Dynamics*. Society of Automotive Engineers Inc., 1992.

[38] H. Godthelp. *Studies on Human vehicle control*. PhD thesis, Institute for perception TNO, Soesterberg, Netherlands, 1984.

[39] M. Gosdin. *Analyse und Optimierung des dynamischen Verhaltens eines PKW-Antriebsstranges*, volume 69 of *VDI Fortschrittberichte, Reihe 11: Schwingungstechnik/Lärmbekämpfung*. VDI Verlag, Düsseldorf, 1985.

[40] E.-H. Hackbath and W. Merhof. *Verbrennungsmotoren*. Vieweg Verlag, 1998.

[41] C. Halfmann. *Adaptive semiphysikalische Echtzeitsimulation der Kraftfahrzeugdynamik im bewegten Fahrzeug*. VDI-Fortschrittsberichte, Reihe 12, Nr. 467. VDI-Verlag, Düsseldorf, 2001.

[42] A.J. Healy, E. Nathman, and C.C. Smith. An analytical and experimental study of automobile dynamics with random inputs. *Transcription of the ASME*, pages 284–292, 1977.

[43] E. Hendricks. Mean value modelling of spark ignition engines. *SAE-Technical Paper Series*, (900616), 1990.

[44] M. Henn. *On-Board-Diagnose der Verbrennung von Ottomotoren*. PhD thesis, Institute of Industrial Information Technology, Universität Karlsruhe, 1995.

[45] P. Herzog. *Möglichkeiten, Grenzen und Vorausberechnung der einspritzspezifischen Gemischbildung bei schnellaufenden Dieselmotoren mit direkter luftverteilender Kraftstoffeinspritzung*, volume 127 of *VDI Fortschrittberichte, Reihe 12: Verkehrstechnik/Fahrzeugtechnik*. VDI Verlag, Düsseldorf, 1989.

[46] John B. Heywood. *Internal Combustion Engine Fundamentals*. McGraw-Hill series in mechanical engineering. McGraw-Hill, 1992.

[47] M. Hiemer and J. Barrho. Observer design for road gradient estimation. *Reports in Industrial Information Technology*, Vol. 7, 2004.

[48] M. Hiemer et al. Cornering stiffness adaptation for improved side slip angle observation. *Proceedings of the First IFAC Symposium on Advances in Automotive Control AAC04*, 2004.

[49] M. Hiemer, S. Lehr, U. Kiencke, and T. Matsunaga. A fuzzy system to determine the vehicle yaw angle. *Transactions of the SAE World Congress*, 2004.

[50] H. Hiroyuki. Diesel engine combustion modelling. *Proceedings of International Symposium on Diagnostics and Modelling of Combustion in Reciprocating Engines*, 1985.

[51] K.L. Höfner and J. Hoskovec. Registrierung der Blickbewegungen beim Autofahren - bisherige Forschungen. *Zeitschrift für Verkehrssicherheit*, 19(4):222–241, 1973.

[52] P.L. Hsu, K.L. Lin, and L.C Shen. Diagnosis of multiple sensor and actuator failures in automative engines. *IEEE trans. on Vehicular Technology*, 44(4):779–789, 1995.

[53] R.D. Huguenin. *Fahrverhalten im Straßenverkehr*. Number 37 in Faktor Mensch im Verkehr. Rot- Gelb- Grün- Verlag, Braunschweig, 1988.

[54] R. Isermann. Process fault detection on modeling and estimation methods - a survey. *Automatica*, 20(4):387–404, 1984.

[55] R. Isermann. *Identifikation dynamischer Systeme I und II*. Springer Verlag, 1991.

[56] R. Isermann and P. Ballé. Trends in the application of model based fault detection and diagnosis of technical processes. *Control Engineering Practice*, 5(5):709–719, 1997.

[57] Ronald Jurgen. *Automotive Electronics Handbook*. McGraw-Hill, 1994.

[58] E. Justi. *Spezifische Wärme, Enthalpie, Entropie und Dissoziation technischer Gase*. Springer-Verlag, Berlin, 1938.

[59] J. Kahlert. *Fuzzy control für Ingenieure*. Vieweg, Braunschweig/Wiesbaden, 1995.

[60] T. Kailath. *Linear systems*. Prentice Hall, 1980.

[61] N. Kalouptsidis and S. Theodoridis. *Adaptive Signal Identification and Signal processing Algorithms*. Prentice Hall, 1993.

[62] Y. Kanayama, Y. Kimura, F. Miyazaki, and T. Nogutchi. A stable tracking control for an autonomous mobile robot. *IEEE International conference on robotics and automation*, pages 384–389, 1982.

[63] U. Kiencke. A view of automotive control systems. *IEEE-Control Systems*, 8(4), 1988.

[64] U. Kiencke. Realtime estimation of adhesion characteristic between tyres and road. *12.th IFAC World Congress of Automatic Control*, 1:15–22, 1993.

[65] U. Kiencke. *Ereignisdiskrete Systeme*. R. Oldenbourg Verlag, 1997.

[66] U. Kiencke and A. Daiß. Estimation of tyre friction for enhanced abs-systems. *AVEC Congress*, 1994.

[67] U. Kiencke, S. Daiß, and M. Litschel. Automotive serial controller area network. *SAE Technical Paper*, (860391), 1986.

[68] U. Kiencke, H. Kronmüller, and R. Eger. *Meßtechnik*. Springer Verlag, 2001.

[69] A. Kracke. *Untersuchung der Gemischbildung durch Hochdruckeinspritzung bei PKW-Dieselmotoren*, volume 175 of *VDI Fortschrittbericht 12*. VDI Verlag, Düsseldorf, 1992.

[70] V. Krishnaswami, G. Chun-Luh, and G. Rizzoni. Diagnosis of exhaust emissions control systems during the E.P.A. tailpipe inspection program. IFAC Fault Detection, Supervision and Safety for Technical Processes, pages 381–386, Espoo, Finland, 1994.

[71] V. Krishnaswami, G.C. Luh, and G. Rizzoni. Fault detection in IC engines using nonlinear parity equations. Proceedings of the American Control Conference, pages 2001–2005, Baltimore, Maryland, 1994.

[72] H. Kronmüller. *Digitale Signalverarbeitung*. Springer-Verlag, Berlin Heidelberg New York, 1991.

[73] A. Laschet. *Simulation von Antriebssystemen*. Springer Verlag, Berlin, Heidelberg, New York, 1988.

[74] K. Ledjeff. *Brennstoffzellen: Entwicklung, Technologie, Anwendung*. Müller, Heidelberg, 1995.

[75] Leins, Meewes, and Gerz. *Zur Beschreibung des Verkehrsablaufes auf Straßen mit und ohne Richtungstrennung*. Forschungsbericht des Landes NRW, 1975.

[76] Ljung and Lennart. *System Identification, Theory for the user*. Prentice Hall, 1987.

[77] G.C. Luh and G. Rizzoni. Identification of a nonlinear mimo ic engine model during I/M240 driving cycle for on-board diagnosis. *Proceedings of the American Control Conference*, pages 1581–1584, Baltimore, Maryland, 1994.

[78] J.M. Maciejowski. *Multivariable feedback design*. Addison-Wesley, 1989.

[79] R. Majjad. *Hybride Modellierung und Identifikation eines Fahrer-Fahrzeug Systems*. PhD thesis, Institute of Industrial Information Technology, Universität Karlsruhe, 1997.

[80] G.F. Mauer. On-line performance diagnostics for internal combustion engines. *Int. Conf. on Ind. Electronics, Control, Instrumentation and Automation, San Diego*, 3:1460–1465, 1992.

[81] J.L. Meriam and L.G. Kraige. *Engineering mechanics, dynamics*. John Wiley & Sons, 1987.

[82] G.-P. Merker and C. Schwarz. *Technische Verbrennung*. Verlag B. G. Teubner, Stuttgart, Leipzig, Wiesbaden, 2001.

[83] P.S. Min and W.B. Ribbens. A vector space solution to incipient sensor failure detection with applications to automotive enviroments. *IEEE trans. on Vehicular Technology*, 38(3):148–158, 1989.

[84] M. Mitschke and H. Wallentowitz. *Dynamik der Kraftfahrzeuge*. Springer Verlag, Berlin, 4. edition, 2004.

[85] C.Y. Mo, A.J. Beaumount, and N.N. Powell. Active control of driveability. *SAE Technical Paper*, (960046), 1996.

[86] K. Mollenhauer. *Handbuch Dieselmotoren*. VDI, Springer Verlag, Berlin, Heidelberg, New York, 1997.

[87] Lars Nielsen and Lars Eriksson. An ion-sense engine-fine-tuner. *IEEE Control Systems*, vol. 18(no. 5):pp. 43–52, October 1998.

[88] K. Nordgård and H. Hoonhorst. Developments in automated clutch management systems. *SAE Technical Paper*, (950896), 1995.

[89] C.O. Nwagboso. *Automotive sensory systems*. Chapman & Hall, 1993.

[90] M. Nyberg and L. Nielsen. Model based diagnosis for the air intake system of the SI-engine. *SAE Paper*, (970209), 1997.

[91] L. Orehall. Scania OptiCruise: Mechanical gearchanging with engine control. *Truck and Commercial Vehicle International*, 1995.

[92] P. Paolella and D. Cho. A robust failure detection and isolation method for automotive power train sensors. IFAC Fault Detection, Supervision and Safety for Technical Processes, pages 509–515, Baden-Baden, Germany, 1991.

[93] R. Patton, P. Frank, and R. Clark, editors. *Fault diagnosis in Dynamic systems*. Systems and Control Engineering. Prentice Hall, 1989.

[94] R.J. Patton. Robust model-based fault diagnosis:the state of the art. IFAC Fault Detection, Supervision and Safety for Technical Processes, pages 1–24, Espoo, Finland, 1994.

[95] M. Pettersson. *Driveline modeling and principles for speed and gear-shift control*. PhD thesis, University of Linsköpping, 1996.

[96] M. Pettersson. *Driveline Modeling and Control*. Linköping Studies in Science and Technology, 1997.

[97] Philips GmbH, Hamburg. *CAN Specification*, 1991. Version 2.0.

[98] R. Pischinger and G. Kraßnig. *Thermodynamik der Verbrennungskraftmaschine*, volume 5 of *Die Verbrennungskraftmaschine*. Springer Verlag, Berlin, Heidelberg, New York, 1989.

[99] L. Rasmussen. *Information processing and human-machine interaction*. North-Holland, New York, 1986.

[100] J. Reimpell. *Fahrwerktechnik, Fahrzeugmechanik*. Vogel Fachbuch, Würzburg, 1992.

[101] J. Reimpell and P. Sponagel. *Fahrwerktechnik: Reifen und Räder*. Vogel Fachbuch, Würzburg, 1995.

[102] R. Reiter. A theory of diagnosis from first principles. *Artificial Intelligence*, 32(1):57–95, April 1987.

[103] A. Riekert and T. Schunk. Zur Fahrmechanik des gummibereiften Kraftfahrzeuges. *Ingenieur Archiv*, 11:210–224, 1940.

[104] R.N. Riggins and G. Rizzoni. The distinction between a special class of multiplicative events and additive events: Theory and application to automotive failure diagnosis. American Control Conf., pages 2906–2911, San Diego, California, 1990.

[105] G. Rizzoni, P.M. Azzoni, and G. Minelli. On-board diagnosis of emission control system malfunctions in electronically controlled spark ignition engines. pages 1790–1795, 1993.

[106] G. Rizzoni and R. Hampo. Real time detection filters for onboard diagnosis of incipient failures. *SAE Paper*, (890763), 1989.

[107] G. Rizzoni and P.S. Min. Detection of sensor failures in automotive engines. *IEEE Trans. on Vehicular Technology*, 40(2):487–500, 1991.

[108] K. Rumar. The role of perceptual and cognitive filters in observed behaviour. *Human behaviour and traffic safety*, pages 151–170, 1985.

[109] K. Rumar. In vehicle information systems. *Proc. 3rd AAVD congress on vehicle design and components*, pages D33–D42, 1995.

[110] M. Sampath, R. Sengupta, S. Lafortune, K. Sinnamohideen, and D. Teneketzis. Diagnosability of discrete-event systems. *IEEE Transactions on Automatic Control*, 40(9):1555–1575, September 1995.

[111] R. Schernewski. *Modellbasierte Regelung ausgewählter Antriebssystemkomponenten im Kraftfahrzeug.* PhD thesis, Institute of Industrial Information Technology, Universität Karlsruhe, 1999.

[112] R. Schernewski and U. Kiencke. Cylinder balancing on common rail diesel engine. Technical report, 12th IAR Annual Meeting, Mulhouse, 1998.

[113] R.A. Schubiger, K. Boulouchos, and M.K. Eberle. Russbildung und Oxidation bei der dieselmotorischen Verbrennung. *MTZ*, pages 342–352, 2002.

[114] T. Schumacher, A. Reitz, and J.-W. Biermann. *Lastwechselschwingungen in Kfz-Antriebssträngen - eine Kompromissauslegung zwischen Komfort und Agilität.* Institut für Kraftfahrwesen, RWTH Aachen, Aachen, 2002.

[115] E. Schwartz. *Erkennung und Regelung querdynamisch kritischer Fahrzustände bei der Kurvenfahrt von PKW.* Dissertation, Universität Braunschweig, Braunschweig, 1990.

[116] SFB224. *Abschlussbericht: Motorische Verbrennung.* Deutsche Forschungsgemeinschaft, Germany, 2001.

[117] L.C. Shen and P.L. Hsu. Design of the robust unknown input observer for automotive engines. pages 129–134, San Francisco, USA, 1996. IFAC.

[118] *Simulink User's Guide.* MathWorks Inc., 1993.

[119] O.J. Smith. A controller to overcome deadtime. *ISA Journal*, 6, 1959.

[120] K. Suzuki and Y. Tozawa. Influence of powertrain torsional rigidity NVH of 6x4 trucks. *SAE Technical Paper*, (922482), 1992.

[121] W. Thiemann. *Messungen und Rechnungen zur Bestimmung der Abhängigkeit des Verbrennungsablaufs vom Einspritzvorgang im schnellaufenden Dieselmotor mit direkter Kraftstoffeinspritzung*, volume 119 of *VDI Fortschrittberichte, Reihe 12: Verkehrstechnik/Fahrzeugtechnik*. VDI Verlag, Düsseldorf, 1989.

[122] D.D. Torkzadeh. *Echtzeitsimulation der Verbrennung und modellbasierte Reglersynthese am Common-Rail-Dieselmotor*. PhD thesis, Institute of Industrial Information Technology, Universität Karlsruhe, 2003.

[123] A. Unger and K. Smith. Misfire detection by evaluating crankshaft speed - a means to comply with obdii. *SAE Paper*, (930399), 1993.

[124] A. Unger and K. Smith. The OBDII system in the Volvo 850 turbo. *SAE Paper*, (932665), 1993.

[125] A. Urlaub. *Verbrennungsmotoren Grundlagen*, volume 1. Springer Verlag, Berlin, Heidelberg, New York, 1987.

[126] H.D. Utzelmann. Tempowahl und -motive. *Faktor Mensch im Verkehr*, (24), 1976.

[127] C.D. Wickens. *Engineering Psychology and human performance*. Harper Collins Publishers Inc., 1992.

[128] W.W. Wierwille. Strategic use of visual resources by the driver while navigating with an in-car navigation display system. *SAE Technical Paper*, 2(885180):2661–2675, 1988.

[129] W.W. Wierwille and L. Tijerina. Darstellung des Zusammenhangs zwischen der visuellen Beanspruchung des Fahrers im Fahrzeug und dem Eintreten eines Unfalls. 47(2):67–74, 1997.

[130] G. Woschni. Verbrennungsmotoren. Skriptum zur Vorlesung, Lehrstuhl und Institut für Verbrennungskraftmaschinen und Kraftfahrzeuge, Technische Univerität München, 1980.

[131] M. Zeitz. *Nichtlineare Beobachter für chemische Reaktoren*, volume 27 of *VDI Fortschrittberichte*. VDI Verlag, Düsseldorf, 1977.

[132] H.J. Zimmermann. *Fuzzy set theory and its applications*. Kluwer Academic Publishers, Boston, 1991.

# Index